ZHANG JING

*Biochemistry of
Lipids and Membranes*

Selected Benjamin/Cummings Titles in the Life Sciences

F. J. Ayala and J. A. Kiger, Jr.
Modern Genetics, second edition (1984)

I. D. Campbell and R. A. Dwek
Biological Spectroscopy (1984)

M. Dworkin
Developmental Biology of the Bacteria (1985)

P. B. Hackett, J. A. Fuchs, and J. W. Messing
*An Introduction to Recombinant DNA Techniques:
Basic Experiments in Gene Manipulation* (1984)

L. E. Hood, I. L. Weissman, W. B. Wood,
and J. H. Wilson
Immunology, second edition (1984)

J. King (ed.)
*Protein and Nucleic Acid Structure
and Dynamics* (1985)

A. Levitzki
Receptors: A Quantitative Approach (1984)

S. E. Luria, S. J. Gould, and S. Singer
A View of Life (1981)

R. L. Rodriguez and R. C. Tait
*Recombinant DNA Techniques:
An Introduction* (1983)

A. S. Spirin
Ribosome Structure and Protein Synthesis (1985)

J. M. Squire
Muscle: Design, Diversity and Disease (1985)

J. D. Watson, N. Hopkins, J. Roberts,
J. Steitz, and A. Weiner
Molecular Biology of the Gene,
fourth edition (1985)

W. B. Wood, J. H. Wilson, R. M. Benbow,
and L. E. Hood
Biochemistry: A Problems Approach,
second edition (1981)

Biochemistry of Lipids and Membranes

EDITED BY

Dennis E. Vance and Jean E. Vance

THE BENJAMIN/CUMMINGS PUBLISHING COMPANY, INC.

Menlo Park, California · Reading, Massachusetts
Don Mills, Ontario · Wokingham, U.K. · Amsterdam · Sydney
Singapore · Tokyo · Mexico City · Bogota · Santiago · San Juan

Sponsoring Editor: *Paul Elias*
Production Supervisor: *Mimi Hills*
Production Coordinator: *Julia Chitwood*
Designer: *Rodelinde Albrecht*

Library of Congress Cataloging in Publication Data
Main entry under title:
Biochemistry of lipids and membranes.
  Includes index.
    1. Lipids.   2. Membranes (Biology)   I. Vance, Dennis E.
II. Vance, Jean E.   [DNLM: 1. Biochemistry.   2. Cell
Membrane.   3. Lipids.   QU 85 B6154]
QP751.B474   1985   574.87′5   84-29777
ISBN 0-8053-9420-6

B C D E F G H I J–H A–8 9 8 7 6

The Benjamin/Cummings Publishing Company, Inc.
2727 Sand Hill Road
Menlo Park, California 94025

For Konrad and Lore Bloch,
with thanks from several generations of lipid biochemists

CONTRIBUTORS

Dr. Konrad Bloch
Harvard University

Dr. Pierre Borgeat
Le Centre Hospitalier de l'Université Laval
Quebec

Dr. David N. Brindley
University of Nottingham

Dr. Harold W. Cook
Dalhousie University

Dr. John E. Cronan, Jr.
University of Illinois, Urbana

Dr. Pieter R. Cullis
University of British Columbia, Vancouver

Dr. Christopher J. Fielding
Dr. Phoebe Fielding
University of California, San Francisco

Dr. Alan G. Goodridge
Case Western Reserve University

Dr. Michael J. Hope
University of British Columbia, Vancouver

Dr. Reinhart A. F. Reithmeier
University of Alberta, Edmonton

Dr. Charles O. Rock
St. Jude's Childrens' Research Hospital
Memphis, Tennessee

Dr. Horst Schulz
City College
   of The City University of New York

Dr. William L. Smith
Michigan State University

Dr. Fred Snyder
Oak Ridge Associated Universities

Dr. Charles C. Sweeley
Michigan State University

Dr. Dennis E. Vance
University of British Columbia, Vancouver

Dr. Dennis R. Voelker
National Jewish Hospital, Denver, Colorado

Dr. Moseley Waite
Wake Forest University

Contents

Preface

This book has two major objectives. One is to provide an advanced textbook in the field of lipid and membrane biochemistry. The second is to provide a clear summary of the field for scientists engaged in research in the area of lipids and membranes and related fields.

Biochemistry has matured to the point that advanced textbooks in the various subcategories are required. This book should satisfy that need for the field of lipid and membrane biochemistry. The chapters are written for students who have taken an introductory course in biochemistry. We assume the students are familiar with the basic principles and concepts of biochemistry and have a general background in lipid and membrane biochemistry.

The second objective relates to the need for a general reference and review book for scientists in the lipid and membrane field. Such a book does not presently exist. Certainly there are many excellent reviews available of the various topics covered by this book, and these reviews are cited in the appropriate chapters. The availability of a current, readable, and critical summary of the biochemistry of lipids and membranes should fill an important gap in scientists' libraries. The literature in the field is vast, and there are usually many constraints on researchers' time. This book should allow these scientists to become more familiar with other areas of lipid metabolism related to their research interests. Finally, this book should help clinical researchers keep abreast of developments in basic science that are important for subsequent clinical advances.

The first chapter was written by Konrad Bloch, who for 50 years has made very important contributions to the lipid and membrane field. His chapter differs from the other chapters in that he summarizes the advances of his major research topics during the past 10 years—the evolutionary and structural functions of cholesterol in membranes. It is a lively and thought-provoking contribution.

The second chapter, Physical Properties and Functional Roles of Lipids in Membranes, introduces advanced information on membrane lipids. A major theme of the chapter reminds us that membrane lipids do more than separate aqueous compartments in cells. The following chapters provide current information on the biochemistry of fatty acids, phospholipids, triacylglycerols, sphingolipids, eicosanoids, cholesterol, and lipoproteins. The book concludes with two chapters that summarize the rapidly developing areas of assembly of lipids and proteins into membranes.

The book does not attempt to cover the general area of structure and function of biological membranes, since that subject has already been covered in a large number of excellent books. Second, the addition of such material would greatly increase the length, and therefore the cost, of this book.

The naming of lipids and enzymes in the book generally adheres to the rules of IUPAC-IUB. For further information on the nomenclature of lipids, please see *Biochemical Journal* 171 (1978): 21–35. A new edition of *Enzyme Nomenclature* has recently been published by Academic Press.

The editors and contributors assume full responsibility for the content of the various chapters. We would be pleased to receive comments and suggestions about this book.

Finally, the editors and contributors are indebted to the many other people who have made this book possible. In particular, we extend our thanks to Judith Smith, Teresa Vollmer, Theresa Fillwoch, Tommyz Campbell, Dawn Oare, Patricia Knight, and Perry d'Obrenan. We also thank the following colleagues who have read parts of the book and made useful suggestions: Dave Severson, Subhash Basu, Carlos Hirschberg, Matt Spence, and Günter Blobel.

Dennis and Jean Vance
Vancouver, Canada
July 1984

CHAPTER 1

Cholesterol: Evolution of Structure and Function

Konrad Bloch

NATURAL OCCURRENCE OF STEROLS

Biochemical unity has been a dominant concept for several decades. Nucleic acids, proteins, carbohydrates, and phospholipids of the same general structure are shared by all forms of life. The genetic code is universal. Darwinian evolution, the common descent of organisms, is manifest at the chemical level. Superimposed on unity, biochemical diversity is phenotypically expressed, at least in part, by organic molecules that are not ubiquitous: they are found in or needed by some cells but not others. Hormones, pigments, sterols, and many other substances concerned with specialized function belong to this category. Cholesterol and the structurally related sterols of fungi and plants are, as far as we know, not universal. We can therefore state with certainty that the sterol structure is not essential for the life process per se. In a more speculative vein, we can say that sterols arrived late in the evolution of organisms. The appearance of oxygen in the biosphere was essential for the biosynthetic pathway of sterols to develop.

Sterols are common in eucaryotic cells but rare in procaryotes. Vertebrates without exception synthesize cholesterol; in no instance is the pathway deleted or incomplete. Most invertebrates, lacking the enzymatic machinery for sterol synthesis, rely on an outside sterol supply. This generalization, valid until recently, may need to be qualified. Drosophila cell lines appear to be viable and exist without measurable endogenous or exogenous sterol (Silberkang et al. 1983). Yeasts and fungi, again with some apparent exceptions, harbor side-chain-alkylated sterols. While in photosynthetic organisms sito-sterol and stigmasterol (see Figure 1.2 for structure) are the most abundant and widely distributed sterols, cholesterol and ergosterol are by no means absent. The classical distinction between animal and plant sterols no longer corresponds to reality.

The occasional presence of sterols in procaryotes is of unknown functional significance. Substantial sterol synthesis (4-methyl sterols) is known with certainty to occur in *Methylococcus capsulatus*, an aerobic methanotroph (Bird et al. 1971). In another well-documented case, the isolation of Δ^8-cholestenol from a myxobacterium has recently been reported (Kohl, Gloe, and Reichenbach 1983). Claims for the presence of sterol traces in other procaryotes need to be substantiated. At any rate, for the vast majority of bacteria and blue-green algae, sterol is not a required molecule. Notable and special cases are the grossly heterotrophic sterol-requiring *Mycoplasma* species, which normally parasitize animal or plant tissues (Edward and Fitzgerald 1951). In a few instances squalene-derived molecules other than those containing the typical tetracyclic steriod nucleus appear to be functionally equivalent to sterol, for example, pentacyclic triterpenes (see the following discussion).

METABOLIC AND PRECURSOR FUNCTIONS OF THE STEROL MOLECULE

Perhaps not surprisingly, the more advanced the organism, the more diverse the role of the sterol molecule. That cholesterol is the essential precursor for bile acids, corticoids, sex hormones, and vitamin D–derived hormones is well established for all vertebrates. These transformations (Figure 1.1) involve partial shortening or complete elimination of the isooctyl side chain as well as a wide variety of ring hydroxylations. Oxygen is an essential reagent for both side-chain and nuclear modifications catalyzed by highly specific mixed-function oxygenases, which often, but not invariably, involve cytochrome P_{450}. It is important to note that in all these transformations but one (vitamin D) the C_{19} carbocyclic ring system remains intact. Changes of the ring conformation from all trans (planar) to A/B cis occur only in the formation of ecdysone, an insect hormone, and in bile acid formation. The position of oxygen functions and the length of the truncated side chain determine hormone specificity.

In invertebrates, with their primitive endocrine system, trans-formations of diet-derived sterol are apparently restricted to hydrox-ylations and transformation from A/B trans to cis of the otherwise intact C_{27} sterol structure (ecdysone, see Figure 1.1). Side-chain shortening does not seem to occur. However, as an interesting example of environmental adaptation, certain insects have evolved a mechanism for converting nonanimal sterols to nutritionally compe-tent cholesterol derivatives by removing C-24 alkyl groups from the

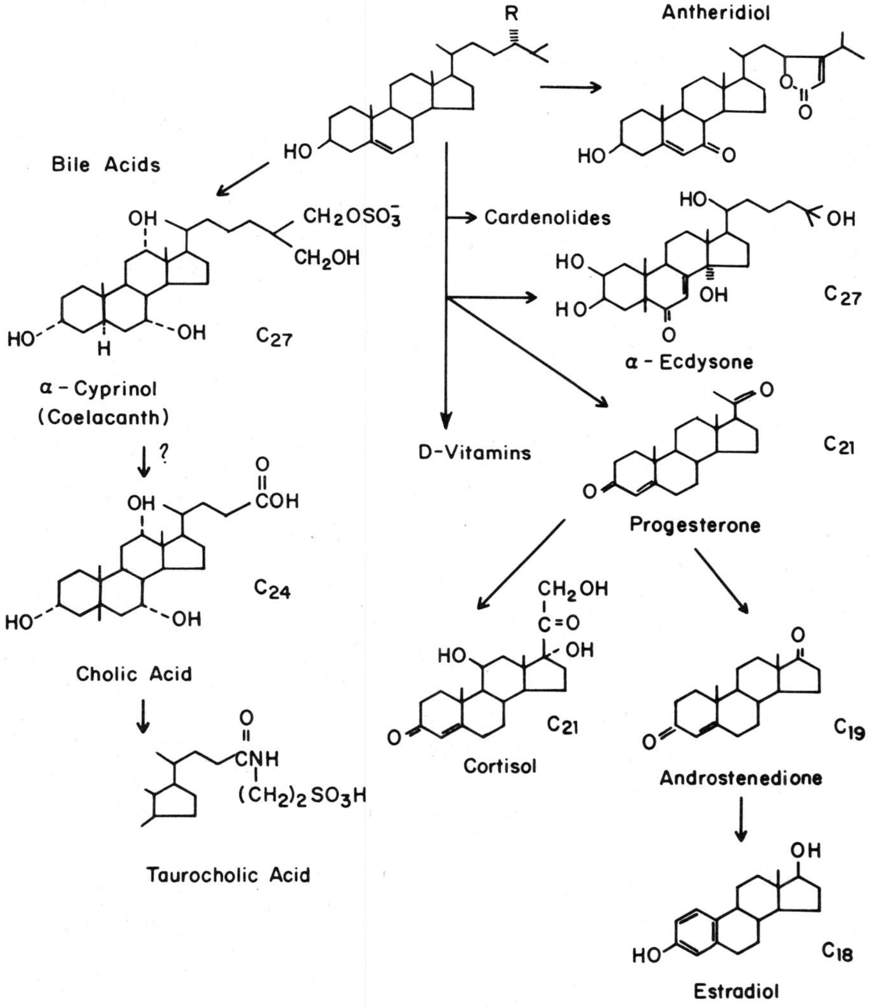

Figure 1.1. Functional evolution of the sterol molecule.

side chain. Thus the omnivorous cockroach dealkylates ergosterol or C_{29} plant sterols to 22-dehydrocholesterol, while *Dermestes vulpinus*, an obligate carnivore, lacks—because it does not need—the requisite dealkylating enzymes (Clark and Bloch 1959).

An extensive examination of marine invertebrates (sponges, gorgonians) has uncovered a bewildering variety of side-chain-modified sterols. Structures bearing additional alkyl or cyclopropane groups at six of the eight isooctyl sterol side-chain positions have been identified (Djerassi et al. 1979). It has been suggested, and to some extent documented, that the phospholipids of such marine organisms contain unique structural features complementary to the side-chain-alkylated sterols.

In lieu of the mammalian bile acids of the cholic acid type, some crustaceans and perhaps also other invertebrates, elaborate aliphatic aminosulfonates as intestinal emulsifiers, presumably for aiding triacylglycerol absorption (van den Oord, Danielsson, and Ryhage 1965).

Metabolites of ergosterol, the prototypical sterol of yeast and fungi, have not been found or adequately characterized. We may conclude that in unicellular eucaryotes only unmodified sterol molecules play an essential or beneficial role.

The C_{24} side-chain-alkylated plant sterols sitosterol and stigmasterol do not appear to undergo functionally essential conversions involving the loss of ring skeletal carbon atoms. Surprisingly, however, numerous plant families produce ecdysone, either identical with the cholesterol-derived invertebrate molting hormones or variants thereof, in quantities exceeding those found in insects by up to five orders of magnitude or more. Plant-feeding insects therefore have the choice of deriving these hormones directly from their diet or by converting the sterols they ingest. The cardioactive digitalis glycosides (cardenolides and bufalins) formed from sterols in *Digitalis* and *Strophanthus* species are probably secondary metabolites, whose physiological function in the organism of origin is unknown.

Physiologically useful modifications of the sterol structure have not been described in the few bacterial sterol producers or sterol auxotrophs. Information on the role of sterols in procaryotes exists only for the wall-less mycoplasmas (see the following discussion).

STEROL PATTERNS   In some animal tissues (liver and brain) cholesterol comprises more than 95% of the sterol fraction. Cholesterol precursors (lanosterol and partially dealkylated lanosterol derivatives) account for the remainder. In some cells, for example, lymphocytes, the concentration of these intermediates may be substantial (Burns et al. 1982). Sterol absorption from the mammalian gastrointestinal tract appears to be specific for cholesterol and its precursors, regardless of diet. Fungal and plant sterols are effectively excluded except in rare hereditary disorders. This remarkable discrimination, which accounts for the homogeneity of tissue sterols, appears to be especially pronounced for the absorption process mediated by the brush border membranes. However, animal cells in culture, for instance, Chinese hamster ovary cells, readily take up plant sterols from the medium and may indeed metabolize them. By contrast marine invertebrates, whether or not they are sterol auxotrophs, appear to discriminate much less against dietary sterols, with the result that their tissue sterol compositions show varying degrees of complexity.

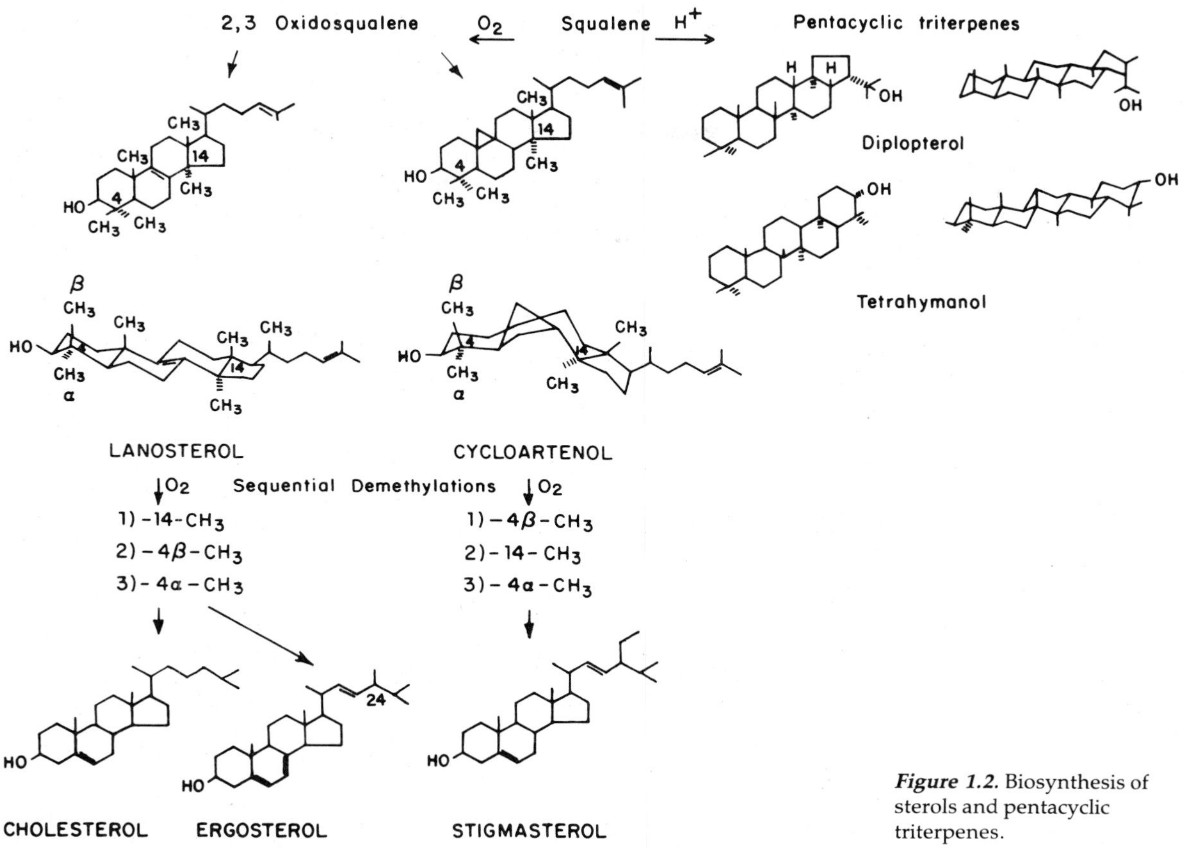

Figure 1.2. Biosynthesis of sterols and pentacyclic triterpenes.

Bewildering sterol mixtures are found in most lower and higher plants. While the C_{24}-ethyl sterols predominate, especially in plant leaves, sterols conventionally regarded as either typical for animals (cholesterol) or yeasts and fungi (ergosterol) (see Figure 1.2 for structure) are often present in substantial amounts. Especially striking is the fact that about a dozen species of red algae contain cholesterol exclusively. The argument made for invertebrates that the complexity of the sterol mixtures is attributable to indiscriminate absorption is not likely to hold in the case of plants. Since side-chain-alkylated sterols exhibit membrane properties quite different from those shown by cholesterol, the possibility that plants produce sterols for diverse functions deserves to be explored.

The amounts of sterol found in various animal tissues greatly exceed their bodily needs for the production of bile acids, steroid hormones, and vitamin D, probably by several orders of magnitude. According to conventional wisdom, the bulk of the tissue sterol is a major modulator of the membrane physical state. This modulation is performed by cholesterol per se, not a metabolite, or sterol ester. Much of the following discussion will deal with the membrane function of cholesterol. At this point the question arises, Why does the sterol content vary widely from tissue to tissue and from one intracellular organelle to another? Membranes of the red cell, the myelin sheath, and the plasma membrane contain the highest sterol: phospholipid ratios, ranging from 0.5 to 1. In most organs, except those concerned with steroid hormone production, mitochondria and nuclei are relatively cholesterol-poor. There is no consensus whether these low levels are real or due to contaminating organelles. However, the metabolic role of membrane sterol to be discussed raises the possibility that the presence of trace amounts is physiologically significant.

REGULATION OF STEROL BIOSYNTHESIS
Regulation of HMG-CoA Reductase

The multistep pathway for de novo sterol synthesis in animal tissues and yeast is treated adequately in most introductory texts. Likewise, the metabolic control of the pathway—the regulation of hydroxymethylglutaryl-CoA (HMG-CoA) reductase mediated by low-density lipoprotein (LDL)—is well established and has been comprehensively reviewed (see Chapter 13) (Goldstein and Brown 1977, 1983). HMG-CoA-reductase-catalyzed mevalonate formation is clearly the committed step in vertebrate cholesterol biosynthesis. However, since the polyprenol side chains of ubiquinone coenzymes and of dolichol are likewise mevalonate-derived, feedback control or alternate control mechanisms of HMG-CoA reductase, possibly by oxygenated sterols, are attracting attention (Sexton et al. 1983). LDL control of cholesterol biosynthesis operates by controlling HMG-CoA reductase synthesis and breakdown at the level of translation or transcription. Whether the reductase activity is also allosterically controlled is not yet clear. Compactin, a naturally occurring mevalonate analogue isolated from fungi, potently inhibits reductases from all sources. Whether a similar inhibitory molecule occurs in animal tissues is not known, but the possibility should not be dismissed. The physiological mechanisms for control of sterol biosynthesis in microorganisms and plants remain totally unexplored.

Regulation of Hepatic Sterol Biosynthesis by Cytosolic Proteins

The late stages of cholesterol biosynthesis appear to be controlled by an unusual mechanism which may operate specifically on membrane-associated enzymes that act on lipophilic substrates. The controlling agents are noncatalytic proteins, which affect neither enzyme synthesis nor enzyme activity by direct contact. The relevant research carried out mainly in the laboratories of M. Dempsey, J. Gaylor, T. Scallen, and the author has revealed the following general phenomenon.

The overall conversion of squalene to cholesterol catalyzed by microsome-embedded integral enzymes is markedly stimulated by rat liver cytosol. Two soluble proteins responsible for this effect, but operating at different stages of the pathway, have been isolated. Supernatant protein factor (SPF), with a molecular weight of 45,000, is a single polypeptide chain and a minor liver protein (0.01% of the total) which promotes the sequential conversion of either endogenous or exogenous squalene to squalene epoxide and also the cyclization of the epoxide to lanosterol. For expression of SPF activity, an anionic phospholipid is required. Sterol carrier protein (SCP), with a molecular weight of 16,000, enhances the rate of the subsequent steps, the overall conversion of lanosterol to cholesterol. A few of the partial reactions in this segment have been examined as isolated events (for example, Δ^7-cholestenol dehydrogenase). SCP, which comprises up to 10% of the soluble hepatic protein also affects a number of reactions apart from those of the cholesterol biosynthetic pathway, for instance, cholesterol esterification. SCP appears to be identical with Z-protein (also known as fatty acid binding protein; see Chapter 4) a molecule with high affinity for lipophilic substances.

The detailed mode of action of either SCP or SPF is unknown. Their effects, which are temperature sensitive, can be demonstrated in three different assay systems: (1) Measuring the uptake of substrate by microsomes and subsequent conversion of substrate to products. (2) Conversion of substrate generated enzymatically in microsomes (for example, squalene from farnesyl-pyrophosphate) or substrate incorporated into particles prior to assay. (3) Intramembrane transfer of substrate from trypsinized (enzymatically inactive) donor microsomes to fresh, untreated microsomes. Neither SPF nor SCP appears to be a conventional substrate carrier. Unlike the phospholipid exchange proteins, SPF and SCP fail to bind the molecules whose metabolism they promote. The current hypothesis is that these stimulatory proteins guide and direct the lipophilic substrates from an inactive membrane pool to the specific enzyme, maintaining high substrate concentrations at the active site. SPF and SCP do not enter the membrane space, and they bind to the membrane only loosely. Their effect appears to be membrane conditioning in a manner that facilitates the continued flow of substrate to its conversion site. This flow most likely depends on the physical state and the organization and dynamics of the endoplasmic reticulum, but how these are modulated by external proteins remains a mystery. The physiological significance of this regulation, which is always positive, is also unknown. Only intact microsomes are sensitive to the proteins; solubilized enzymes do not respond to SPF or SCP.

Conceivably the integral membrane proteins that catalyze the conversion of squalene to cholesterol are organized in a loose multienzyme complex rather than being distributed randomly and individually throughout the bilayer matrix. The fact that trypsin treatment of microsomes inactivates squalene epoxidase and Δ^7-cholestenol dehydrogenase but not squalene epoxide cyclase suggests that the active sites of some of the enzymes are nearer the cytosolic membrane surface, while others may be oriented toward the luminal membrane space. Differences in enzyme topology may require specific factors such as SPF and SCP for regulating intramembrane transport of substrates and products.

STEROLS AND MEMBRANE FUNCTION
Function of Sterols in Animal Membranes

The literature dealing with sterol effects on both artificial and natural membranes is vast. It leaves little doubt that cholesterol and a few of its derivatives can effectively modulate the physical state of phospholipid bilayers. In warm-blooded animals the bulk membrane phospholipid is probably in the liquid-crystalline phase. Under these conditions, that is, above the transition temperature, cholesterol lowers membrane fluidity by restricting the motion of fatty acyl chains. A large arsenal of physical methods is available for testing the reality of this effect: solute permeability, natural abundance ^{13}C-NMR, fluorescence depolarization of suitable probes, ESR signals with spin-labeled (nitroxyl) fatty acids, and scanning calorimetry. Sterol-induced fluidity changes begin to manifest themselves at molar cholesterol-phospholipid ratios above 1:10. The methods available inherently monitor only bulk phase fluidities. Few techniques exist for detecting fluidities above or below average in isolated regions or specific domains of biomembranes.

One elegant approach to the problem of distinguishing sterol-rich and sterol-poor membrane domains takes advantage of the sterol-complexing capacity of fungal antibiotics (such as filipin). Such regions can be visualized as distinctive knobs by freeze-fracture electron microscopy of cells. In fibroblasts exposed to filipin and radioactively labeled (^{125}I) LDL, the observed electron microscopy patterns have been interpreted as showing domains of lower-than-average cholesterol content. They are coincident with coated pits— the specific LDL receptor sites (Orci et al. 1978). Since fibroblast receptor-bound LDL is subsequently internalized, a low cholesterol content—and consequently high fluidity or plasticity—may be characteristic for membrane sites concerned with endocytosis of ligand-receptor complexes.

Solubilization, purification, and reconstitution of functional receptors for neurotransmitters has been achieved in several instances. Activity or ligand response in some of the reconstituted systems studied requires the presence of cholesterol (see the following discussion); ligand internalization is not involved in these instances. The question deserves to be raised: Do receptors which fall into one or the other category—internalization of the signal-producing ligand or immobilization of the ligand on the receptor surface—operate generally in an environment which is more fluid in the former case and less in the latter? Whether cholesterol levels influence the activity of the ligand-receptor protein system by direct contact or by modulating the fluidity of embedding lipid matrix is not known. The possible role of sterol as an effector with hormonal attributes will be discussed next.

Sterol Auxotrophs   In principle, natural sterol auxotrophs are ideal for investigating the role of sterols in membranes. Such auxotrophs are rare, however, and difficult to raise on synthetic media; moreover they furnish useful information only if the test sterol fails to be metabolized by the recipient cell. The two organisms that come closest to meeting these requirements are *Mycoplasma* species and the yeast mutant GL-7. It is an added advantage that both are fatty acid auxotrophs as well; therefore their membrane lipid composition can be experimentally manipulated. Studies with these two auxotrophs have helped to define the structural features of the sterol molecules that are essential for membrane function. In addition, they have provided evidence that sterols play a metabolic as well as a structural role in membranes (Dahl, Dahl, and Bloch 1980b; Dahl and Dahl 1983).

The growth rates of these two sterol auxotrophs respond synergistically rather than additively to pairs of exogenous sterols. *Mycoplasma capricolum*, will grow optimally when supplied with 10 μg/mL of cholesterol but only sluggishly when given either the same quantity of the cholesterol precursor lanosterol or limiting amounts (0.2 μg) of cholesterol. However, 10 μg/mL of lanosterol combined with 0.2 μg/mL of cholesterol affords nearly optimal growth rates. Cells grown on lanosterol (10 μg/mL) have high membrane fluidity, while cells supplied with nonlimiting cholesterol have much less. Supplementing high-fluidity lanosterol cells with cholesterol (1:20) does not change the bulk fluidity. The conclusion is that the sterol required by *Mycoplasma* serves in more than one capacity: Membrane fluidity may be one (the bulk function), and the second, involving only a small fraction of the total sterol required, may be control of membrane-associated metabolic processes (Dahl, Dahl, and Bloch 1980b). In *Mycoplasma*, this metabolic cholesterol component optimizes rates of phospholipid synthesis, specifically some step involving the utilization of exogenous unsaturated fatty acids. Increased protein synthesis is not the controlled process, since chloramphenicol does not inhibit the cholesterol effect. Moreover, enhanced phospholipid synthesis occurs within a few minutes of cholesterol supplementation (Dahl and Dahl 1983), while rate increases in the synthesis of RNA and protein and of growth follow after a 1–2 h lag. Cholesterol therefore may be viewed as a signal for membrane assembly, which in turn is essential for macromolecular synthesis and associated growth. There is some evidence for cholesterol binding sites on the plasma membrane of *M. capricolum* (Efrati et al. 1982).

The same general phenomenon—sterol synergism as expressed by promotion of phospholipid synthesis and growth—has been demonstrated in eucaryotic cells. GL-7, a yeast mutant deficient in squalene epoxide cyclase (Gollub et al. 1977) grows very much faster on combinations of ergosterol and cholesterol (1:3) than on suboptimal amounts of single sterol supplements (Ramgopal and Bloch 1983). For yeast, ergosterol is the native and cholesterol the foreign sterol. More strikingly, mutant yeast cells that fail to grow altogether when supplied with highly purified cholestanol grow vigorously when provided with trace amounts of ergosterol in addition to cholestanol. The phenomenon has been termed the *sparking effect* (Rodriguez, Taylor, and Parks 1982).

One metabolic process which is affected under conditions of sterol synergism in yeast is the synthesis of phosphatidylcholine. GL-7 microsomes catalyze the (probably sequential) methylation of phosphatidylethanolamine to phosphatidylcholine by *S*-adenosylmethionine. Methylation rates are enhanced by ATP and are faster in microsomes derived from cells grown on either ergosterol alone or on the synergistic ergosterol:cholesterol 1:3 combination than in microsomes of cells grown on cholesterol alone. Thus small amounts of ergosterol appear to promote transmethylation and possibly enzyme phosphorylation (Ramgopal and Bloch, unpublished). This type of metabolic triggering is reminiscent of events that follow the interaction of some hormones with receptors. Thus the sterol molecule appears to have hormonelike properties.

For DNA synthesis to occur, cultured animal cells rely on cholesterol supplied either with the medium or by endogenous synthesis (Goldstein and Brown 1977, 1983). When cholesterol is withheld or synthesis prevented by the potent HMG-CoA reductase inhibitor 25-hydroxycholesterol, DNA synthesis is arrested—after a lag. From the previous discussion it may be inferred that the cholesterol effect on macromolecular synthesis in animal cells is not necessarily direct but may well serve as a signal for membrane bilayer assembly, a prerequisite for macromolecular synthesis.

A notable feature of the metabolic sterol effects in the membranes of sterol auxotrophs is the small quantity required, certainly much less than the amount needed to raise membrane fluidity significantly. Metabolic cholesterol or ergosterol could, of course, be localized in specific domains and could control fluidity locally in the vicinity of nonrandomly distributed membrane enzymes; however, in at least one instance, fluidity control of either bulk membrane or microphase does not appear to be the operating mechanism. For *M. capricolum* 3α-methylcholesterol is an effective single sterol source. Membranes isolated from cells raised on this sterol are as fluid as those obtained from cells growing much more sluggishly on lanosterol (Dahl, Dahl, and Bloch 1980a), yet 3α-methylcholesterol in combination with lanosterol is also effective as a synergistic sterol (J. Dahl, unpublished). In this instance, sterol may interact directly with membrane enzyme, with attendant changes in protein conformation.

Metabolic Regulation in the Membrane Environment

Effector molecules that modulate the kinetic parameters of a responding system may do so in one of several ways: (1) The rate of protein synthesis or degradation may be altered. A variant of this mechanism is posttranslational or cotranslational protein modification. In many instances, the stimulatory molecules, notably hormones, operate by way of secondary messengers, for example, cAMP protein kinase systems. (2) Because direct effects of ligands on isolated catalytic proteins are generally allosteric, kinetically more or less competent conformational states of the macromolecule may result. For technical reasons, allosteric effects can be demonstrated only with cytosolic enzymes or enzymes solubilized from membranes.

The control of membrane-associated events, whether solute transport, the activity of integral enzymes, or responses of surface receptors, operates in a heterogeneous phase and therefore poses special problems. Almost certainly, the nature of the physical environment is a major influence. Since phospholipids—and in few instances glycolipids—provide the essential milieu, their chemical structure (head groups, fatty acid composition) will necessarily determine the activity of those membrane-associated events that are shared by all cells. Cholesterol, which is not a universal membrane constituent, will contribute in some instances to this environmental control but probably only associated with or mediated by phospholipids. Why sterols are essential for additional control of membrane events in some cells and organelles and not in others is not known. Solubilized membrane enzymes which can be activated or controlled by cholesterol alone have not been described.

Changes of physical state—increased or lowered fluidity of the membrane—arising from changes in the sterol:phospholipid ratio or specific sterol-phospholipid interactions are in many instances adequate, but not necessarily sufficient, explanations for sterol effects. However, the synergistic effect of traces of cholesterol on phospholipid synthesis in mycoplasma and of ergosterol in phosphatidyl-ethanolamine-phosphatidylcholine transmethylase of yeast suggests, at the very least, modes of sterol action that are not fluidity-related but more like hormonal signals.

Fluidity Control   In response to environmental challenges (nutritional, temperature) microorganisms employ devices for adjusting membrane fluidity by modulating the degree of phospholipid unsaturation. The underlying mechanism is not well understood but presumably involves changes in the rates of enzyme synthesis. Such adaptive responses are likely to be relatively slow. In higher multicellular organisms, especially those that maintain constant internal temperatures, certain stimuli—for example, hormones—may require a more rapid response of membrane-associated events in addition to long-range or adaptive control. Cholesterol may be uniquely suited to serve as a short-term regulator because it can move relatively rapidly either to and from the inner and outer half of the membrane bilayer or from one organelle to another (for example, endoplasmic reticulum ↔ plasma membrane) perhaps assisted by transport proteins. Many of the surface receptors for rapidly acting hormone signals are sensitive to sterol content. Another possible reason for the selection of cholesterol as a short-term effector is that fluctuations in sterol concentrations are not likely to perturb membrane integrity. In several membrane systems, artificial cholesterol depletion and the resultant change in a given biological activity are reversible. On the other hand, the phospholipid content (though not necessarily the composition) of a functional membrane probably remains constant at all times. The tight association of phospholipids with membrane protein, which can be disrupted only by detergents or phospholipases, probably ensures the structural integrity of membranes.

The function commonly ascribed to membrane-resident cholesterol is to order or decrease acyl chain mobility above and to raise mobility below the phospholipid transition temperature. At physiological temperatures and balanced phospholipid compositions, the ordering effect on acyl chain mobility probably predominates, at least as far as the bulk phase is concerned. In those membranes or membrane regions rich in sphingomyelin, a phospholipid of high transition temperature, cholesterol may however have fluidizing effects even at 37°C.

Much of the research on the role of cholesterol in biological systems, that is, with intact cells or cell-derived membranes, has been carried out and interpreted against the background of the responses of artificial membranes (liposomes). A variety of cells can either be artificially depleted or enriched with cholesterol. These manipulations produce the bulk fluidity changes expected from the behavior of cholesterol-rich or cholesterol-poor liposomes. Understandably, therefore, concomitant changes in the rate of the biological event under study tend to be interpreted as fluidity-mediated under physiological conditions as well. It seems reasonable to assume that the *bulk* of membrane cholesterol with its high preference for associating with a lipid phase is intercalated between phospholipid acyl chains and not in direct contact with membrane proteins. As a consequence, sterol effects, whether positive or negative, would be indirect, that is, mediated by modulating the physical state of the phospholipid bilayer, yet there are no a priori reasons for ruling out sterol-protein interactions as well, for example, in hydrophobic regions of polypeptide chains. The well-known propensity of bovine serum albumin to bind cholesterol tightly is an example, though the physiological significance of this interaction is unknown.

The following section describes some selected examples of sterol effects on cellular phenomena and membrane systems. It will be noted that invariably the issue is whether membrane fluidity control accounts for the sterol response or whether sterol interacts directly with the functional macromolecule. In no instance so far is the evidence sufficiently conclusive to decide one way or another.

REGULATORY ROLES FOR STEROLS IN MEMBRANES
Fungi and Invertebrates

Evidence that the intact sterol molecule may have regulatory properties has already been presented. Other relevant examples at the cellular level include certain fungi deficient in the sterol biosynthetic pathway. They require exogenous sterol for completion of the reproductive cycle (Warner, Sovocool, and Domnas 1983). The mosquito-parasitizing Oomycete *Lagenidium giganteum* fails to produce infective zoospores unless supplemented with sterol. By contrast the related fungus *Lagenidium callinectes*, which synthesizes traces of sterol (6–7 ng/mg dry weight), produces zoospores without sterol addition. It may be inferred that whatever their biochemical role in the two Oomycetes, the quantities of sterol required for eliciting the biological response are minute and insufficient for modulating the physical state of the target receptor. In these studies a conversion of sterol to steroids (oxygenated sterols) was not detected. A related phenomenon, formation of oospores capable of germination in the fungal plant pathogen *Phytophora cactorum* is also sterol-dependent (Elliot 1977; Nes and Stafford 1983). Again, the intact sterol and not an oxygenated metabolite appears to be responsible for the biological effect.

Hormonally active sterols that are oxidatively modified but retain the intact carbon skeleton occur in the saprophytic fungus *Achlya*. Fucosterol (24-ethylidenecholest-5-en-3β-ol)-derived antheridiol (Figure 1.1) promotes the development of antheridial hyphae, apparently by stimulating RNA synthesis and related events of the type involved in the action of mammalian sex hormones (McMorris 1978).

Sterol ring and side-chain oxygenations with retention of the intact carbon skeleton, perhaps the next step in the evolution of the sterol molecule, are typical for the developmental hormones of invertebrates (ecdysones, molting hormones). The sites of oxygen introduction, however, are not those found in the fungal hormones. Vicinal (2-, 3-) hydroxyl groups and oxygen functions at C-6 and C-14 are characteristic features of the ecdysones. Presumably they evolved from cholesterol independently along phylogenetic lines separate from those leading to fungal hormones.

Sterol Effects on Hormone and Neurotransmitter Receptors

An early report (Puchwein, Pfeuffer, and Helmreich 1973) deals with the potential role of sterol in the control of a hormone-sensitive adenylate cyclase. Exposure of pigeon erythrocyte membranes to the sterol-complexing antibiotic filipin drastically reduced the catecholamine (isoproterenol) activation of this cyclase. β-Sitosterol reversed the filipin effect. GTP stimulation of catecholamine-responsive adenylate cyclase was even more sensitive to filipin. The sterol-sequestering antibiotic did not alter the activity of unstimulated enzyme nor did it affect membrane microviscosity. The conclusion is that filipin in some manner interferes with signal transduction from hormone receptor to the catalytic subunit of adenylate cyclase. Sterol maintains structural order in the lipid matrix without measurable effects on fluidity. Results leading to the same conclusions for glucagon-stimulated adenylate cyclase have been obtained with rat liver plasma membranes (Whelton, Gordon, and Houslay 1983).

With the aid of a nonspecific lipid transfer protein (Crain and Zilversmit 1980) and liposomes, the cholesterol:phospholipid ratio in rat forebrain synaptosomes or cell-derived membranes can be substantially reduced (North and Fleischer 1983). When this ratio is lowered from 0.5 to 0.2, the rate of uptake of the neurotransmitter γ-aminobutyric acid (GABA) falls 3- to 6-fold but remains relatively constant when the ratio is artificially raised from 0.5 to 1.0. Qualitatively, these responses correlate with attendant changes in membrane fluidity, but inspection of the data obtained shows a disproportionately large rise in neurotransmitter uptake when the cholesterol:phospholipid ratio is raised only slightly, for instance, from 0.2 to 0.3. While it is considered plausible that GABA uptake is a function of membrane fluidity, "the possibility of a specific requirement for cholesterol by the GABA transporter" is not ruled out (North and Fleischer 1983). These and related studies reflect the uncertainties in distinguishing between the role of cholesterol in modulating membrane fluidity and perhaps more specifically affecting receptor activity. The same report (North and Fleischer 1983) notes that several receptors, for example, β-adrenergic receptor in Chang liver cells, and also serotonin and opiate receptors respond negatively to increased membrane fluidity, while others, for example, the β-adrenergic receptor in turkey erythrocytes and β-adrenergic receptors in human platelets, are insensitive to such changes. In some, but not all, of these studies, the effect of sterol on hormone binding to surface receptors has been distinguished from cholesterol-promoted changes on signal transduction, that is, control of cAMP-dependent protein kinase.

The isolated acetylcholine receptor from the electric organ of *Torpedo marmorata* has a cholesterol:phospholipid ratio close to unity; moreover, purified receptor protein interacts more readily with sterol than with phospholipid. While the sterol specificity is not especially high, the protein shows a marked preference for cholesterol over ergosterol (Popot et al. 1978).

An essential or highly beneficial role of cholesterol during detergent solubilization and reincorporation of solubilized acetylcholine receptor protein into lipid vesicles (cholate dialysis) has been observed (Criado, Eibl, and Barrantes, 1982). Cholesterol appears to be critical for preserving the agonist-induced state transitions both during detergent solubilization and reincorporation into lipid vesicles. Here again the observed sterol effects cannot be interpreted unequivocally. A related report also stresses the essential need for cholesterol in the reconstitution of the acetylcholine receptor. In this instance, recovery of agonist-induced ion flux was shown to be cholesterol-dependent as well (Dalziel, Robbins, and McNamee 1980).

Studies on acetylcholine receptor solubilization and reconstitution provide the best evidence so far for specific sterol-protein interactions that have functional consequences.

Sterol and Fusion of Viral Membranes

Fusion between viral particles and a target membrane—either artificial vesicles or a cellular host—necessarily requires the physical contact between the interacting entities. The specificity of the interaction appears to be largely determined by virus-specific glycoprotein. Fusion of Semliki Forest virus with vesicles is relatively independent of the phospholipid composition of the target liposome but requires cholesterol (White, Kielian, and Helenius 1983) (cholesterol:phosphatidylcholine ratios of 0.5 or higher). Curiously, fusion occurs also with cholesterol analogues that are usually inert as membrane components (coprostanol or androstanediol). However, a free 3β-hydroxy group (equatorial) is an essential structural feature. The suggestion that cholesterol in this instance, as in many others, modifies the physical properties of the target membrane is difficult to reconcile with the known inability of coprostanol or androstanediol to alter membrane bulk membrane fluidity. A sterol requirement for virus fusion with liposomes is not a general phenomenon. Influenza virus, for example, fuses with sterol-free vesicles. Nevertheless, when a virus fuses with a target animal cell instead of liposomes, it may be assumed that the surface membrane of the latter ordinarily contains accessible cholesterol.

Studies with vesicular stomatitis virus suggest that cholesterol, a major component of the viral envelope, may effect the organization (aggregation) of the viral glycoprotein spikes (Pal et al. 1983) and thereby play a key role in promoting viral infectivity (Moore et al. 1978).

Capping of Surface Immunoglobulin

Capping, the clustering of antibody receptors in lymphocytes over one of the cell poles, is one of numerous cellular processes which respond to alterations in the lipid environment of the cell surface. *Cis*-olefinic fatty acids, but not saturated fatty acids, interfere with the translocation of IgG in murine lymphocytes. The extent of capping is also cholesterol-dependent. Cholesterol depletion by incubation of the lymphocytes with phospholipid vesicles reduces and repletion with cholesterol raises IgG movement to the cell poles (Hoover et al. 1983). These cholesterol effects are attributed to the stabilization of gel-like surface domains or microenvironments in which the protein involved in capping is located.

Sterol Effects on Enzyme Activities

Human red cell Na^+, K^+-ATPase activity responds to artificially induced changes in the cholesterol:phospholipid ratio. Relative to normal, low cholesterol levels activate and high levels inhibit ATPase activity. The various potential mechanisms for this phenomenon are (1) a simple change in membrane order, that is, the physical state; (2) a change in the lateral distribution of membrane lipids; and (3) a direct cholesterol protein interaction resulting in conformational changes. The third is considered to be most likely (Yeagle 1983).

Similarly, sarcoplasmic Ca^{2+}-ATPase is sensitive to cholesterol. An interesting correlation between this enzyme activity, cholesterol content, and speed of twitch has been established for different anatomically distinct sarcoplasmic reticulum (SR) fractions of rabbit muscle. Slow-twitch soleus SR contains two to three times as much cholesterol as fast-twitch caudofemoralis SR. The higher cholesterol (and also sphingomyelin) content reduces bilayer fluidity, which in turn lowers Ca^{+2}-ATPase activity, which regulates muscle contraction and relaxation rates (Borchman, Simon, and Bicknell-Brown 1982).

A positive cholesterol effect has been reported for the stimulation of ADP–ATP exchange catalyzed by reconstituted submitochondrial particles and isolated inner mitochondrial membranes (mitoplasts). Enhanced exchange activity is seen when the membrane fluid contains as little as 7–13% of cholesterol (Kramer 1982). In this range, cholesterol alters the physical state of artificial phospholipid vesicles only marginally, and therefore a sterol-induced membrane ordering effect is not a likely explanation for the effect on ADP–ATP exchange. Again, the sterol may affect the molecular activity of the protein directly, but other explanations are not ruled out. Since the inner mitochondrial membrane normally contains very little, if any, cholesterol, the physiological significance of this phenomenon is unclear.

Several reports deal with the question of whether the transfer of acyl residues from lecithin to cholesterol catalyzed by lecithin:cholesterol acyltransferase (LCAT) is sensitive to the concentration of unesterified cholesterol. Specifically, the question has been raised whether substrate (cholesterol) availability fully accounts for the sterol-dependent acyl transferase activity. A recent study concludes that with highly purified LCAT and defined complexes of phosphatidylcholine and apolipoprotein A-I, enzyme activity is independent of cholesterol concentration, cholesterol:phospholipid ratios, or lipid-domain fluidity (Jonas and McHugh 1983).

Microsomal acyl-CoA:cholesterol acyltransferase (ACAT) is believed to be responsible for the bulk of intracellular cholesterol esterification activity. ACAT therefore plays an important role in the coordinate control of cholesterol homeostasis. Membrane-associated enzymes in general are subject to a variety of controls including substrate availability at the active site, the lipid environment, and possibly substrate carrier proteins. Such multiple controls appear to exist also for ACAT. Some published reports suggest that cholesterol stimulation of ACAT is only partially accounted for by substrate availability. Also, the rate enhancement of cholesterol esterification by 25-hydroxycholesterol (Field and Mathur 1983) raises the possibility that ACAT activity is subject to allosteric regulation by cholesterol, one of its substrates. The effect may be exerted directly on enzyme protein or via control of posttranslational enzyme modification such as phosphorylation (Suckling, Stange, and Dietschy 1983). ACAT properties and its control mechanism have recently been reviewed (Chang and Doolittle 1983).

Effects of Cholesterol on Phospholipid Synthesis

The phenomenon of sterol synergism and the dependence of phospholipid biosynthesis on specific sterol structures in *Mycoplasma* and in the yeast mutant GL-7 appear to be the most direct evidence for a regulatory role of sterol per se in membrane-associated processes. Evidence for a similar role of cholesterol in animal cells is less direct but suggestive. Stimulated by the observation that in human atheromata, phospholipid accumulates in foam cells, probably as the result of de novo synthesis, Day, Fidge, and Wilkinson (1966) have examined the effect of cholesterol on phospholipid synthesis in macrophages which can be precursors of foam cells. Synthesis of all phospholipids (from [^{32}P]phosphate) was stimulated substantially as a function of the external cholesterol concentration, with significant effects seen after 1 h. The cholesterol-affected step was not identified. Lipid vesicle-associated cholesterol has also been shown to stimulate phosphatidylcholine synthesis in cultured arterial smooth muscle cells (Slotte and Lundberg 1983). In studies on the coordination of lipid synthesis with the cell cycle in L$_6$-myoblasts, the cholesterol-synthesis inhibitors 25-hydroxycholesterol and compactin rapidly reduced phospholipid synthesis and caused a slower decline in the synthesis of DNA, RNA, and protein. The coordinated inhibition of lipid synthesis appears to be responsible for an arrest in the G$_1$ phase of the cell cycle (Cornell and Horwitz 1980). One of the cholesterol-sensitive steps in L$_6$-myoblast phospholipid synthesis appears to be CTP:phosphocholine cytidylyltransferase (Cornell and Goldfine 1983).

Similar conclusions were reached and amplified on the basis of cholesterol-feeding experiments. Rat liver CTP:phosphocholine cytidylyltransferase activity from animals on the cholesterol regime was twice as high as the enzyme activity obtained from animals on a cholesterol-free diet. Moreover, the cholesterol-enriched diet appears to cause translocation of the cytidylyltransferase from cytosol to microsomes (Lim et al. 1983).

EVOLUTION OF THE STEROL STRUCTURE

The great achievement of classical biochemistry has been to establish the molecular events of intermediary metabolism, the modes of formation and transformation of cellular constituents. Moreover, the isolation of numerous enzymes has provided detailed information on the highly sophisticated chemistry of biocatalysis. In essence, the questions traditionally asked and answered were how biological systems bring about chemical reactions. We know much less, and rarely ask, about the motives of nature—why certain molecules and reactions and not others have evolved and became established. This question of choice or selection is of course a central issue that evolution poses at the molecular as well as at the organismic level. The remaining section of this chapter addresses one aspect of this issue. It summarizes and attempts to rationalize the chemical structure of a biomolecule in terms of biological function.

According to the hypothesis of Oparin and Haldane, the prebiotic atmosphere on earth was reducing and probably entirely free of molecular oxygen. While direct proof will always remain beyond reach, the argument that few prebiogenic organic compounds could have survived in an aerobic environment seems compelling. In support of the hypothesis organisms regarded as the most primitive, the *Archaebacteria* and anaerobic *Eubacteria*, derive energy from fermentation, not respiration. They lack porphyrins , quinone coenzymes, and in general, all molecules which contemporary cells form by oxygen-requiring processes. The tetracyclic sterol nucleus belongs in this category. While the isoprene-derived polyprenols, including the acyclic sterol precursor squalene, could have been formed abiotically in the absence of oxygen, the sterol pathway surely arose after aerobic cells had evolved.

Early trial-and-error experiments of nature to create molecules with sterol-like properties are reflected in the structures of the various squalene-derived pentacyclic triterpenes (Figure 1.2) (Bloch 1983). Proton-initiated, rather than oxidative, squalene cyclization produces the hopanes, for example, diplopterol found in numerous bacteria (Rohmer, Bouvier, and Ourisson 1980) and tetrahymanol, which replaces cholesterol in some ciliated protozoans (tetrahymena species) (Mallory, Gordon, and Conner 1963). In their effects on model membranes and also nutritionally (in *Tetrahymena* and *Mycoplasma*), pentacyclic triterpenes substitute to some extent for the more advanced squalene epoxide–derived sterol structure. This functional equivalence is remarkable because the fully cyclized tetrahymanol and diplopterol lack any structural feature comparable to the conformationally unrestrained sterol side chain. At the same time, the three-dimensional structure of the biologically active pentacyclic terpenes is sterol-like with respect to bulk and rigidity. Notably, these molecules display two planar regions favorably disposed for interaction with phospholipid acyl chains. These features of pentacyclic triterpenes are not due to processing, that is, by demethylation, but result from the specific folding of squalene during cyclization. It is conceivable that tetrahymanol and diplopterol retained all alkyl branches because their removal would fail to produce molecules functionally superior to the intact C_{30} compound. By the same token, the failure of tetrahymanol-producing protists to elaborate steroids by oxidative modifications may be due to structural restraints. Because they are anaerobically formed, pentacyclic triterpenes probably preceded the tetracyclic sterols in evolution. It can further be argued that the route to pentacyclic triterpenes was only a limited success. While their structures are suitable for modulating the physical properties of membranes by acyl chain ordering, they appear to be metabolically inert. Oxidative modifications of pentacyclic terpenes of the type that lead to the formation of bile acids and steroid hormones in higher organisms have not been observed. This may be one of the reasons why the eucaryotic tetrahymena failed to evolve further.

Relevant to the divergent routes of squalene cyclizations is the observation that the aerobic procaryote *Methylococcus capsulatus* produces both pentacyclic triterpenes and sterols, side by side (Rohmer, Bouvier, and Ourisson 1980). Notably, the pathway to C_{27} sterols in this bacterium is incomplete. Demethylation of lanosterol terminates at the stage of 4-monomethyl sterols, suggesting that *Methylococcus* was an early, if not the earliest, sterol producer.

For both invertebrate and vertebrate function, cholesterol appears to be optimally designed. Any modifications, no matter where in the molecule or of what kind, diminish its functional competence in membranes. Structural features responsible for this "perfection" are indicated in Figure 1.3. They include the following: (1) All-trans fusion of the ring system results in the all-chair conformation which equips the tetracyclic moiety with two planar regions, the lower α- and the upper β-face. (2) The equatorial 3-hydroxy function located at the membrane-water interface allows the sterol molecule to orient itself most stably in the phospholipid bilayer. (3) In naturally occurring sterols, the right-handed structure (17β, 20R) with respect to the isooctyl side chain is the thermodynamically preferred conformer. This fully extended structure optimizes sterol-phospholipid interactions. (4) The bridgehead methyl groups at C-10 and C-18 are retained. (5) The unmodified isooctyl side chain renders the bilayer core relatively fluid. Sterols containing unbranched, shortened, or lengthened side chains are less effective in modulating bilayer fluidity. However, the physiological advantage of a higher fluidity in the bilayer interior remains to be rationalized. (6) The tetracyclic ring

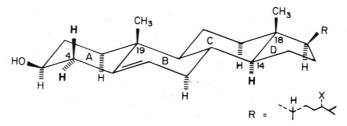

Figure 1.3. Structural features of importance for the functioning of the intact sterol molecule: (1) All-trans ring fusions, (2) 3-(equatorial) hydroxyl group, (3) intact isooctyl side chain (R), (4) right-handed conformation of side chain (R), (5) absence of methyl groups at C-14 and C-4, (6) retention of methyl groups at C-18 and C-19, (7) 5,6-double bond, (8) extra alkyl groups (X) in side chain, X = H in animal sterols, X = CH_3 in sterols of yeast and fungi, X = C_2H_5 in "plant" sterols. Ring B is a modified chair conformation, but this is not shown in this diagram.

system is uniquely compact, bulky, and rigid. Under physiological conditions the energy barrier for conformational changes is too large. Except for the pentacyclic triterpenes, other molecules of comparable hydrophobicity (polyprenols, fatty acid derivatives) are much less restrained conformationally. Because of their bulk, sterol inserts will not only separate or laterally displace fatty acyl chains but may also separate polar phospholipid head groups. Charge densities at the membrane surface may therefore be altered.

The planarity-producing modifications of cholesterol precursors eliminate methyl groups from the lanosterol α-face in the sequence (1) 14α-methyl, (2) 4β-methyl, and (3) 4α-methyl. This invariant order* produces intermediates of progressively greater competence, as shown by the induced fluidity changes in artificial membranes and by the growth response and membrane viscosities of the sterol auxotroph *Mycoplasma capricolum* (Dahl, Dahl, and Bloch 1980a). Elimination of the axial 14-methyl group, which protrudes from the α-face, causes the largest changes in the fluidity parameter and in the biological response. For this reason it may be the first of the three alkyl groups to be removed. It has been argued that this segment of the sterol pathway evolved not synchronously but step by step in response to evolutionary pressures (Bloch 1976).

The pathway to sterols employed by plants is characterized by a major variant of the squalene epoxide–sterol cyclization (Figure 1.2). Squalene epoxide cyclizes not to lanosterol but to the isomeric cycloartenol. The final proton elimination during cyclization produces a three-membered ring (between rings B and C) with profound conformational consequences. Restraints imposed by the cyclopropane structure forces ring B into a boat conformation and results in a bent rather than planar tetracyclic ring system. The same axial hydrogen atoms which cause the two faces of an all-chair system (cholestan) to be planar lie instead in a semicircular arc or belt. As a further consequence of the conformational change, the 14α-methyl group no longer protrudes but becomes embedded in the arclike α-face. The adverse effect of this bulky substituent on hydrophobic interactions is thereby diminished if not entirely relieved. Structural analysis therefore rationalizes why the behavior of cycloartenol in both artificial and natural membrane systems is much more cholesterol- than lanosterol-like (Bloch 1983). Burial rather than exposure of the 14α-methyl group can also be invoked to explain why in the cycloartenol demethylation sequence employed by plants, the 14α-methyl group is not the first to be removed (Heintz and Benveniste 1974). Since it is reorientated in space, the 14α-methyl group no longer interferes with acyl chain packing in the phospholipid bilayer.

*The apparent paradox that the first of the two methyl groups removed at C-4 is α-oriented (equatorial) and that the product is the 4α- and not the 4β-monomethyl derivative is due to an intervening epimerization. For a discussion see Bloch (1983).

The selection pressures to remove it are no longer paramount. Instead, α-face demethylation begins with removal of the 4β-methyl group at C-4. It has been proposed that this step has the higher priority because it is a steric prerequisite for opening of the cyclopropane ring. Once ring opening has occurred, the tetracyclic system assumes the normal all-trans conformation and further demethylation proceeds as expected: (1) 14α-methyl and (2) 4α-methyl, leading eventually to the typical phytosterols (sitosterol, stigmasterol).

Of the two isomeric squalene epoxide cyclization products, cycloartenol is less favored thermodynamically than lanosterol. Why then is it an intermediate in the plant sterol pathway, given the fact that the cyclopropane ring is a transitory feature which disappears again farther along in the pathway? One may speculate that cycloartenol is both an intermediate and an end product. In certain plant tissues it may have biological functions of its own, distinct from those of the major side-chain-alkylated sterols.

The preceding analysis considers the sterol molecule not in isolation but as one of two lipid components—phospholipid being the other—interacting in a cellular system, the membrane bilayer. Perhaps a similar approach will help to explain the origins and structure of other biological molecules that are taken for granted but have not yet been rationalized.

This chapter does not attempt a systematic treatment of the diverse aspects of sterol biochemistry. Rather, roles of sterols other than those long recognized are emphasized and the possible direction of future research indicated. Original papers and citations have been arbitrarily selected, and many important references may have been overlooked.

BIBLIOGRAPHY

Entries preceded by an asterisk are suggested for further reading.

Bird, C. W.; Lynch, J. M.; Port, F. J.; Reid, W. W.; Brooks, C. J. W.; and Middleditch, B. S. 1971. Steroids and squalene in *Methylococcus capsulatus* grown on methane. *Nature* 230:473.

*Bloch, K. 1976. "On the evolution of a biosynthetic pathway." In *Reflections on biochemistry*, ed. A. Kornberg, 143–50. New York: Pergamon.

*Bloch, K. 1983. Sterol structure and membrane function. *Crit. Rev. Biochem.* 14:47–92.

Borchman, D.; Simon, R.; and Bicknell-Brown, E. 1982. Variation in the lipid composition of rabbit muscle sarcoplasmic reticulum membrane with muscle type. *J. Biol. Chem.* 257:14136–39.

Burns, C. P.; Welshman, I. R.; Scallen, T. J.; and Spector, A. A. 1982. Mechanism of defective sterol synthesis in human leukocytes. *Biochim. Biophys. Acta* 713:519–28.

Chang, T. Y., and Doolittle, G. M. 1983. Acyl coenzyme A:Cholesterol O-acyltransferase. In *The enzymes*, ed. P. D. Boyer, vol. 16, 523–39. New York: Academic Press.

Clark, A. J., and Bloch, K. 1959. Conversion of ergosterol to 22-dehydrocholesterol in *Blattella germanica*. *J. Biol. Chem.* 234.2589–94.

Cornell, R. B., and Goldfine, H. 1983. The coordination of sterol and phospholipid synthesis in cultured myogenic cells. Effect of cholesterol synthesis inhibition on the synthesis of phosphatidylcholine. *Biochim. Biophys. Acta* 750:504–20.

Cornell, R. B., and Horwitz, A. F. 1980. Apparent coordination of the biosynthesis of lipids in cultured cells: Its relationship to the regulation of the membrane sterol: phospholipid ratio and cell cycling. *J. Cell Biol.* 86:810–19.

Crain, R. C., and Zilversmit, D. 1980. Two nonspecific phospholipid exchange proteins from beef liver. 1. Purification and Characterization. *Biochemistry* 19:1433–39.

Criado, M.; Eibl, H.; and Barrantes, F. J. 1982. Effects of lipids on acetylcholine receptor. Essential need of cholesterol for maintenance of agonist-induced state transition lipid vesicles. *Biochemistry* 21:3622–29.

Dahl, C.; Dahl, J.; and Bloch, K. 1980b. Effect of alkyl-substituted precursors of cholesterol on artificial and natural membranes and on the viability of *Mycoplasma capricolum*. *Biochemistry* 19:1462–67.

Dahl, J., and Dahl, C. 1983. Coordinate regulation of unsaturated phospholipid, RNA, and protein synthesis in *Mycoplasma capricolum* by cholesterol. *Proc. Natl. Acad. Sci. USA* 80:692–96.

Dahl, J.; Dahl, C.; and Bloch, K. 1980a. Sterols in membranes: Growth characteristics and membrane properties of *Mycoplasma capricolum* cultured on cholesterol and lanosterol. *Biochemistry* 19:1467–72.

Dalziel, A. W.; Robbins, E. S.; and McNamee, M. G. 1980. The effect of cholesterol on agonist-induced flux in reconstituted acetylcholine receptor vesicles. *FEBS Lett.* 122:193–96.

Day, A. J.; Fidge, N. H.; and Wilkinson, G. N. 1966. Effect of cholesterol in suspension on the incorporation of phosphate into phospholipid by macrophages in vitro. *J. Lipid Res.* 7:132–40.

Djerassi, C.; Theobald, N.; Kokke, W. C.; Pak, C. S.; and Carlson, R. M. K. 1979. Recent progress in the marine sterol field. *Pure Appl. Chem.* 51:1815–28.

Edward, D. G., and Fitzgerald, W. A. 1951. Cholesterol in the growth of organisms of the Pleuropneumonia group. *J. Gen. Microbiol.* 5:576–86.

Efrati, H.; Oschry, Y.; Eisenberg, S.; and Razin, S. 1982. Preferential uptake of lipids by *Mycoplasma*

membranes from human plasma low density lipoproteins. *Biochemistry* 21:6477–82.

Elliot, C. G. 1977. Sterols in fungi: Their functions in growth and reproduction. *Adv. Microl. Physiol.* 15:121–73.

Field, F. J. and Mathur, S. N. 1983. Regulation of acyl CoA-cholesterol acyltransferase by 25-hydroxycholesterol in rabbit intestinal microsomes and absorptive cells. *J. Lipid Res.* 24:1049–59.

*Fieser, L. F., and Fieser, M. 1957. *Steroids*, New York: Reinhold.

*Goldstein, J. L., and Brown, M. S. 1977. The low density lipoprotein pathway and its relation to atherosclerosis. *Ann. Rev. Biochem.* 46:897–930.

*———. 1983. Lipoprotein metabolism in the macrophage. *Ann. Rev. Biochem.* 52:223–61.

Gollub; E. G.; Liu K.-P.; Dayan, J.; Adlersberg, M.; and Sprinson, D. 1977. Yeast mutants deficient in heme biosynthesis and a heme mutant additionally blocked in cyclization of 2,3-oxidosqualene. *J. Biol. Chem.* 252:2846–54.

Heintz, R., and Benveniste, P. 1974. Plant sterol metabolism. *J. Biol. Chem.* 249:4267–74.

Hoover, R. L.; Dawidowicz, E. A.; Robinson, J. M.; and Karnovsky, M. J. 1983. Role of cholesterol in the capping of surface immunoglobulin receptors on murine lymphocytes. *J. Cell Biol.* 97:73–80.

Jonas, A., and McHugh, H. T. 1983. Reaction of lecithin: cholesterol acyltransferase with micellar complexes of apolipoprotein A-1 and phosphatidylcholine, containing variable amounts of cholesterol. *J. Biol. Chem.* 258:10335–40.

Kohl, W.; Gloe, A.; and Reichenbach, H. 1983. Steroids from *Mycobacterium nannocystis exedens*. *J. Gen. Microbiol.* 129:1629–35.

Kramer, R. 1982. Cholesterol as an activator of ADP–ATP exchange in reconstituted liposomes and in mitochondria. *Biochim. Biophys. Acta* 693:296–304.

Lim, P. H.; Pritchard, P. H.; Paddon, H. B.; and Vance, D. E. 1983. Stimulation of hepatic phosphatidylcholine biosynthesis in rats fed a high cholesterol and cholate diet correlates with translocation of CTP:phosphocholine cytidyltransferase from cytosol to microsomes. *Biochim. Biophys. Acta* 753:74–82.

Mallory, F. B.; Gordon, J. T.; and Conner, R. L. 1963. The isolation of a pentacyclic triterpenoid alcohol from a protozoan. *J. Am. Chem. Soc.* 85:1362–63.

McMorris, T. C. 1978. Sex hormones of the aquatic fungus *Achlya*. *Lipids* 13:716–22.

Moore, N. F.; Patzer, E. J.; Shaw, J. M.; Thompson, T. E.; and Wagner, R. R. 1978. Interaction of vesicular stomatitis virus and effect of virion membrane fluidity and infectivity. *J. Virol.* 27:320–29.

Nes, W. D., and Stafford, A. E. 1983. Evidence for metabolic and functional discrimination of sterols by *Phytophthora cactorum*. *Proc. Natl. Acad. Sci. USA* 80:3227–31.

*Nes, W. R., and McKean, M. L. 1977. *Biochemistry of steroids and other isopentoids*, University Park Press.

North, P., and Fleischer, S. 1983. Alteration of synaptic membrane cholesterol/phospholipid ratio using a lipid transfer protein. *J. Biol. Chem.* 258:1242–53.

Orci, L.; Carpenter, J. L.; Perelet, A.; Anderson, R. G. W.; Goldstein, J. L.; and Brown, M. S. 1978. Occurrence of low density lipoprotein receptors within large pits on the surface of human fibroblasts as demonstrated by freeze-etching. *Exp. Cell Res.* 113:1–13.

Pal, R.; Wiener, J. R.; Barenholz, Y.; and Wagner, R. R. 1983. Influence of the membrane glypoprotein and cholesterol of vesicular stomatitis virus on the dynamics of viral and model membranes. *Biochemistry* 22:3624–30.

Popot, J. L.; Demel, R. A.; Sobel, A.; van Deenen, L. L. M.; and Changeux, J. P. 1978. Interaction of the acetylcholine (nicotinic) receptor protein from *Torpedo marmorata* electric organ with monolayers of pure lipids. *Eur. J. Biochem.* 85:27–42.

*Porter, J. W., and Spurgeon, S. L. 1981. *Biosynthesis of isoprenoid compounds*, Vol. I, New York: Wiley Interscience.

Puchwein, G.; Pfeuffer, T.; and Helmreich, E. J. M. 1973. Uncoupling of catecholamine activation of pigeon erythrocyte membrane adenylate cyclase by filipin. *J. Biol. Chem.* 249:3232–40.

Ramgopal, M., and Bloch, K. 1983. Sterol synergism in yeast. *Proc. Natl. Acad. Sci. USA* 80:712–15.

Rodriguez, R. J.; Taylor, F. R.; and Parks, L. 1982. A requirement for ergosterol to permit growth of yeast sterol auxotrophs on cholestanol. *Biochem. Biophys. Res. Comm.* 106:435–41.

Rohmer, M.; Bouvier, P.; and Ourisson, G. 1980. Non-specific lanosterol and hopanoid biosynthesis by a cell-free system from the bacterium *Methylococcus capsulatus*. *Eur. J. Biochem.* 112:557–60.

*Sabine, J. R. 1977. *Cholesterol*, New York: Marcel Dekker.

Schindler, H., and Quast, U. 1980. Functional acetylcholine receptor from *Torpedo marmorata* in planar membranes. *Proc. Natl. Acad. Sci. USA* 77:3052–56.

Sexton, R. C.; Panini, S. R.; Azran, F.; and Rudney, H. 1983. Effects of 3β-[2-(diethylamino)ethoxyl]-androst-5-en-17-one on the synthesis of cholesterol and ubiquinone in rat intestinal epithelial cell cultures. *Biochemistry* 22:5687–92.

Silberkang, M.; Havel, C. M.; Friend, D. S.; McCarthy, B. J.; and Watson, J. A. 1983. Isoprene synthesis in isolated embryonic *Drosophila* cells. *J. Biol. Chem.* 258:8503–11.

Slotte, J. P., and Lundberg, B. 1983. Effects of cholesterol surface transfer on cholesterol and phosphatidylcholine syntheses in cultured rat arterial smooth muscle cells. *Med. Biol.* 61:223–27.

Suckling, K. E.; Stange, E. F.; and Dietschy, J. M. 1983. In vivo modulation of rat liver acyl-coenzyme A:cholesterol acyltransferase by phosphorylation and substrate supply. *FEBS Lett.* 158:29–32.

van den Oord, A.; Danielsson, H.; and Ryhage, R. C. 1965. On the structure of the emulsifiers in gastric juice from the crab. *J. Biol. Chem.* 240:2242–47.

Warner, S. A.; Sovocool, G. W.; and Domnas, A. J. 1983. Comparative utilization of sterols by the oomycetes *Lagenidium giganteum* and *Lagenidium callinectes*. *Exp. Mycol.* 7:227–32.

Whelton, A. D.; Gordon, L. M.; and Houslay, M. D. 1983. Adenylate cyclase is inhibited upon depletion of plasma-membrane cholesterol. *Biochem. J.* 212:331–38.

White, J.; Kielian, M.; and Helenius, A. 1983. Membrane fusion proteins of enveloped animal viruses. *Quart. Rev. Biophys.* 16(2):151–95.

Yeagle, P. 1983. Cholesterol modulation of $(Na^+ + K^+)$-ATPase ATP hydrolyzing activity in the human erythrocyte. *Biochim. Biophys. Acta* 727:39–44.

CHAPTER 2

Physical Properties and Functional Roles of Lipids in Membranes

Pieter R. Cullis
Michael J. Hope

INTRODUCTION AND OVERVIEW

Biological membranes contain an astonishing variety of lipids. As detailed throughout this book, generation of this diversity requires elaborate metabolic pathways. The lipid compounds representing the end products of these pathways must bestow significant evolutionary advantages to the cellular or multicellular systems in which they reside, implying particular functional roles for each component. However, clarification of the functional roles of individual lipid species has proven a difficult problem. Here we present a synopsis of the physical properties of lipid systems and indicate how they may relate to the functional capacities of biological membranes.

The major role of membrane lipids has been understood in broad outline since the early experiments of Gorter and Grendell (1925), who extracted lipids from the erythrocyte membrane and measured the area these lipids were able to cover as a monolayer at an air-water interface. Although a number of unwarranted assumptions were made in the analysis of these data, the errors fortunately compensated for one another and led to the correct conclusion that the erythrocytes contained sufficient lipid to provide a bilayer lipid matrix surrounding the red blood cell. This bilayer lipid organization, which provides a permeability barrier between exterior and interior compartments, was further characterized by Danielli and Davson (1935) and has remained a dominant theme in our understanding of the organization and function of biological membranes. Subsequent observations that such bilayers are fluid, allowing rapid lateral diffusion of lipid and protein in the plane of the membrane, and that membrane proteins are often inserted into and through the lipid matrix have further contributed to our present understanding of membranes, resulting in the Singer and Nicolson (1972) *fluid mosaic model*, a refined version of which is shown in Figure 2.1.

The ability of lipids to assume the basic bilayer organization is dictated by a unifying characteristic of membrane lipids—namely, their *amphipathic* character, which is indicated by the presence of a

Plasma membrane

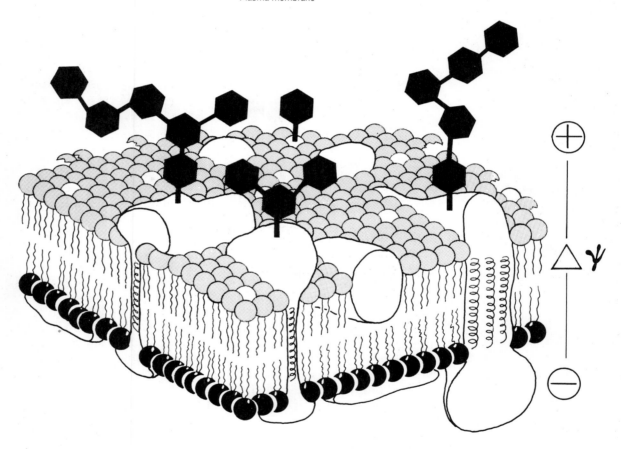

Cytosol

Figure 2.1. The topography of membrane protein, lipid, and carbohydrate in the fluid mosaic model of a typical eucaryotic plasma membrane. Phospholipid asymmetry results in the preferential location of phosphatidylethanolamine and phosphatidylserine in the cytosolic monolayer. Carbohydrate moieties on lipids and proteins face the extracellular space. $\Delta \psi$ represents the transmembrane potential, negative inside the cell.

polar or hydrophilic (water loving) head group region and nonpolar or hydrophobic (water hating) region. The chemical nature of these hydrophilic and hydrophobic sections can vary substantially. However, the lowest-energy macromolecular organizations assumed in the presence of water have similar characteristics, where the polar regions tend to orient toward the aqueous phase, while the hydrophobic sections are sequestered from water. In addition to the familiar bilayer phase, a number of other macromolecular structures are compatible with these constraints, as indicated later in this chapter. It is of particular interest that many naturally occurring lipids prefer nonbilayer structures in isolation.

The fluidity of membranes depends on the nature of the acyl chain region comprising the hydrophobic domain of most membrane lipids. Most lipid species in isolation can undergo a transition from a very viscous gel (frozen) state to the fluid (melted) *liquid-crystalline* state as the temperature is increased. This transition has been studied intensively, since the local fluidity, as dictated by the gel or liquid-crystalline nature of membrane lipids, may regulate membrane-mediated processes. However, at physiological temperatures most, and often all, membrane lipids are fluid; thus, the major emphasis of this chapter will concern the properties of liquid-crystalline lipid systems. As indicated later, the melted nature of the acyl chains depends on the presence of cis double bonds, which can dramatically lower the transition temperature from the gel to the liquid-crystalline state for a given lipid species.

The ability of lipids to self-assemble into fluid bilayer structures is consistent with two major roles in membranes: establishing a permeability barrier and providing a matrix with which membrane proteins are associated. Roles of individual lipid components may therefore relate to establishing appropriate permeability characteristics, satisfying insertion and packing requirements in the region of integral proteins (which penetrate into or through the bilayer), as well as allowing the surface association of peripheral proteins via electrostatic interactions. All these demands are clearly critical. An intact permeability barrier to small ions such as Na^+, K^+, and H^+, for example, is vital for establishing the electrochemical gradients which give rise to a membrane potential and drive other membrane-mediated transport processes. In addition, the lipid in the region of membrane protein must seal the protein into the bilayer so that nonspecific leakage is prevented and an environment appropriate to a functional protein conformation is provided.

In summary, membrane lipids satisfy demands related to membrane structure, fluidity, and permeability, as well as protein association and function. These aspects will be dealt with at length; however, before a coherent discussion is possible, a basic overview of lipid diversity in membranes, a study of the methods employed to isolate individual components, and a discussion of the physical properties of lipids are essential. These will comprise the bulk of the next three sections.

LIPID DIVERSITY AND DISTRIBUTION

The general definition of a *lipid* is a biological material soluble in organic solvents, such as ether or chloroform. Here we shall discuss the diverse chemistry of the subclass of lipids which are found in membranes. This excludes other lipids which are poorly soluble in bilayer membrane systems, such as fats (triglycerides) and cholesterol esters.

Chemical Diversity of Lipids

The major classes of lipids found in biological membranes are summarized in Figure 2.2. We shall discuss most of these compounds in depth at various points in this book; we present only a brief synopsis here. In eucaryotic membranes the glycerol-based phospholipids are predominant, including phosphatidylcholine, phosphatidylethanolamine, phosphatidylserine, phosphatidylinositol, and cardiolipin. Sphingosine-based lipids, including sphingolipids and the glycosphingolipids, also constitute a major fraction. The glycolipids, which can also include carbohydrate-containing glycerol-based lipids (found particularly in plants), play major roles as cell-surface-associated antigens and recognition factors in eucaryotes. The physical properties of glycolipids have not been extensively characterized and will not be discussed in this chapter. Cholesterol is also a major component of eucaryotic membranes, particularly in mammalian plasma membranes, where it may be present in equimolar proportions with phospholipid.

In most procaryotic membranes, phosphatidylcholine is not usually present; the major phospholipids observed are phosphatidylethanolamine, phosphatidylglycerol, and cardiolipin. In plant membranes on the other hand, lipids such as monogalactosyl and digalactosyl diglycerides can form the majority components of membranes such as the chloroplast membrane.

These observations give some impression of the lipid diversity in membranes, but it must be emphasized that this diversity is much more complex. Minority species such as sulfolipids, phospholipids with phosphorylated head groups, and lysolipids abound. Furthermore, each lipid species exhibits a characteristic fatty acid composition. In the case of glycerol-based phospholipids, for example, it is usual to find a saturated fatty acid esterified at the 1-position of the glycerol backbone and an unsaturated fatty acid at the 2-position. Also, in eucaryotic membranes it is usual to find that phosphatidyl-

Major classes of Phospholipid

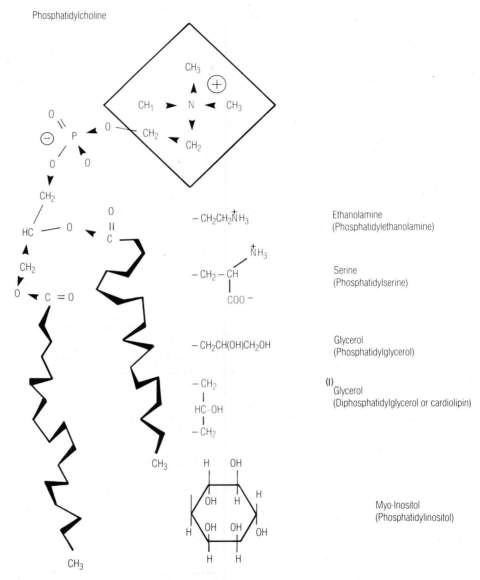

Figure 2.2. The structure of the phospholipid molecule distearoylphosphatidylcholine in the liquid-crystalline state is represented schematically. Head groups for the other major classes of phospholipid are also shown. [1] The glycerol moiety of a cardiolipin is esterified to two phosphatidic acid molecules.

Table 2.1. Gas Chromatographic Analyses of the Fatty Acid Chains in Human Red Cell Phospholipid

Chain length and unsaturation	Total phospholipids	Sphingomyelin	Phosphatidyl-choline (lecithin)	Phosphatidyl-ethanolamine	Phosphatidyl-serine
16:0*	20.1%	23.6 %	31.2 %	12.9 %	2.7 %
18:0	17.0	5.7	11.8	11.5	37.5
18:1	13.3	+	18.9	18.1	8.1
18:2	8.6	+	22.8	7.1	3.1
20:0	+†	1.9	+	+	+
20:3	1.3	—	1.9	1.5	2.6
22:0	1.9	9.5	1.9	1.5	2.6
20:4	12.6	1.4	6.7	23.7	24.2
23:0	+	2.0	+	+	+
24:0	4.7	22.8	+	+	+
22:4	3.1	—	+	7.5	4.0
24:1	4.8	24.0	+	+	+
22:5	2.0	—	+	4.3	3.4
22:6	4.2	—	2.1	8.2	10.1

Note: The data are expressed as weight % of the total.
* This code indicates the number of carbon atoms in the chain and the number of double bonds.
† Denotes that the concentration did not exceed 1% of the total.
Reproduced with permission of van Deenen and de Gier (1974).

ethanolamine and phosphatidylserine, for example, are more unsaturated than other phospholipids. In order to give a true impression of the molecular diversity of phospholipids in a single membrane, we list in Table 2.1 the fatty acid composition of phospholipids found in the human erythrocyte membrane. From this table and other analyses (van Deenen and de Gier 1974) it is clear that the number of different molecular species of phospholipids in a membrane can easily exceed 100.

The lipid composition of membranes can vary dramatically among different cells or organelles. In addition, different sides or monolayers of the same membrane can contain different lipid species. These different compositions are indicated in the following sections.

Membrane Lipid Compositions

The lipid compositions of several mammalian membrane systems are given in Table 2.2. Dramatic differences are observed for the cholesterol contents. Plasma membranes such as those of myelin or the erythrocyte contain equimolar quantities of cholesterol and phospholipid, whereas the organelle membranes of endoplasmic reticulum or the inner mitochondrial membrane contain little or no cholesterol. This cholesterol distribution correlates well with the distribution of

Table 2.2. The Lipid Composition of Various Biological Membranes

Lipid	Erythrocyte*	Myelin*	Mitochondria† (inner and outer membrane)	Endoplasmic reticulum†
Cholesterol	23	22	3	6
Phosphatidylethanolamine	18	15	35	17
Phosphatidylcholine	17	10	39	40
Sphingomyelin	18	8	—	5
Phosphatidylserine	7	9	2	5
Cardiolipin	—	—	21	—
Glycolipid	3	28	—	—
Others	13	8	—	27

Note: The data are expressed as weight % of total lipid.
* Human sources.
† Rat liver.

sphingomyelin. Cholesterol may have a "fluidizing" role in membranes containing sphingomyelin, which is relatively saturated.

Cardiolipin is almost exclusively localized to the inner mitochondrial membrane, and it has been suggested that cardiolipin is required for the activity of cytochrome *c* oxidase, the terminal member of the respiratory electron-transfer chain. In general, the lipids of more metabolically active membranes are considerably more unsaturated, as indicated in Table 2.3.

It is interesting to note that the lipid composition of the same membrane system in different species can also vary significantly. The rat erythrocyte membrane, for example, contains low levels of sphingomyelin and elevated levels of phosphatidylcholine with respect to the human erythrocyte. In the bovine erythrocyte, this distribution is reversed, with high sphingomyelin, and low phosphatidylcholine, contents.

Table 2.3. Double-Bond Composition of Phospholipids of Various Membranes

Membrane	Number of double bonds per acyl chain
Myelin	0.5
Erythrocyte	1.0
Sarcoplasmic reticulum	1.4
Mitochondria (inner)	1.5
Nerve synapse	>2

Transbilayer Lipid Asymmetry

A major discovery of recent years has been the observation that the inner and outer leaflets of membrane bilayers may exhibit different lipid compositions (Op den Kamp 1979). Several different species of membranes have been investigated with respect to lipid asymmetry; however, the plasma membrane of human erythrocytes has been the most thoroughly investigated.

The results obtained indicate that most membranes display some degree of lipid asymmetry. The use of impermeable probes that react with the primary amines of phosphatidylethanolamine and phosphatidylserine on only one side of the membrane has shown that the majority of the amino-containing phospholipids of the erythrocyte are located on the inner monolayer. Combinations of chemical probes and phospholipase treatments indicate that in a normal red blood cell all the phosphatidylserine is located in the inner monolayer, whereas approximately 20% of the phosphatidylethanolamine can be detected at the outer surface, with 80% confined to the inner monolayer. The outer monolayer consists predominantly of phosphatidylcholine, sphingomyelin, and glycolipids. Figure 2.3 summarizes the transbilayer lipid distributions obtained for various mammalian cell membranes and viral membranes derived from animal-cell plasma membranes. A common feature is that the amino-containing phospholipids are chiefly limited to the cytosolic side of plasma membranes. It is interesting that the little information available for organelle membranes suggests that phosphatidylethanolamine and phosphatidylserine are also oriented toward the cytosol.

A general feature of plasma membrane asymmetry is that the majority of phospholipids that exhibit a net negative charge at physiological pH (phosphatidylserine and phosphatidylinositol—phosphatidylethanolamine is only weakly anionic) are limited to the cytosolic half of the bilayer. Certain proteins appear to be involved in maintaining this asymmetry. Treatment of erythrocytes with diamide, which induces cross-linking of the cytoskeletal protein spectrin, results in the appearance of phosphatidylserine in the outer monolayer. Pathological red blood cells known to have lesions associated with cytoskeletal proteins also exhibit a partial breakdown of asymmetry, with an increased exposure of phosphatidylserine and phosphatidylethanolamine on the outer half of the bilayer and an equivalent transfer of phosphatidylcholine to the inner monolayer (Lubin and Chiu 1982).

These experiments suggest a possible interaction between cytoskeletal proteins and membrane phospholipids. The functional importance of lipid asymmetry is not clear but could be related to prevention of exposure of phosphatidylserine at the outer surface of a normal cell (which has been suggested to be a signal of senescence (Tanaka and Schroit 1983)). Alternatively, phosphatidylserine may

Outer Monolayer

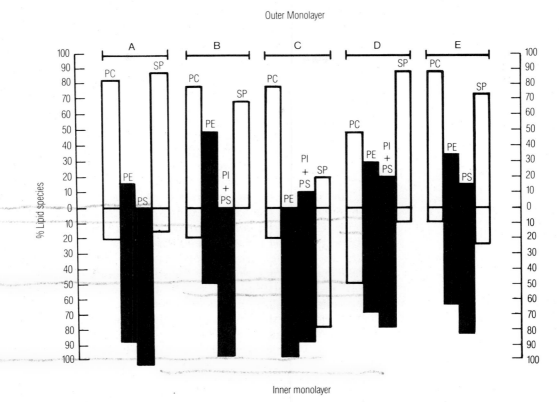

Figure 2.3. Phospholipid asymmetry in plasma membranes. (A) Human erythrocyte membrane, (B) rat liver blood sinusoidal plasma membrane, (C) rat liver continuous plasma membrane, (D) pig platelet plasma membrane, (E) VSV envelope derived from hamster kidney BHK-21 cells. The data were adapted from Houslay and Stanley (1982). See Table 2.6 for phospholipid nomenclature.

be required at the cytosolic surface to maintain a functioning cytoskeleton.

MODEL MEMBRANE SYSTEMS

The physical properties and functional roles of individual lipid species in membranes are exceedingly difficult to ascertain in an intact biological membrane due to the complex lipid composition. In order to gain insight into the roles of individual components, it is necessary to construct *model membrane* systems that contain the lipid species of interest. This requires three steps, namely, isolation or chemical synthesis of a given lipid, construction of an appropriate model system containing that lipid, and subsequent incorporation of a particular protein if understanding the influence of a particular lipid on protein function is desired. By this method specific models of biological membranes can be achieved in which the properties of individual lipid components can be well characterized.

Lipid Isolation and Purification

A variety of techniques has been developed for isolation of lipids from membranes (Kates 1975). These differ according to the particular source and type of lipid being isolated. A procedure commonly employed for the preparation of erythrocyte phospholipids is illustrated in Figure 2.4. A first step common to most procedures is to disrupt the membrane in a solvent system which denatures and precipitates most of the protein and solubilizes the lipid component. The Bligh and Dyer procedure is perhaps most often employed and involves incubation of the membrane system in a chloroform-methanol-water (1:2:0.8) (v:v:v) mixture, which forms a one-phase system. The subsequent addition of chloroform and water to the mixture containing the extracted lipids results in a two-phase system where the lower (chloroform) phase contains most membrane lipids.

Column chromatography is usually subsequently employed for isolation of individual lipid species. A solid phase such as silicic acid, DEAE cellulose, aluminum oxide, or carboxymethyl cellulose is used, depending upon the lipid being isolated, and lipids are eluted using mixtures of solvents with different polarities, such as chloroform and methanol. *Thin-layer chromatography* is generally used for lipid identification and for ascertaining purity. All these separation techniques rely upon the different partitioning characteristics of lipids between the stationary phase surface and mobile solvent phase for different solvent polarities. The exact nature of the binding of lipid to the solid phase is not well understood but appears to involve both electrostatic and hydrophobic interactions. Carboxymethyl cellulose and DEAE cellulose are often used for separation of anionic lipids.

Modern developments include *high-pressure preparative liquid chromatography*, which enables the rapid purification of large quantities of natural lipids (Patel and Sparrow 1978). Techniques for separating phospholipids according to the degree of acyl chain unsaturation are, as yet, tedious and expensive and normally utilize the tendency of unsaturated lipids to form complexes with certain metals; silver ions are commonly employed. Silica gel impregnated with silver nitrate can be used to prepare appropriate columns or thin-layer plates.

Reversed-phase chromatography (Skipski and Barclay 1969), where the stationary phase is hydrophobic and the mobile phase hydrophilic, is becoming more popular for separation of membrane lipids. The solid support is usually coated with hydrocarbon chains of a defined length (and consequently of regulated hydrophobicity), and the mobile phase is hydrophilic. This technique is particularly useful for separating single lipid classes according to their acyl chain length.

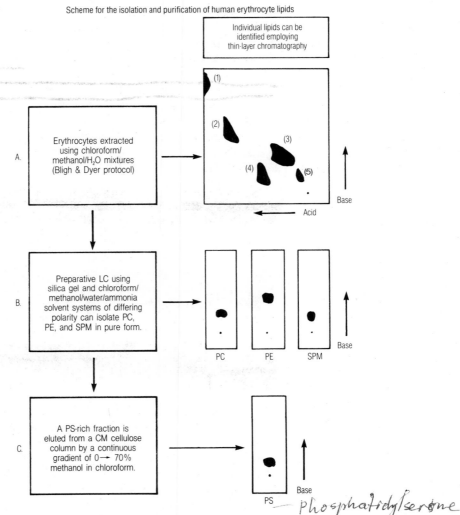

Scheme for the isolation and purification of human erythrocyte lipids

Individual lipids can be
identified employing
thin-layer chromatography

A. Erythrocytes extracted
using chloroform/
methanol/H₂O mixtures
(Bligh & Dyer protocol)

B. Preparative LC using
silica gel and chloroform/
methanol/water/ammonia
solvent systems of differing
polarity can isolate PC,
PE, and SPM in pure form.

C. A PS-rich fraction is
eluted from a CM cellulose
column by a continuous
gradient of 0 → 70%
methanol in chloroform.

phosphatidylserone

Figure 2.4. An outline of the procedure for extracting and purifying the phospholipid species of human erythrocytes. In step A, red cells are extracted using the Bligh and Dyer protocol. Denatured hemoglobin precipitates at this stage and is readily removed by centrifugation. Two-dimensional thin-layer chromatography is used to identify all the phospholipid species in the total lipid extract: (1) cholesterol, (2) PE, (3) PC, (4) PS, and (5) SPM. Step B utilizes preparative liquid chromatography (LC) to obtain pure PC, PE, and SPM. The PC and SPM fractions are readily separated using chloroform/methanol/water (60:30:4, v/v) to elute the lipid from the silica gel column. PE is further purified by passing the lipid once more through the column using chloroform/methanol/water/25% ammonium hydroxide (60:30:1:1, v/v). In step C an impure PS fraction, obtained from the passes outlined above, is purified by elution from carboxymethyl (CM) cellulose using a continuous gradient of 0 to 70% methanol in chloroform. For phospholipid nomenclature, see Table 2.6. Acid refers to the thin-layer plate running solvent chloroform/methanol/acetic acid/water (60:30:8:3, v/v) and base, to chloroform/methanol/25% ammonium hydroxide/water (90:54:6:5, v/v).

Techniques for Making Model Membrane Vesicles

Once lipids have been isolated, purified, and chemically characterized, their properties as membrane components can be studied. For this purpose a number of techniques have been developed for producing model membranes from lipids. Preparation of the simplest model system involves the straightforward hydration of a lipid film by mechanical agitation, such as vortex mixing. In the case of bilayer-forming lipids, this hydration results in a macromolecular structure which is composed of a series of concentric bilayers interspersed by narrow aqueous spaces (Bangham, Standish, and Watkins 1965). Such structures are usually referred to as *liposomes* or *multilamellar vesicles* (MLVs) and have been used for many years as models for the bilayer matrix of biological membranes. Their use is mostly restricted to physical studies on bilayer organization and the motional properties of individual lipids within a membrane structure. MLVs are not ideal models for the study of other aspects of lipids in membrane structure and function, mainly because as little as 10% of the total lipid of a MLV is contained in the outermost bilayer. As a result, methods have been sought by which unilamellar (single bilayer) model membranes can be obtained either directly or from MLVs.

Small unilamellar vesicles (SUVs) can be made from MLVs by subjecting the MLVs to ultrasonic irradiation (Huang 1969) or by passage through a French press (Barenholtz, Amselem, and Lichtenberg 1979). However, their small size limits their use in model membrane studies. Typically, diameters in the range 25–40 nm are observed. The radius of curvature experienced by the bilayer in SUVs is so small (Figure 2.5) that the ratio of lipid in the outer monolayer to lipid in the inner monolayer can be as large as 2:1. As a result of this curvature, the packing constraints experienced by the lipids perturb their physical properties in comparison with less highly curved systems. This restricts the use of SUVs for physical studies on the properties of membrane lipid. Moreover, the aqueous volume enclosed by the SUV membrane is often too small to allow studies of permeability or ion distributions between the internal and external aqueous compartments.

A more useful membrane model is the *large unilamellar vesicle* (LUV) system, where the mean diameter is larger, and the distribution of lipid between the outer and inner monolayers is closer to 1:1. The most common procedures for producing LUVs (outlined in Table 2.4) result in unilamellar vesicles with diameters ranging from 50 to 500 nm. These preparative procedures usually include the use of detergents (Mimms et al. 1981) or organic solvents (Szoka and Papahadjopoulos 1980), although LUVs can be produced directly from MLVs (Hope et al. 1985).

Procedures that employ detergents vary depending upon the type of detergent; however, the principle is the same. Lipids are solubi-

Curvature and some characteristics of large unilamellar and small unilamellar vesicles

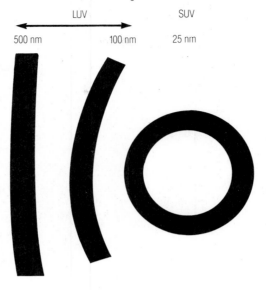

Diameter (nm)	IM/OM (mole ratio)	Trap (μl per μmol)	No. phospholipid molecules per vesicle	No. vesicles per μmol of lipid
25	0.36	0.2	3.6×10^3	1.7×10^{14}
100	0.81	2.7	8.0×10^4	7.6×10^{13}
500	0.96	17	2.2×10^6	2.7×10^{11}

Figure 2.5. The curvature and some characteristics of large unilamellar vesicles (LUV) and small unilamellar vesicles (SUV). LUVs typically have diameters in the range 100–500 nm. SUVs prepared by sonication can be as small as 25 nm in diameter. The radius of curvature for each vesicle size is shown in proportion. The ratio of lipid in the inner monolayer (IM) compared with lipid in the outer monolayer (OM) gives an indication of the packing restrictions in bilayers with a small radius of curvature. The trapped volume refers to the volume of aqueous medium enclosed per micromole of phospholipid. The calculations were made assuming a bilayer thickness of 4 nm and a surface area per phospholipid molecule of 0.6 nm^2.

lized by the detergent of choice (such as cholate or octylglucoside); then the detergent is removed either rapidly by dilution or gel filtration, or slowly by dialysis. As the detergent concentration decreases, the lipids adopt unilamellar vesicular structures. The

Table 2.4. Common Procedures for the Generation of Large Unilamellar Vesicles

Category	Technique	Trapped Volumes (μL/μmol lipid)	Advantages	Disadvantages
Detergent dialysis	Cholate dialysis	0.5–5	Large trapped volumes.	Detergents difficult to remove completely; procedures lengthy; low trapping efficiency; limited to certain lipid mixtures.
	Octylglucoside dialysis	~10		
Hydration from organic solvent	Reverse-phase evaporation	~10	High trapping efficiency for reverse-phase technique only; large trapped volumes.	Technically complex; limited to certain lipid mixtures; vaporization and injection techniques have low trapping efficiency; residual organic solvent.
	Ether vaporization	~10		
	Ethanol injection	0.5–5		
Direct from MLVs	MLVs extruded through 0.1-μm polycarbonate filters	1–2	High trapping efficiency for extrusion techniques only; no detergents or solvents used; fast procedures.	Trapped volumes low unless freeze-thaw protocol is employed.
	Plus freeze-thaw	1–10		
	Sonication	0.2–0.5		
	Plus freeze-thaw	1–10		

vesicle size can be controlled to some extent by the rate at which detergent is removed.

A number of methods exist for preparation of LUVs employing organic solvents (Szoka and Papahadjopoulos 1980). The lipid is first solubilized in an organic solvent which is subsequently diluted by aqueous buffer. The largest unilamellar vesicles are produced by injection procedures whereby lipid is dissolved in ether or ethanol, then slowly injected into aqueous buffer. An alternative protocol employing organic solvent is called the *reverse phase evaporation* procedure, which involves making an emulsion of lipid (dissolved in ether or mixtures of other organic solvents) and aqueous buffer. The organic solvent is carefully removed under partial vacuum, which gives rise to hydrated lipid in the form of a thick gel. This gel can be diluted and sized by extrusion through polycarbonate filters of defined pore sizes to give LUVs. Finally, it is possible to form LUVs by repeated extrusion of MLVs through polycarbonate filters with pore sizes of 100 nm or less (Hope et al. 1985). An advantage of this procedure is that it does not require detergents or solvents, which are difficult to remove completely.

Techniques for Making Planar Bilayers and Monolayers

Planar bilayers (also known as *black lipid membranes*) are favorite model membranes of electrophysiologists interested in current flow across a bilayer (Fettiplace et al. 1974). They are formed by dissolving phospholipids in a hydrocarbon solvent and painting them across a small aperture (approximately 2 mm in diameter) which separates two aqueous compartments. The solvent tends to collect at the perimeter of the aperture, leaving a bilayer film across the center. The electrical properties of the barrier are readily measured employing electrodes in the two buffered compartments. It is also possible to incorporate some membrane proteins into the film, if the protein can be solubilized by the hydrocarbon. With this technique, ion channels have been reconstituted and voltage-dependent ion fluxes recorded. The most serious problem of black lipid membranes is the presence of the hydrocarbon solvent, which may change the normal properties of the lipid bilayer being studied.

With regard to monolayer systems, amphipathic lipids orient at an air-water interface. The result is a monolayer film which, in the case of phospholipids, represents half of a bilayer, where the polar regions are in the aqueous phase and the acyl chains extend above the buffer surface. Such films can be compressed and their resistance to compression measured. The study of compression pressure versus surface area (occupied by the film) yields information on molecular packing of lipids and lipid-protein interactions. Perhaps the best-known result of monolayer studies is the *condensation effect* of cholesterol and phospholipid, in which the area occupied by a typical membrane phospholipid molecule and a cholesterol molecule in a monolayer is less than the sum of their molecular areas in isolation. This phenomenon provides a strong indication of a specific interaction between this sterol and membrane phospholipids (Demel and de Kruijff 1976).

Reconstitution of Integral Membrane Protein into Vesicles

An important step, both for the study of membrane protein function and for the building of simple but more representative biological membranes, is the insertion of purified integral membrane proteins into well-defined lipid model membranes. A large variety of membrane proteins have been reconstituted (Racker 1973). For the purpose of discussing the salient features of reconstitution techniques, we shall use the example of cytochrome c oxidase from bovine heart mitochondria. This integral membrane protein, which has been purified and is relatively well characterized, spans the inner mitochondrial membrane and oxidizes cytochrome c in the terminal reaction of the electron-transfer chain.

Purified integral proteins such as cytochrome oxidase maintain a functional conformation when solubilized in detergents. The goals of reconstitution can be summarized as follows. First, the protein must be inserted into a bilayer of desired lipid composition. This insertion is commonly achieved by solubilizing the lipid in detergent, mixing the solubilized lipid and protein together, and then removing the detergent by dialysis. This method produces LUVs containing various amounts of protein. Second, the reconstituted systems must have constant lipid to protein ratios between vesicles. Most reconstitution procedures give rise to heterogeneous systems, where vesicles contain various amounts of protein. Column chromatography techniques can be employed to obtain systems exhibiting uniform lipid to protein ratios (Madden, Hope, and Cullis 1984). Finally, the systems should have asymmetric protein orientation. In contrast with the intact biological membrane, the protein in reconstituted systems is not necessarily inserted with a well-defined asymmetric orientation. In the case of reconstituted cytochrome oxidase systems, for example, oxidase-containing vesicles can exhibit protein orientations in which the cytochrome c binding sites are on the outside or the inside. Asymmetric protein orientation can be achieved by reconstitution at low protein to lipid ratios such that most vesicles contain one or zero protein molecules. Populations containing only one oxidase molecule per vesicle with well-defined transmembrane orientations of the oxidase can subsequently be achieved by ion-exchange or affinity column chromatography, as illustrated in Figure 2.6.

In some cases asymmetric incorporation of other proteins can be achieved by different procedures. Erythrocyte glycophorin, for example, has a large carbohydrate-containing region which is normally localized on the exterior of the red cell. Reconstituted systems can be obtained by hydrating a dried film of lipid and glycophorin (McDonald and McDonald 1975), resulting in asymmetric vesicles in which more than 80% of the carbohydrate groups are on the vesicle exterior. This is presumably due to the small size of the reconstituted vesicle, which limits the fraction of the bulky carbohydrate-containing groups that can pack into the interior volume.

Alternative reconstitution techniques involving protein insertion into preformed vesicles have achieved some success in obtaining

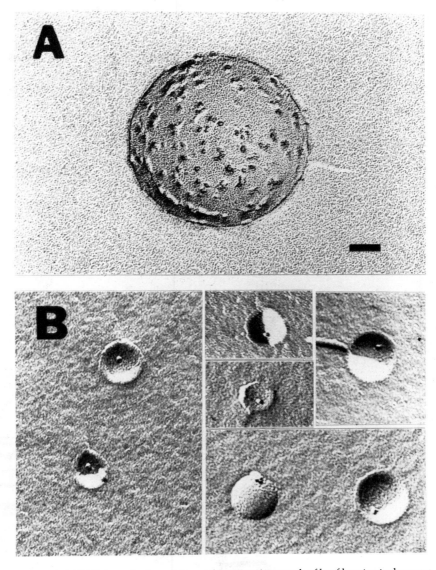

Figure 2.6. (A) Rotary-shadowed freeze-fracture micrograph of beef heart cytochrome *c* oxidase reconstituted into a vesicle of dioleoylphosphatidylcholine by the cholate dialysis procedure, at a protein to lipid ratio of 1:15 (w/w).

(B) Unidirectionally shadowed freeze-fracture micrographs of cytochrome *c* oxidase reconstituted at protein to lipid ratios of <1:5000 (w/w). Each particle has been shown to represent one dimer of cytochrome *c* oxidase and is approximately 10 nm in diameter. The bar represents 100 nm (see Madden, Hope, and Cullis [1984] for details of the reconstitution procedure).

asymmetric incorporation. One of these asymmetric insertion techniques utilizes the detergent octylglucoside. It is possible to form vesicles in the presence of relatively high detergent concentrations (approximately 20mM) which are sufficient to solubilize the spike protein of Semliki Forest virus (Helenius, Sarvas, and Simons 1981). The spike protein consists of a hydrophilic spike and a smaller hydrophobic anchor portion of the molecule. The anchor portion is solubilized by a coat of detergent, and this half of the molecule can insert into the preformed bilayer on dialysis.

In summary, a large variety of sophisticated and well-defined model membrane systems are becoming available. The simplest model systems consist of aqueous dispersions of lipid which can be converted to LUV forms by a variety of techniques. The subsequent incorporation of protein, with well-defined lipid to protein ratios and asymmetric transmembrane protein orientations, is becoming more feasible. Problems remain, however, both in removing the last traces of detergent in reconstituted systems and in generating the lipid asymmetry observed in biological membrane systems.

PHYSICAL PROPERTIES OF LIPIDS
Membrane Fluidity

It is common practice to characterize membranes in terms of an ill-defined parameter known as *fluidity*. Unfortunately, the concept of membrane fluidity can be most misleading. For example, it is commonly assumed that a more saturated lipid composition, or the presence of cholesterol, makes membranes less fluid. This is not necessarily the case. Strictly speaking, membrane fluidity is the reciprocal of membrane viscosity, which in turn is inversely proportional to rotational and lateral diffusion rates of membrane components. Thus, a linear relation between membrane fluidity and rotational and lateral diffusion rates would be expected with increasing cholesterol content, for example. However, incorporation of cholesterol into liquid-crystalline phosphatidylcholine model membranes has little or no influence on the phospholipid lateral diffusion rates (Lindblom, Johansson, and Arvidson 1981), and can actually increase rotational diffusion rates. The major influence of cholesterol or decreased unsaturation is to increase the order in the hydrocarbon matrix. It is this increase or decrease in order, which is a measurable quantity expressed by NMR or ESR order parameters (Davis 1983), for example, that should be correlated with such changes as increased or decreased membrane permeabilities.

Gel–Liquid-Crystalline Phase Behavior

As indicated earlier, membrane lipids can exist in a frozen gel state or fluid liquid-crystalline state, depending on the temperature (Silvius 1982), as illustrated in Figure 2.7. Transitions between the gel and

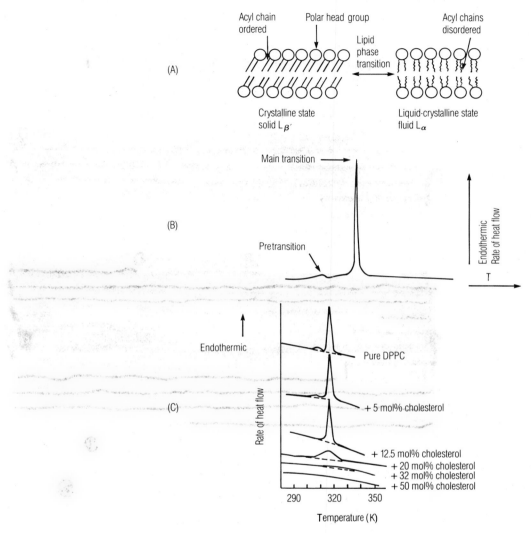

Figure 2.7. The phospholipid gel–liquid-crystalline phase transition and the effect of cholesterol. (A) Phospholipids, when fully hydrated, can exist in the gel or crystalline form (L_β) or in the fluid or liquid-crystalline state (L_α). In bilayers of gel-state phosphatidylcholine, the molecules can be packed such that the acyl chains are tilted with respect to the bilayer normal ($L_{\beta'}$) state. Raising the temperature converts the crystalline state into the liquid-crystalline state. (B) The exothermic gel–liquid-crystalline phase transition as detected by DSC. T represents temperature. For dipalmitoylphosphatidylcholine (DPPC) the onset of the main transition occurs at approximately 41°C. The pretransition represents a small endothermic reorganization in the packing of the gel-state lipid molecules prior to melting. (C) Influence of cholesterol. The enthalpy of the phase transition (represented by the area under the endotherm) is dramatically reduced. At greater than 30 mol % cholesterol, the lipid phase transition is effectively eliminated. Adapted from Houslay and Stanley (1982).

liquid-crystalline phases can be monitored by a variety of techniques, including nuclear magnetic resonance (NMR), electron spin resonance (ESR), and fluorescence, among others. Perhaps the most direct technique is *differential scanning calorimetry* (DSC), which measures the heat absorbed (or released) by a sample as it undergoes an endothermic (or exothermic) phase transition. A representative DSC scan of dipalmitoylphosphatidylcholine, which exhibits a gel to liquid-crystalline transition temperature (T_c) of 41°C, is illustrated in Figure 2.7. Three parameters of interest in such traces are the area under the transition peak, which is proportional to the enthalpy of the transition; the width of the transition, which gives a measure of the "cooperativity" of the transition; as well as the transition temperature T_c itself. The enthalpy of the transition reflects the energy required to melt the acyl chains, whereas cooperativity reflects the number of molecules that undergo a transition simultaneously.

Before describing the calorimetric behavior of various phospholipid systems, we emphasize two general points. First, gel-state lipids always assume an overall bilayer organization, presumably because the interactions between the crystalline acyl chains are then maximized. Thus, the nonbilayer hexagonal (H_{II}) or other phases discussed in the following section are not available to gel-state systems. Second, species of naturally occurring lipids exhibit broad noncooperative transitions due to the heterogeneity in the acyl chain composition. Thus, sharp gel–liquid-crystal transitions, indicating highly cooperative behavior, are observed only for aqueous dispersions of molecularly well-defined species of lipid. These can presently be obtained only by synthetic routes.

The calorimetric behavior of a variety of synthetic phospholipids is given in Table 2.5. There are three points of interest. First, for the representative phospholipid species, phosphatidylcholine, there is an increase in T_c by approximately 20°C as each two-carbon unit is added and a corresponding increase in enthalpy (2–3 kcal/mol). Second, inclusion of a cis double bond at C-9 results in a remarkable decrease in T_c, which is further lowered as the degree of unsaturation is increased. It is interesting to note that inclusion of only one cis-unsaturated fatty acid at the C-1 or C-2 position of the glycerol backbone is sufficient to lower T_c from 41°C for dipalmitoyl phosphatidylcholine to −5°C for the palmitoyl-oleoyl species, a major molecular subspecies of phosphatidylcholine in biological membranes. A final point is that the T_c and enthalpy are also sensitive to the head-group constituent. For example, molecular species of phosphatidylethanolamine commonly exhibit T_c values 20°C higher than corresponding species of phosphatidylcholine. The data of Table 2.5 have some predictive value in that approximate values of T_c can be estimated for other molecular species of lipids.

Table 2.5. Temperature (T_c) and Enthalpy (ΔH) of the
Gel–Liquid-Crystalline Phase Transition of Phospholipids
(in Excess Water)

Lipid species*		$T_c \pm 2°C$	$\Delta H \pm 1\ kcal/mol$
12:0/12:0	PC	−1	3
14:0/14:0	PC	23	6
16:0/16:0	PC	41	8
16:0/18:1cΔ^9	PC		
16:1cΔ^9/16:1cΔ^9	PC	−36	9
18:0/18:0	PC	54	10
18:1cΔ^9/18:1cΔ^9	PC	−20	9
16:0/16:0	PE	63	9
16:0/16:0	PS	55	9
16:0/16:0	PG	41	9
16:0/16:0	PA	67	5

* The code denotes the number of carbons per acyl chain and the number of double bonds. Δ gives the position of the double bond. The abbreviations are PC, phosphatidylcholine; PE, phosphatidylethanolamine; PS, phosphatidylserine; PG, phosphatidylglycerol; PA, phosphatidic acid; c indicates cis.

The calorimetric behavior of individual lipid species cannot be directly related to the behavior of the complex lipid mixtures found in biological systems; therefore, considerable attention has been devoted to the properties of mixtures of pure lipid species. Two general features have emerged. First, when all component lipids are liquid crystalline (that is, $T > T_c$), the lipid systems exhibit characteristics consistent with complete mixing of the various lipids. Second, at temperatures below the T_c of one of the constituents, separation of the component with the highest melting temperature into crystalline domains (*lateral phase separation*) can occur under certain conditions. For example, equimolar mixtures of two saturated phosphatidylcholines differing by four carbon units or more ($\Delta T_c > 20°C$) can exhibit lateral phase separation (indicated by calorimetric and freeze-fracture studies). On the other hand, mixtures of two phosphatidylcholines differing in their length by only two carbon units are miscible and co-crystallize at a temperature intermediate between the component T_c values. Similar behavior is observed for binary mixtures of synthetic lipids with different head group compositions—if the T_c values of the components differ by more than 20°C, lateral phase separation phenomena can occur. It should be emphasized that lateral segregation of a particular lipid species in the plane of the membrane has been observed only for model systems in which a sizable proportion of the lipids exhibit a T_c which is not only well above the temperature at which the experiment is performed but is also significantly greater than the T_c of other component lipids.

Further studies of the calorimetric behavior of lipid systems have emphasized the remarkable physical properties of cholesterol (Demel and de Kruijff 1976), which are detailed in part in the previous chapter. This lipid has the ability to inhibit the crystallization of lipids to form gel-state systems, as illustrated for dipalmitoylphosphatidylcholine in Figure 2.7. The enthalpy of the transition is progressively reduced as the cholesterol content is increased. For phosphatidylcholine to cholesterol molar ratios of 2:1, no transition is observable. This lipid mixture exhibits basically liquid-crystalline characteristics, since, as indicated previously, equimolar cholesterol levels do not reduce the lateral diffusion rates of (liquid-crystalline) phosphatidylcholine significantly (the lateral diffusion rates of gel-state lipids are at least two orders of magnitude slower). Furthermore, at temperatures below the T_c of the phospholipid, NMR characteristics indicative of rapid axial rotation of the phospholipid molecule are observed in the presence of cholesterol, behavior similar to that observed for liquid-crystalline phospholipids.

The relation between the gel–liquid-crystalline properties of lipids and the roles of lipids in biological membranes remains obscure. The observation that individual lipid components can adopt gel or liquid-crystalline arrangements has led to the suggestion that segregation of particular lipids into a local gel-state environment may occur within a biological membrane. This segregation could affect protein function by restricting protein mobility within the bilayer matrix or could provide packing defects resulting in permeability changes. There are two major difficulties encountered with these concepts, however. First, while certain procaryotic systems can exhibit characteristics consistent with the presence of gel-state lipids at temperatures which allow growth, such observations are by no means universal. In eucaryotic membranes, for example, there is no evidence for the presence of gel-state lipid components at physiological temperatures. The second difficulty concerns the way in which lateral segregation of lipid into crystalline domains might be regulated. Clearly, an organism cannot regulate fluidity by regulating temperature; thus, physiological factors are required which can induce isothermal modulation of the local lipid composition. The presence of factors capable of segregating lipids into local crystalline domains in a biological membrane has not been unambiguously demonstrated.

The theme that membranes do not require the presence of gel-state lipids is easily developed for eucaryotic membrane systems, such as the well-characterized erythrocyte membrane. Of the erythrocyte membrane lipids, only sphingomyelin exhibits a T_c close to physiological temperatures, with the attendant possibility of forming local crystalline domains. However, this possibility is seriously compromised by the presence in the membrane of equimolar levels of cholesterol, which would be expected to inhibit such formation, in agreement with the observation that no reversible phase transition is observable in the intact erythrocyte (ghost) membrane by calorimetric or other techniques. In other membranes which contain little or no cholesterol, such as the membranes of various subcellular organelles, the absence of gel-state domains is indicated by the absence of relatively saturated lipid species, such as sphingomyelin, as well as by the increased unsaturation of other lipids present. Table 2.3 gives the unsaturation index (number of unsaturated bonds per phospholipid) for the lipid component of a variety of membranes. In general, more metabolically active membranes, such as the inner mitochondrial membrane, contain a higher fraction of more unsaturated lipid species (whose T_c values may be 60°C or more below physiological temperatures).

In summary, available evidence indicates that membranes require a fluid bilayer matrix for function and that modulation of local fluidity and function by formation of crystalline domains is unlikely to be a general phenomenon.

Lipid Polymorphism   In addition to an ability to adopt a gel or liquid-crystalline bilayer organization, lipids can also adopt entirely different liquid-crystalline structures on hydration (Cullis et al. 1983). The major structures assumed are illustrated in Figure 2.8. These structures have three general features. First, the predominant structures assumed by isolated species of membrane lipids on hydration in excess aqueous buffer are the familiar bilayer organization and the *hexagonal* H_{II} structure. Lipids which form micellar structures, such as lysophosphatidylcholine, are minority components of membranes. Second, the H_{II} phase, which consists of a hydrocarbon matrix penetrated by hexagonally packed aqueous cylinders with diameters of about 20 Å, is not compatible with maintenance of a permeability barrier between external and internal compartments. This immediately raises questions concerning the functional role of lipids in membranes which preferentially adopt this structure in isolation. Finally, in contrast with the situation for gel-state (crystalline) lipids, it now appears that all biological membranes contain an appreciable fraction (up to 40 mol %) of lipid species which prefer the H_{II} arrangement, as well as lipids which prefer bilayer structure.

The ability of lipids to adopt different structures on hydration is commonly referred to as *lipid polymorphism*. Three techniques which have been extensively employed to monitor lipid polymorphism are X-ray diffraction, ^{31}P- and ^{2}H-NMR, and freeze-fracture procedures. X-ray diffraction is the classical technique, allowing the detailed nature of the phase structure to be elucidated. The use of ^{31}P-NMR for

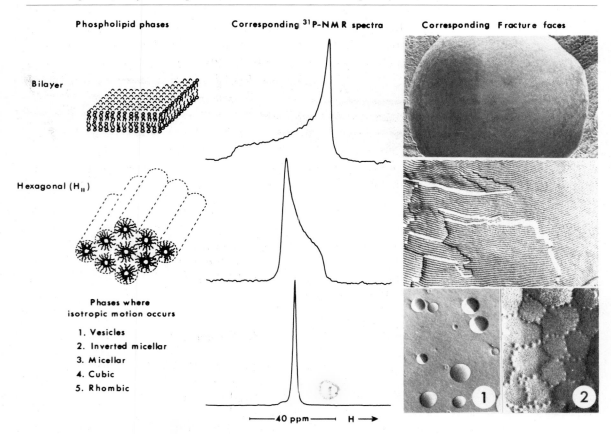

Figure 2.8. ^{31}P-NMR and freeze-fracture characteristics of phospholipids in various phases. The bilayer spectrum was obtained from aqueous dispersions of egg yolk phosphatidylcholine, and the hexagonal (H$_{II}$) phase spectrum from phosphatidylethanolamine (prepared from soybean phosphatidylcholine). The ^{31}P-NMR spectrum representing isotropic motion was obtained from a mixture of 70 mol % soy phosphatidylethanolamine and 30 mol % egg yolk phosphatidylcholine. All preparations were hydrated in 10mM Tris–acetic acid (pH 7.0) containing 100mM NaCl, and the spectra were recorded at 30°C in the presence of proton decoupling. The freeze-fracture micrographs represent typical fracture faces obtained. The bilayer configuration (total erythrocyte lipids) gives rise to a smooth fracture face, whereas the hexagonal (H$_{II}$) configuration is characterized by ridges displaying a periodicity of 6–15 nm. Two common conformations that give rise to isotropic motion are represented in the bottom micrograph: (1) bilayer vesicles (less than 200 nm diameter) of egg phosphatidylcholine prepared by extrusion techniques and (2) large lipid structures containing lipidic particles (egg phosphatidylethanolamine containing 20 mol % egg phosphatidylserine at pH 4.0).

identification of polymorphic phase characteristics of phospholipids relies on the different motional averaging mechanisms available to phospholipids in different structures and provides a convenient and reliable diagnostic technique. Finally, freeze-fracture electron microscopy allows visualization of local structure which need not be arranged in a regular lattice, yielding information not available from X-ray or NMR techniques.

The ^{31}P-NMR and freeze-fracture characteristics of bilayer and H_{II} phase phospholipid systems are illustrated in Figure 2.8. Bilayer systems exhibit broad, asymmetric ^{31}P-NMR spectra with a low-field shoulder and high-field peak separated by about 40 ppm, whereas H_{II} phase systems exhibit spectra with reversed asymmetry which are narrower by a factor of two. The difference between bilayer and H_{II} phase ^{31}P-NMR spectra arises from the ability of H_{II} phase phospholipids to diffuse laterally around the aqueous channels (Cullis et al. 1983). Freeze-fracture techniques show flat, featureless fracture planes for bilayer systems, whereas H_{II} phase structures give rise to a regular corrugated pattern as the fracture plane cleaves between the hexagonally packed cylinders.

The polymorphic phase preferences of a large variety of synthetic and naturally occurring phospholipids have been investigated, and the results obtained for eucaryotic lipid species are summarized in Table 2.6. It is immediately apparent that a significant proportion of membrane lipids adopt or promote H_{II} phase structure under appropriate conditions. Phosphatidylethanolamine, which commonly comprises up to 30% of membrane phospholipids, is perhaps

Table 2.6. Phase Preferences of Membrane Lipids from Eucaryotes

Bilayer	Hexagonal H_{II}
PC	
SPM	
	PE
PS	PS (pH < 3)
PG	
PI	
PA	PA ($+Ca^{2+}$)
	PA (pH < 3)
CL	CL ($+Ca^{2+}$)
	Cholesterol*
	Fatty acids

Note: The abbreviations are PC, phosphatidylcholine; SPM, sphingomyelin; PS, phosphatidylserine; PG, phosphatidylglycerol; PI, phosphatidylinositol; PA, phosphatidic acid; CL, cardiolipin; PE, phosphatidylethanolamine.
* Cholesterol and long-chain unsaturated fatty acids can induce the hexagonal (H_{II}) phase in some lipid mixtures.

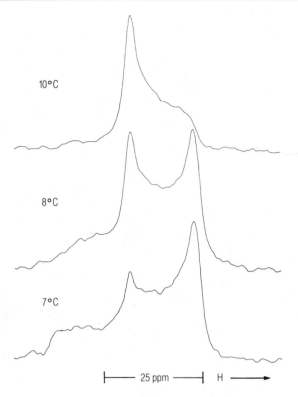

Figure 2.9. 36.4-MHz ^{31}P-NMR spectra of an aqueous dispersion of human erythrocyte phosphatidylethanolamine dispersed in 25mM Tris-acetic acid (pH 7.0) and 2mM EDTA. These spectra were obtained employing broad-band proton decoupling.

hexagonal
2 phts 266

the most striking example, and particular effort has been devoted to understanding the factors which result in a predilection for the H_{II} arrangement. As illustrated in Figure 2.9, phosphatidylethanolamine isolated from erythrocytes can adopt both the bilayer and H_{II} arrangements, depending on the temperature. The H_{II} structure is formed above a characteristic bilayer to hexagonal (H_{II}) transition temperature (T_{BH}) of about 10°C. Similar or lower values of T_{BH} have been observed for phosphatidylethanolamine isolated from endoplasmic reticulum and the inner mitochondrial membrane. Lower T_{BH} values are observed for more unsaturated species. This dependence of T_{BH} on acyl chain unsaturation has been characterized more definitively, employing synthetic species of phosphatidylethanolamine, as summarized in Table 2.7. This table illustrates that a minimal degree of unsaturation of the acyl chains is required for H_{II} structure to be adopted and that increased unsaturation progressively favors the H_{II} arrangement.

Biological membranes contain mixtures of lipids which individually prefer bilayer or H_{II} structures; therefore, the properties of mixed systems are of considerable interest. Studies on model systems

Table 2.7. The Temperature (T_{BH}) of the
Bilayer–Hexagonal H_{II} Phase Transition for
Some Phosphatidylethanolamines

Phosphatidylethanolamine	T_{BH} (°C)
18:0/18:0	>105
18:1tΔ^9/18:1tΔ^9	60 to 63
18:1cΔ^9/18:1cΔ^9	10
18:2c$\Delta^{9,12}$/18:2c$\Delta^{9,12}$	−15 to −25
18:3c$\Delta^{9,12,15}$/18:3c$\Delta^{9,12,15}$	−15 to −30

show that mixtures of an H_{II} phase lipid (for example, phosphatidyl-ethanolamine) with a bilayer phospholipid (such as phosphatidyl-choline) result in a progressive stabilization of net bilayer structure for the whole mixture as the percentage of bilayer lipid increases, as illustrated in Figure 2.10. This is a general feature of mixtures of

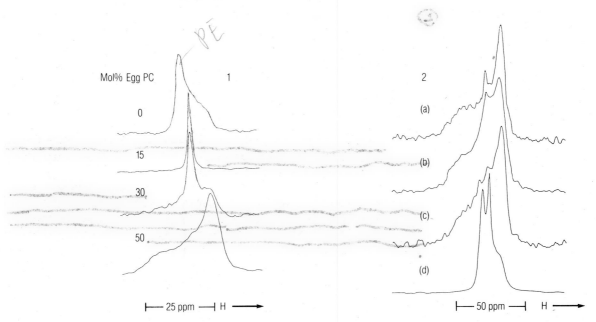

Figure 2.10. Phase behavior of phosphatidylcholine (PC) and phosphatidylethanol-amine (PE) mixtures and the effects of cholesterol. (1) 36.4-MHz ^{31}P-NMR spectra of aqueous dispersions of mixtures of soy phosphatidylethanolamine and egg phosphatidylcholine. The amount present is expressed as a percentage of the total phospholipid. (2) 36.4-MHz ^{31}P-NMR spectra obtained at 30°C from equimolar mixtures of dioleoyl-PE with dioleoyl-PC in the presence of (a) 0 mol %; (b) 15 mol %; (c) 30 mol %, and (d) 50 mol % cholesterol.

bilayer and H_{II} lipids. Depending on the acyl chain composition, temperature, and head group size and charge, complete bilayer stabilization can be achieved by the addition of 10 to 50 mol % of the bilayer species. These systems appear to retain the ideal mixing behavior characteristic of liquid-crystalline systems. For example, in phosphatidylethanolamine-phosphatidylcholine mixtures containing intermediate amounts of the bilayer-stabilizing species, situations can arise where H_{II} phase and bilayer phase components coexist in the same sample. ^{2}H-NMR studies of ^{2}H-labeled varieties of these lipids indicate a homogeneous lipid composition, with no preference of the H_{II}-preferring phosphatidylethanolamine species for the H_{II} component or of phosphatidylcholine for the bilayer component (Tilcock et al. 1982).

There are two other features of these mixed systems which are of particular interest. The first concerns cholesterol, which has the remarkable ability to induce H_{II} phase structure for phosphatidyletha-nolamine-containing systems where bilayer structure has been stabilized by phosphatidylcholine (Figure 2.10). This effect of cholesterol is also observed in other mixed-lipid systems (Tilcock et al. 1982). The second point concerns the narrow ^{31}P-NMR peak occasionally observed in the mixed-lipid systems of Figure 2.10. Such a spectral feature arises from phospholipids, which experience *isotropic motional averaging* over all possible orientations. Such a resonance cannot arise from phospholipids in H_{II} or large (diameter $\geq$ 200 nm) bilayer structures, where the motion is restricted. Freeze-fracture studies suggest that this isotropic peak corresponds to a novel particulate feature observed on the fracture face of these systems, as illustrated in Figure 2.11. These "lipidic particles" are a general feature of mixtures of bilayer- and H_{II}-preferring lipids (Verkleij 1984), and there is a growing consensus that they correspond to inverted micellar interbilayer structures, as indicated in Figure 2.11. These structures appear to represent intermediaries between the bilayer and H_{II} phases and may be of particular importance, as such nonbilayer structures can be localized to a particular region of the membrane. Such structures may have functional utility, since their formation would not result in a large-scale disruption of the bilayer permeability barrier that necessarily accompanies generation of macroscopic H_{II} phase lipid structure.

The functional roles of nonbilayer lipid structures in membranes have been investigated by characterizing the influence of divalent cations, ionic strength, pH, and membrane protein on lipid polymorphism (Cullis et al. 1983). These factors can strongly influence the structural preferences of appropriate lipid systems. In the case of pure lipid systems, for example, reduction of the pH results in H_{II} phase structure for (unsaturated) phosphatidylserine and

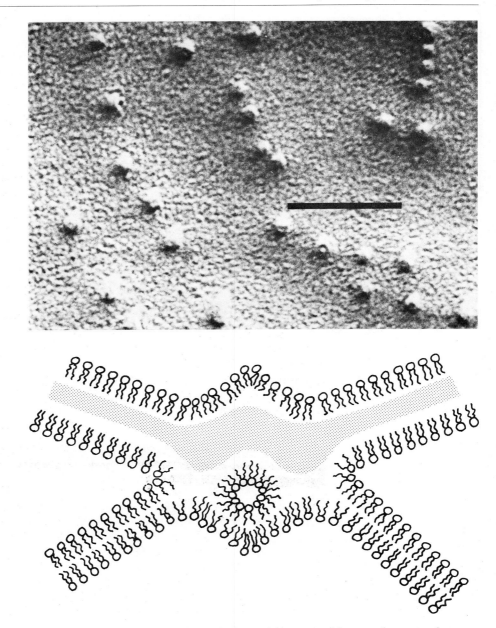

Figure 2.11. A schematic representation of the proposed fracture plane around an inverted micelle formed at the contact point between two bilayers. The corresponding lipidic particles observed by freeze-fracture electron microscopy are shown in the micrograph of a mixture of soy phosphatidylinositol (7.5 mol %) with soy phosphatidylethanolamine. Each particle is approximately 10 nm in diameter. The scale represents 100 nm.

phosphatidic acid systems, and the addition of Ca^{2+} to cardiolipin triggers bilayer-H_{II} transitions (Table 2.6). Similar observations extend to mixed-lipid systems, where the addition of Ca^{2+} to bilayer systems containing phosphatidylethanolamine and various acidic phospholipids can also trigger H_{II} phase formation.

Phosphatidylserine-phosphatidylethanolamine systems are perhaps the best characterized in this regard (Tilcock et al. 1984), and certain features of the mechanisms involved deserve emphasis. First, in some binary phospholipid mixtures containing phosphatidylserine, Ca^{2+} can segregate the phosphatidylserine component into a crystalline (gel-phase) structure with a characteristic morphology described as *cochleate* (as observed by freeze-fracture). In the case of phosphatidylserine-phosphatidylethanolamine systems, the bilayer-stabilizing influence of phosphatidylserine is thus removed, allowing the phosphatidylethanolamine to adopt the H_{II} organization it favors in isolation. When cholesterol is present, however, Ca^{2+}-dependent generation of H_{II} structure proceeds by a different mechanism which does not involve lateral segregation phenomena— rather, all lipid components, *including* phosphatidylserine, adopt the H_{II} organization. The potential relevance of these observations is illustrated by Figure 2.12, where it is shown that Ca^{2+} can trigger H_{II} formation in a mixture of lipids isolated from human erythrocytes, with a composition corresponding to that of the erythrocyte inner monolayer (which contains predominantly phosphatidylethanolamine and phosphatidylserine).

The ability of lipids to adopt different macroscopic structures on hydration has stimulated studies aimed at understanding the physical properties of lipids which dictate these preferences. These studies have given substantial support to a simplistic hypothesis that a generalized *shape property* of lipids determines the phase structure adopted (Cullis et al. 1983). This concept is illustrated in Figure 2.13, where bilayer phase lipids are proposed to exhibit cylindrical geometry compatible with that organization, while H_{II} phase lipids have a cone shape where the acyl chains subtend a larger cross-sectional area than the polar head group region. Detergent-type lipids which form micellar structures are suggested to have reversed geometry corresponding to an inverted cone shape. It should be noted that "shape" is an inclusive term reflecting the effects of the size of polar and apolar regions, head group hydration and charge, hydrogen-bonding processes, and effects of counterions, among other possibilities. The cone shape of unsaturated phosphatidylethanolamines, for example, can be ascribed to a smaller, less-hydrated head group (in comparison with phosphatidylcholine). There may also be intermolecular hydrogen bonding between phosphatidylethanolamines, which would further reduce the area per molecule in the head group region.

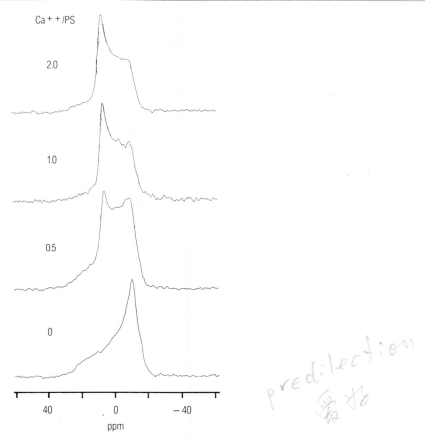

Figure 2.12. Effect of calcium on "inner monolayer" phospholipids: 81.0-MHz
^{31}P-NMR spectra at 37°C arising from an aqueous dispersion of reconstituted inner
monolayer lipid isolated from human erythrocyte membrane. The lipid
composition is PE:PS:PC:SPM (in the ratios 0.5:0.25:0.13:0.12) and contains
equimolar cholesterol with respect to total phospholipid. The ratio Ca^{2+}/PS refers to
the molar ratio of Ca^{2+} to PS.

Alternatively, the increased predilection of more unsaturated species
of phosphatidylethanolamine for the H_{II} arrangement (Table 2.7) may
be attributed to the increased cross-sectional area of the unsaturated
(compared with saturated) acyl chains. A striking observation supporting the shape concept is that lipid mixtures containing detergents
(inverted cone shape) and unsaturated phosphatidylethanolamines
(cone shape) can adopt bilayer structure, which may be attributed to
shape complementarity (Cullis et al. 1983).

In summary, studies on model systems show that lipids found in
biological membranes can exist in a variety of structures in addition to

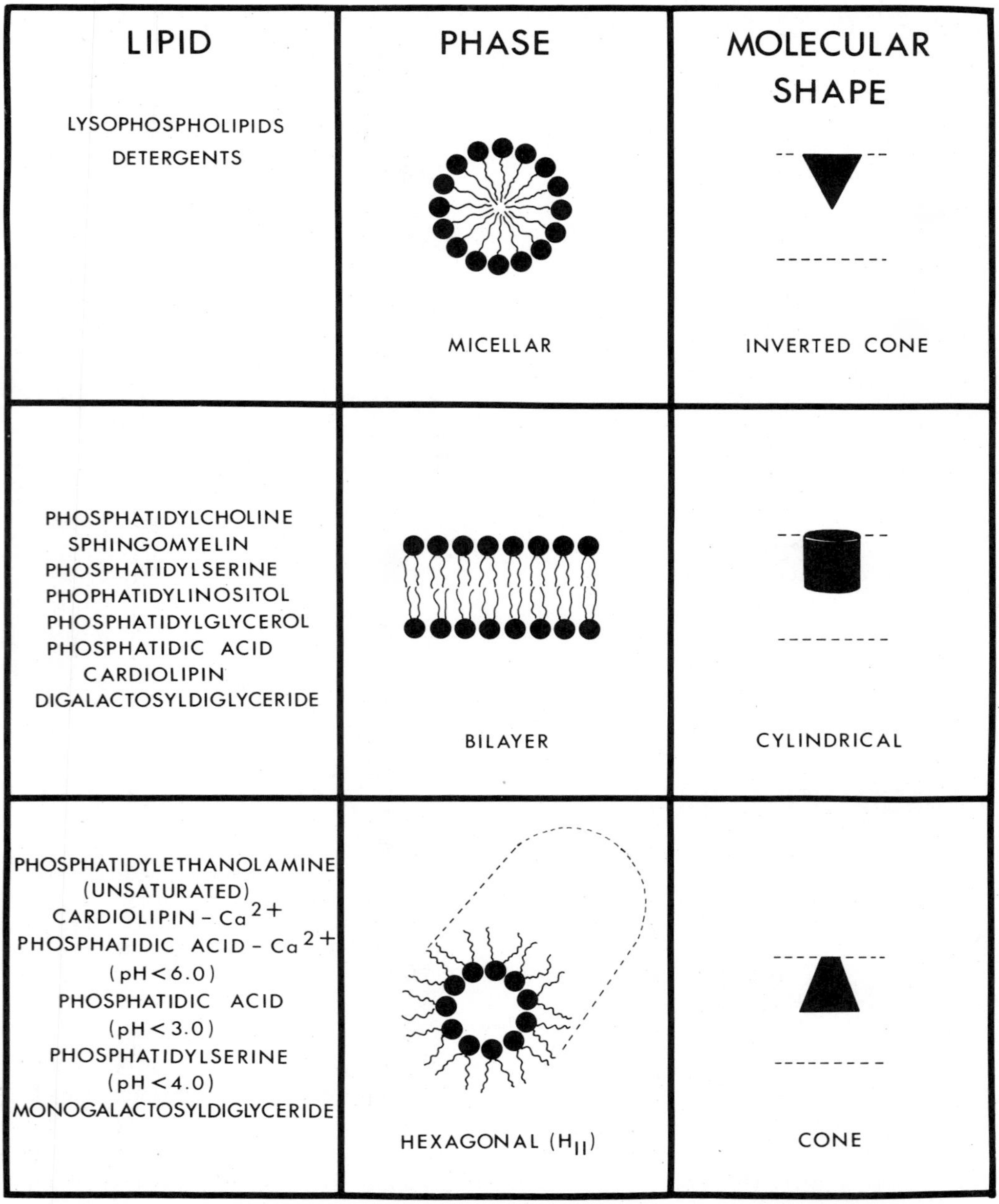

LIPID	PHASE	MOLECULAR SHAPE
LYSOPHOSPHOLIPIDS DETERGENTS	 MICELLAR	 INVERTED CONE
PHOSPHATIDYLCHOLINE SPHINGOMYELIN PHOSPHATIDYLSERINE PHOPHATIDYLINOSITOL PHOSPHATIDYLGLYCEROL PHOSPHATIDIC ACID CARDIOLIPIN DIGALACTOSYLDIGLYCERIDE	 BILAYER	 CYLINDRICAL
PHOSPHATIDYLETHANOLAMINE (UNSATURATED) CARDIOLIPIN – Ca^{2+} PHOSPHATIDIC ACID – Ca^{2+} (pH < 6.0) PHOSPHATIDIC ACID (pH < 3.0) PHOSPHATIDYLSERINE (pH < 4.0) MONOGALACTOSYLDIGLYCERIDE	 HEXAGONAL (H_{II})	 CONE

Figure 2.13. Polymorphic phases and corresponding dynamic molecular shapes of lipids.

the bilayer phase. These structural preferences can be modulated by many biologically relevant variables, supporting the possibility that nonbilayer lipid structures play roles in membrane-mediated phenomena requiring local departures from bilayer organization. As indicated later in this chapter, membrane fusion may be a most important example.

LIPIDS AND THE PERMEABILITY PROPERTIES OF MEMBRANES

The ability of lipids to provide a bilayer permeability barrier between external and internal environments constitutes one of their most important functions in a biological membrane. The nature and selectivity of this barrier to various molecules and ions of biological interest (water, uncharged water-soluble nonelectrolytes, and ionized solutes of varying hydrophobicity) have been extensively investigated. A succinct review of this data is difficult to achieve, due in part to the different model systems employed and the discrepancies among experiments. One major problem concerns the use of the black lipid membrane or LUV systems where residual solvents such as decane, detergent, ether, or ethanol may influence the permeability properties of the ion of interest. Here we present a summary of salient general principles of membrane permeability in relation to properties of component lipids such as fluidity, polar head group charge, and phase structure. A synopsis of the permeability coefficients observed for different solutes for a variety of membranes is presented in Table 2.8.

Table 2.8. Permeability Coefficients (cm/s) for Some Common Polar Solutes Across Model and Biological Membranes (at 20°C unless otherwise indicated)

Membrane	Na^+	K^+	Cl^-
Egg phosphatidylcholine (4°C)	$<1.2 \times 10^{-14}$	—	5.5×10^{-11}
Soy phosphatidylcholine*	4.0×10^{-13}	—	7.0×10^{-11}
Equimolar soy phosphatidyl-* choline and cholesterol (40°C)	$<2.0 \times 10^{-13}$	—	—
Dilinoleoylphosphatidylserine*	$<7.0 \times 10^{-13}$	—	—
Human red blood cell* phosphatidylserine	$<7.0 \times 10^{-13}$	—	—
Human red blood cell[†]	1.0×10^{-10}	2.4×10^{-10}	1.4×10^{-4}
Squid axon (resting)[†]	1.5×10^{-8}	5.6×10^{-8}	1.0×10^{-8}
Squid axon (excited)[†]	5.0×10^{-6}	1.7×10^{-4}	1.0×10^{-8}

* The data were obtained from large unilamellar vesicles prepared by extrusion through filters according to Hope et al. (1985).
[†] The data were adapted from Jain and Wagner (1980).

Theoretical Considerations

In order to appreciate the meaning of the permeability coefficient parameter for a given lipid system, some understanding of the underlying theory is required. A basic phenomenological treatment of diffusion begins with *Fick's law*, which states that the diffusion rate of a given substance (number of molecules per unit time, dn/dt) through a membrane is directly proportional to the area A of the membrane and the difference in the concentration $\Delta C(t)$ of the material across the membrane. Thus, $dn/dt \propto A\Delta C(t)$, which may be rewritten as $dn/dt = -PA\Delta C(t)$, where P, which has the units of length over time (for example, cm/s), is the permeability coefficient and t is time. If we consider the special case of a LUV of radius R containing an initial concentration of solute $C_I(0)$, where the initial external concentration of this solute is zero, it is straightforward to show that $\Delta C(t) = C_I(0)\exp(-3Pt/R)$. Under conditions where the external volume is much greater than the internal trapped volume, $\Delta C(t) \cong C_I(t)$ (where $C_I(t)$ is the internal concentration at time t); thus, $C_I(t) = C_I(0)\exp(-3Pt/R)$. For a 100-nm diameter LUV it may therefore be calculated that the time required for release of one-half the entrapped material ($t_{1/2}$) is 0.1 s for $P = 10^{-5}$ cm/s, whereas for $P = 10^{-10}$ cm/s, $t_{1/2} = 3.2$ h.

It should be emphasized that the preceding example, while illustrative, neglects several important factors which can strongly influence the net flux of molecules through membranes. These include the effects associated with the aqueous layer (more than 20 nm thick) that extends from the lipid-water interface, in which solute molecules are not mixed to the same extent as in the bulk solution (Fettiplace and Haydon 1980). Such unstirred layers can effectively reduce the solute concentration difference ΔC across the membrane itself, giving rise to a smaller measured value of P. For charged molecules, the efflux can be strongly limited by generation of a membrane potential, as will be discussed later. Finally, the permeability of various solutes through membranes is strongly temperature dependent, with activation energies E_0 in the range of 8–20 kcal/mol. A measure of the influence of temperature is given by the observation that an activation energy of 12 kcal/mol will increase the permeability coefficient by a factor of two for every 10°C increase in temperature.

Permeability of Water and Nonelectrolytes

Liquid-crystalline lipid bilayers are remarkably permeable to water, which exhibits permeability coefficients in the range of 10^{-2} to 10^{-4} cm/s (Fettiplace and Haydon 1980). Membrane systems enclosing high concentrations of a relatively impermeable solute will swell when placed in an aqueous medium containing little or no solute, due to a net influx of water to achieve osmotic balance. Conversely, the reverse conditions will lead to shrinkage. As a result, the relative

permeability of different membrane systems to water can be monitored by measuring swelling rates (employing light scattering techniques, for example) when osmotic gradients are applied (Blok,
van Deenen, and de Gier 1976). Results obtained from such studies
indicate that increased unsaturation of the fatty acids of the membrane causes increases in water permeability. Similarly, the inclusion
of cholesterol reduces water permeability, leading to the general
conclusion that factors contributing to increased order in the hydrocarbon region decrease water permeability. It is of interest to note
that whereas MLVs and LUVs in the size range of 1 μm or larger
demonstrate these osmotic properties, the smaller SUV systems do
not.

The diffusion properties of nonelectrolytes (uncharged polar
solutes) appear to depend on the properties of the lipid matrix in much
the same manner as does the diffusion of water. In general, the
permeability coefficients observed are at least two orders of magnitude smaller. For example, the permeability coefficient of urea across
egg phosphatidylcholine bilayers is approximately 4×10^{-6} cm/s at
25°C (Poznansky et al. 1976). Furthermore, for a given homologous
series of compounds, the permeability increases as the solubility in a
hydrocarbon environment increases, indicating that the rate-limiting
step in diffusion is the initial partitioning of the molecule into the lipid
bilayer (Poznansky et al. 1976). With regard to the influence of lipid
composition on the permeability of nonelectrolytes, the order in the
acyl chain region has the same qualitative effects as in the case of
water. Thus, decreased unsaturation of lipids or increased cholesterol
content results in lower permeability coefficients (Fettiplace and
Haydon 1980). Gel-phase systems are particularly impermeable.
However, in systems exhibiting lateral phase separation of gel and
liquid-crystalline domains, the permeability can be higher than for
liquid-crystalline systems. This increased permeability can be attributed to packing defects at the crystalline–liquid-crystalline hydrophobic interface.

Permeability of Ions   Lipid bilayers are remarkably impermeable to most small ions (Table
2.8). Permeability coefficients of less than 10^{-10} cm/s are commonly
observed, and they can be as small as 10^{-14} cm/s for Na$^+$ and K$^+$. For
the example of a 100-nm diameter LUV, this would correspond to a
half-life for release of entrapped Na$^+$ of approximately 3.6 years. In
contrast, lipid bilayers appear to be much more permeable to H$^+$ or
OH$^-$ ions, which have been reported to have permeability coefficients in the range of 10^{-4} cm/s (Deamer 1982). The Cl$^-$ anion also
exhibits anomalous permeability behavior, with permeability coefficients up to 300 times greater than those observed for Na$^+$ in similar
systems.

Measures of the permeability of membranes to small ions are complicated, since for free permeation to proceed, a counterflow of other ions of equivalent charge is required; otherwise, a membrane potential is established which is equal and opposite to the chemical potential of the diffusing species. As an example, consider the 100-nm diameter LUV which has a well-buffered interior pH of 4.0 and an exterior pH of 7.0 in a Na^+ buffer. The relatively permeable H^+ ions can diffuse out, but Na^+ ions cannot move in. Thus, a membrane potential ($\Delta\psi$) is established (interior negative), where

$$\Delta\psi = -59 \log \frac{[H^+]_i}{[H^+]_0} = -177 \text{ mV}$$

and the subsequent efflux of protons is coupled to the much slower influx of Na^+ ions. Assuming a membrane thickness of 4 nm and interior dielectric constant of 2, the capacitance of the vesicle membrane can be calculated as $C = 0.5 \ \mu F/cm^2$; thus, the number of protons that diffuse out to set up $\Delta\psi$ can be calculated to be about 150. Subsequent H^+ efflux will occur only as Na^+ ions permeate in.

The relation between the physical properties of lipids and the permeability properties of membranes to small ions is not understood in detail. Difficulties in understanding this relationship arise from the different model systems employed, the various impurities present, and complexities due to ion counterflow and related membrane potential effects. Vesicles prepared by techniques involving detergents or organic solvents contain residual detergent or solvent which can strongly influence the permeabilities observed, and the presence of n-decane or other long chain alkanes in black lipid membrane systems may also influence permeability. In general, however, the permeability of a given ion appears to be related to the order in the hydrocarbon region, where increased order leads to a decrease in permeability.

The charge on the phospholipid polar head group can also strongly influence permeability by virtue of the resulting surface potential ϕ. For example, approximately 30% of the lipid of the inner monolayer of the erythrocyte membrane is the negatively charged lipid phosphatidylserine. If we assume an area per lipid molecule of 0.6 nm^2, the resultant surface charge density σ is 8 $\mu C/cm^2$ (where C is coulombs). The resulting surface potential ϕ can be calculated from Gouy-Chapman theory (McLaughlin 1977) for a 150mM monovalent salt buffer according to the relation $\phi = 0.052 \times \sinh^{-1}(\sigma/4.5)$. This gives a negative surface potential of $\phi = -69$ mV. This potential will repel anions from and attract cations to the lipid-water interface. For example, the H^+ concentration at the inner monolayer interface will be increased by the Boltzmann factor $\exp(e\phi/kT) = 14.5$ in comparison with the bulk solution, resulting in a significantly lower pH at the membrane interface and correspondingly higher H^+ efflux rates.

LIPID-PROTEIN INTERACTIONS

Any complete understanding of biological membrane systems necessitates a detailed understanding of the nature and influence of lipid-protein interactions. Such interactions can be divided into two classes. The first concerns proteins with hydrophobic segments which penetrate into or through the lipid bilayer (*intrinsic,* or *integral, proteins*), whereas the second concerns water soluble proteins which interact electrostatically with negatively charged groups at the lipid-water interface (*extrinsic,* or *peripheral, proteins*). The effects of intrinsic and extrinsic proteins on membrane lipid fluidity (for example, the gel or liquid-crystalline nature of associated lipids) or lipid polymorphism will provide the primary focus of this section. However, it should be noted that studies of lipid-protein interactions (Jost and Griffith 1982) have generated a large and often confusing literature which has not yet led to a generally accepted understanding. We emphasize here only those points which we believe provide the most important insight.

Extrinsic Proteins

The interaction of extrinsic proteins with lipids has been studied using a variety of proteins, including polylysine, cytochrome *c*, the A_1 basic protein from myelin, and spectrin from the red blood cell. In order for these basic (positively charged) molecules to interact extensively with lipid systems, the presence of acidic (negatively charged) lipids is required, consistent with an electrostatic protein-membrane association. Two general points can be made. First, while it is possible that such surface interactions may induce a time-averaged enrichment of the negatively charged lipid in the region of the protein, there is presently no unambiguous evidence to suggest that such clustering can induce a local fluidity decrease via formation of crystalline domains. Indeed, in model membrane systems containing acidic phospholipids, such extrinsic proteins as cytochrome *c*, the A_1 basic protein, and spectrin induce a decreased T_c and enthalpy of the lipid gel–liquid-crystalline transition, indicating an increased disorder in the acyl chain region. This effect has been related to an ability of such proteins to partially penetrate the hydrophobic region, as indicated by increases in permeability and monolayer surface pressure on binding. The second point is that there is evidence of competition between divalent cations and extrinsic proteins for binding to membranes. Thus, spectrin can shield the effects of Ca^{2+} on the gel-liquid phase transition properties of systems containing negatively charged lipids.

Studies on the influence of extrinsic protein on the polymorphic properties of lipids (de Kruijff et al. 1984) also yield results consistent with a competition between the protein and divalent cations. For example, polylysine, which is highly positively charged, can to some extent destabilize the bilayer structure of cardiolipin-phosphatidylethanolamine systems and strongly protects against the ability of Ca^{2+} to induce complete H_{II} organization in the pure lipid system. A particularly interesting observation is that cytochrome c can induce nonbilayer structures in cardiolipin-containing systems. This observation may be related to an apparent ability of cytochrome c to translocate rapidly across bilayers that contain cardiolipin, possibly including the inner mitochondrial membrane.

Intrinsic Proteins   Intrinsic or integral membrane proteins cannot be solubilized without detergent and contain one or more hydrophobic sequences which span the lipid bilayer one or more times in α-helical structures. Studies on the interactions of lipids with such proteins have resulted in a particularly large literature. This work has mainly focused on the specificity of such lipid-protein interactions and on the physical state of the lipid (Jost and Griffith 1982). In particular, it has been shown that lipids residing at the lipid-protein interface of intrinsic proteins experience a different environment than do bulk bilayer lipids. It has been speculated that such *boundary lipids* may be specific to a given protein and provide environments that are appropriate to, and that possibly regulate, function. These theories were supported by early ESR studies of spin-labeled lipids in reconstituted systems which demonstrated that such lipids, when in the vicinity of integral proteins, exhibited increased order parameters (that is, restricted motion of the lipid) in the acyl chain region. Other studies indicating the importance of the physical state of boundary lipids demonstrated that gel-state boundary lipids inhibited the function of the sarcoplasmic reticulum Ca^{2+} ATPase and other membrane-bound enzymes in reconstituted systems.

More recent work is pointing to a rather different picture, however. First, with the exception of a possible requirement for one or two molecules of a particular lipid, lipid-protein interactions appear relatively nonspecific, in that a large variety of different (liquid-crystalline) lipids can usually support protein activity. The sarcoplasmic reticulum ATPase, for example, has excellent activity when reconstituted with a variety of phospholipids as well as detergents (Dean and Tanford 1977). Similar observations have been made for many other integral proteins, including cytochrome oxidase. A second point is that, in general, a long-lived boundary layer of lipid does not appear to exist at the lipid-protein interface. For example, whereas ESR spin-label studies indicate long-lived boundary components, ^{2}H-NMR studies on analogous systems containing ^{2}H-labeled lipids do not reveal such components. This apparent discrepancy has been reconciled, since ESR and NMR report on phenomena occurring during different time scales. Boundary-bulk lipid exchange rates in the region 10^{-6}–10^{-8} s would appear slow on the ESR time scale but fast on the NMR time scale. These observations, together with NMR and calorimetric results indicating that integral proteins can have disordering effects on adjacent lipids, suggest that lipids in the region of intrinsic protein are in relatively rapid exchange ($T_{ex} \sim 10^{-7}$ s) and do not have gel-state characteristics. This does not mean that the lipid composition in contact with the protein is necessarily the same as the bulk composition, as effects such as electrostatic lipid-protein interactions may enhance the local concentration of a particular lipid species on a time-averaged basis. Furthermore, such generalizations may not hold for particular situations. The purple membrane fragments of *Halobacterium halobium*, which contain bacteriorhodopsin, for example, exhibit a unique lipid composition distinct from the rest of the membrane (Stoekenius 1976).

The influence of intrinsic proteins on lipid polymorphism has received less detailed attention; however, some interesting features are emerging (de Kruijff et al. 1984). First, the hydrophobic peptide antibiotic *gramicidin*, which spans the membrane as a dimer, has a very strong bilayer destabilizing capacity and even induces H$_{II}$ phase structure in phosphatidylcholine systems. On the other hand, *glycophorin*, the major asialoglycoprotein from the erythrocyte, stabilizes the bilayer structure for unsaturated phosphatidylethanolamines. Thus, the message from these initial studies is clear—membrane proteins may well play an active role in determining local membrane structure.

In summary, our understanding of lipid-protein interactions in biological membranes remains relatively unsophisticated. It may be that some fraction of lipid diversity satisfies relatively nonspecific requirements and provides an appropriate solvent for the optimal function of integral proteins. Alternatively, specific functions of lipids may be more related to other membrane properties, such as permeability, than to protein function per se. In addition, many fundamental questions have not yet been adequately addressed, including the role of various lipids in insertion of protein, in sealing proteins within the bilayer matrix, and in providing an interface appropriate for membrane protein-substrate interactions.

LIPIDS AND MEMBRANE FUSION

Membrane fusion (in various expressions) is one of the most ubiquitous membrane-mediated events, occurring in processes of fertilization, cell division, exo- and endocytosis, infection by membrane-bound viruses, and intracellular membrane transport, to name but a few.

There are strong experimental and theoretical indications that the lipid components of membranes are directly involved in such fusion processes. For example, model membrane systems such as LUVs can be induced to fuse in the absence of any protein factors. In addition, it is topologically impossible for two membrane-bound systems to fuse together to achieve mixing of internal compartments without a local transitory departure from the normal lipid bilayer structure at the fusion interface. We shall discuss the possible nature of the fusion intermediates, as indicated by studies on model membrane systems. The fusion intermediates are subsequently related to fusion behavior observed in biological membrane systems.

Fusion of Model Systems

For fusion events to proceed in vivo the presence of Ca^{2+} is often required. As a result, numerous studies have been concerned with the induction of Ca^{2+}-stimulated fusion between vesicle systems and analysis of the lipid factors involved. We shall discuss in turn the modulation of gel-liquid-crystalline properties of lipids and the modulation of the polymorphic properties of lipids in relation to membrane fusion.

It has been recognized for some time that model membrane SUV systems will undergo fusion when incubated at temperatures in the region of their gel–liquid-crystalline transition temperature T_c. Continued recycling of sonicated dipalmitoylphosphatidylcholine vesicles through $T_c = 41°C$, for example, results in fusion and formation of larger systems. Isothermal induction of crystalline structure by the addition of Ca^{2+} to phosphatidylserine systems results in fusion to form the large crystalline cochleate structures noted previously (Papahadjopoulos, Poste, and Vail 1978). Given the involvement of Ca^{2+} in biological fusion events, the latter observation suggests that Ca^{2+} may induce lateral segregation of negatively charged phospholipids, such as phosphatidylserine, in vivo, which may act as local crystalline nucleation points for fusion. However, phosphatidylserine is not always present in membranes which undergo fusion, nor is Ca^{2+} able to induce crystalline cochleate-type structures for other species of (unsaturated) negatively charged phospholipids. Furthermore, in more complex lipid mixtures containing phosphatidylethanolamine and cholesterol, for example, there are strong indications that Ca^{2+} is not able to induce segregation of unsaturated phosphatidylserines (Tilcock et al. 1984). Finally, the concentration of Ca^{2+} required to induce crystalline phosphatidylserine-Ca^{2+} complexes is $2mM$ or larger, a concentration much higher than could occur in the cell cytoplasm, for example.

The hypothesis that membrane fusion proceeds by taking advantage of the polymorphic capabilities of component lipids is more viable, but not proven. Three important observations have been made which support this hypothesis. First, it has been shown that lipid-soluble *fusogens* (such as glycerolmonooleate, which induces cell fusion in vitro), induce H_{II} phase structures in model and biological membranes (Cullis et al. 1983), which is consistent with a role of nonbilayer structure during fusion. Second, MLV systems composed of lipid mixtures such as phosphatidylethanolamine and phosphatidylserine form H_{II} structures on the addition of Ca^{2+}. SUV or LUV systems with this lipid composition first fuse to form larger lamellar systems exhibiting lipidic particle structures (as shown in Figures 2.11 and 2.14), before assuming the H_{II} arrangement. Finally, a variety of factors which engender H_{II} organization, such as pH variation or increased temperatures, can induce fusion of vesicle systems with appropriate lipid compositions (Tilcock et al. 1982; Verkeleij 1984; de Kruijff et al. 1984).

These observations have lead to a general hypothesis that factors which tend to induce nonbilayer (H_{II} phase) structure will also induce fusion between membrane-bound systems. There are many attractive features to this hypothesis. In particular, lipids which adopt H_{II}

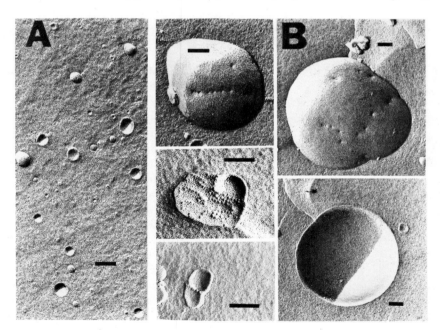

Figure 2.14. (A) Freeze-fracture micrographs of vesicles composed of soy phosphatidylethanolamine containing 20 mol % soy phosphatidylserine.

(B) The vesicles undergo fusion to large bilayer structures containing lipidic particles following the addition of $2mM$ Ca^{2+} at room temperature. Each bar represents 100 nm.

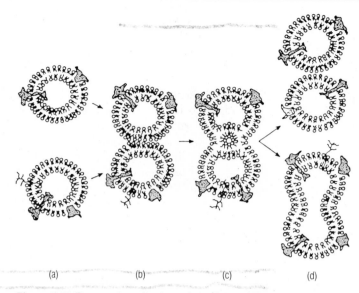

Figure 2.15. Proposed mechanism of membrane fusion proceeding via an inverted cylinder or inverted micellar intermediate. The process whereby the membranes come into close apposition (a)-(b) is possibly protein mediated, whereas the fusion event itself (b)-(c) is proposed to involve formation of an inverted lipid intermediate.

organization hydrate poorly in comparison with bilayer lipids and thus allow the close apposition of membranes required for fusion. In addition, the ability of such lipids to adopt inverted structures, such as inverted micelles or inverted cylinders, clearly provides an attractive intermediate structure for fusion. Furthermore, all membranes appear to contain lipids that can adopt nonbilayer structures, and a large number of biologically relevant variables can modulate the structural preferences of these lipids. These facts support the proposition that fusion proceeds via a nonbilayer intermediate, as shown in Figure 2.15. Difficulties with this model are that lipidic particle structures observed on fusion (Figure 2.14) appear to be generated subsequent to the actual fusion event and that for Ca^{2+}-induced fusion, high Ca^{2+} levels (greater than $0.5mM$) are required. However, the nonbilayer fusion intermediates may be extremely short-lived and thus difficult to observe by freeze-fracture. Other factors, such as Mg^{2+} or protein, may act synergistically with Ca^{2+} to induce fusion in vivo at lower Ca^{2+} levels.

Fusion of Biological Membranes

Extension of the preceding observations on fusion of model systems to fusion processes in vivo is difficult to show directly. However, work on several experimental systems has provided circumstantial evidence in support of the hypothesis that fusion processes rely on the polymorphic capabilities of lipids. One system studied was the fusion process involved in the exocytotic events occurring during release of the contents of secretory vesicles such as the chromaffin granules of the adrenal medulla (Cullis et al. 1983). Such exocytosis is dependent on the influx of Ca^{2+}, which stimulates fusion between the granule and the cytosolic side of the plasma membrane. By analogy to the erythrocyte membrane, the inner (cytosolic) monolayer is probably composed primarily of phosphatidylserine and phosphatidylethanolamine. Studies have shown that chromaffin granules will undergo Ca^{2+}-stimulated fusion with SUVs of inner-monolayer lipid composition. Such fusion appears to depend on the ability of Ca^{2+} to promote nonbilayer structures. In another system, myoblast cells (which fuse to form the multinucleated muscle fibers) have been studied (Sessions and Horwitz 1981). Such fusion, which is also Ca^{2+} dependent, may rely on a different transmembrane distribution of phosphatidylethanolamine and phosphatidylserine, which appear to reside mainly in the outer monolayer of the myoblast plasma membrane.

Yet another system concerns the tight junction network formed by epithelial and endothelial cells to separate *apical* (membrane facing the lumen) and *basolateral* (surface opposite the lumen) domains. Such networks may correspond to a situation of arrested fusion. Recent freeze-fracture work suggests that the striated patterns characteristic of tight junction assemblies (Figure 2.16) may correspond to long, inverted lipid cylinders similar to those comprising the H_{II} phase structure (Kachar and Reese 1982). Similar states of arrested

Figure 2.16. Diagram of a cross section of a tight junction strand combined with freeze-fracture micrographs. Reproduced with permission from Kachar and Reese (1982).

fusion may correspond to the contact sites between the inner and outer membranes of mitochondria and *E. coli*.

MODEL MEMBRANES AND DRUG DELIVERY

The preceding sections have dealt primarily with the use of lipids in various model membrane systems to gain insight into the physical properties and relative functional roles of individual lipid components in biological membranes. However, these model membrane systems have important potential uses in their own right, as carriers of biologically active agents such as drugs, enzymes, and DNA vectors for clinical application (Poste 1980). Natural membrane lipid components such as phosphatidylcholine are remarkably nontoxic and nonimmunogenic and can therefore provide benign carriers for more toxic or labile agents encapsulated within lipid vesicles. An important aim, which has not yet been realized, is to target liposomal systems containing drugs such as anticancer agents to specific tissues via antibodies attached to the vesicle surface, as indicated in Figure 2.17.

The many difficulties involved in drug delivery via liposomal systems may be summarized as follows. First, vesicle systems must be employed which exhibit an adequate trapped volume to entrap sufficient drug, and a mode of preparation must be used which allows a high trapping efficiency. Several such procedures now exist, including the reversed-phase evaporation protocol (Szoka and Papahadjopoulos 1980) and the extrusion protocol (Hope et al. 1985) outlined previously (Table 2.4), which allow maximum trapping efficiencies in the range of 30–50% of available drug. The second difficulty concerns the phenomenon of serum-induced leakage of the liposomes due to interaction with serum components such as lipoproteins. This problem can be significantly alleviated by inclusion of lipids that are more saturated and/or cholesterol in the carrier vesicle. A third and major difficulty for liposomal delivery systems involves uptake of the liposomes by the fixed and free macrophages of the reticuloendothelial system, which are primarily localized to the liver (Kupffer cells) and spleen. This problem has not yet been circumvented, and even if such uptake could be avoided, other significant problems would remain. For instance, although several procedures exist for coupling antibodies to vesicles, it is unlikely that such targeted systems will be able to cross the endothelial barrier to gain access to extravascular tissue.

Despite these problems, the attractive nature of vesicle-mediated drug delivery has engendered increasing interest and effort which have already resulted in protocols of potential clinical importance. These advances have taken advantage of the natural targeting to macrophages, with two distinct aims. The first involves parasites which reside in the macrophages and which are difficult to eliminate

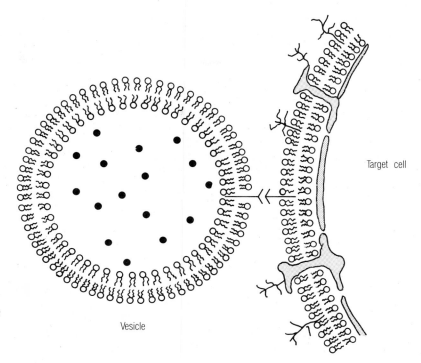

Figure 2.17. The delivery of biologically active materials encapsulated in membrane vesicles. Tissue-specific antibodies are covalently attached to the surface of the vesicle and enable the targeting of entrapped material.

Target cell

Vesicle

by conventional means. However, encapsulation of an appropriate drug into a vesicle carrier which is subsequently taken up by the macrophages can result in elimination of parasites such as *Lieshmania* (Alving 1982). An advantage of this method of treatment is that the dose levels needed are at least an order of magnitude lower than otherwise required. A second clinical application of vesicle-mediated drug delivery is the incorporation of macrophage-activating factors into the vesicles which, when endocytosed, result in macrophage activation. Such activated macrophages appear to be remarkably effective for recognizing and destroying diseased tissue, including transformed cells (Poste 1980).

PROBLEMS

1. The phospholipids of the erythrocyte membrane are asymmetrically distributed across the membrane. On the basis of the lipid composition of the two monolayers, which monolayer would you expect to be the most permeable to Na^+ or to Cl^- ions? How would you verify this experimentally?

2. The apical and basolateral domains of polarized epithelial and endothelial cells exhibit different lipid and protein compositions. It has been observed that lipids introduced into the outer monolayer of the apical region remain there, whereas lipids which can redistribute across the bilayer are subsequently found in both apical and

basolateral domains. Indicate how the barrier presented by the arrested fusion model of Figure 2.16 is compatible with these properties.

3. The microorganism *Acholeplasma laidlawii* can be manipulated so that its fatty acid composition is essentially homogeneous. It is observed that as the acyl chain unsaturation is increased, the ratio of endogenous monoglucosyldiglyceride (MGluDG) to diglucosyldiglyceride (DGluDG) decreases dramatically. MGluDG is an H_{II} phase lipid, whereas DGluDG is a bilayer lipid. Indicate how the change in ratios of MGluDG to DGluDG could be rationalized in terms of lipid shape properties.

4. Valinomycin is a relatively specific K^+ *ionophore*, which translocates K^+ ions across membranes. LUV systems are prepared so that a K^+ buffer is on the inside and a Na^+ buffer is on the outside. How would you expect the efflux of the K^+ ions to vary with time when valinomycin is added? How would you measure this?

5. Indicate four procedures which you would expect to induce fusion between LUVs composed of a mixture of unsaturated phosphatidylserine and phosphatidylethanolamine.

BIBLIOGRAPHY

Entries preceded by an asterisk are suggested for further reading.

Alving, C. R. 1982. "Therapeutic potential of liposomes as carriers in leishmaniasis, malaria, and vaccines." In *Targeting of Drugs*, ed. G. Gregoriadis and A. Tromet, 337–53. New York: Plenum.

Bangham, A. D.; Standish, M. M.; and Watkins, J. D. 1965. Diffusion of univalent ions across the lamellae of swollen phospholipids. *J. Mol. Biol.* 13:238–51.

Barenholtz, Y.; Amselem, S.; and Lichtenberg, D. 1979. A new method for preparation of phospholipid vesicles (liposomes)—French press. *FEBS Lett.* 99:210–13.

Blok, M. C.; van Deenen, L. L. M.; and de Gier, J. 1976. Effect of the gel to liquid crystalline phase transition on the osmotic behaviour of phosphatidylcholine liposomes. *Biochim. Biophys. Acta* 433:1–12.

*Cullis, P. R.; de Kruijff, B.; Hope, M. J.; Verkleij, A. J.; Nayar, R.; Farren, S. B.; Tilcock, C. P. S.; Madden, T. D.; and Bally, M. B. 1983. "Structural properties of lipids and their functional roles in biological membranes." In *Membrane fluidity in biology*, ed. R. C. Aloia, vol. 1, 39–81. New York: Academic Press.

Danielli, J. F., and Davson, H. 1935. *J. Cell. Comp. Physiol.* 5:495–502.

Davis, J. H. 1983. The description of membrane lipid conformation, order, and dynamics by ^{2}H NMR. *Biochim. Biophys. Acta* 737:117–71.

Deamer, D. W. 1982. "Proton permeability in biological and model membranes." In *Intracellular pH: Its measurement, regulation, and utilization in cellular functions*, 173–87. New York: Alan Liss.

Dean, W. L., and Tanford, C. 1977. Reactivation of a lipid-depleted Ca^{2+}-ATPase by a non-ionic detergent. *J. Biol. Chem.* 252:3351–53.

*de Kruijff, B.; Cullis, P. R.; Verkleij, A. J.; Hope, M. J.; van Echteld, C. J. A.; and Toraschi, T. F. 1984. "Lipid polymorphism and membrane function." In *Enzymes of biological membranes*, ed. A. Martinosi, 131–204. New York: Plenum.

Demel, R. A., and de Kruijff, B. 1976. The function of sterols in membranes. *Biochim. Biophys. Acta* 457:109–32.

Fettiplace, R.; Gordon, I. G. H.; Hladky, S. B.; Requens, J.; Zingshen, H. B.; and Haydon, D. A. 1974. "Techniques in the formation and examination of black lipid bilayer membranes." In *Methods in membrane biology*, ed. E. D. Korn, vol. 4, 1–75. New York: Plenum.

Fettiplace, R., and Haydon, D. A. 1980. Water permeability of lipid membranes. *Physiol. Rev.* 60:510–50.

Gorter, E., and Grendel, F. 1925. On bimolecular layers of lipids on the chromocytes of the blood." *J. Exp. Med.* 41:439–43.

Helenius, A.; Sarvas, M.; and Simons, K. 1981. Asymmetric and symmetric membrane reconstitution by detergent elimination. *Eur. J. Biochem.* 116:27–31.

*Hope, M. J.; Bally, M. B.; Webb, G.; and Cullis, P. R. 1985. Production of large unilamellar vesicles by a rapid extrusion procedure. Characterization of size distribution, trapped volume, and ability to maintain a membrane potential. *Biochim. Biophys. Acta.* 812:55–65.

Houslay, M. D., and Stanley, K. K. 1982. *Dynamics of biological membranes*, New York: Wiley.

Huang, C. 1969. Studies on phosphatidylcholine vesicles. Formation and physical characteristics. *Biochemistry* 8:344–49.

Jain, M. K., and Wagner, R. C. 1980. *Introduction to biological membranes*, New York: Wiley.

*Jost, P. C., and Griffith, O. H., eds. 1982. *Lipid-protein interactions*, vols. 1 and 2. New York: Wiley.

Kachar, B., and Reese, T. S. 1982. Evidence for the lipidic nature of tight junction strands. *Nature* 296:464–67.

Kates, M. 1975. "Techniques of lipidology." In *Isolation, analysis, and identification of lipids*, edited by T. S. Work, and E. Work. New York: American Elsevier.

Lindblom, G.; Johansson, L. B. A.; and Arvidson, G. 1981. Effect of cholesterol in membranes. *Biochemistry* 20:2204–9.

Lubin, B., and Chiu, D. 1982. "Membrane phospholipid organization in pathologic human erythrocytes." In *Membranes and genetic disease*, ed. J. R. Sheppard, V. E. Anderson, and J. W. Eaton, 137–64. New York: Alan R. Liss.

*Madden, T. D.; Hope, M. J.; and Cullis, P. R. 1984. Influence of vesicle size and oxidase content on respiratory control in reconstituted cytochrome oxidase vesicles. *Biochemistry* 23:1413–18.

McDonald, R. J., and McDonald, R. C. 1975. Assembly of phospholipid vesicles bearing sialoglycoprotein from the erythrocyte membrane. *J. Biol. Chem.* 250:9206–11.

*McLaughlin, S. 1977. "Electrostatic potentials at membrane-solution interfaces." In *Current topics in membranes and transport*, vol. 9, 71–144. New York: Academic Press.

Mimms, L. T.; Zampighi, G.; Nozaki, Y.; Tanford, C.; and Reynolds, J. A. 1981. Formation of large unilamellar vesicles by detergent dialysis employing octylglucoside. *Biochemistry* 20:833–40.

*Op den Kamp, J. A. F. 1979. Lipid asymmetry in membranes. *Ann. Rev. Biochem.* 48:47–71.

*Papahadjopoulos, D.; Poste, G.; and Vail, W. J. 1978. "Studies on membrane fusion with natural and model membranes." In *Methods in membrane biology*, ed. E. D. Korn, vol. 10, 1–121. New York: Plenum.

Patel, K. M., and Sparrow, J. T. 1978. Rapid large scale purification of crude egg phospholipids using radially compressed silica gel columns. *J. Chromatogr.* 150:542–47.

Poste, G. 1980. "The interaction of lipid vesicles (liposomes) with cultured cells and their uses as carriers for drugs and macromolecules." In *Liposomes in biological systems*, ed. G. Gregoriadis and A. C. Allison, 101–51. New York: Wiley.

Poznansky, M.; Tang, S.; White, P. C.; Milgram, J. M.; and Selenen, M. 1976. Non-electrolyte diffusion across lipid bilayer systems. *J. Gen. Physiol.* 67:45–66.

Racker, E. 1973. Reconstitution of cytochrome oxidase vesicles and conferral of sensitivity to energy transfer inhibitors. *J. Membr. Biol.* 10:221–33.

Sessions, A., and Horwitz, A. F. 1981. Myoblast aminophospholipid asymmetry differs from that of fibroblasts. *FEBS Lett.* 134:75–79.

*Silvius, J. R. 1982. "Thermotropic phase transitions of pure lipids in model membranes and their modification by membrane proteins." In *Lipid-protein interactions*, ed. P. C. Jost, and O. H. Griffith, vol. 2, ch. 7. New York: Wiley.

*Singer, S. J., and Nicolson, G. L. 1972. The fluid mosaic model of the structure of cell membranes. *Science* 175:720–31.

Skipski, V. P., and Barclay, M. 1969. "Thin layer chromatography of lipids." In *Methods in enzymology*, ed. J. M. Lowenstein, vol. 14, 530–98. New York: Academic Press.

*Stoekenius, W. 1976. The purple membrane of salt loving bacteria. *Scientific American*, June, 38–44.

*Szoka, F., and Papahadjopoulos, D. 1980. Comparative properties and methods of preparation of lipid vesicles (liposomes). *Ann. Rev. Bioeng.* 9:467–508.

Tanaka, Y., and Schroit, A. J. 1983. Insertion of fluorescent phosphatidylserine into the plasma membrane of red blood cells: Recognition by autologous macrophages. *J. Biol. Chem.* 258:11335–43.

Tilcock, C. P. S.; Bally, M. B.; Farren, S. B.; and Cullis, P. R. 1982. Influence of cholesterol on the structural preferences of dioleoylphosphatidyl-ethanolamine-dioleoylphosphatidylcholine systems: A phosphorus-31 and deuterium magnetic resonance study. *Biochemistry* 21:4596–4601.

Tilcock, C. P. S.; Bally, M. B.; Farren, S. B.; Cullis, P. R.; and Gruner, S. M. 1984. Cation-dependent segregation phenomena and phase behaviour in model membrane systems containing phosphatidylserine. Influence of cholesterol and acyl chain composition. *Biochemistry* 23:2696–2703.

*van Deenen, L. L. M., and de Gier, J. 1974. "Lipids of the red cell membrane." In *The red blood cell*, ed. D. Surgenor, 147–213. New York: Academic Press.

Verkleij, A. J. 1984. Lipidic intramembranous particles. *Biochim. Biophys. Acta* 779:43–64.

CHAPTER 3

Lipid Metabolism in Procaryotes

Charles O. Rock
John E. Cronan, Jr.

THE STUDY OF BACTERIAL LIPID METABOLISM

There are several major advantages of using bacteria to study the regulation of phospholipid biosynthesis and the role of phospholipids in membrane function. The main advantage bacteria offer over other systems is the use of the genetic approach to manipulate the experimental system. This approach has allowed bacterial physiologists to vary the components of a metabolic network in the cell in order to test critically hypotheses about the role of these components in vivo. Other distinct advantages are that (1) the experimenter has complete control over the conditions of cell growth, (2) large quantities of precisely grown cells for biochemical analysis are readily obtained, (3) most bacteria have a simple lipid composition and lack intracellular membrane systems, and (4) there is a wealth of general information on bacterial metabolism and biochemistry. The biological organism most thoroughly understood at the molecular level is the Gram-negative bacterium *Escherichia coli*, and this generality carries over to lipid metabolism. Most of our knowledge of the molecular aspects of procaryotic lipid metabolism is based on studies of *E. coli*; hence, research on this bacterium will dominate the discussions in this chapter. However, the diversity in life styles, lipid structures, and metabolic pathways represented by bacteria is so large that *E. coli* physiology should not be considered typical of the procaryotic kingdom.

Historical Background   The modern era of *E. coli* phospholipid enzymology began in the early 1960s when Vagelos and his colleagues discovered that the intermediates of fatty acid biosynthesis were bound to a heat-stable cofactor termed *acyl carrier protein* (ACP). The realization that the reactions in fatty acid biosynthesis could be separated and studied individually precipitated a flurry of activity, primarily by the laboratories of Bloch, Vagelos, and Wakil, and within a few years the structures of all the intermediates in fatty acid biosynthesis had been elucidated. These experiments had a great deal of impact on the lipid metabolism field as a whole because the fatty acid synthases of higher organisms are multifunctional protein complexes, and the individual reactions could not be isolated. In the late 1960s, the enzymatic steps leading to the major phospholipid classes were established by the classical demonstration, primarily by Kennedy and co-workers, of the enzymes in crude cell extracts that catalyze these reactions. Later radiolabeling studies established the rapid turnover of the anticipated intermediates in the pathway such as phosphatidic acid, CDP-diacylglycerol, and phosphatidylserine. Armed with the knowledge of the biochemical pathways and the intermediates, selection schemes were devised to obtain mutants in specific pathway enzymes, thus ushering in the most recent work on the regulation of the phospholipid metabolic network that forms the focus of the present chapter.

An Overview of Phospholipid Metabolism in *E. coli*   In *E. coli*, phospholipids are synthesized exclusively for use in the biogenesis of membranes, and there does not appear to be any significant alternative fate for these lipids. The steps in the biosynthetic pathway leading to the major phospholipid classes of *E. coli* have been established, and the biochemistry of these enzymes is covered in most basic biochemistry texts. The enzymes are isolated as individual protein species, and to date there is no convincing evidence for the existence of multienzyme complexes in the lipid biosynthetic pathway. The exception to this rule is the high molecular weight fatty acid synthases of phylogenetically advanced bacteria (see the following discussion). The enzymes of fatty acid biosynthesis are located in the cytoplasmic compartment, and the enzymes that metabolize phospholipids are bound to the inner face of the cytoplasmic membrane. Each mole of phospholipid requires 32 mol of ATP for its biosynthesis from acetyl-CoA and *sn*-glycerol-3-phosphate, and since approximately 10% of the dry weight of the cell is phospholipid, a significant amount of energy expended in the biogenesis of a new cell is used in the production of phospholipid. The advantage to the cell of maintaining fine control over the biosynthesis of phospholipid is evident from these numbers, and the more recent work on lipid biogenesis has focused on uncovering these control mechanisms. The thrust of this chapter will also be on the regulation of this branch of intermediary metabolism.

Brief Background on Bacterial Genetics

$$CH_2-OH$$
$$HO-CH$$
$$CH_2-O-\overset{O}{\underset{O^-}{\overset{\|}{P}}}-O^-$$

E. coli has a single chromosome composed of sufficient DNA to encode about 3000 proteins. Two general types of mutants can be obtained. *Auxotrophic mutants* are strains that require a growth supplement that the organism isolated from nature (called the wild-type strain) does not require. The auxotrophic lipid metabolic mutants are those that require the addition of fatty acids or other lipid precursors (for example, *sn*-glycerol-3-phosphate) to the growth medium. The reason for the nutrient supplement is generally that the mutant organism has lost the function of a key enzyme required in the biosynthesis of the compound; however, the reason for the requirement can be more complex (for example, the *plsB* and *acpS* mutants).

Conditional lethal mutations represent a second class of mutants. Many compounds, such as phospholipids, are not readily taken up from the growth medium by bacteria. Therefore, mutants unable to synthesize a phospholipid required for membrane function would be nonviable (dead) and thus could not be isolated and studied using the approach just described. However, the isolation of conditionally lethal mutants allows the study of such pathways. Conditionally lethal mutants function normally under one set of conditions but become defective when shifted to a second set of conditions. Although several types of conditionally lethal mutants have been used in bacterial genetics, only temperature-sensitive mutations have been used in the study of lipid metabolism. *Temperature-sensitive mutants* are strains that grow at a low temperature (for instance, 30°C), but not at 42°C (wild-type *E. coli* grows from 8–44°C). The mutants owe this restricted growth temperature range to a mutation in the gene encoding a required protein such that the mutant protein denatures at an abnormally low temperature. Therefore, the protein is functional in cells incubated at 30°C, but at 42°C, the protein is nonfunctional and growth ceases. The mutational alteration in protein structure is usually a single amino acid change, and the increased thermal lability of the protein can often be demonstrated in vitro. Such a demonstration is strong evidence that the mutation is within the gene encoding the protein.

Three general approaches have been used to isolate the *E. coli* mutants listed in Table 3.1. In all three approaches, the wild-type strain is treated with a strongly mutagenic chemical to increase the rate of mutation, after which mutants are sought. The first technique is to isolate auxotrophic mutants that require either fatty acids or

(handwritten margin notes:)
① The phospholipids are not readily taken up from the growth medium by bacteria.

② Temperature sensitive mutation can be used in the study of lipid metabolism –

Table 3.1. Mutants in *E. coli* Phospholipid Biosynthesis

Gene	Enzyme affected	Structural gene	Phenotype	Cloned
fabA	3-Hydroxydecanoyl-ACP dehydrase	Yes	Unsaturated fatty acid auxotroph	Yes
fabB	3-Ketoacyl-ACP synthase I	Yes	Unsaturated fatty acid auxotroph	Yes
fabD	Malonyl-CoA:ACP transacylase	Yes	Ts* mutant requires both saturated and unsaturated fatty acids	No
fabE	Acetyl-CoA carboxylase	Probably	Same as *fabD*	No
fabF	3-Ketoacyl-ACP synthase II	Yes	Altered thermal regulation	No
acpS	[ACP] synthase	Probably	Requires high intracellular CoA levels	No
cfa	Cyclopropane fatty acid synthase	Yes	Grows normally but lacks cyclopropane fatty acids	Yes
plsB	*sn*-glycerol-3-phosphate acyltransferase	Yes	*sn*-glycerol-3-phosphate or glycerol auxotroph	Yes
cds	CDP-diacylglycerol synthase	?	pH sensitive	No
cdh	CDP-diacylglycerol hydrolase	?	None	Yes
pss	Phosphatidylserine synthase	Yes	Ts mutant	Yes
psd	Phosphatidylserine decarboxylase	Yes	Ts mutant	Yes
pgsA	Phosphatidylglycerol phosphate synthase	Probably	None	Yes
pgsB	Disaccharide-1-phosphate synthase	Probably	None	Yes
pgsA, pgsB	Double mutant	Probably	Ts mutant	No
mdoB	Transfer of *sn*-glycerol-1-phosphate to MDO† precursor	?	None	No
cls	Cardiolipin synthase	?	None	No
dgk	Diacylglycerol kinase	Yes	Osmotically sensitive	Yes
pgpAB	Phosphatidylglycerol phosphate phosphatase	?	None	No
pldA	Detergent-resistant phospholipase A	Probably	None	Yes

Regulatory Genes				
pssR	Overproduction of phosphatidylserine synthase		None	No
dgkR	Overproduction of diacylglycerol kinase		None	No
adk	Adenylase kinase and *sn*-glycerol-3-phosphate acyltransferase		Ts mutant; *sn*-glycerol-3-phosphate acyltransferase also thermolabile	Yes
cdsS	Stabilizes mutant CDP-diacylglycerol synthase		Suppresses pH sensitivity of *cds* strain	No
fabAUp	Overproduction of 3-hydroxydecanoyl-ACP dehydrase		Overproduction of saturated fatty acids	No

* Ts represents temperature sensitive.   † MDO represents membrane-derived oligosaccharides.

Handwritten margin notes:

① general bacterial genetic selection method.

② Radio precursor
↓
mutant + normal
↓
mutant + (normal Radio precursor) → die
↓
mutant

③ colony autoradiography

sn-glycerol-3-phosphate by use of standard bacterial genetic selection methods. A second type of selection is to incorporate large quantities of highly radioactively labeled lipid precursors into the mutagenized cells at the nonpermissive growth temperature (usually 42°C) and then store the labeled cells. During storage, those cells competent in lipid synthesis are irradiated by the disintegration of the incorporated radioactive precursor, and they die, whereas those cells mutationally defective in lipid synthesis survive due to their lack of incorporation of the radioactive precursor (Figure 3.1). The third method is to screen

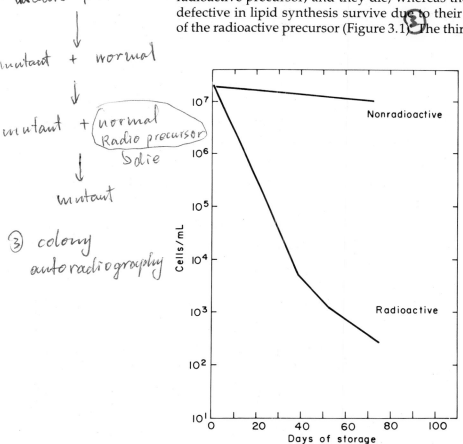

Figure 3.1. Procedure for the isolation of mutants by tritium suicide. Mutagenized cells are grown in the presence of a highly radioactive tritiated compound which is specifically incorporated into phospholipid (this often requires blocking other metabolic pathways which would utilize the precursor). The cells are washed free of unincorporated radioactive compound, suspended in buffer, and stored in the refrigerator. At various intervals, the number of viable cells is determined. Cells having a normal pathway of incorporation (normal lipid synthesis) become highly radioactive and are killed by radioactive disintegration (probably through generation of free radicals), whereas mutant strains unable to incorporate the precursor survive (plateau of the curve marked radioactive). Thus, the rare mutant cells can be selected from the population. If a nonradioactive precursor is used, only the slow rate of death due to storage is seen.

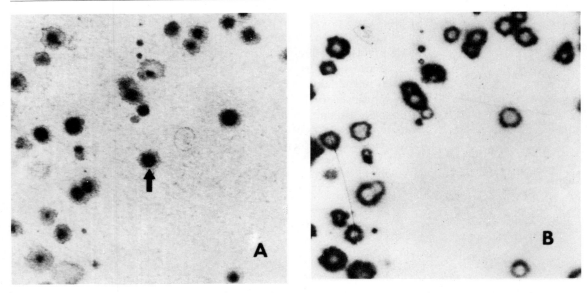

Figure 3.2. Isolation of mutants by colony autoradiography. Mutagenized colonies growing on the surface of an agar petri plate are blotted onto a sheet of filter paper. Much of the colony adheres to the paper. The paper is then immersed in a solution, which lyses the colonies (the membranes of the lysed cells remain bound to the paper), and is next immersed in a second solution containing the precursors and buffer necessary for the enzyme reaction. A water-soluble substrate of the reaction is included in a radioactively labeled form. The immobilized lysed cells are allowed to incorporate the radioactive precursor for a short time; then the filter is washed repeatedly with a solution in which lipids are insoluble but in which water-soluble compounds remain in solution. The washed paper is stained with a specific protein stain, dried, and exposed to a sheet of X-ray film. When the resulting autoradiogram (panel B) is compared with the stained filter paper (panel A), colonies that lack enzyme activity (arrow) can be identified. The mutant colony is then isolated from the original agar plate.

colonies of mutagenized cells either by performing a given lipid synthetic enzyme assay on isolated bacterial colonies (Figure 3.2) or by screening such colonies for the inability to synthesize a given lipid. The use of such brute-force colony autoradiography schemes has proven extremely valuable in isolating phospholipid mutants (Raetz 1982).

The chromosome of *E. coli* is circular and is divided into 100 minutes (map units) as determined by the time-of-entry in inter-rupted-conjugation experiments. The entire chromosome is also linked by cotransduction data using bacteriophage P1. Each minute of the map corresponds to about 45 kilobases of DNA. The locations of the known genetic loci of the enzymes involved in phospholipid metabolism are shown in Figure 3.3, with the 100/0 point of the map located at the top center of the figure. The most important point to be

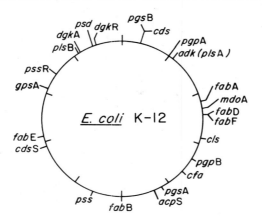

Figure 3.3. Location of the known genes of lipid metabolism on the *E. coli* chromosome.

① The gene are not clustered into operon
② regulation of phospholip is exerted at the level of individual enzyme activities and not at the level of gene expression

made from this figure is that the genes governing phospholipid metabolism are scattered throughout the circular map and are not clustered into operons. In fact, there are no examples of two genes in phospholipid biosynthesis that form a transcriptional unit. These enzymes are indispensable for cell proliferation; therefore, these enzymes are not inducible, but rather are expressed at the same relative level regardless of the growth condition. This means that the regulation of the phospholipid biosynthetic pathway is exerted at the level of individual enzyme activities and not at the level of gene expression. In addition to the genes listed in Table 3.1, there are many other loci that have been extremely useful in studying phospholipid metabolism. A listing of these mutations and their uses can be found in Clark and Cronan (1981).

It should be noted that mutants are defined by two characteristics: The *phenotype* of a mutant is its outward manifestation, whereas the *genotype* is the genetic alteration which causes the phenotype. Examples are the *fabA* and *fabB* mutants. Both the *fabA* and *fabB* mutants (Table 3.1) have the same phenotype, a requirement for unsaturated fatty acids for growth. However, the two mutants have different genotypes, since their phenotype is due to lesions in two different genes (Figure 3.3) which encode different fatty acid synthetic enzymes.

Membrane Systems of *E. coli*

The only known metabolic fate for phospholipids is in the formation of the two membrane systems of *E. coli* (Figure 3.4). Like other Gram-negative bacteria, *E. coli* has an inner cytoplasmic membrane that contains the enzymes of phospholipid biosynthesis, electron transport, metabolite and ion transport, and other metabolic processes. Between the inner and outer membranes there is an osmoti-

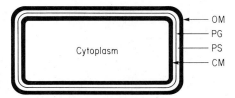

Figure 3.4. Membrane systems of *E. coli*. The major structures indicated are OM, outer membrane; PG, peptidoglycan; PS, periplasmic space; and CM, cytoplasmic (inner) membrane.

cally active compartment called the *periplasmic space.* Membrane-derived oligosaccharides, peptidoglycan, and binding proteins involved with metabolite transport are found in this compartment. The outer membrane is considerably different from the inner membrane. Pores exist in this structure that allow the passage of molecules having a molecular weight less than 600, whereas the inner membrane is impermeable to solutes unless specific transport systems are present. The outer layer of the outer membrane is composed primarily of lipopolysaccharides rather than phospholipid. The outer membrane is rich in structural lipoproteins and proteins involved in the transport of high molecular weight compounds. Some of these proteins also function as receptors for bacteriophages that infect *E. coli*.

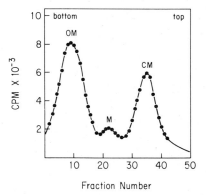

Figure 3.5. Separation of the cytoplasmic (inner) membrane from the outer membrane by sucrose density gradient centrifugation. The two membrane systems are separated by isopycnic centrifugation on linear sucrose gradients (30–55%, w/w). The distribution of [³H]glycerol-labeled phospholipids in the gradient is shown in the figure. The abbreviations are CM, cytoplasmic membrane; OM, outer membrane; and M, an intermediate fraction composed of a mixture of the two membrane systems.

The marked difference in the composition of the inner and outer membrane systems of *E. coli* has been established due to the development of a reliable method for the separation of these two structures (Osborn and Munson 1974). If the associations between the two membrane systems are broken, the two structures can be clearly resolved using density-gradient centrifugation (Figure 3.5). The presence of lipopolysaccharides and the higher protein:lipid ratio in the outer membrane results in the banding of the outer membrane at a higher density (1.22 g/mL) than the inner membrane (1.16 g/mL). Virtually everything we know about the subcellular distribution of phospholipids and phospholipid enzymes in *E. coli* is derived from the analysis of such gradients.

PHOSPHOLIPID BIOSYNTHESIS IN *E. COLI*

Initiation

The precursors for fatty acid biosynthesis are derived from the acetyl-CoA pool. Four enzyme reactions are involved (Figure 3.6). Malonyl-CoA is required for all the elongation steps and is formed by the carboxylation of acetyl-CoA by acetyl-CoA carboxylase. Acetyl-CoA carboxylase is composed of three individual proteins: biotin carboxylase, biotin carboxyl carrier protein, and carboxyltransferase. Acetyl-ACP is the primer of fatty acid biosynthesis; it is formed from acetyl-CoA by the action of acetyl transacylase. This is the only step where an acetate unit is used in fatty acid biosynthesis. Similarly, malonyl-ACP is formed from malonyl-CoA by an analogous transacylase.

[handwritten notes in margin: acetyl CoA carboxylase; ① biotin carboxylase; ② biotin carboxyl carrier protein; ③ carboxyl transferase]

2 Elongation

The initial condensation of malonyl-ACP and acetyl-ACP is catalyzed by the 3-ketoacyl-ACP synthase. There are two condensing enzymes in *E. coli*, and the genetic evidence indicates that either of these two

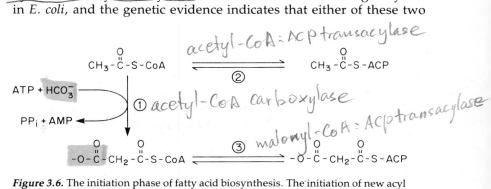

[handwritten annotations on figure: acetyl-CoA : ACP transacylase; ① acetyl-CoA carboxylase; ③ malonyl-CoA : ACP transacylase]

Figure 3.6. The initiation phase of fatty acid biosynthesis. The initiation of new acyl chains is accomplished by the action of four enzymes: (1) acetyl-CoA carboxylase, (2) acetyl-CoA:ACP transacylase, and (3) malonyl-CoA:ACP transacylase.

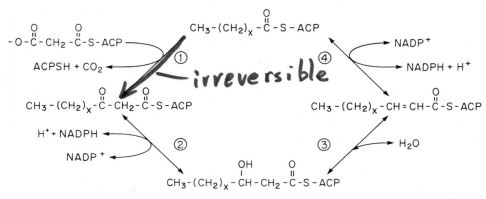

Figure 3.7. The elongation cycle of fatty acid biosynthesis. The elongation of a growing acyl chain is accomplished by the action of four enzymes: (1) 3-ketoacyl-ACP synthase, (2) 3-ketoacyl-ACP reductase, (3) 3-hydroxyacyl-ACP dehydrase, and (4) *trans*-2-acyl-ACP reductase (enoyl reductase).

enzymes can catalyze the first condensation reaction to form ace-toacetyl-ACP. There are four enzymes that participate in each cycle of chain elongation; the general reaction scheme is shown in Figure 3.7. First, the condensing enzymes add an additional two-carbon unit from malonyl-ACP. The resulting ketoester is reduced by an NADPH-dependent 3-ketoacyl-ACP reductase, and a water molecule is then removed by the 3-hydroxyacyl-ACP dehydrase. The last step is catalyzed by enoyl-ACP reductase to form a saturated acyl-ACP, which in turn can serve as the substrate for another condensation reaction. NADPH is probably the preferred cofactor for the enoyl-ACP reductase, but there may be two enzymes involved—one specific for NADPH and the other specific for NADH. An important point to remember about the elongation phase of fatty acid biosynthesis is that the condensing enzymes catalyze the only irreversible steps in the elongation cycle.

3. Product Diversification   There are three major fatty acids produced by the *E. coli* fatty acid synthase system, namely, palmitic, 16:0, palmitoleic, 16:1(n − 7), and *cis*-vaccenic, *c*-18:1(n − 7), acids. A specific dehydrase enzyme, 3-hydroxydecanoyl-ACP dehydrase, first described by Bloch and co-workers, catalyzes a key reaction at the point where the biosynthesis of saturated fatty acids diverges from unsaturated fatty acids (Figure 3.8). This dehydrase is a distinctly different enzyme from the 3-hydroxyacyl-ACP dehydrase that participates in the elongation cycle

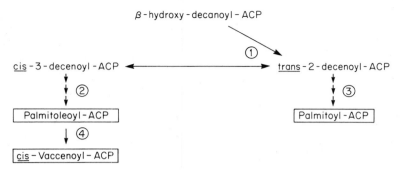

Figure 3.8. Product diversification in fatty acid biosynthesis. There are three main fatty acids produced by the *E. coli* fatty acid synthase system. The ratio of these fatty acids is controlled by the activity of three enzymes: (1) 3-hydroxydecanoyl-ACP dehydrase is a specific dehydrase that introduces the double bond into the acyl chain, (2) 3-ketoacyl-ACP synthase I catalyzes an essential step in the unsaturated fatty acid elongation pathway, (3) both 3-ketoacyl-ACP synthases I and II can elongate saturated fatty acids, and (4) 3-ketoacyl-ACP synthase II is responsible for the elongation of 16:1(n-7) to 18:1(n-7).

in that it is also capable of isomerizing *trans*-2- to *cis*-3-decenoyl-ACP. Genetic studies have also shown that the two condensing enzymes, I and II, are responsible for different aspects of the elongation reactions in the unsaturated branch of the pathway. 3-Ketoacyl-ACP synthase I is absolutely required for one of the elongation reactions in this branch, and 3-ketoacyl-ACP synthase II is responsible for the elongation of palmitoleate to *cis*-vaccenate. Control of product distribution is one of the most important adaptive responses in bacterial physiology. The regulation of the enzymes shown in Figure 3.8 will be covered in detail in later sections.

4. Transfer to the Membrane

The transformation of the water-soluble acyl-ACP end products of fatty acid biosynthesis into a membrane phospholipid is accomplished by the glycerolphosphate acyltransferase system (Figure 3.9). This enzyme system transfers an acyl moiety from acyl-ACP to the 1-position of glycerolphosphate followed by the esterification of the 2-position of the glycerol backbone with a second acyl chain. Like most other phospholipids in nature, bacterial phospholipids have an asymmetric distribution of fatty acids between the 1- and 2-positions of the glycerolphosphate backbone (Figure 3.9), and the acyl chain specificity of the glycerolphosphate acyltransferase system is generally considered to account for this important aspect of membrane

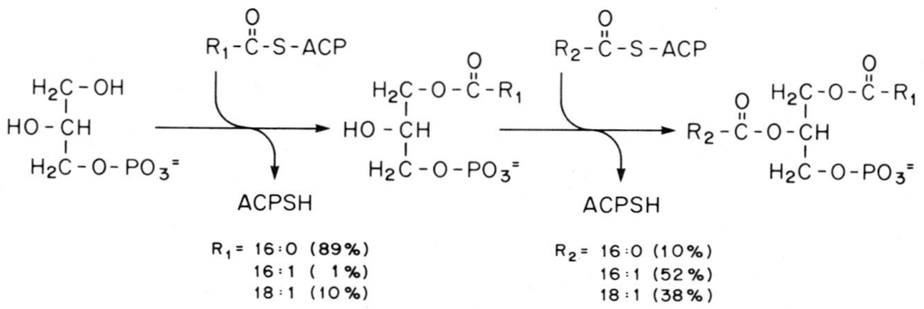

Figure 3.9. Transfer to the membrane. The utilization of fatty acids to form the first membrane phospholipid in the pathway is catalyzed by the glycerol-3-phosphate acyltransferase system, which consists of two enzymes: (1) *sn*-glycerol-3-phosphate acyltransferase and (2) 1-acyl-*sn*-glycerol-3-phosphate acyltransferase. The positional distribution of fatty acids between the 1- and 2-positions of the glycerol backbone found in vivo is indicated below each step.

phospholipid structure. However, some bacteria do not have diacylphospholipids. *Archaebacteria* contain glycerolipids containing phytanoyl ether groups at the 1- and 2-positions of glycerol, and *Clostridia* contain an abundant amount of alk-1′-enyl groups at the 1-position of their glycerolphosphatides. The details of the biosynthesis of these more unusual phospholipid structures remain to be elucidated.

5. Diversification of Polar Head Groups

Phosphatidic acid formed by the glycerolphosphate acyltransferase is utilized by phosphatidate cytidylyltransferase (Figure 3.10) along with CTP to form CDP-diacylglycerol, the key intermediate in the formation of the diverse phospholipid species found in bacterial systems. The diversity of polar head groups in the procaryotic kingdom defies adequate description in this short space; the reader is referred to Goldfine's review (1982) for a more comprehensive treatment of bacterial phospholipid structures.

Phospholipid biosynthesis has been most carefully studied in *E. coli*. This organism possesses one of the simplest phospholipid compositions, consisting primarily of phosphatidylethanolamine (75%), phosphatidylglycerol (15–20%), and cardiolipin (5–10%). The two more abundant phospholipids are synthesized from CDP-diacylglycerol in two steps (Figure 3.10). Phosphatidylserine synthase exchanges L-serine for CMP to form phosphatidylserine, which is subsequently decarboxylated by phosphatidylserine decarboxylase to form the most abundant phospholipid, phosphatidylethanolamine. Similarly, phosphatidylglycerol arises from the exchange of CMP for *sn*-glycerol-3-phosphate to form phosphatidylglycerolphosphate, which is subsequently dephosphorylated to form phosphatidyl-

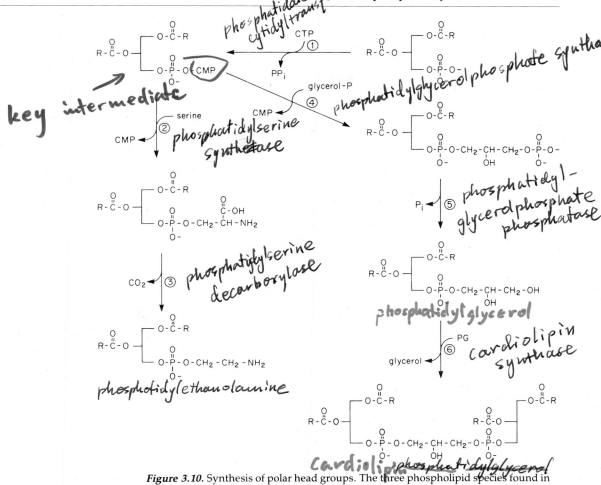

Figure 3.10. Synthesis of polar head groups. The three phospholipid species found in *E. coli* are synthesized by a series of reactions utilizing six enzymes: (1) phosphatidate cytidylyltransferase, (2) phosphatidylserine synthase, (3) phosphatidylserine decarboxylase, (4) phosphatidylglycerolphosphate synthase, (5) phosphatidyl-glycerolphosphate phosphatase, and (6) cardiolipin synthase. PG is an abbreviation for phosphatidylglycerol.

glycerol. Cardiolipin is synthesized by the condensation of two molecules of phosphatidylglycerol (Pieringer 1983).

The Central Role Played by Acyl Carrier Protein in Bacterial Lipid Biosynthesis

A quick examination of Figures 3.6–3.10 shows that ACP is a critical cofactor in four of the five phases of membrane phospholipid biogenesis; consequently, this protein has received considerable experimental attention. ACP is one of the most abundant proteins in *E. coli* (0.25% of the total soluble protein). It has been purified to

```
      +    1                6              10
     NH₃-Ser-Thr-Ile-Glu-Glu-Arg-Val-Lys-Lys-Ile-Ile-Gly-Glu-

                      20
     Gln-Leu-Gly-Val-Lys-Gln-Glu-Glu-Val-Thr-Asp-Asn-Ala-Ser-

            30              34        (P)-Pantetheine
                                           |
     Phe-Val-Glu-Asp-Leu-Gly-Ala-Asp-Ser-Leu-Asp-Thr-Val-Glu-

            44                   50                  55
     Leu-Val-Met-Ala-Leu-Glu-Glu-Glu-Phe-Asp-Thr-Glu-Ile-Pro-

                  60
     Asp-Glu-Glu-Ala-Glu-Lys-Ile-Thr-Thr-Val-Gln-Ala-Ala-Ile-

     70                      77
     Asp-Tyr-Ile-Asn-Gly-His-Gln-Ala-COO⁻
```

Figure 3.11. The amino acid sequence of acyl carrier protein.

homogeneity and its complete amino acid sequence has been determined (Figure 3.11). ACP, whose molecular weight is 8847, has a number of characteristic structural features. It has a preponderance of acidic residues, resulting in an isoelectric point of pH 4.1, and a high content of α-helical secondary structure. Hydrodynamic studies have shown that ACP is an asymmetrically shaped protein with a major axis:minor axis ratio of 1.4. The intermediates of fatty acid biosynthesis are bound to the protein as thioesters attached to the terminal sulfhydryl of the phosphopantetheine prosthetic group (Figure 3.12). The prosthetic group is in turn attached to the protein via a phosphodiester linkage to Ser-36 of the protein. A β-turn is commonly found between structural domains of proteins and consists of a group of four amino acids that constitute an approximate 180° reversal in the direction of the amino acid backbone. Ser-36 is predicted to be located in the fourth position of such a β-turn structure located between the second and third α-helical segments of ACP. The prosthetic group sulfhydryl is the only thiol group in *E. coli* ACP. Sequence data are

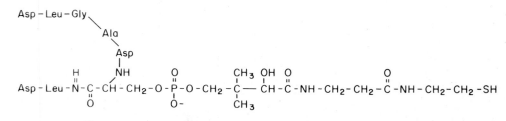

Figure 3.12. Attachment of the phosphopantetheine prosthetic group to Ser-36 of acyl carrier protein.

available for ACP in fungi, plants, and animals; all of these ACPs have a helical secondary structure and a primary sequence that is highly conserved in the region of the protein on either side of the attachment site for the phosphopantetheine prosthetic group.

ACP undergoes a reversible conformational change from its compact helical form at neutral pH to a random-coil structure at elevated pH. Do the enzymes of fatty acid metabolism specifically recognize the native helical form of the protein? The clearest answer to this question has come from studies with glycerolphosphate acyltransferase. This enzyme has a pH optimum in vitro of pH 8.5 when acyl-CoAs are used as the acyl donor; however, acyltransferase activity measured using palmitoyl-ACP as the acyl donor is almost nonexistent at this pH unless divalent cations such as magnesium or spermidine are included in the assay. The reason for this curious observation was established when the solution structure of acyl-ACP was examined under three different assay conditions. The technique chosen was determination of the hydrodynamic radius of ACP by gel-filtration chromatography (Figure 3.13). This chromatographic medium fractionates molecules according to their molecular volume. Therefore, denatured acyl-ACP elutes faster than the native, more compact form of the protein. At pH 8.5, acyl-ACP exists in its expanded, denatured conformation and is correspondingly inactive in the acyltransferase assay. On the other hand, at either neutral pH or at pH 8.5 in the presence of divalent cations, acyl-ACP adopts its native α-helical structure (Figure 3.13) and consequently is an excellent substrate for the glycerolphosphate acyltransferase. Although this is the best described example, it seems reasonable that all the enzymes that utilize ACP derivatives are specific for the native solution structure of the protein.

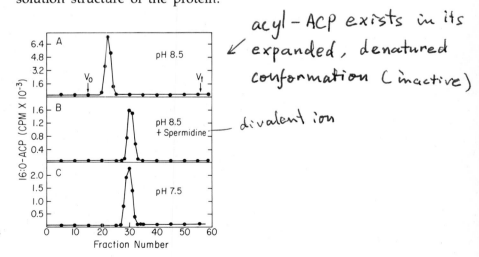

Figure 3.13. Effect of pH and cations on the solution structure of acyl carrier protein. Gel-filtration chromatography was performed using Bio-Gel P-150 equilibrated in Tris buffer at the indicated pH, containing either no additions or 5 mM spermidine.

(Handwritten annotations:) acyl-ACP exists in its ← expanded, denatured conformation (inactive); — divalent ion

Figure 3.14. Turnover of the phosphopantetheine prosthetic group of acyl carrier protein. The hydrolysis of the prosthetic group is accomplished by the enzyme [ACP]phosphodiesterase, and the phosphopantetheine is added to the apo-protein by the enzyme [ACP]synthase.

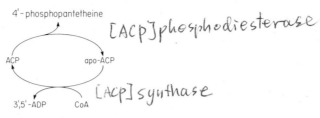

Prosthetic Group Turnover

The protein portion of ACP is metabolically stable in *E. coli*, but its phosphopantetheine prosthetic group is metabolically active and undergoes considerable turnover in vivo. The prosthetic group turnover cycle is mediated by the action of two enzymes: [ACP]synthase, which is responsible for the transfer of the phosphopantetheine portion from CoA to Ser-36 of apo-ACP, and [ACP]phosphodiesterase, which initiates the turnover cycle by cleaving the prosthetic group from ACP (Figure 3.14). Each round of prosthetic group turnover results in the expenditure of two molecules of ATP that are used in the conversion of phosphopantetheine to CoA.

The measurement of the rate of prosthetic group turnover in vivo presents an interesting biochemical challenge. A conventional pulse-chase experiment cannot be used to study ACP turnover because the CoA pool is both large and metabolically stable and thus cannot be effectively chased. Two approaches have been used to circumvent this difficulty. In the first, a mutant unable to synthesize the pantothenate portion of phosphopantetheine is supplemented with [³H]pantothenate to uniformly label the CoA and ACP pools. Next, the CoA pool is severely depleted by depriving the mutant of the pantothenate supplement required for the production of CoA. After the intracellular CoA supply is virtually exhausted, the cells are resuspended in media containing [¹⁴C]pantothenate, thereby initiating growth and CoA synthesis. The rate of prosthetic group turnover is determined by measuring the flow of tritium from [pantothenate-³H]ACP into the CoA pool which has been depleted of tritium during the starvation procedures (Powell, Elovson, and Vagelos 1969).

A second approach is to exploit the effect of heavy isotopes on the solution structure of ACP (the *constitutional isotope effect*) to differentiate between newly synthesized and preexisting ACP in a pulse-chase experiment (Jackowski and Rock 1984) (Figure 3.15). The incorporation of deuterium into nonexchangeable positions (for example, —CH₂—) of the ACP amino acid backbone results in the destabilization of the protein and increased sensitivity to pH-induced hydrodynamic expansion. Therefore, deuterio-ACP can be separated from the normal protio-ACP by conformationally sensitive gel electrophoresis (Rock and Cronan 1981). This technique separates ACP species

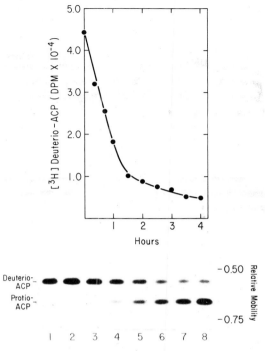

Figure 3.15. Determination of the rate of prosthetic group turnover by heavy-isotope labeling. The intracellular CoA pool was minimized by first growing an *E. coli* strain (*panD*) that requires either a β-alanine or pantethenate supplement to permit the synthesis of the phosphopantetheine portion of CoA and ACP in D_2O medium with a low concentration ($0.5\mu M$) of β-[^{3}H]alanine. The culture was washed and shifted to H_2O medium with a higher concentration ($4\mu M$) of β-[^{14}C]alanine, which reinitiated growth. Cell samples were taken during the recovery from β-alanine starvation and into the logarithmic stage of growth, and the distribution of tritium between preexisting (deuterio) and newly synthesized (protio) ACP was determined by separating the two species by conformationally sensitive gel electrophoresis. The upper panel shows the loss of tritium from deuterio-ACP following the switch from D_2O to H_2O medium, and the lower panel shows a fluorogram of the original gel. Lanes 1 through 8 at the bottom correspond to the first eight time points in the top panel. The actual rate of turnover was calculated from the change in the ^{3}H/^{14}C ratio in deuterio-ACP as a function of time.

according to their stability to pH-induced denaturation. In practice, the gels are cast at pH 9.0 and run at 37°C, conditions that partially denature ACP. ACP derivatives that are more stable than normal ACP run faster in this gel system due to their smaller hydrodynamic radii, whereas ACP derivatives that are less stable than ACP migrate more slowly due to their increased molecular volume. After the ACP prosthetic group is prelabeled with tritiated prosthetic group precursor (β-alanine) during growth on deuterium oxide medium, the cells are chased with ^{14}C-labeled β-alanine. The rate of prosthetic group turnover is determined by measuring the loss of tritium and its replacement with ^{14}C in preexisting deuterio-ACP (Figure 3.15). At low intracellular CoA levels, ACP prosthetic group turnover was four times faster than the rate of new ACP synthesis, but at higher CoA concentrations characteristic of logarithmic growth, turnover was an order of magnitude lower, amounting to 25% of the ACP pool per generation. Despite continued interest in this topic, the role of ACP prosthetic group turnover in bacterial lipid metabolism remains an enigma.

Although ACP plays an indispensable role in fatty acid biosynthesis, recent work suggests that the size of the ACP pool can be severely depleted without significantly affecting phospholipid production. This point can be investigated by the analysis of [^{3}H]leucine-labeled cell extracts by conformationally sensitive gel electrophoresis (Figure 3.16), since the removal of the prosthetic group to form apo-ACP also destabilizes the protein to pH-induced denaturation. Normally a significant pool of inactive apo-ACP does not exist in vivo (lanes A and B, Figure 3.16), but in *E. coli* strains containing the *acpS* mutation (abnormal ACP synthase function), apo-ACP accumulates and becomes the major form of ACP in vivo (lanes C and D, Figure 3.16). Although the *acpS* mutant contains much less ACP than wild-type strains, it still possesses a normal lipid:protein ratio, indicating that ACP concentration per se is not a factor in determining the overall rate of phospholipid biosynthesis.

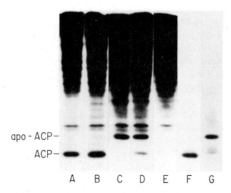

apo-ACP—
ACP—

A   B   C   D   E   F   G

Figure 3.16. Active and inactive forms of acyl carrier protein in vivo. The occurrence of inactive apo-ACP in vivo was investigated by labeling the cells with leucine and separating the proteins by conformationally sensitive gel electrophoresis. Strain SJ16 (*panD*) was labeled in the presence of either $0.25\mu M$ β-alanine (lane A) or $25\mu M$ β-alanine (lane B), and strain MP4 (*panB, acpS*) was labeled with either $0.25\mu M$ pantothenate (lane C) or $25\mu M$ pantothenate (lane D). Lane E is the same as lane D except that the strain MP4 extract was preabsorbed with specific anti-ACP IgG prior to electrophoresis. Standards are ACP (lane F) and apo-ACP (lane G).

**REGULATION OF FATTY ACID COMPOSITION IN *E. COLI*
Role of 3-hydroxydecanoyl-ACP Dehydrase**

The reaction in which the double bond is introduced into the growing fatty acid chain is catalyzed by 3-hydroxydecanoyl-ACP dehydrase (Figure 3.8). This enzyme, a homodimer of an 18,000 molecular weight subunit, catalyzes the specific dehydration of 3-hydroxydecanoyl-ACP to a mixture of *trans*-2-decenoyl-ACP and *cis*-3-decenoyl-ACP (Figure 3.17). The double bond of the *trans*-2 intermediate is reduced to decanoyl-ACP by enoyl-ACP reductase, whereas the *cis*-3 double bond is retained through subsequent elongation steps and becomes the double bond of the unsaturated fatty acids of *E. coli*, palmitoleic acid, and *cis*-vaccenic acid. The dehydration reaction proceeds first by formation of the *trans*-2-decenoyl-ACP as an enzyme-bound intermediate. A portion of this intermediate can dissociate from the enzyme and be converted to saturated fatty acids by enoyl-ACP reductase, whereas the remaining *trans*-2 isomer (still enzyme-bound) is isomerized to *cis*-3-decenoyl-ACP by the dehydrase. The phenotype of mutants lacking this dehydrase demonstrates that the isomerase reaction is the activity required for unsaturated fatty acid synthesis.

Mutants lacking the dehydrase (called *fabA*) were the first mutants isolated in the fatty acid biosynthetic pathway. These mutants require

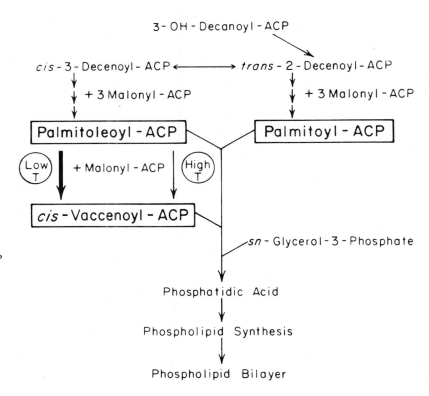

Figure 3.17. Thermal regulation of fatty acid biosynthesis. 3-Ketoacyl-ACP synthase II is primarily responsible for the temperature control of *E. coli* fatty acid composition by being more active in the conversion of palmitoleate to *cis*-vaccenate at lower temperatures than at higher temperatures.

unsaturated fatty acids for growth but synthesize saturated fatty acids normally. In vitro, the mutant enzymes are unable to catalyze the formation of their *trans*-2-decenoyl-ACP or *cis*-3-decenoyl-ACP. This finding, together with continued synthesis of saturated fatty acids observed in vivo, indicates that *trans*-2-decenoyl-ACP can be synthesized by a dehydrase other than 3-hydroxydecanoyl-ACP dehydrase. However, this second dehydrase is unable to catalyze the isomerization to *cis*-3-decenoyl-ACP. This second dehydrase is presumably the enzyme that catalyzes the dehydration of the 3-hydroxyacyl-ACPs of shorter and longer chain length (Figure 3.8).

Role of 3-ketoacyl-ACP Synthase I

Another class of fatty acid mutants (*fabB*) has the same phenotype as the *fabA* strains. Both mutants synthesize saturated fatty acids normally, but require for growth the addition of unsaturated fatty acid to the medium. However, *fabB* strains contain normal levels of 3-hydroxydecanoyl-ACP dehydrase, and genetic analysis shows that the *fabA* and *fabB* mutations are in two different genes (Figure 3.3). When first isolated, the *fabB* mutants seemed an enigma, since the same set of fatty acid biosynthetic enzymes (except 3-hydroxydecanoyl-ACP dehydrase) was thought to function in both unsaturated and saturated fatty acid synthesis. Hence, a mutation resulting in the loss of one of these enzymes was expected to block the synthesis of saturated as well as unsaturated fatty acids.

The enigma was resolved by enzymological studies showing that two distinct enzymes catalyze the 3-ketoacyl-ACP synthase reaction in *E. coli* (D'Agnolo, Rosenfeld, and Vagelos 1975). Therefore, the loss of one enzyme could give a *fabB* mutant, since the second enzyme was available for the elongation steps required in saturated fatty acid synthesis. Indeed, it was soon shown that *fabB* mutants lacked synthase I and that the *fabB* gene encodes the synthase I enzyme. Therefore, it is clear that synthase I catalyzes a key reaction in unsaturated fatty acid synthesis in which synthase II is unable to function. Both synthases I and II are capable of elongating saturated fatty acids.

Role of 3-ketoacyl-ACP Synthase II

The discovery of the role of 3-ketoacyl-ACP synthase II in unsaturated fatty acid synthesis was an outcome of the investigation of a phenomenon called *thermal control* of fatty acid synthesis. *E. coli*, in common with most (if not all) other organisms, synthesizes phospholipids with a greater proportion of unsaturated fatty acids when grown at low, rather than high, temperatures. This regulatory system is believed designed to ameliorate the effects of temperature change on the physical state of the membrane phospholipids. As discussed in

Chapter 2, the proportion of fluid (disordered) lipid to nonfluid (ordered) lipid in cell membranes plays a major role in membrane function. Increased incorporation of unsaturated fatty acids decreases the melting temperature of the membrane phospholipids, whereas increased incorporation of saturated fatty acids has the opposite effect. The thermal regulatory system can thus adapt the membrane lipids for optimal functioning at new growth temperatures.

A key finding in unraveling the mechanism of thermal regulation in *E. coli* was that protein synthesis was not required for the synthesis of new phospholipids with a fatty acid composition characteristic of the new temperature. This result indicated that thermal regulation was exerted by a protein synthesized at all growth temperatures but active only at low temperatures. A clue to the identity of this protein was the observation that the unsaturated fatty acid synthesized in greater quantity at low temperature was *cis*-vaccenic acid; the amount of palmitoleic acid did not vary with temperature. This result suggested that thermal regulation involved the conversion of palmitoleic acid to *cis*-vaccenic acid. This suggestion was buttressed by the isolation of a mutant (called *fabF*) defective both in the conversion of palmitoleic acid to *cis*-vaccenic acid and in thermal regulation. The *fabF* strain had only traces of *cis*-vaccenic acid in its phospholipids, and its fatty acid composition (chiefly palmitic and palmitoleic acids) did not change with growth temperature.

The search for the enzyme missing in the *fabF* strains paralleled that described for the *fabB* mutant and resulted in the demonstration that *fabF* strains are defective in the gene encoding 3-ketoacyl-ACP synthase II. Moreover, genetic analysis of strains carrying mutations in both *fabB* (a temperature-sensitive mutant was used) and *fabF* showed that this double mutant was completely defective in fatty acid synthesis; thus, no putative third synthase is present in *E. coli*.

The analysis of the *fabF* mutants suggests that 3-ketoacyl-ACP synthase II functions much better than synthase I in the elongation of palmitoleoyl-ACP to the 3-keto precursor of *cis*-vaccenoyl-ACP and that the difference is accentuated at lower temperatures (Figure 3.17). Since mutants containing either synthase I or synthase II synthesize saturated fatty acids normally, either enzyme can catalyze all the elongation steps required in saturated fatty acid synthesis. These predictions were borne out by in vitro studies on the purified enzymes. A simple model emerged in which temperature alters the activity of synthase II, which in turn regulates the fatty acid composition by producing more *cis*-vaccenoyl-ACP for incorporation into phospholipid (deMendoza and Cronan 1983).

Though all the available data were consistent with this model, one dilemma remained. It could be argued either that the lack of synthase II resulted in the loss of thermal regulation or, conversely, that the

Table 3.2. Thermal Regulation in *E. coli*

Genotype	Relative intracellular synthase levels I	II	Growth temperature (°C)	18:1/16:1 Ratio
Wild-type	1	1	30	0.7
			37	0.4
			42	0.3
fabB clone	10	1	30	2.0
			37	1.2
			42	0.8
fabF	1	0	30	0.03*
			37	0.03
			42	0.02
fabB clone, *fabF*	10	0	30	0.7
			37	0.6
			42	0.5

* The overall ratio of unsaturated to saturated fatty acids does not change with growth temperature in these strains.

lack of *cis*-vaccenoyl-ACP, the product of elongation of palmitoleoyl-ACP by synthase II, caused the lack of the temperature regulation (thermal regulation would, in this case, be exerted at a later step). Although indirect evidence indicated that the first argument was correct, direct proof was needed. This was provided by an experiment in which normal amounts of *cis*-vaccenic acid were synthesized in the absence of 3-ketoacyl-ACP synthase II using a cloned synthase I gene (deMendoza, Ulrich, and Cronan 1983). Although synthase I elongates palmitoleoyl-ACP more poorly than does synthase II, the synthase I reaction proceeds at a measurable rate with this substrate. Thus, the investigators reasoned that if synthase I could be overproduced by molecular cloning onto a multicopy plasmid, appreciable *cis*-vaccenic acid synthesis should occur. The results of this experiment (Table 3.2) show that the presence of a plasmid carrying the *fabB* gene results in the synthesis of a normal level of *cis*-vaccenic acid in the absence of synthase II. More important, the fatty acid composition of these strains is not altered by growth temperature. This experiment therefore demonstrates that 3-ketoacyl-ACP synthase II is the only protein responsible for thermal regulation of membrane lipid composition in *E. coli* (Figure 3.17).

Factors Affecting Fatty Acid Chain Length

Palmitate, palmitoleate, and *cis*-vaccenate comprise the bulk of the fatty acids found in *E. coli* membranes. One likely candidate for the site of chain-length regulation is at the level of the reactions of 3-ketoacyl-ACP synthases I and II. Substrate specificity studies on these enzymes in vitro support the view that the reason membrane phospholipids are devoid of chain lengths longer than 18 carbons is due in part to the reduced activity of these enzymes on 18-carbon substrates. The fatty acid composition of mutants lacking one or the other of the condensing enzymes has also given some clues to their substrate specificity in vivo (see preceding discussion). The *fabB* mutants lack all unsaturated fatty acids; therefore, it is not possible to assign an essential role for synthase I in controlling chain length. Synthase II mutants (*fabF*) are defective in the elongation of palmitoleate to *cis*-vaccenate; therefore, synthase II plays a critical role in determining the amount of 18-carbon fatty acids in the membrane. Although these data suggest that the condensing enzymes play a significant role in determining chain length, additional physiological experiments indicate that the activity of the glycerolphosphate acyltransferase is also important. When phospholipid biosynthesis is arrested at the acyltransferase step (by glycerol starvation of a *plsB* mutant), the fatty acids synthesized in the absence of their utilization by the acyltransferase have abnormally long chain lengths compared with the normal distribution of fatty acids synthesized in the presence of acyltransferase activity. These data suggest that competition between the rate of elongation and utilization of the acyl-ACPs by the acyltransferase may be a significant determinant of fatty acid chain length in *E. coli*.

Synthesis of Cyclopropane Fatty Acids

Cyclopropane fatty acids are a third major type of fatty acid found in *E. coli* and many other bacteria. These acids are formed by a postsynthetic modification of the unsaturated fatty acids of membrane phospholipids. In vitro, neither free fatty acids nor their thioesters are substrates for cyclopropane fatty acid synthase, only phospholipid dispersed in a micelle will suffice. The reaction involves the addition of a CH_2 group from *S*-adenosylmethionine to the double bond to form a cyclopropane ring. The cis double bond is converted to a cis cyclopropane group.

This novel reaction has several interesting aspects. For example, cyclopropane fatty acid synthesis occurs primarily as bacterial cultures enter the stationary growth phase (that is, as the cultures cease growth due to oxygen or nutrient limitation); rapidly growing cells accumulate little cyclopropane fatty acids. However, the levels of cyclopropane fatty acid synthase vary little with growth phase; thus, the timing of cyclopropane fatty acid synthesis is not due to a change in enzyme production. We have no good rationale for the timing of synthesis. Recent work shows that cells carrying a cloned segment of DNA containing the cyclopropane fatty acids synthase gene synthesize the acids throughout log phase. Despite this modification, the cells grow normally; therefore, the presence of these acids does not

[handwritten marginalia:]

antagonize

2f --- 효소의 양이 증가

Cyclopropane F.A synthase is large and hydrophobic. So How does the cyclopropane F.A synthase ~~access~~ have access to the bilayer lipid? it is unknown.

antagonize normal growth. The continuous synthesis of cyclopropane acids in strains carrying a cloned *cfa* gene indicates that whatever the regulatory process that inhibits cyclopropane fatty acid synthesis in rapidly growing cells, it can be overcome by increased production of the enzyme. This result suggests a stoichiometric inhibition, such as a protein-protein interaction, but no further data are available.

A second novel aspect of cyclopropane fatty acid synthesis is the topological problem. Cyclopropane fatty acid synthase is primarily a soluble enzyme, although traces of activity are found loosely bound to the inner membrane. How then does the enzyme have access to virtually all of the phospholipids of both the inner and the outer membranes? The enzyme is large (molecular weight between 80,000 and 90,000) and hydrophilic; thus, it seems unlikely that cyclopropane fatty acid synthase can cross lipid bilayers. It seems more probable that the lipid somehow gains access to the enzyme. There is some evidence for a flow of lipid between the inner and outer membranes, but neither the mechanism of the flow nor that of the flip-flop of phospholipid molecules from one face of the bilayer to the other is known, although the former process seems to require an electrochemical gradient (see following discussion). Molecular cloning of the cyclopropane synthase gene should allow complete purification of this enzyme and hence allow a detailed analysis of this novel reaction.

Another intriguing question is the physiological rationale for the modification, since the physical and chemical properties of the cyclopropane fatty acids are rather similar to those of the unsaturated fatty acids from which they are derived. Many functions have been proposed for cyclopropane fatty acids. However, mutants of *E. coli* (called *cfa*) have been isolated which completely lack cyclopropane fatty acids (fewer than 100 molecules per cell) due to disruption of the gene encoding the enzyme. These mutants grow and survive various environmental stresses (such as stationary phase and high and low oxygen tension) as well as strains which accumulate cyclopropane acids in a normal manner. We are left with an enigma: Cyclopropane fatty acids are widely conserved among bacteria but seem to play no essential role in cellular metabolism. We can conclude only that these acids play a role in the natural environment which has not yet been duplicated in the laboratory.

Importance of the Fatty Acid Composition of the Membrane Phospholipids

An important lesson learned from studies of the various mutants of *E. coli* blocked in fatty acid synthesis is that the organism tolerates a wide variation in the fatty acid composition of the membrane phospholipids. The unsaturated fatty acid auxotrophs (*fabA* and *fabB*) can grow if the medium is supplemented with any of a large number of fatty acids. Although saturated fatty acids alone will not support growth, a wide variety of cis unsaturated fatty acids (mono-, di-, or tri-unsaturated) will support growth of *fabA* and *fabB* mutants. Indeed, even unsaturated fatty acids with a centrally located trans double bond (a type of fatty acid not found in *E. coli* and very rarely found in nature) will suffice. It is clear that the double bond per se plays no chemical role in metabolism. The role of the double bond is only to decrease the temperature of the phase transition of the phospholipid in which it resides, since a number of fatty acids lacking double bonds also support growth. These acids (*cis*- or *trans*-cyclopropane, branched, centrally brominated) do, however, share with double bonds the ability to disrupt the close packing of phospholipid acyl chains and lower the phase transition (Chapter 2). This property is purely physical. The presence of a substituent or a double bond in the middle of the hydrocarbon chain sterically disrupts strong hydrophobic interactions with other acyl chains.

However, there are limits to the fatty acid compositions which allow growth. The finding that *fabA* and *fabB* mutants require an unsaturated or equivalent fatty acid for growth indicates that a membrane composed of phospholipids containing only saturated fatty acids is nonfunctional. Indeed, these mutants undergo cell lysis when deprived of the unsaturated fatty acid supplement. Thus, *E. coli* requires some fluid lipid for a functional membrane. Several laboratories have reported that if less than half of the membrane lipid was in the fluid state, the *E. coli* membrane became nonfunctional. A similar argument can be made for the importance of some nonfluid lipid for a functional membrane. Mutants (*fabD*, *fabE*, or *fabB* plus *fabF*) which block fatty acid synthesis at an early step require both a saturated and an unsaturated fatty acid for growth. If only an unsaturated fatty acid were added, the cells would leak internal components and eventually lyse and die.

The conclusions from these experiments are straightforward. A functional *E. coli* membrane requires that the composition of the membrane phospholipids be within the limits of the phase transition. If all the phospholipids were in either the ordered state or the disordered state, the membrane would be nonfunctional. However, quite wide variations in fluidity are tolerated; that is, the cells do not have to maintain a precise ratio of fluid to nonfluid lipid to have functional membranes. However, there does seem to be an optimal fluidity at which cell growth is most rapid, and the thermal control regulatory systems seem designed for optimizing the fluidity within the tolerated range, rather than to extending the range.

REGULATION OF PHOSPHOLIPID BIOSYNTHESIS IN *E. COLI*
Role of the *sn*-glycerol-3-phosphate Acyltransferase

The *sn*-glycerol-3-phosphate acyltransferase step in the pathway for phospholipid synthesis (Figure 3.9) represents the transition point from soluble to membrane-bound enzymes and intermediates and has received considerable experimental attention. The acyltransferase is an integral cytoplasmic membrane protein, which makes it intrinsically more difficult to work with than the soluble enzymes of fatty acid biosynthesis. Primarily through the use of gene-cloning procedures, the glycerolphosphate acyltransferase has become one of the best characterized membrane-bound enzymes of phospholipid biosynthesis. Following the finding that the acyltransferase could be solubilized from the membrane with the detergent Triton X-100, the most important breakthrough in dealing with this enzyme was the isolation of hybrid plasmids harboring the glycerolphosphate acyltransferase (*plsB*) structural gene. Strains containing this plasmid overproduce the acyltransferase 10-fold. Solubilization of the membrane and three column-chromatographic steps produce a single protein species having a molecular weight of 83,000 (Lightner et al. 1980). Each step of the purification is carried out in detergent-containing buffers, but as is common for membrane enzymes, reconstitution of the protein with phospholipids is required for enzymatic activity. The complete primary sequence (806 residues) of the acyltransferase has been determined by a combination of protein- and DNA-sequencing techniques (Larson et al. 1980). The single polypeptide catalyzes the formation of 1-acylglycerolphosphate from either acyl-CoA or acyl-ACP acyl donors. This lack of substrate specificity is not shared by *Rhodopseudomonas sphaeroides* and *Clostridium butyricum* glycerolphosphate acyltransferases, which utilize acyl-ACP thioesters exclusively. These data suggest that the utilization of both acyl-CoA and acyl-ACP by the *E. coli* acyltransferase may be a special adaptation of this organism rather than a general rule for bacterial acyltransferases.

The 1-acyl-*sn*-glycerol-3-phosphate acyltransferase is poorly characterized compared with the glycerolphosphate acyltransferase just described. A biochemical assay for 1-acylglycerolphosphate acyltransferase activity in isolated membranes is available. This enzyme will also utilize acyl-CoA and acyl-ACP as the acyl donor in vitro. No mutants are available that are defective in 1-acylglycerolphosphate acyltransferase, accounting in part for our lack of knowledge about this acyltransferase.

Naturally occurring phospholipids are generally characterized as having a saturated fatty acid at the 1-position and an unsaturated fatty acid at the 2-position of the glycerol backbone. The substrate specificity of the glycerolphosphate acyltransferase system is considered the most likely origin of acyl-group asymmetry in bacterial phospholipids. Accordingly, higher V_{max} and lower K_m values are

[handwritten margin note: glycerol-3-phosphate acyltransferase in E. coli can use both acyl-ACP and acyl-CoA this is probably a special cases]

found for saturated rather than unsaturated acyl donors when either the purified or membrane-bound forms of *sn*-glycerol-3-phosphate acyltransferase are used as the enzyme source. These data demonstrate that the acyltransferase has a substrate specificity that is consistent with the role of this enzyme in controlling the positional placement of fatty acids on the glycerol backbone. Acyltransferase specificity can also be investigated in vivo using unsaturated fatty acid auxotrophs. When *E. coli* mutants that are unable to synthesize unsaturated fatty acids (either *fabA* or *fabB*) are deprived of their exogenous unsaturated fatty acid supplement, a significant accumulation of disaturated molecular species of phospholipid is observed. Restoration of unsaturated fatty acids to the medium results in the synthesis of molecular species having the typical fatty acid positional asymmetry. These data demonstrate that the acylation specificity of the glycerolphosphate acyltransferase is not absolute in vivo, but is controlled in part by the supply of fatty acids. This is another good example of how the genetic approach was used to dissect the interrelationships between enzymes in a metabolic pathway that were not directly evident from in vitro studies.

Control of Phospholipid Biosynthesis

The biosynthesis of phospholipid is an energy-intensive process, and there is abundant evidence that phospholipid production is tightly regulated in vivo. There are two possible sites where regulation of the pathway could occur. Phospholipid production could be controlled either at the level of fatty acid supply or at the level of fatty acid incorporation into phospholipid. This question has been addressed in vivo by measuring the pool size of long-chain acyl-ACP substrates for the acyltransferase during balanced growth and after the cessation of phospholipid synthesis (Rock and Jackowski 1982). To accomplish this goal, the genetic approach was used to construct strains of *E. coli* containing both the *plsB* acyltransferase mutation and the *panD* defect in the CoA biosynthetic pathway. The *plsB* mutation results in a reduced affinity of the glycerolphosphate acyltransferase for glycerol-3-phosphate and therefore allows phospholipid biosynthesis to be turned on and off by the presence or absence of the glycerolphosphate supplement in the culture medium. The *panD* allele renders the cell auxotrophic for β-alanine, a precursor to the phosphopantetheine prosthetic group of ACP, thus allowing the ACP pool to be uniformly and specifically labeled. Cell samples were then labeled with β-[^{3}H]alanine, the ACP and acyl-ACP were extracted, and the concentration of acyl-ACP substrates for the acyltransferase determined by reversed-phase high-pressure liquid chromatography. If the utilization of acyl-ACP were not rate limiting in phospholipid production, a large pool of acyl-ACP molecules awaiting acyltransfer would be anticipated. On the other hand, if the supply of acyl-ACP were rate limiting, the concentration of long-chain acyl-ACP species would be low. In these experiments, the acyl-ACP concentration was approximately 10% of the total ACP pool, and the majority of these acyl-ACPs were not of chainlengths that are substrates for the acyltransferase. Hence, it would appear that the supply of acyl-ACP does limit the rate of phospholipid synthesis in E. coli. When phospholipid biosynthesis was inhibited by the removal of glycerolphosphate, the acyl-ACP pool rose to about 34% of the total ACP, thereby establishing that acyl-ACPs are the acyl donors for phospholipid synthesis. As a check on this point, the long-chain acyl-CoA pool was also measured before and after inhibition of phospholipid synthesis. Long-chain acyl-CoAs were not detected under either of these conditions. These data strongly point to a step in the initiation phase of fatty acid biosynthesis (Figure 3.6) as the primary rate-controlling step in the pathway of phospholipid synthesis. More work is needed to establish which enzyme is the pacemaker of phospholipid synthesis. Control of the pathway at the level of initiation appears to be a logical hypothesis, since 94% of the ATP used in the biosynthesis of a phospholipid is expended in the synthesis of the fatty acid components.

Phospholipid Production is Coordinated with Macromolecular Biosynthesis

The exponential phase of bacterial growth is typically referred to as balanced since the myriad of biochemical pathways are well coordinated. These regulatory mechanisms have proven difficult to isolate experimentally during balanced growth, but some progress has been made using amino acid starvation to perturb the metabolic system. Wild-type cells respond to the decreased availability of any aminoacyl-tRNA species by a dramatic reduction in stable RNA accumulation in addition to several other metabolic adjustments that down-regulate cellular metabolism. A single site mutation, relA, abolishes this entire set of adjustments, conferring a phenotype called relaxed. These regulatory effects are mediated by a family of nucleotides, most notably guanosine-5'-diphosphate-3'-diphosphate (ppGpp) that accumulate in rel^+ but not relA strains. Several laboratories have reported that phospholipid production decreases dramatically following the starvation of rel^+ but not relA strains (Rock and Cronan 1982). Careful measurement of both the ppGpp concentration and the rate of phospholipid biosynthesis shows a quantitative correlation between ppGpp concentration and phospholipid accumulation. Although the in vivo evidence strongly supports the role of ppGpp in the regulation of total phospholipid production, the identity of the enzyme or enzymes affected by ppGpp remains to be firmly established.

Regulation of Phospholipid Head Group Composition

The biochemical mechanisms that control the ratio of phosphatidylethanolamine, phosphatidylglycerol, and cardiolipin in the membranes of *E. coli* are less well understood than the factors that control the phospholipid fatty acid composition. Mutants in the enzymes of phospholipid biosynthesis can be used to render any of the steps in Figure 3.10 rate limiting; however, these data do not shed much light on the mechanisms that operate in vivo. Similarly, gene-cloning experiments provide little insight. For example, a 10-fold overproduction of phosphatidylserine synthase in strains harboring a hybrid plasmid containing the *pss* gene does not significantly affect the ratio of polar head groups in the membrane. It therefore seems that either the phospholipid head group composition is not regulated by the levels of the synthetic enzymes or the enzyme levels measured in vitro are not an accurate indication of the activity present in vivo (Raetz 1982). The isolation of mutants that cause overproduction of diacylglycerol kinase (*dgkR*) and phosphatidylserine synthase (*pssR*) indicate that regulatory mechanisms do exist (Raetz 1982). However, both the method whereby the synthesis of these enzymes is controlled and the reasons for this control (since overproduction does not affect phospholipid composition) are unknown. Since these enzymes are generally integral membrane proteins, their activity could be modified by the phospholipid environment in vivo, although there is little evidence for this type of mechanism either in vivo or in vitro. Selection schemes for the isolation of strains with abnormal phospholipid compositions and second-site revertants of strains possessing lesions in phospholipid enzymes may provide additional insight into this important problem in the future.

E. coli Tolerates Variations in Phospholipid Composition

Studies of mutants blocked early in the phospholipid biosynthetic pathway (for example, *plsB*) show that cell growth and macromolecule synthesis continue for about a cell generation following severe inhibition of phospholipid synthesis. Consequently, the cell membranes become unusually dense due to continued membrane protein synthesis in the absence of lipid synthesis. Thus, other metabolic processes are not tightly coupled to membrane phospholipid synthesis. Analysis of mutants blocked later in the phospholipid pathway shows that the membrane lipid composition of *E. coli* (although closely maintained in normally growing cells) can be significantly altered without abrupt effect on cell growth and membrane function (Table 3.3). For example, *cls* strains have less than 10% of the normal level of cardiolipin but grow normally. However, it should be noted that cardiolipin is a major membrane component only in cells grown into stationary phase. Mutants blocked in the other late steps of phospholipid synthesis (*pss*, *psd*, or *psgA* plus *pgsB*) do show severe disturbances in growth, however, these problems do not arise until the synthesis of a given lipid has been inhibited for one to two generations. At this time, cell division is often inhibited, and disruption of the barrier properties of the membrane also becomes evident. These experiments show that a precisely aligned phospholipid composition is not required for growth and membrane function. Functional membranes can also contain large amounts of phospholipids which are only trace components of wild-type strains in *E. coli*, such as phosphatidylserine, a major component of *psd* strains and phosphatidic acid, a major component of *cds* strains (Table 3.3). Indeed, by fusion of artificial liposomes with *Salmonella typhimurium* cells, such abnormal lipids as cholesterol and phosphatidylcholine have been introduced into bacterial membranes without affecting bacterial physiology (Chapter 14).

It is not clear why cells synthesize a diversity of phospholipid species when a single species would suffice to form a lipid bilayer (Chapter 2). Some evidence suggests that the differing physical properties of bilayers composed of various lipid mixtures could be important, but definitive evidence is lacking.

Table 3.3. *E. coli* Mutants Used for the Modification of Membrane Lipid Composition

Defective gene	Phenotype	Condition*	Membrane lipid modification
plsB	Glycerol auxotroph	Remove glycerol, 37°C, 60 min	Lipid:protein ratio reduced 40%
dgk	Osmotic sensitivity	Log phase, 37°C	Accumulation of diacylglycerol and other neutral glycerides
cds	pH-sensitive	pH 8, 37°C, 2 h	Reduced phosphatidylethanolamine; phosphatidic acid accumulation
pss	Ts (42°C)[†]	42°C, 4 h	Reduced phosphatidylethanolamine; increased phosphatidylglycerol plus cardiolipin
psd	Ts (42°C)	42°C, 4 h	Accumulation of phosphatidylserine at expense of phosphatidylethanolamine
cls	None	37°C, log phase	Lack of cardiolipin
pgsA, pgsB	Ts (42°C)	42°C, 3 h	Lack of phosphatidylglycerol; accumulation of two lipid A precursors
fabB or *fabA*	Unsaturated fatty acid auxotroph	Remove fatty acid supplement	Accumulation of disaturated phospholipid molecular species
pyrG	Cytidine auxotroph	37°C, deplete cytidine for 2 h	Accumulation of phosphatidic acid; lipid:protein ratio increased threefold

* Here condition means the growth conditions required to produce the indicated membrane lipid modification.
[†] Ts represents temperature-sensitive.

The Diacylglycerol Cycle

(handwritten margin notes:)

① phosphatidylglycerol

↓

cardiolipin   ↓

sn-glycerol-1-phosphate.

② MDO is osmotic pressure sensitive substance

Early observations on phospholipid metabolism showed that the polar head group of phosphatidylglycerol was lost in a pulse-chase experiment, whereas that of phosphatidylethanolamine was quite stable. At first it was thought that the phosphatidylglycerol was being degraded. Upon the discovery that *E. coli* contains cardiolipin, it was realized that some of the phosphatidylglycerol "turnover" was actually the conversion of phosphatidylglycerol to cardiolipin catalyzed by cardiolipin synthase. However, cardiolipin synthesis did not account for all the loss of ^{32}P-labeled phosphatidylglycerol observed in pulse-chase experiments nor did it explain why the nonacylated glycerol of phosphatidylglycerol was labeled (and chased) more rapidly than the acylated glycerol moiety. A nonlipid phosphate-containing compound derived from the head group of phosphatidylglycerol was sought, and a family of molecules called *membrane-derived oligosaccharides* (MDO) was discovered (van Golde, Schulman, and Kennedy 1973). These molecules are composed of *sn*-glycerol-1-phosphate (derived from phosphatidylglycerol), glucose, and (usually) succinate moieties, have molecular weights of 4000–5000 and are found in the periplasm of Gram-negative bacteria. The *periplasm*, the space between the cytoplasmic and outer membranes of these organisms (Figure 3.4), is an osmotically sensitive compartment. The synthesis of the MDO compounds is regulated by the osmotic pressure of the growth medium (decreased osmotic pressure gives an increased rate of MDO synthesis); thus, MDO seem to be involved in osmotic regulation.

The discovery of the MDO compounds provided a function for the well-studied, but enigmatic, enzyme diacylglycerol kinase. In the synthesis of MDO (Figure 3.18), the *sn*-glycerol-1-phosphate polar group of phosphatidylglycerol is transferred to the oligosaccharide, with 1,2-diacylglycerol as the other product. Diacylglycerol kinase will phosphorylate the diacylglycerol to phosphatidic acid, which can reenter the phospholipid biosynthetic pathway (Figure 3.10) to complete the diacylglycerol cycle (Figure 3.18). In the overall reaction only the *sn*-glycerol-1-phosphate portion of the phosphatidylglycerol molecule is consumed; the lipid portion of the molecule is recycled back into phospholipid. It is clear that MDO synthesis is responsible for most of the metabolic instability of the polar group of phosphatidylglycerol, since this turnover is greatly decreased if MDO synthesis is blocked at the level of oligosaccharide synthesis (by lack of UDP-glucose). Moreover, the rate of accumulation of diacylglycerol in strains lacking diacylglycerol kinase (*dgk*) depends on both the presence of the oligosaccharide acceptor and the osmotic pressure of the growth medium. It should be noted that some species of MDO contain phosphoethanolamine. Although direct proof is lacking, it is likely that the ethanolamine moiety is derived from phosphatidyl-ethanolamine, as this is the only known source of ethanolamine.

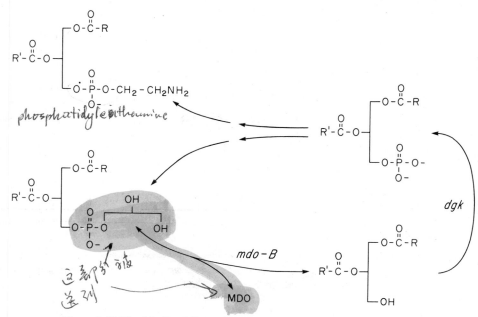

Figure 3.18. The 1,2-diacylglycerol cycle. First, the *sn*-glycerol-1-phosphate moiety is removed from phosphatidylglycerol for the biosynthesis of membrane-derived oligosaccharides (MDO). The resulting 1,2-diacylglycerol is then phosphorylated by 1,2-diacylglycerol kinase to phosphatidic acid, which is subsequently reutilized in the synthesis of membrane phospholipids (see Figure 3.10).

The 2-acylglycerol-phosphoethanolamine Cycle

2-Acylglycerolphosphoethanolamine acyltransferase is another inner membrane enzyme that participates in a metabolic cycle (Figure 3.19). This acyltransferase esterfies the 1-position of 2-acyl-lysophospholipids utilizing acyl-ACP as the acyl donor (Figure 3.19). Unlike the *sn*-glycerol-3-phosphate acyltransferase, 2-acylglycerolphosphoethanolamine acyltransferase does not utilize acyl-CoA thioesters. 2-Acylglycerolphosphoethanolamine has been identified as a minor membrane lipid in *E. coli* and is apparently the substrate for this enzyme in vivo. There is a small amount of fatty acid turnover at the 1-position of phosphatidylethanolamine, although the metabolic fate of the 1-position fatty acids and the physiological significance of this cycle remain to be established.

Normally, acyl-ACP intermediates are not observed in the 2-acylglycerol phosphoethanolamine acyltransferase assay system (Figure 3.19); however, after the addition of chaotropic salts (such as LiCl), acyl-ACP accumulates. The enzyme which catalyzes this partial reaction of the acyltransferase (acyl-ACP synthetase) has been purified and is most active with saturated fatty acids as substrates, although the enzyme ligates a broad range of fatty acids to ACP.

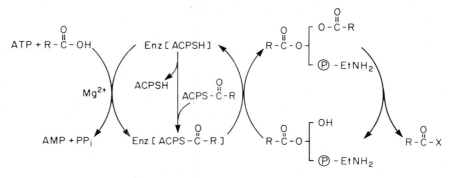

Figure 3.19. The 2-acylglycerol-3-phosphoethanolamine cycle. First, the 1-position fatty acid of phosphatidylethanolamine is removed by a process that has not been identified. The resulting 2-acylglycerolphosphoethanolamine is then reacylated by 2-acylglycerolphosphoethanolamine acyltransferase. This enzyme either can utilize acyl-ACP as the acyl donor or can acylate fatty acids to an enzyme-bound acyl-ACP intermediate in the presence of ATP and Mg^{2+}, thus completing the cycle.

This enzymatic activity is of considerable practical significance since acyl-ACP synthetase can be used to prepare native acyl-ACPs for use as substrates in the study of other fatty acid and phospholipid biosynthetic enzymes.

DEGRADATION OF PHOSPHOLIPIDS AND FATTY ACIDS
Phospholipid Degradation

There are 10 reported enzymatic activities that degrade phospholipids, intermediates in the phospholipid biosynthetic pathway, or triacylglycerol (Table 3.4). The best characterized of these is the detergent-resistant phospholipase A of the outer membrane. This enzyme is unusually resistant to inactivation by heat and ionic detergents and requires calcium for maximal activity. The phospholipase has been purified to homogeneity and exists as a single subunit with a molecular weight of 28,000. Hydrolysis of fatty acids from the 1-position of phospholipids is the most rapid reaction, but the enzyme will also hydrolyze 2-position fatty acids, as well as both isomeric forms of lysophosphatides. A detergent-sensitive phospholipase A has also been described. This enzyme differs from the detergent-resistant protein in that it is located in the soluble fraction of the cell, is inactivated by heat and ionic detergents, and in contrast with the broad substrate specificity of the outer membrane phospholipase, has a high degree of specificity for phosphatidylglycerol. The cytoplasmic phospholipase A also requires calcium for activity.

The physiological role of these degradative enzymes remains unknown. Mutants lacking the detergent-resistant phospholipase, the detergent-sensitive phospholipase, or both enzymes·do not have

Table 3.4. Degradative Enzymes in *E. coli*

Enzyme	Location	Substrates
Phospholipase A	Outer Membrane	Phosphatidylethanolamine, phosphatidylglycerol, cardiolipin + lyso derivatives
Phospholipase A	Cytoplasm	Phosphatidylglycerol
Lysophospholipase	Inner Membrane	Lyso-phosphatidylethanolamine
Lysophospholipase	Cytoplasm	Lyso-phosphatidylethanolamine, lyso-phosphatidylglycerol
Phospholipase C	Unknown	Phosphatidylethanolamine
Phospholipase D	Cytoplasm	Cardiolipin
Phospholipase D	Cytoplasm	Phosphatidylserine
Lipase	Membrane	Triacylglycerol
CDP-diacylglycerol hydrolase	Inner Membrane	CDP-diacylglycerol
Phosphatidic acid phosphatase	Membrane	Phosphatidic acid
Thioesterase I	Cytoplasm	Acyl-CoA
Thioesterase II	Cytoplasm	Acyl-CoA

any obvious defects in growth, phospholipid composition, or turnover. Moreover, strains that overproduce the detergent-resistant enzyme (constructed by molecular cloning) also grow normally. It has been established that the detergent-resistant phospholipase is responsible for the release of fatty acids from phospholipids that occurs during infection with T4 and λ phages. However, phospholipid hydrolysis is not essential for the life cycle of these bacteriophages. One possible function for these enzymes is that they are actually biosynthetic proteins that act as hydrolases in the absence of suitable acceptor molecules in the assay systems employed. Examples of such enzymes are the phospholipase D and CDP-diacylglycerol hydrolase activities that are associated with phosphatidylserine synthase. Phosphatidylserine synthase appears to function via a phosphatidyl-enzyme intermediate, and in the absence of a suitable acceptor such as serine or CMP, the phosphatidyl-enzyme complex can be hydrolyzed by water; thus, the enzyme exhibits either phospholipase D or CDP-diacylglycerol hydrolase activity. Recently, CDP-diacylglycerol hydrolase of the inner membrane (a different enzyme than phosphatidylserine synthase) has been shown to be a cytidylyl donor to inorganic phosphate and other phosphomonoester acceptors, which suggests that this enzyme is a biosynthetic cytidylyltransferase, although the identity of the acceptor molecule in vivo has not been determined. Finally, some of these enzyme activities may reflect a broad substrate specificity of a single enzyme rather than the presence of several distinct protein species. For example, the lipase activity that cleaves the l-position fatty acids from triacylglycerols (a lipid usually not found in *E. coli*) is probably due to the presence of the detergent-resistant phospholipase A acting on triacylglycerol as an alternate substrate. In conclusion, much remains to be learned about the role of phospholipid degradative enzymes in the lipid metabolism of bacteria.

E. coli contains two thioesterases that catalyze the hydrolysis of acyl-CoA molecules (*Rhodopseudomonas sphaeroides* has two similar enzymes). Thioesterase I is a small (molecular weight of 22,000) serine esterase that hydrolyzes only long-chain acyl-CoAs. Thioesterase II is a much larger protein (molecular weight of 122,000) that is insensitive to serine esterase inhibitors and cleaves a broad range of acyl-CoA chain lengths. Both enzymes are much more active on acyl-CoA molecules than on acyl-ACPs because the native solution structure of ACP acts to protect the thioester bond from attack by these enzymes. No function for either enzyme is known. Acyl-CoAs are found in *E. coli* only when the β-oxidation pathway (Chapter 5) is operative. These thioesterases could play a role in the synthesis of various fatty-acid-containing molecules in the cell other than phospholipids, as described in the next section. It is possible that these thioesterases actually are acyltransferases that in the absence of the unknown physiological acceptor molecules transfer the acyl chain to water. A recently isolated mutant deficient in thioesterase II should give valuable information on the in vivo role of this enzyme. A thioesterase that specifically hydrolyzes acyl-ACP has not been identified.

Fatty Acid Oxidation in Bacteria

E. coli has an inducible system for the uptake and oxidation of fatty acids as a carbon source for growth. The genes comprising the regulon for fatty acid degradation are scattered throughout the bacterial chromosome, and the expression of these genes is controlled by a single genetic locus (*fadR*). An important point to keep in mind is that acyl-CoAs serve as the substrates for the enzymes of β-oxidation, whereas the fatty acid biosynthetic enzymes utilize acyl-ACPs. The biochemistry of the enzymes of fatty acid oxidation in *E. coli* is covered in detail in Chapter 4.

OTHER LIPIDS IN *E. COLI* 3-Hydroxymyristic Acid and Lipopolysaccharides

The outer leaflet of the outer membrane of *E. coli* contains little phospholipid but is instead made up of lipopolysaccharide. Lipopolysaccharides consist of three regions of contrasting chemical and biological properties. The outermost region consists of the *O*-specific polysaccharide and forms the basis for the serological differences between closely related bacteria. The *O*-antigen region is linked to a core polysaccharide region, which is common to groups of bacteria. This region is in turn attached via a 2-keto-3-deoxyoctonate trisaccharide to the lipid component termed *lipid A*. Lipid A serves to anchor the lipopolysaccharide to the outer membrane and also functions as an endotoxin and a mitogen during bacterial infections.

The only major *E. coli* fatty acid that is not a component of the phospholipids is 3-hydroxymyristic acid. Rather, this fatty acid is attached by both ester and amide linkages to the saccharide residues of the lipid A portion of the outer membrane lipopolysaccharide (Figure 3.20). The available evidence suggests that 3-hydroxymyristic acid is derived from the central fatty acid biosynthetic machinery. The enzymes responsible for the transfer of 3-hydroxymyristic acid to lipid A and the mechanism that determines whether the 3-hydroxymyristoyl-ACP is channeled to lipopolysaccharide biosynthesis rather than elongation to palmitic acid and hence to phospholipid are unknown.

Some confusion has been created in this area, since the originally published structure of the lipopolysaccharide was in error with regard to the placement of the 3-hydroxymyristoyl moieties. Interestingly, the genetic approach provided the key to unraveling the chemical structure of lipopolysaccharide (Nishijima and Raetz 1979). Using the colony autoradiography approach, strains defective in phosphatidylglycerolphosphate synthase were isolated that contain

Smallest Lipid A Unit

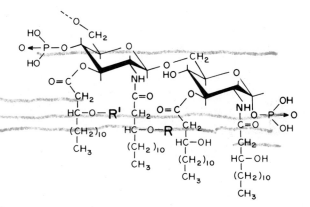

Figure 3.20. The structure of the lipid A portion of the outer membrane lipopolysaccharides. R represents predominantly a lauroyl, and R' a myristoyl, moiety. In some molecules, R and R' may be H. The 3-deoxy-D-manno-octulosanate portion of the lipopolysaccharide is attached to the hydroxyl group on the C-6' position located in the upper left of the figure.

[Handwritten margin notes:]
mutant ① phosphatidylglycerolphosphate synthase mutant is not temperature sensitive and difficult to isolate

mutant ② use mutant ① to generate defective psgB gene. So easy to isolate

less than 5% of the wild-type enzymatic activity. Interestingly, these mutants (*pgsA*) were not temperature-sensitive for growth and contained normal levels of phosphatidylglycerol. In order to obtain mutants with reduced phosphatidylglycerol content, second-step mutants were generated by starting with a parent harboring a defective *psgA* gene. This approach yielded a temperature-sensitive organism that contains less than 5% of the normal amount of phosphatidylglycerol after a 3-h incubation at 42°C. This second mutation was designated *pgsB*; it is far removed from *pgsA* on the bacterial chromosome. It was then discovered that two intermediates in the lipid A biosynthetic pathway accumulate in strains harboring both the *pgsA* and *pgsB* mutations. The structure of these simple intermediates suggested a lipid A structure different from the prevailing model, thus providing the impetus to reevaluate lipopolysaccharide structure and biosynthesis. The currently accepted lipid A structure (Figure 3.20) is synthesized by the condensation of UDP-2,3-diacylglucosamine with 2,3-diacylglucosamine-1-phosphate to form a (β1, $\rightarrow$6)tetraacyldisaccharide-1-phosphate, which is subsequently phosphorylated at the C-4' position. The acyltransferase reactions that attach 3-hydroxymyristic acid to the glucosamine residue have not been characterized. A curious finding is that 3-hydroxymyristic acid is taken up by the bacteria and can be used as a carbon source via the β-oxidation pathway; however, exogenous 3-hydroxymyristate is not incorporated into lipid A.

Outer Membrane Lipoproteins

Lipids are also used in bacteria in the synthesis of outer membrane lipoproteins. The most well known of these is *Braun's lipoprotein*. This lipoprotein is one of the most abundant proteins in the cell, and the mature form possesses an amino terminal cysteine that is modified with both a diacylglycerol and an amide-linked fatty acid (Figure 3.21). The lipid modification of this protein is presumably important to its proper attachment to the membrane. The glycerol portion and the acyl moieties bound to the lipoprotein are derived from the

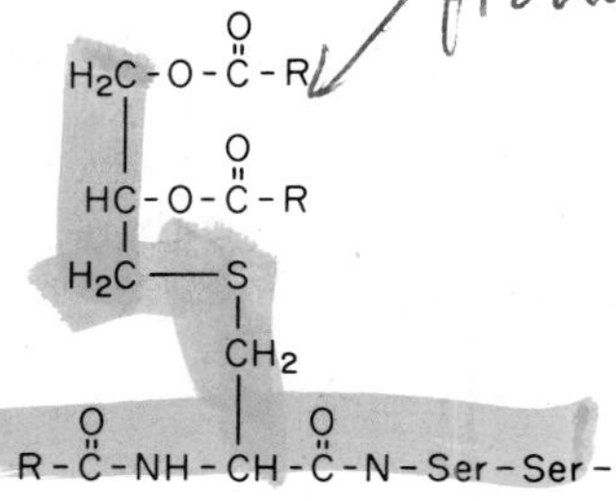

[Handwritten note: from phospholipid pool]

Figure 3.21. Location of acyl chains on the amino terminus of the major lipoprotein of *E. coli*.

phospholipid pool. An important structural point is that the diacylglycerol is linked via a thioether to the C-1 carbon of *sn*-glycerol and not to the C-3 position. This finding indicates that the precursor is not a diacylglycerol moiety of a phospholipid (linkage to C-3 would be expected in this case). There is strong evidence that the source of the *sn*-1-glycerol is the nonacylated polar head group of phosphatidylglycerol. The ester-linked fatty acids are then added to the protein, followed by the amide-linked fatty acid. The exact source of these fatty acids has not been determined, but the data strongly argue that they are derived from the phospholipid pool and not from the fatty acid biosynthetic pathway. Likewise, the nature of the transacylase enzymes responsible for the placement of these fatty acids on the lipoprotein remains to be elucidated. Therefore, phospholipids are used as intermediates in the biosynthesis of membrane proteins in addition to fulfilling their structural role.

LIPID METABOLISM IN BACTERIA OTHER THAN *E. COLI*

It must be emphasized that the lipid metabolism of *E. coli* differs greatly from that of some other bacteria. Although many bacteria follow the *E. coli* paradigm, others ignore it. What is most striking about procaryotic lipids is their incredible diversity.

Bacteria Lacking Unsaturated Fatty Acids

Many bacteria, such as the very successful *Bacillus* genus, possess only very low levels of unsaturated fatty acids under most growth conditions. Instead of unsaturated fatty acids, the major fatty acids imparting membrane fluidity are terminally branched chain fatty acids, which have physical properties similar to those of unsaturated fatty acids. Indeed, it has been shown that *E. coli* can use such acids as unsaturated fatty acid substitutes. The terminally branched chain acids are made by substituting isobutyryl-CoA or 2-methylvaleryl-CoA for acetyl-CoA as the primer of fatty acid biosynthesis. The branched chain acyl-CoA primer is transacylated to ACP and is then used as the acceptor of malonyl groups in processive cycles of fatty acid synthesis.

Bacteria Containing Phosphatidylcholine

Most bacteria lack phosphatidylcholine. However, a few bacteria possess this lipid (for example, *Rhodopseudomonas spheroides*), and these tend to be rather highly specialized or highly evolved bacteria such as photosynthetic or nitrogen-fixing bacteria. Bacterial phosphatidylcholine is synthesized by three successive methylations of phosphatidylethanolamine (Chapter 8). These organisms seem to lack the ability to incorporate choline directly into phospholipid.

Bacteria Synthesizing Unsaturated Fatty Acids by an Aerobic Pathway

The pathway used by plants and animals to synthesize monounsaturated fatty acids involves formation of a double bond in an oxygen-requiring step (Chapter 6). Although most bacteria such as *E. coli* use the anaerobic pathway, some obligately aerobic bacteria (for instance, *Bacilli*) synthesize unsaturated fatty acids by an oxygen-requiring reaction which resembles that of higher cells. In the *Bacilli*, significant amounts of unsaturated fatty acid are synthesized only at low growth temperatures, a situation reminiscent of the thermal control of *E. coli*. However, it should be noted that new protein synthesis seems to be required for commencement of unsaturated fatty acid synthesis upon shift to low temperature; thus, synthesis of some new protein(s) at the lower temperature is probably required. However, it has not been possible to study the desaturation reaction in vitro, so a detailed analysis of this system is not yet available. This pathway is also used by various *Bacilli* to synthesize diunsaturated fatty acids. Polyunsaturated fatty acids are abundant components of all eucaryotic membranes, but few procaryotes synthesize these acids. However, the diunsaturated fatty acids of *Bacilli* have very different double-bond positions from those commonly found in eucaryotes (for example, 5,10-hexadecadienoic acid in *Bacillus licheniformis*).

Bacteria with a Multifunctional Fatty Acid Synthase

The fatty acid synthase complex of *Mycobacterium smegmatis* represents an exception to the general rule that the enzymes of bacterial fatty acid synthesis do not form multifunctional complexes. It was discovered that this bacterium has a fatty acid synthase complex composed of six identical subunits, each having a molecular weight of 290,000. Each of these subunits is a multifunctional polypeptide that contains all six of the reaction centers required for saturated fatty acid synthesis, similar to the liver enzyme described in Chapter 5. This fatty acid synthase also differs from the *E. coli* system in that the products of the synthase are acyl-CoAs having chain lengths ranging from 16 to 24 carbons. Another unusual feature is that the fatty acid synthase system is markedly stimulated by methylated polysaccharides that are polymeric forms of either 3-*O*-methylglucose or 3-*O*-methylmannose. These polysaccharide structures have hydrophobic domains that bind long-chain acyl-CoAs and stimulate fatty acid production by relieving the synthase system from feedback inhibition by acyl-CoA. Interestingly, it appears that the diffusion of the acyl-CoA from the enzyme surface is the rate-limiting step for the synthesis of fatty acids in *M. smegmatis*. *Corynebacterium diphtheriae* has a fatty acid synthase similar to that in *M. smegmatis*, although the aggregate molecular weight is somewhat larger (2.5×10^{-6}). *Brevibacterium ammoniagenes* has an unusual multienzyme complex (molecular weight of 1.23×10^{-6}) that synthesizes both saturated and unsaturated fatty acids. The unsaturated acids are produced in the absence of oxygen and, therefore, appear to be synthesized by a modification of the anaerobic pathway used by *E. coli*. These organisms are members of a highly developed group of bacteria that is thought to be the progenitor of fungi; thus, the finding that the organization of their fatty acid synthase resembles that of the fungi is not surprising.

Bacteria with Intracytoplasmic Membranes

Some specialized bacteria elaborate intracytoplasmic membrane systems that harbor specific metabolic processes in response to changes in the environment. *Rhodopseudomonas sphaeroides* is an example of the type of system used to study the production and differentiation of intracytoplasmic membranes. When this organism is grown phototrophically, the cytoplasmic membrane invaginates and differentiates into an intracytoplasmic membrane that contains the reaction centers required for photosynthetic growth. The quantity of intracytoplasmic membrane produced is inversely related to the intensity of the incident light. The phospholipid components of the intracytoplasmic membrane are acquired discontinuously during the cell cycle, resulting in cyclic alterations in the composition of the intracytoplasmic membranes in synchronously dividing cell populations. The enzymes responsible for the biosynthesis of intracytoplasmic membrane phospholipids are located in the cytoplasmic membrane compartment and the phospholipids are then translocated to the intracytoplasmic membrane.

Other Bacterial Oddities

One bacterium is known (*Bacteroides*) which synthesizes sphingolipids. Although not yet studied in detail, the biosynthetic pathway seems very similar to that used in mammals. A number of bacterial species (for example, *Clostridium*) synthesize 1-alk-1'-enyl lipids (plasmalogens) and in some cases further modify the ether group. The synthetic pathways of these lipids are unknown but bacterial plasmalogens are clearly synthesized by a pathway different from that in mammals. The bacterial plasmalogens are made by strictly anaerobic bacteria. Since the mammalian pathway requires oxygen (Chapter 9), a markedly different pathway must be used by these bacteria. Several bacteria synthesize methyl-branched fatty acids with the methyl group located in the center (rather than at the end) of the acyl chain. These acids are synthesized by a reaction that resembles cyclopropane fatty acid synthesis in that the donor of the $-CH_3$ group is *S*-adenosylmethionine, and the substrate is the fatty acid residue of an intact phospholipid molecule.

Lipids of Nonbacterial (But Related) Organisms

It has recently been realized that there is a group of what were considered to be bacteria but which actually form a group of organisms distinct from the common bacteria (eubacteria) and eucaryotes. These organisms, called the *Archaebacteria*, have a very unusual lipid composition in that the building block of these lipids is the five-carbon unit, mevalonic acid, rather than the two-carbon unit, acetic acid. The result is that phytanyl chains are bound to the glycerol moieties of the complex lipids by ether linkages; thus, these lipids differ in several basic features from those of the common bacterial and eucaryotic cells.

FUTURE DIRECTIONS

Although our knowledge of the individual enzymatic steps in the phospholipid biosynthetic pathway and their regulation has increased dramatically over the last 20 years, there are still a number of fundamental questions that are unanswered. The identity of the enzyme or enzyme system that functions as the pacemaker of phospholipid biosynthesis is not known. Similarly, the mechanisms that operate to produce the observed distribution of phospholipid polar head groups has not been uncovered, and the details of how phospholipid and macromolecular biosynthesis are coordinated during normal growth and nutritional stress remain to be worked out. The regulatory point that controls the divergence of 20% of the acyl-ACP to lipid-A biosynthesis rather than to phospholipid production is a mystery, and the steps in the biosynthetic pathway leading to lipid A are also unknown. Further experiments to measure the intracellular level of intermediates in the phospholipid biosynthetic pathway in vivo along with an in vitro biochemical analysis of isolated regulatory enzymes will be needed to complete our understanding of the rate-determining steps in the biosynthesis of bacterial phospholipids. The genetic approach will continue to be an important tool, and recent advances in genetic-cloning procedures have begun to have a profound effect on our understanding of both the biochemistry and physiology of *E. coli*. The selection of regulatory mutants holds the key to testing our hypothesis about the regulation of the pathway in vivo. Designing selection schemes to isolate regulatory mutants will not be straightforward and represents one of the most challenging aspects of our future work.

PROBLEMS

1. Cerulenin is a fungal antibiotic that is a covalent, active-site-directed inhibitor of both 3-ketoacyl-ACP synthases I and II. How would treatment of *E. coli* cells with cerulenin affect the synthesis of fatty acids, phospholipids, lipid A, and lipoproteins? The amount of cerulenin used in this experiment is sufficient to inhibit all the endogenous 3-ketoacyl-ACP synthase activity.

2. *E. coli* strains harboring the *pgi* mutation contain a defective phosphoglucose isomerase and consequently cannot synthesize glucose. How would adding glucose to a culture of a *pgi* mutant growing on glycerol as a carbon source affect membrane-derived oligosaccharide biosynthesis and membrane phospholipid turnover? What would be the effect on membrane lipid composition if the experiment were performed using a *pgi, dgk* double mutant? How would the ionic strength of the growth medium influence these results?

3. The *gpsA* mutant has a defective biosynthetic *sn*-glycerol-3-phosphate dehydrogenase and therefore requires a glycerolphosphate supplement in order to grow. At the midlogarithmic phase of growth, this strain is shifted from glucose minimal medium containing glycerolphosphate to glucose minimal medium without glycerolphosphate. What would be the effect of glycerolphosphate deprivation on phospholipid biosynthesis, the long-chain acyl-ACP pool, and membrane composition?

4. A new mutant has been isolated that overproduces *cis*-vaccenic acid and does not regulate the synthesis of this acid as a function of temperature. What enzyme activity has most likely been affected? How could this hypothesis be tested?

5. A thioesterase exhibits a pH optimum of 9.5 when acyl-ACP is used as a substrate. What could be predicted about the solution structure of acyl-ACP under these assay conditions? How could the structural specificity of the thioesterase be investigated?

6. 3-Decynoyl-*N*-acetylcysteamine is a covalent, active-site-directed inhibitor of 3-hydroxydecanoyl-ACP dehydrase. How would the addition of this compound to a growing culture of *E. coli* affect membrane phospholipid composition and cell growth? How would the addition of oleic acid to the growth medium affect the results? How would using a *fabAUp* mutant for these studies affect the results?

BIBLIOGRAPHY

Entries preceded by an asterisk are suggested for further reading.

*Clark, D. P., and Cronan, J. E., Jr. 1981. Bacterial mutants for the study of lipid metabolism. *Methods Enzymol.* 72:693–707.

D'Agnolo, G.; Rosenfeld, I. S.; and Vagelos, P. R. 1975. Multiple forms of β-ketoacyl-acyl carrier protein synthetase in *Escherichia coli. J. Biol. Chem.* 250:5289–94.

*deMendoza, D., and Cronan, J. E., Jr. 1983. Thermal regulation of membrane lipid fluidity in bacteria. *Trends Biochem. Sci.* 8:49–52.

deMendoza, D.; Ulrich, A. K.; and Cronan, J. E., Jr. 1983. Thermal regulation of membrane fluidity in *Escherichia coli. J. Biol. Chem.* 258:2098–2101.

*Goldfine, H. 1982. "Lipids of prokaryotes: Structure and distribution." In *Current topics in membranes and transport*, ed. S. Razin, and S. Rottem, vol. 17, 1–43. New York: Academic Press.

Jackowski, S., and Rock, C. O. 1983. Ratio of active to inactive forms of acyl carrier protein in *Escherichia coli. J. Biol. Chem.* 259:15186–91.

Jackowski, S., and Rock, C. O. 1984. Turnover of the 4'-phosphopantetheine prosthetic group of acyl carrier protein. *J. Biol. Chem.* 259:1891–95.

Larson, T. J.; Lightner, V. A.; Green, P. R.; Modrich, P.; and Bell, R. M. 1980. Membrane phospholipid synthesis in *Escherichia coli*. Identification of the *sn*-glycerol-3-phosphate acyltransferase polypeptide as the *plsB* gene product. *J. Biol. Chem.* 255:9421–26.

Lightner, V. A.; Larson, T. J.; Tailleur, P.; Kantor, G. D.; Raetz, C. H. R.; Bell, R. M.; and Modrich, P. 1980. Membrane phospholipid synthesis in *Escherichia coli*. Cloning of a structural gene (*plsB*) of the *sn*-glycerol-3-phosphate acyltransferase. *J. Biol. Chem.* 255:9413–20.

Nishijima, M., and Raetz, C. R. H. 1979. Membrane lipid biogenesis in *Escherichia coli*: Identification of genetic loci for phosphatidylglycerophosphate synthetase and construction of mutants lacking phosphatidylglycerol. *J. Biol. Chem.* 254:7837–44.

Osborn, M. J., and Munson, R. 1974. Separation of the inner (cytoplasmic) and outer membranes of gram-negative bacteria. *Methods Enzymol.* 31:642–52.

*Pieringer, R. A. 1983. "Formation of Bacterial Glycerolipids." In *The Enzymes*, ed. P. D. Boyer, vol XVI, 255–306. New York: Academic Press.

Powell, G. L.; Elovson, J.; and Vagelos, P. R. 1969. Acyl carrier protein. XII. Synthesis and turnover of the prosthetic group of acyl carrier protein in vivo. *J. Biol. Chem.* 244:5616–24.

*Raetz, C. R. H. 1982. "Genetic control of phospholipid bilayer assembly." In *Phospholipids*, ed. J. N. Hawthorne and G. B. Ansell, 435–77. Amsterdam:Elsevier.

Rock, C. O., and Cronan, J. E., Jr. 1981. Acyl carrier protein from *Escherichia coli*. *Methods Enzymol.* 71:341–51.

*Rock, C. O., and Cronan, J. E., Jr. 1982. "Regulation of bacterial membrane lipid synthesis." In *Current topics in membranes and transport*, ed. S. Razin and S. Rottem, vol. 17, 207–27. New York: Academic Press.

Rock, C. O., and Jackowski, S. 1982. Regulation of phospholipid synthesis in *Escherichia coli*. Composition of the acyl-acyl carrier protein pool in vivo. *J. Biol. Chem.* 257:10759–65.

van Golde, L. M. G.; Schulman, H.; and Kennedy, E. P. 1973. Metabolism of membrane phospholipids and its relation to a novel class of oligosaccharides in *Escherichia coli*. *Proc. Natl. Acad. Sci. USA* 70:1368–72.

CHAPTER 4

Oxidation of Fatty Acids

Horst Schulz

THE PATHWAY OF β-OXIDATION: A HISTORICAL ACCOUNT Fatty acids are a major source of energy in animals. The study of their biological degradation began in 1904, when Knoop (1904) performed the classical experiments which led him to formulate the theory of β-oxidation. In his experiments Knoop used fatty acids that contained a phenyl residue in place of the terminal methyl group. The phenyl residue served as a *reporter group* because it was not metabolized but instead was excreted in the urine. When Knoop fed phenyl-substituted fatty acids with an odd number of carbons, like phenyl-propionic acid (C_6H_5—CH_2—CH_2—COOH) or phenylvaleric acid (C_6H_5—CH_2—CH_2—CH_2—CH_2—COOH), to dogs, he isolated from their urine hippuric acid (C_6H_5—CO—NH—CH_2—COOH), the conjugate of benzoic acid and glycine. In contrast, phenyl-substituted fatty acids with an even number of carbons, such as phenylbutyric acid (C_6H_5—CH_2—CH_2—CH_2—COOH), were degraded to phenyl-acetic acid (C_6H_5—CH_2—COOH) and excreted as phenylaceturic acid (C_6H_5—CH_2—CO—NH—CH_2—COOH). These observations led Knoop to propose that the oxidation of fatty acids begins at C-3, the β-carbon, and that the resulting β-keto acids are cleaved between the α-carbon and β-carbon to yield fatty acids shortened by two carbons. Knoop's experiments, and later ones performed by Dakin, prompted the idea that fatty acids are degraded in a stepwise manner by successive β-oxidation. In the years following Knoop's initial study, Dakin (1909) performed similar experiments with phenyl-propionic acid. He was able to isolate besides hippuric acid the glycine conjugates of the following β-oxidation intermediates:

phenylacrylic acid (C_6H_5—CH≡CH—COOH), β-phenyl-β-hydroxypropionic acid (C_6H_5—CHOH—CH_2—COOH), and benzoylacetic acid (C_6H_5—CO—CH_2—COOH). At the same time Embden and co-workers demonstrated by use of perfused livers that unsubstituted fatty acids are degraded by β-oxidation to ketone bodies. Thus, by 1910, the basic information necessary for formulating the pathway of β-oxidation was available.

After a 30-year period of little progress, Munoz and Leloir in 1943, and Lehninger in 1944, demonstrated the oxidation of fatty acids in cell-free preparations from liver. Their work set the stage for the complete elucidation of β-oxidation. Detailed investigations with cell-free systems, especially the studies of Lehninger, demonstrated the need for energy to "spark" the oxidation of fatty acids. ATP was shown to meet this requirement and to be essential for the activation of fatty acids. Activated fatty acids were shown by Wakil and Mahler, as well as by Kornberg and Pricer, to be thioesters formed from fatty acids and coenzyme A. This advance was made possible by earlier studies of Lipmann and co-workers, who isolated and characterized coenzyme A, and Lynen (1952–53) and co-workers, who proved the structure of "active acetate" to be acetyl-CoA. Acetyl-CoA was found to be identical with the two-carbon fragment removed from fatty acids during their degradation. The subcellular location of the β-oxidation system was finally established by Kennedy and Lehninger, who showed that mitochondria were the cellular components most active in fatty acid oxidation. The mitochondrial location of this pathway agreed with the observed coupling of fatty acid oxidation to the citric acid cycle and to oxidative phosphorylation. The most direct evidence for the proposed β-oxidation cycle, as shown in Figure 4.1, emerged from enzymological studies carried out in the 1950s primarily in the laboratories of Green in Wisconsin, Lynen in Munich, Germany, and Ochoa in New York. Their studies were

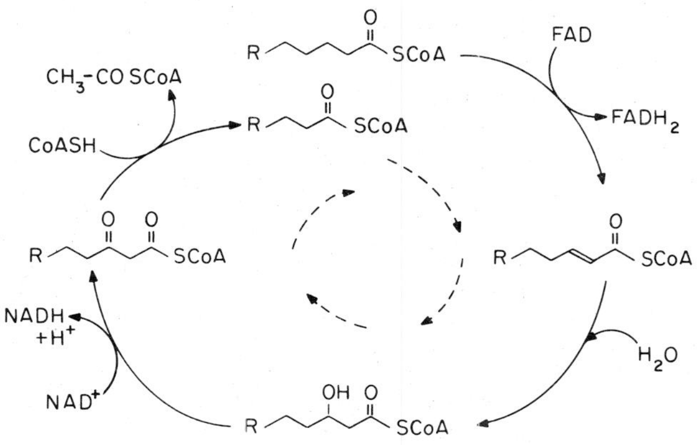

Figure 4.1. The β-oxidation cycle.

greatly facilitated by newly developed methods of enzyme purification and by the use of spectrophotometric enzyme assays that utilized chemically synthesized intermediates of β-oxidation.

TRANSPORT AND ACTIVATION OF FATTY ACIDS IN ANIMAL CELLS   Fatty acids are transported between organs either in the form of unesterified fatty acids complexed to serum albumin or in the form of triacylglycerols associated with lipoproteins. Triacylglycerols are hydrolyzed outside of cells by lipoprotein lipase to yield free fatty acids. The mechanism by which free fatty acids enter cells remains poorly understood despite a number of studies performed with isolated cells from heart, liver, and adipose tissue. Kinetic evidence has been obtained for both a saturable and nonsaturable uptake of fatty acids. The saturable uptake predominates at low concentrations of free fatty acids and reflects either a carrier-mediated uptake or a rate-limiting metabolic step subsequent to the uptake of fatty acids. The nonsaturable uptake, which becomes significant at higher concentrations of free fatty acids, has been attributed to nonspecific diffusion across the membrane.

Once long-chain fatty acids have crossed the plasma membrane, they either diffuse or are transported to the outer mitochondrial membrane and to the endoplasmic reticulum. At both locations they are activated by conversion to their CoA thioesters. Whether this transfer of fatty acids between membranes is a facilitated process or occurs by simple diffusion continues to be a matter of speculation. The identification of a low molecular weight (approximately 14,000) fatty acid binding protein (FABP) in the cytosol of various animal tissues has led to the suggestion that this protein may function as a carrier of fatty acids in the cytosolic compartment (Ockner et al. 1972). However, FABP also binds fatty acyl-CoA, sterols, and other hydrophobic molecules. Hence, this protein may function as a carrier of any of these compounds or may simply provide a temporary place of storage for damaging compounds like long-chain acyl-CoA thioesters. The tissue distribution of FABP is interesting and may provide some insight into the physiological function of this protein. The best characterized FABP is the Z-protein of liver, which is also present in small intestine and possibly in adipose tissue. The epithelial cells of small intestine contain a second FABP which shows some sequence homology with Z-protein. A third FABP, which is immunologically unrelated to the liver protein, has been identified in heart and a related but nonidentical FABP was found in red skeletal muscle. The presence of distinctive FABPs in different tissues raises doubts about a common physiological function of this group of proteins.

The metabolism of fatty acids requires their prior activation by conversion to fatty acyl-CoA thioesters. Most of the activating enzymes are ATP-dependent acyl-CoA synthetases which catalyze the formation of acyl-CoA by the following two-step mechanism (Groot, Scholte, and Hülsmann 1976):

ATP-dependent acyl CoA synthetase

The evidence for this mechanism was derived primarily from a study of acetyl-CoA synthetase. Although the postulated intermediate acetyl-AMP does not accumulate in solution, and therefore only exists bound to the enzyme, the indirect evidence for this intermediate is very convincing. Synthetic acetyl-AMP is a donor of acetate in the reaction with CoASH to form acetyl-CoA and is a donor of AMP in the pyrophosphate-dependent formation of ATP. All exchange reactions agree with this two-step mechanism, which is also supported by the observed incorporation of one oxygen atom of acetate into AMP. Other fatty acids are believed to be activated by a similar mechanism, even though less evidence to support this hypothesis has been obtained. The activation of fatty acids is catalyzed by a group of acyl-CoA synthetases which differ in their subcellular locations and in their specificities with respect to the chain length of their fatty acid substrates. The chain-length specificities are the basis for classifying these enzymes as short-chain acyl-CoA synthetases, medium-chain acyl-CoA synthetases, and long-chain acyl-CoA synthetases.

A short-chain specific acetyl-CoA synthetase has been isolated in a highly purified form from beef heart mitochondria. The enzyme, which is stimulated by monovalent ions like K^+, is most active with acetate as a substrate ($K_m = 0.8mM$), but exhibits some activity toward propionate ($K_m = 11mM$). Acetyl-CoA synthetase has been found in mitochondria of heart, skeletal muscle, kidney, adipose tissue, and intestine, but not in liver mitochondria. A cytosolic

classification base on chain length
① short chain acyl CoA synthetase
② medium-chain acyl CoA synthetase
③ long-chain acyl CoA synthetase

short chain acyl CoA synthetase exhibits some activity toward propionate

acetyl-CoA synthetase was identified in liver, intestine, adipose tissue, and mammary gland, all of which have high lipogenic activities. It is possible that the cytosolic enzyme synthesizes acetyl-CoA for lipogenesis, whereas the mitochondrial acetyl-CoA synthetase activates acetate headed for oxidation. Evidence for the presence of a distinct propionyl-CoA synthetase in liver mitochondria has been obtained. This enzyme is active with acetate, propionate, and butyrate, but the K_m value for propionate is much lower than the K_m values for the other two substrates.

Medium-chain acyl-CoA synthetases are present in the mitochondria of various mammalian organs. The partially purified enzyme from beef heart mitochondria acts on fatty acids with 3 to 7 carbon atoms, but is most active with butyrate, for which a K_m value of $1.5 mM$ was determined. In contrast, a partially purified enzyme from bovine liver mitochondria activates fatty acids with 4 to 12 carbon atoms. Of the fatty acids with an even number of carbon atoms, octanoate is the best substrate, with a K_m value of $0.15 mM$. This enzyme also activates branched, unsaturated, and hydroxy-substituted medium-chain carboxylic acids and, more surprisingly, acts on aromatic carboxylic acids like benzoic acid and phenylacetic acid.

Long-chain acyl-CoA synthetase is a membrane-bound enzyme located in the endoplasmic reticulum and in the outer mitochondrial membrane. This enzyme has been isolated from both organelles and purified to homogeneity. The purified rat liver enzymes from both intracellular locations are identical as judged by several molecular and catalytic properties (Tanaka et al. 1979). The molecular weight of the enzyme was estimated by polyacrylamide gel electrophoresis in the presence of sodium dodecylsulfate to be 76,000. The enzyme acts efficiently on saturated fatty acids containing 10 to 18 carbon atoms and on unsaturated fatty acids containing 16 to 20 carbon atoms. The dual subcellular location of this enzyme may reflect its functions in both lipid synthesis and fatty acid oxidation. If so, the mitochondrial enzyme would synthesize fatty acyl-CoA thioesters for oxidation, whereas the enzyme of the endoplasmic reticulum would provide substrates for lipid (for example, glycerolipid) synthesis.

In addition to ATP-dependent acyl-CoA synthetases, a group of GTP-dependent acyl-CoA synthetases has been described. The best known of these is succinyl-CoA synthetase, which cleaves GTP to GDP and phosphate and functions in the tricarboxylic acid cycle. A mitochondrial GTP-dependent acyl-CoA synthetase, which activates fatty acids ranging from butyrate to oleate, has been described. However, its existence as a distinct enzyme has been questioned.

Since the inner mitochondrial membrane is impermeable to CoA

[handwritten margin note: the mitochondrial enzyme would synthesize fatty acyl-CoA thioesters for oxidation whereas the enzyme of endoplasmic reticulum would provide substrates for lipid synthesis.]

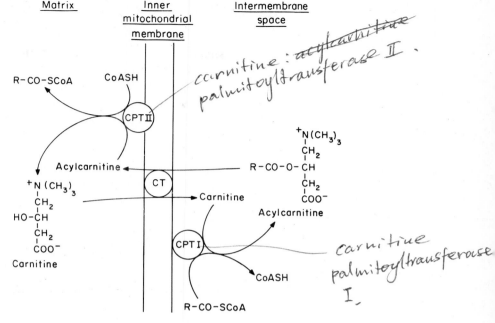

Figure 4.2. Carnitine-dependent uptake of fatty acids by mitochondria. CPT I, carnitine palmitoyltransferase I; CPT II, carnitine palmitoyltransferase II; CT, carnitine:acylcarnitine translocase.

and its derivatives, fatty acyl-CoA thioesters formed at the outer mitochondrial membrane cannot directly enter the mitochondrial matrix where the enzymes of β-oxidation are located. Instead, the acyl residues of acyl-CoA thioesters are carried across the inner mitochondrial membrane by L-carnitine. This carnitine-dependent translocation of fatty acids across the inner mitochondrial membrane is schematically shown in Figure 4.2. The reversible transfer of fatty acyl residues from CoA to carnitine is catalyzed by carnitine palmitoyltransferase (CPT). Two forms of this enzyme have been identified (Hoppel 1982). Both are associated with the inner mitochondrial membrane, but one form, CPT I or CPT A, faces the intermembrane space, whereas the other, CPT II or CPT B, is directed toward the matrix space. During the translocation of fatty acid residues from the cytosol into the mitochondrial matrix, CPT I catalyzes the transfer of acyl residues from CoA to carnitine. The acylcarnitines cross the inner mitochondrial membrane via the carnitine:acylcarnitine translocase (Pande 1975). This enzyme catalyzes a slow unidirectional diffusion of carnitine both in and out of the mitochondrial matrix in addition to a much faster mole-to-mole exchange of acylcarnitine for carnitine, carnitine for carnitine, and acylcarnitine for acylcarnitine. The unidirectional transfer of carnitine may be an important mechanism by which mitochondria acquire carnitine, which is synthesized in the liver. The fast exchange, especially of acylcarnitine for carnitine, is believed to be essential for the translocation of long-chain fatty acids from the cytosol into mitochondria. In the mitochondrial matrix, CPT II catalyzes the transfer of acyl residues from carnitine back to CoA to form acyl-CoA thioesters, which then enter the β-oxidation cycle. The possible identity of CPT I and CPT II with respect to their primary structures has been considered but not been established. Kinetic differences between these two enzymes may be a consequence of their different environments. Carnitine palmitoyltransferase has been extracted from bovine heart mitochondria and purified to near homogeneity. This purified preparation catalyzes the reversible transfer of acyl residues with 8 to 16 carbon atoms between CoA and carnitine.

In addition to CPT I and II, mitochondria contain a carnitine acetyltransferase which has been purified to homogeneity. The bovine heart enzyme has an estimated molecular weight of 60,000 and is composed of a single polypeptide chain. It catalyzes the transfer of acyl groups with 2 to 10 carbon atoms. The function of this enzyme has not been conclusively established. Possibly, the enzyme regenerates free CoA in the mitochondrial matrix by transferring acetyl groups and other short-chain or medium-chain acyl residues from CoA to carnitine. The resulting acylcarnitines can leave the mitochondria, possibly via the carnitine:acylcarnitine translocase, and move to other organelles or tissues. Although carnitine acetyltransferase and medium-chain carnitine acyltransferase activities have been detected in peroxisomes and microsomes, their functions remain to be elucidated.

Short-chain and medium-chain fatty acids with fewer than 10 carbon atoms are taken up by mitochondria as free acids independent of carnitine. They are activated in the mitochondrial matrix, where short-chain and medium-chain acyl-CoA synthetases are located.

ENZYMES OF β-OXIDATION IN MITOCHONDRIA
Acyl-CoA Dehydrogenase

The first step of β-oxidation is the dehydrogenation of an acyl-CoA to 2-*trans*-enoyl-CoA according to the following equation:

$$R—CH_2—CH_2—CO—SCoA + FAD \longrightarrow$$
$$R—CH{=}CH—CO—SCoA + FADH_2$$

This reaction is catalyzed by acyl-CoA dehydrogenases, which have tightly, but noncovalently, bound flavin adenine nucleotides (FAD) as prosthetic groups. Three different acyl-CoA dehydrogenases function in the β-oxidation of fatty acids, namely, short-chain acyl-CoA or butyryl-CoA dehydrogenase, general or medium-chain acyl-CoA dehydrogenase, and long-chain acyl-CoA dehydrogenase. The same set of acyl-CoA dehydrogenases is present in liver and heart mitochondria. Isolation and purification of these enzymes has permitted detailed studies of their molecular and kinetic properties. All three dehydrogenases have similar molecular weights of approximately 180,000 and are composed of four, possibly identical, subunits. The three enzymes differ in their specificities for substrates of various chain lengths. Short-chain acyl-CoA or butyryl-CoA dehydrogenase is active only with hexanoyl-CoA and shorter-chain substrates. General or medium-chain acyl-CoA dehydrogenase acts on a broad spectrum of acyl-CoA thioesters ranging from hexanoyl-CoA to palmitoyl-CoA. The enzyme is most active with the medium-chain substrates hexanoyl-CoA, octanoyl-CoA, and decanoyl-CoA. Interestingly, this dehydrogenase is the only one that will act on 4-cis-decenoyl-CoA, an intermediate in the oxidation of linoleic acid. Long-chain acyl-CoA dehydrogenase is active with hexanoyl-CoA and longer-chain substrates, but exhibits optimal activity with dodecanoyl-CoA. It dehydrogenates stearoyl-CoA more effectively than does the medium-chain enzyme. Kinetic measurements with all three dehydrogenases have yielded K_m values of 1–10μM for good substrates. The dehydrogenation of acyl-CoA thioesters is possibly initiated by the removal of a proton from the carbon atom (α-carbon) adjacent to the carbonyl group of the thioester. The resulting carbanion may then donate electrons directly to the FAD of the enzyme to yield 2-enoyl-CoA and enzyme-bound $FADH_2$. From $FADH_2$ of the dehydrogenase, electrons are transferred to the FAD prosthetic group of a second flavoprotein named electron-transferring flavoprotein (ETF) which donates electrons to an iron-sulfur flavoprotein named ETF:ubiquinone oxidoreductase. The latter enzyme, a component of the inner mitochondrial membrane, feeds electrons into the mitochondrial electron transport chain via ubiquinone. The flow of electrons from acyl-CoA to oxygen is schematically shown in the following flow chart:

$$R—CH_2—CH_2—CO—SCoA \longrightarrow FAD \text{ (Acyl-CoA dehydrogenase)} \longrightarrow FAD \text{ (ETF)} \longrightarrow$$

$$FeS \text{ (ETF:ubiquinone oxidoreductase)} \longrightarrow \text{ubiquinone} ----\rightarrow \text{oxygen}$$

ETF is a soluble matrix protein with a molecular weight close to 60,000. It is composed of two nonidentical subunits of similar molecular weights, and it contains only one FAD per dimer.

In addition to the three acyl-CoA dehydrogenases involved in fatty acid oxidation, two acyl-CoA dehydrogenases specific for metabolites of branched-chain amino acids have been isolated and purified. They are isovaleryl-CoA dehydrogenase and 2-methyl-branched-chain acyl-CoA dehydrogenase.

Enoyl-CoA Hydratase   The second reaction in β-oxidation is the hydration of 2-*trans*-enoyl-CoA to L-3-hydroxyacyl-CoA catalyzed by enoyl-CoA hydratase, as shown in the following equation:

$$R—CH{=}CH—CO—SCoA + H_2O \rightleftharpoons R—CH(OH)—CH_2—CO—SCoA$$

The reaction is reversible, and at equilibrium the ratio of L-3-hydroxy-acyl-CoA to 2-enoyl-CoA is close to 2. Enoyl-CoA hydratase also catalyzes the hydration of 2-*cis*-enoyl-CoA to D-3-hydroxyacyl-CoA.

Two enoyl-CoA hydratases have been identified in heart mitochondria. The better characterized of the two is crotonase, or short-chain enoyl-CoA hydratase, which has been purified to homogeneity and crystallized. Its molecular weight is 165,000, and it is composed of six, most likely identical, subunits. A long-chain enoyl-CoA hydratase, which appears to be at least partially membrane-associated, has been separated from crotonase. The pig heart long-chain enoyl-CoA hydratase does not act on crotonyl-CoA $(CH_3—CH{=}CH—CO—SCoA)$, but is active with all longer-chain enoyl-CoA thioesters. In contrast, pig heart crotonase is most active with crotonyl-CoA and exhibits a decreasing activity with increasing chain lengths of its substrates. Its activity with 2-hexadecenoyl-CoA is only one-fortieth of the activity observed with crotonyl-CoA as a substrate. Kinetic measurements with both hydratases have yielded K_m values between $10\mu M$ and $30\mu M$ for virtually all substrates. Since most of the long-chain enoyl-CoA hydratase activity of pig heart muscle is due to the long-chain enzyme, the suggestion has been made that crotonase and long-chain enoyl-CoA hydratase cooperate in β-oxidation to ensure a high rate of hydration of all enoyl-CoA intermediates in fatty acid degradation. However, the total crotonase activity of some tissues, especially of liver, is so high that the rate of the crotonase-catalyzed hydration of long-chain enoyl-CoA thioesters exceeds the β-oxidation capacity of the tissue. Consequently, long-chain enoyl-CoA hydratase may not be required in the β-oxidation of regular fatty acids but may function in fatty acid elongation.

L-3-Hydroxyacyl-CoA   The third reaction in the β-oxidation cycle is the reversible dehy-
Dehydrogenase   drogenation of L-3-hydroxyacyl-CoA catalyzed by L-3-hydroxyacyl-CoA dehydrogenase, as shown in the following equation:

$$R—CH(OH)—CH_2—CO—SCoA + NAD^+ \rightleftharpoons R—CO—CH_2—CO—SCoA + NADH + H^+$$

L-3-Hydroxyacyl-CoA dehydrogenase is a soluble matrix enzyme which has been obtained in crystalline form from pig heart muscle.

The enzyme is specific for NAD$^+$, but acts with comparable effectiveness on all L-3-hydroxyacyl-CoA thioesters formed during the oxidation of fatty acids. The K_m values for L-3-hydroxybutyryl-CoA and L-3-hydroxydecanoyl-CoA are $14\mu M$ and $3\mu M$, respectively. Recently, a second mitochondrial L-3-hydroxyacyl-CoA dehydrogenase has been described. This dehydrogenase is associated with the inner mitochondrial membrane and exhibits a preference for long-chain substrates. The function of the membrane-bound L-3-hydroxyacyl-CoA dehydrogenase remains to be established.

Thiolase   The last reaction in the β-oxidation cycle is the cleavage of 3-ketoacyl-CoA catalyzed by thiolase, as shown in the following equation:

$$R—CO—CH_2—CO—SCoA + CoASH \rightleftharpoons R—CO—SCoA + CH_3—CO—SCoA$$

The products of the reaction are acetyl-CoA and an acyl-CoA that is two carbon atoms shorter than the substrate. The equilibrium of the reaction is far to the side of the thiolytic cleavage products. All thiolases that have been studied in detail contain an essential sulfhydryl group which participates directly in the carbon-carbon bond cleavage, as outlined in the following equations:

$$E—SH + R—CO—CH_2—CO—SCoA \rightleftharpoons R—CO—S—E + CH_3—CO—SCoA$$

$$R—CO—S—E + CoASH \rightleftharpoons R—CO—SCoA + E—SH$$

According to this mechanism, 3-ketoacyl-CoA binds to the enzyme and is cleaved between its α- and β-carbon atoms. An acyl residue, which is two carbons shorter than the substrate, remains covalently linked to the enzyme via a thioester bond, while acetyl-CoA is released from the enzyme. Finally, the acyl residue is transferred from the sulfhydryl group of the enzyme to CoASH to yield acyl-CoA.

Several types of thiolases have been identified, some of which exist in multiple forms. Mitochondria contain two classes of thiolases: ①acetoacetyl-CoA thiolase or acetyl-CoA acetyltransferase, specific for acetoacetyl-CoA, ②and 3-ketoacyl-CoA thiolase or acetyl-CoA acyltransferase, which acts on 3-ketoacyl-CoA thioesters of various chain lengths. The latter enzyme functions in β-oxidation, whereas acetoacetyl-CoA thiolase is believed to be involved in ketone body synthesis and degradation. Both mitochondrial thiolases have been purified to homogeneity. They are composed of four, possibly identical, subunits with molecular weights between 42,000 and 46,000. Kinetic measurements with 3-ketoacyl-CoA thiolase from pig heart have yielded K_m values that decrease from $17\mu M$ to $2\mu M$ as the acyl chain length of the substrate increases from four to eight carbon atoms. This enzyme acts equally well on all substrates tested except for acetoacetyl-CoA, which is cleaved at half the maximal rate observed with longer-chain substrates. Some tissues, such as liver, also contain a cytosolic acetoacetyl-CoA thiolase which functions in cholesterol biosynthesis.

DEGRADATION OF UNSATURATED AND ODD-CHAIN FATTY ACIDS

Unsaturated fatty acids, which usually contain cis double bonds, also are degraded by β-oxidation. However, their double bonds must be either moved or removed during the degradation process. All double bonds found in unsaturated and polyunsaturated fatty acids can be classified either as double bonds extending from odd-numbered carbon atoms, like the 9-*cis* double bond present in oleic acid, linoleic acid, and many other polyunsaturated fatty acids or as double bonds extending from even-numbered carbon atoms, like the 12-*cis* double bond of linoleic acid. Since both classes of double bonds are present in linoleic acid, its degradation exemplifies the breakdown of all other unsaturated fatty acids. A summary of the oxidation of linoleic acid is presented in Figure 4.3. Linoleic acid after conversion to its CoA thioester (I), undergoes three cycles of β-oxidation to yield 3-*cis*,6-*cis*-dodecadienoyl-CoA (II), which is isomerized to 2-*trans*,6-*cis*-dodecadienoyl-CoA (III) by *cis*-Δ^3-*trans*-Δ^2-enoyl-CoA isomerase, an auxiliary enzyme of β-oxidation (Stoffel, Ditzer, and Caesar 1964). This enzyme can also isomerize 3-*trans*-enoyl-CoA thioesters to their 2-*trans* isomers, although at lower rates. 2-*trans*,6-*cis*-Dodecadienoyl-CoA (III) is a substrate of β-oxidation and can pass once through the cycle to yield 4-*cis*-decenoyl-CoA (IV), which is dehydrogenated to 2-*trans*,4-*cis*-decadienoyl-CoA (V) by acyl-CoA dehydrogenase. Only medium-chain dehydrogenase is capable of catalyzing this reaction, even though long-chain acyl-CoA dehydrogenase will act on decanoyl-CoA. According to Stoffel and Caesar (1965), the further metabolism of 2-*trans*,4-*cis*-decadienoyl-CoA (V) requires completion of the β-oxidation cycle (not shown in Figure 4.3) to yield 2-*cis*-octenoyl-CoA, which can be hydrated by enoyl-CoA hydratase (crotonase) to D-3-hydroxyoctanoyl-CoA. D-3-Hydroxyoctanoyl-CoA is not a substrate of L-3-hydroxyacyl-CoA dehydrogenase of the β-oxidation pathway, but can be converted to the L-isomer by 3-hydroxyacyl-CoA epimerase. The resulting L-3-hydroxy-octanoyl-CoA would be completely degraded by passing three times through the β-oxidation cycle. However, the identification of a NADPH-dependent 2,4-dienoyl-CoA reductase in mitochondria led Kunau and Dommes (1978) to suggest an alternate sequence of steps by which 2-*trans*,4-*cis*-decadienoyl-CoA (V) can be degraded (shown in Figure 4.3). According to their proposed pathway, 2-*trans*,4-*cis*-

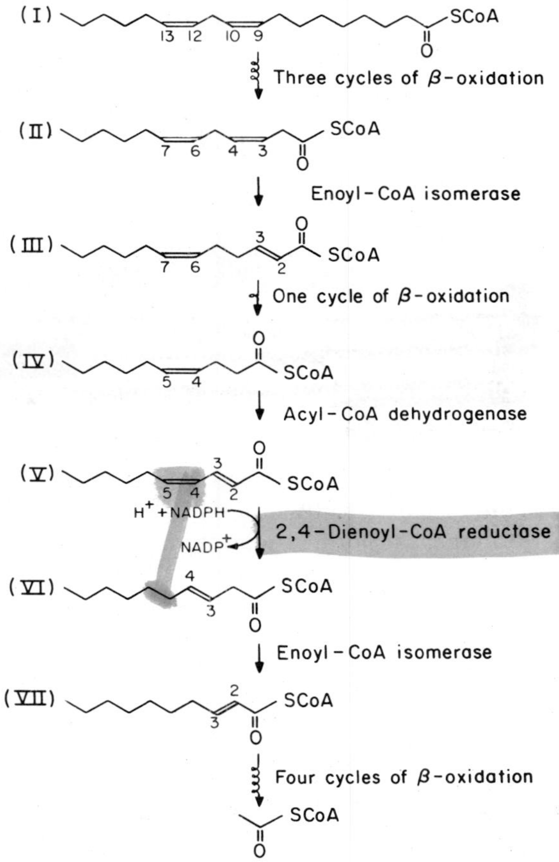

Figure 4.3. β-Oxidation of linoleoyl-CoA.

decadienoyl-CoA (V) is first reduced by NADPH in a reaction catalyzed by 2,4-dienoyl-CoA reductase. The product of this reduction, 3-*trans*-decenoyl-CoA (VI), is isomerized by enoyl-CoA isomerase to 2-*trans*-decenoyl-CoA (VII), which can be completely degraded by cycling four times through the β-oxidation pathway. This modified pathway of linoleate degradation, which requires 2,4-dienoyl-CoA reductase instead of 3-hydroxyacyl-CoA epimerase as an auxiliary enzyme of β-oxidation, is strongly supported by the following results obtained in the laboratories of Kunau (Kunau and Dommes 1978) and Schulz (Cuebas and Schulz 1982): (a) A NADPH-dependent dienoyl-CoA reductase is present in liver and heart mitochondria. (b) Metabolites isolated after incubating 4-*cis*-decenoyl-CoA (IV in Figure 4.3) with mitochondrial extracts are indicative of their formation via the pathway shown in Figure 4.3. (c) 2-*trans*,4-*cis*-Decadienoyl-CoA (V) is not degraded by the β-oxidation system of heart mitochondria nor is it degraded by a β-oxidation system reconstituted from purified enzymes, but it is rapidly reduced by NADPH in the presence of heart mitochondria. (d) Linoleoyl-CoA is effectively degraded by rat heart mitochondria, which are virtually devoid of 3-hydroxyacyl-CoA epimerase activity, but which contain 2,4-dienoyl-CoA reductase. The conclusion is that the degradation of unsaturated fatty acids requires cis-Δ^3-$trans$-Δ^2-enoyl-CoA isomerase and 2,4-dienoyl-CoA reductase in addition to the regular enzymes of β-oxidation. The isomerase is required for the degradation of all fatty acids with double bonds extending from odd-numbered carbon atoms, whereas the reductase is necessary for the removal of all double bonds extending from even-numbered carbon atoms. The physiological function of 3-hydroxyacyl-CoA epimerase remains to be established.

Δ^3-cis-Δ^2-$trans$-Enoyl-CoA isomerase has been isolated and purified from hog liver as well as rat liver. The hog liver enzyme has a molecular weight of 90,000 and is composed of two apparently identical subunits. In addition to catalyzing the isomerization of CoA derivatives of 3-cis-enoic acids and 3-$trans$-enoic acids with 6 to 16 carbon atoms to the corresponding 2-$trans$-enoyl-CoAs, the enzyme catalyzes the conversion of 3-acetylenic acyl-CoA to 2,3-dienoyl-CoA.

2,4-Dienoyl-CoA reductase from bovine liver has been purified to homogeneity. Its molecular weight is estimated to be 124,000. The enzyme is composed of four apparently identical subunits. The reductase has a specific requirement for NADPH. NADH neither substitutes for NADPH nor inhibits the enzyme. K_m values for NADPH and 2-$trans$-4-cis-decadienoyl-CoA are $94\mu M$ and $3\mu M$, respectively.

The oxidation of fatty acids with an odd number of carbons proceeds by β-oxidation and yields in addition to acetyl-CoA 1 mol of propionyl-CoA per mole of fatty acid. Propionyl-CoA, which is further metabolized to succinate, is also formed during the degradation of some amino acids. Propionyl-CoA is carboxylated by the biotin-containing propionyl-CoA carboxylase to D-methylmalonyl-CoA.

$$CH_3-CH_2-CO-SCoA + HCO_3^- + ATP \longrightarrow {}^-OOC-\overset{\overset{\displaystyle CH_3}{|}}{CH}-CO-SCoA + ADP + P_i$$

The D-isomer is isomerized to the L-isomer by methylmalonyl-CoA racemase. In the final step of this pathway, L-methylmalonyl-CoA is isomerized to succinyl-CoA, an intermediate of the tricarboxylic acid cycle.

$$\begin{array}{c} \overset{\displaystyle CH_3}{\underset{\displaystyle |}{}} \\ {}^-OOC-CH-CO-SCoA \end{array} \xrightarrow{\text{\it methylmalonyl-CoA mutase}} {}^-OOC-CH_2-CH_2-CO-SCoA$$

This reaction is catalyzed by methylmalonyl-CoA mutase, one of the few enzymes requiring cobalamin as a cofactor. All reactions of propionyl-CoA catabolism occur within mitochondria.

REGULATION OF MITOCHONDRIAL FATTY ACID OXIDATION

The rate of fatty acid oxidation is determined by the availability of fatty acids and by the rate of utilization of β-oxidation products. Plasma free fatty acids formed by lipolysis in adipose tissue are readily available to most tissues. The concentration of unesterified fatty acids in plasma is regulated by the hormones glucagon, which stimulates, and insulin, which inhibits the breakdown of triacylglycerols in adipose tissue (Chapter 7). The stimulatory effect of glucagon is due to its activation of adenylate cyclase and the resulting increase in the concentration of cellular cyclic AMP, which activates protein kinase. One of the substrates of protein kinase in adipose tissue is the hormone-sensitive lipase, which is activated by phosphorylation and inactivated by dephosphorylation. Thus, when the concentration of glucose is low, as in the fasting animal, a high glucagon to insulin ratio results in an increased plasma concentration of unesterified free fatty acids. These fatty acids will enter cells, where they can be either degraded to acetyl-CoA or incorporated into other lipids. The utilization of fatty acids for either oxidation or lipid synthesis depends on the nutritional state of the animal, more specifically on the availability of carbohydrates. Because of the close relationship between lipid metabolism, carbohydrate metabolism, and ketogenesis, the regulation of fatty acid oxidation in liver differs from and is more complex than the regulation of β-oxidation in tissues like heart and skeletal muscle which have an overwhelming catabolic function. For this reason, the regulation of fatty acid oxidation in liver and heart will be discussed separately.

The direction of fatty acid metabolism in liver depends on the nutritional state of the animal. In the fed animal the liver breaks down carbohydrates to synthesize fatty acids, while in the fasted animal fatty acid oxidation, ketogenesis, and gluconeogenesis are the more active processes. Clearly, there exists a reciprocal relationship between fatty acid synthesis and fatty acid oxidation. Although it is well established that lipid and carbohydrate metabolism are under hormonal control, it has been more difficult to identify the specific sites at which fatty acid synthesis and oxidation are regulated and to elucidate those regulatory mechanisms. McGarry and Foster (1980) have

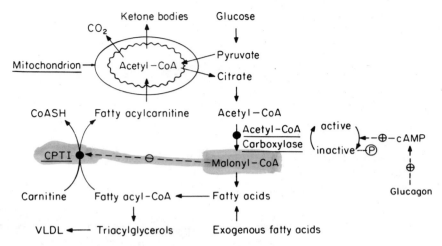

Figure 4.4. Proposed regulation of fatty acid oxidation in liver. (-⊕-) Stimulation; (-⊖-) inhibition; (●) enzymes subject to regulation.

suggested that the concentration of malonyl-CoA, the first committed intermediate in fatty acid biosynthesis, determines the rate of fatty acid oxidation. The essential features of their hypothesis are presented in Figure 4.4. In the fed animal, where glucose is actively converted to fatty acids, the concentration of malonyl-CoA is elevated. Malonyl-CoA is a reversible inhibitor of carnitine palmitoyltransferase I (CPT I). As a result of the inhibition of CPT I by malonyl-CoA, fatty acyl residues are not transferred from acyl-CoA to carnitine and therefore not translocated into mitochondria. Consequently, β-oxidation is depressed. When the animal changes from the fed to the fasted state, hepatic metabolism shifts from glucose breakdown to gluconeogenesis, with a resulting decrease in fatty acid synthesis. The concentration of malonyl-CoA decreases, and the inhibition of CPT I is relieved. As acylcarnitines are formed and translocated into mitochondria, both β-oxidation and ketogenesis are stimulated.

It appears that the cellular concentration of malonyl-CoA is directly related to the activity of acetyl-CoA carboxylase, which is hormonally regulated (Chapter 5). The short-term regulation of acetyl-CoA carboxylase involves the glucagon-dependent phosphorylation and dephosphorylation of the enzyme. Phosphorylated acetyl-CoA carboxylase is the less active form of the enzyme. In the fasted animal, a high glucagon:insulin ratio causes an increase in cellular cAMP, which is responsible for the phosphorylation and inactivation of acetyl-CoA carboxylase. As a consequence, the concentration of malonyl-CoA and the rate of fatty acid synthesis decrease, while the rate of β-oxidation increases. A decrease of the glucagon:insulin ratio reverses these effects. Thus, both fatty acid synthesis and fatty acid oxidation are regulated by the ratio of glucagon to insulin.

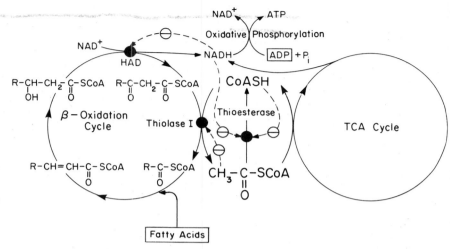

Figure 4.5. Proposed regulation of fatty acid oxidation in heart muscle. (-⊖-) Inhibition; (●) enzymes subject to regulation; HAD, L-3-hydroxyacyl-CoA dehydrogenase.

The rate of fatty acid oxidation in heart is dependent on the free fatty acid concentration of plasma as well as on the energy demand of the tissue except at sufficiently high levels of free fatty acids, where the rate is dependent only on the energy demand of the tissue. Studies with perfused hearts and isolated mitochondria under conditions of reduced energy demand have demonstrated that the concentrations of NADH and acetyl-CoA increase and those of NAD^+ and CoASH decrease. The reduction in the energy demand is either a consequence of lower workloads imposed on the isolated heart or a result of limiting amounts of ADP available to isolated mitochondria (Figure 4.5). An increase in the ratio of NADH to NAD^+ in mitochondria causes inhibition of the tricarboxylic acid cycle and consequently an increase in the ratio of acetyl-CoA to CoASH. Acetyl-CoA, NADH, and $FADH_2$ are the products of fatty acid oxidation, and their accumulation may inhibit β-oxidation. Since 3-hydroxy fatty acids accumulate when oxidative phosphorylation is inhibited in isolated mitochondria, the first dehydrogenation step in β-oxidation is apparently not the primary site of regulation. Kinetic studies with purified 3-hydroxyacyl-CoA dehydrogenase and 3-ketoacyl-CoA thiolase have revealed their severe inhibitions by NADH and acetyl-CoA, respectively (Figure 4.5). The inhibition of 3-hydroxyacyl-CoA dehydrogenase by NADH is competitive with respect to NAD^+, with a K_i value for NADH of $25\mu M$. The inhibition of 3-ketoacyl-CoA thiolase by acetyl-CoA ($K_i = 4\mu M$) is competitive with respect to CoASH. When the mitochondrial activities of these two enzymes are calculated at low and high concentrations of NADH and acetyl-CoA, it appears that 3-ketoacyl-CoA thiolase may be the primary site at which β-oxidation is regulated. However, the influx of reducing equivalents into mitochondria could also cause an inhibition of β-oxidation. The suggested regulation of β-oxidation in heart in response to the energy demand of the tissue is shown in Figure 4.5. According to this hypothesis, a decrease in the energy demand of heart muscle leads to an increase in acetyl-CoA and to a parallel decrease in CoASH. This change in the acetyl-CoA to CoASH ratio causes the inhibition of 3-ketoacyl-CoA thiolase and thereby of β-oxidation. Intermediates of β-oxidation accumulate and prevent the entry of more acyl-CoA into the β-oxidation cycle. A kinetic study of a mitochondrial thioesterase (not thiolase) revealed it to be so strongly inhibited by NADH and CoASH that it would be inactive under all normal conditions. However, should the energy demand be high (NADH is low) and should the tricarboxylic acid cycle operate insufficiently (CoASH is low), the thioesterase would be de-inhibited and would hydrolyze acetyl-CoA to CoASH and thereby stimulate β-oxidation. By this mechanism β-oxidation could be decoupled from the tricarboxylic acid cycle in an emergency situation when, for example, intermediates of the tricarboxylic acid cycle are not available.

β-OXIDATION IN PEROXISOMES AND GLYOXYSOMES

Peroxisomes and *glyoxysomes*, collectively referred to as *microbodies* (de Duve 1983), are subcellular organelles capable of respiration. They do not contain an energy-coupled electron transport system like mitochondria, but instead have flavin oxidases which catalyze substrate-dependent reductions of oxygen to H_2O_2. Since catalase is present in these organelles, H_2O_2 is rapidly reduced to water. Thus, peroxisomes and glyoxysomes are organelles with a primitive respiratory chain where energy released during the reduction of oxygen is lost as heat. Glyoxysomes are peroxisomes that contain the enzymes of the glyoxylate pathway in addition to flavin oxidases and catalase. Peroxisomes or glyoxysomes are found in all major groups of eucaryotic organisms including yeasts, protozoa, plants, and animals.

The presence of an active β-oxidation pathway in microbodies similar to that in mitochondria was first detected in glyoxysomes of germinating seeds. When rat liver peroxisomes were shown to contain a β-oxidation system (Lazarow and de Duve 1976), originally thought to be limited to mitochondria, the interest in this pathway was greatly stimulated. Within a short period of time, all peroxisomal β-oxidation enzymes from rat liver had been purified and characterized (Hashimoto 1982).

The first step in peroxisomal β-oxidation is the dehydrogenation of acyl-CoA to 2-enoyl-CoA catalyzed by acyl-CoA oxidase. This enzyme, in contrast with the mitochondrial dehydrogenases, trans-

$R-CH_2-CH_2-CH_2-\overset{O}{\overset{\|}{C}}-CoA$

H_2O_2

?

$2\,H_2O$

oxidase

$R-CH=CH-CH_2-\overset{O}{\overset{\|}{C}}-CoA$

enoyl CoA hydratase

$R-\underset{H}{\overset{H}{\underset{|}{\overset{|}{C}}}}-\underset{H}{\overset{OH}{\underset{|}{\overset{|}{C}}}}-CH_2-\overset{O}{\overset{\|}{C}}-CoA$

NAD^+

$NADH$

L-3-hydroxyacyl CoA dehydrogenase

$R-CH_2-\overset{O}{\overset{\|}{C}}-CH_2-\overset{O}{\overset{\|}{C}}-CoA$

3-ketoacyl CoA thiolase

$CH_3-\overset{O}{\overset{\|}{C}}-CoA$

$R-CH_2-\overset{O}{\overset{\|}{C}}-CoA$

fers two hydrogens from the substrate to oxygen, which is thereby reduced to H_2O_2. The purified rat liver acyl-CoA oxidase contains FAD and has a molecular weight of close to 140,000. This enzyme is inactive with butyryl-CoA and hexanoyl-CoA as substrates, but dehydrogenates all longer-chain substrates with similar maximal velocities. The K_m values for long-chain substrates are approximately $10\mu M$. The next two reactions in β-oxidation, the hydration of 2-enoyl-CoA to 3-hydroxyacyl-CoA and the NAD-dependent dehydrogenation of L-3-hydroxyacyl-CoA to 3-ketoacyl-CoA, are catalyzed in peroxisomes by a bifunctional polypeptide which harbors both enoyl-CoA hydratase and L-3-hydroxyacyl-CoA dehydrogenase activities. This bifunctional enzyme consists of a single polypeptide chain with a molecular weight close to 80,000. The enoyl-CoA hydratase associated with the bifunctional enzyme is most active with crotonyl-CoA and exhibits decreasing activities with increasing chain lengths of the substrates. The K_m values for crotonyl-CoA is $83\mu M$, while that for all other substrates is approximately $10\mu M$. The NAD⁺-specific L-3-hydroxyacyl-CoA dehydrogenase of the bifunctional enzyme is almost equally active with substrates of various chain lengths. However, the K_m values decrease with increasing chain length, from $42\mu M$ for 3-hydroxybutyryl-CoA to $4\mu M$ for 3-hydroxydecanoyl-CoA.

The last reaction of β-oxidation, the CoA-dependent cleavage of 3-ketoacyl-CoA, is catalyzed by 3-ketoacyl-CoA thiolase. The peroxisomal enzyme has a molecular weight of 89,000 and is composed of two subunits of equal size. This thiolase exhibits little activity toward acetoacetyl-CoA, but is highly active with all longer-chain substrates. The K_m values for all substrates are in the low micromolar range. Glyoxysomes from germinating seeds contain β-oxidation enzymes similar to those of rat liver peroxisomes. The β-oxidation enzymes from mitochondria and peroxisomes of rat liver are different proteins, as judged by their structural, catalytic, and immunological properties.

The mechanism by which fatty acids are taken up by peroxisomes has not been fully elucidated. Either free fatty acids enter peroxisomes and are converted to their CoA thioesters by peroxisomal acyl-CoA synthetase, or fatty acids cross the peroxisomal membrane as their CoA thioesters. Carnitine does not stimulate the initiation of fatty acid oxidation in peroxisomes.

The cofactors required for peroxisomal β-oxidation, NAD⁺ and CoASH, have been found in peroxisomes. The reoxidation of NADH to NAD⁺ may occur via a glycerol phosphate shuttle or may involve the direct translocation of NADH to the cytosol and of NAD⁺ back into peroxisomes.

The inactivity of acyl-CoA oxidase toward butyryl-CoA and hexanoyl-CoA prevents rat liver peroxisomes from completely degrading fatty acids. Peroxisomes can only chain-shorten long-chain fatty acids to medium-chain acyl-CoA thioesters. The medium-chain acyl residues and acetyl groups formed in peroxisomes can be transferred to carnitine by peroxisomal medium-chain carnitine acyltransferase and carnitine acetyltransferase, respectively. The resulting acylcarnitines, including acetylcarnitine, may move from peroxisomes to mitochondria for further oxidation. In contrast with rat liver peroxisomes, glyoxisomes of germinating seeds can degrade fatty acids completely to acetyl-CoA.

Studies of peroxisomal fatty acid oxidation have been greatly aided by the proliferation of peroxisomes in livers of rats fed a diet containing clofibrate or other hypolipidemic drugs. Most importantly, clofibrate causes a 10-fold increase in the specific activities of the enzymes of peroxisomal β-oxidation due to increased enzyme synthesis. In addition to hypolipidemic drugs, high-fat diets, starvation, and diabetes reportedly also increased peroxisomal β-oxidation. The induction of peroxisomal fatty acid oxidation in response to feeding a high-fat diet rich in erucic acid 22:1(n − 9), prompted the proposal that an important function of peroxisomal β-oxidation is the chain-shortening of very long chain fatty acids, like erucic acid, which are poorly metabolized by mitochondria. The cooperation between mitochondria and peroxisomes in fatty acid oxidation suggested by this proposal has been debated since peroxisomes were shown to be capable of β-oxidation. The contribution of peroxisomal β-oxidation to fatty acid oxidation in rat liver has been estimated by some investigators to be insignificant, whereas others have maintained that it may represent half of the total fatty acid oxidation capacity of liver. Further studies are needed to settle this issue and to provide more definitive evidence for the physiological function of peroxisomal β-oxidation.

INHIBITION OF FATTY ACID OXIDATION AND ITS CONSEQUENCES

Although fatty acid oxidation is essential for the survival of animals during prolonged fasting, the importance of this metabolic pathway in the fed animal is more difficult to assess. The study of *Jamaican vomiting sickness* has served to emphasize the importance of fatty acid oxidation in humans. Jamaican vomiting sickness is a life-threatening illness that afflicts mostly malnourished people of Jamaica. The illness is a consequence of eating the unripe fruit of the ackee tree (*Blighia sapida*). Unlike the ripe fruit, the unripe fruit contains a water-soluble toxin, which causes the sickness. Jamaican vomiting sickness is characterized by a sudden onset of vomiting and violent retching 2 to 3 hours after the fruit is eaten. The most striking condition associated with this illness is a severe hypoglycemia. Children and malnourished adults are the most likely victims of this type

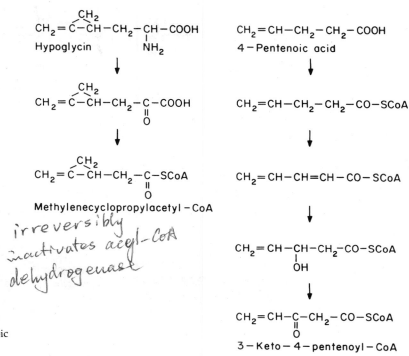

Figure 4.6. Metabolism of hypoglycin and 4-pentenoic acid.

of food poisoning. The isolation of the toxic principle, named *hypoglycin*, and its structural characterization have facilitated in vitro studies which have helped to elucidate the molecular basis of this disease (Billington and Sherratt 1981). Hypoglycin, an amino acid, is metabolized in animals as outlined in Figure 4.6. Transamination of hypoglycin yields methylenecyclopropylpyruvic acid, which is converted to methylenecyclopropylacetyl-CoA. The latter compound irreversibly inactivates acyl-CoA dehydrogenase, especially butyryl-CoA dehydrogenase, and thereby inhibits β-oxidation. The dehydrogenase is inactivated due to the formation of a covalent adduct between the inhibitor and the FAD cofactor of the enzyme. An impairment of β-oxidation results in an increased utilization of glucose and an inhibition of gluconeogenesis due to limiting amounts of β-oxidation products. Together, these effects cause a rapid depletion of available carbohydrates, especially in people with limited glycogen deposits.

The search for structurally simpler analogs of hypoglycin led to the identification of 4-pentenoic acid as a β-oxidation inhibitor that causes hypoglycemia in animals. However, 4-pentenoic acid inhibits β-oxidation by inactivating 3-ketoacyl-CoA thiolase, but not acyl-CoA dehydrogenase (Schulz and Fong 1981). 4-Pentenoic acid must be metabolized via β-oxidation (Figure 4.6) before it inhibits thiolase. 3-Keto-4-pentenoyl-CoA, a metabolite of 4-pentenoic acid, is both a reversible and an irreversible inhibitor of 3-ketoacyl-CoA thiolase and acetoacetyl-CoA thiolase. In addition to 4-pentenoic acid, 4-bromocrotonic acid and 2-bromooctanoic acid are metabolized by β-oxidation to their 3-keto derivatives, which inactivate 3-ketoacyl-CoA thiolase and thereby inhibit β-oxidation. In addition to these compounds, 2-tetradecylglycidic acid,

$$CH_3(CH_2)_{13}-\overset{\displaystyle CH_2-O}{\underset{}{C}}-COOH$$

inhibits fatty acid oxidation by interfering with the entry of fatty acids into mitochondria due to the inactivation of carnitine palmitoyltransferase I.

Deficiencies of enzymes and cofactors of fatty acid oxidation have been identified as the causes of episodic hypoglycemia, nonketotic aciduria, hepatic insufficiency, and myopathy. Many of the patients afflicted with these diseases were children, who died at an early age. In some patients the defect in fatty acid oxidation was traced to a carnitine deficiency (Di Mauro, Trevisan, and Hays 1980). In myopathic carnitine deficiency the carnitine concentration is normal in plasma, but low in muscle, possibly due to an abnormal carnitine uptake by muscle. Carnitine deficiency in muscle results in weakness and abnormal accumulation of lipids in muscle, but generally is not life-threatening. In patients with systemic carnitine deficiency, the carnitine concentration is low in plasma and liver, possibly due to an impaired biosynthesis of carnitine. All patients with systemic carnitine deficiency showed progressive neuromuscular disorders and episodic hepatic insufficiency.

The most frequently observed enzyme deficiency related to fatty acid oxidation is that of carnitine palmitoyltransferase (CPT) (Di Mauro, Trevisan, and Hays 1980). Patients with low levels (5–45% of normal) of CPT have recurrent muscle weakness and myoglobinuria, often precipitated by prolonged exercise, fasting, or a combination of the two. CPT deficiency is paralleled by impaired fatty acid oxidation and an increased dependence on carbohydrate metabolism.

Dicarboxylic aciduria is a metabolic disorder characterized by episodic hypoglycemia and urinary excretion of dicarboxylic acids with 6 to 10 carbon atoms. Mitochondria isolated from fibroblasts of patients with this disease exhibit only 5% of the normal medium-chain acyl-CoA dehydrogenase activity, but have near-normal levels of butyryl-CoA dehydrogenase and long-chain acyl-CoA dehydrogenase (Rhead et al. 1983). Generally, an impairment of β-oxidation

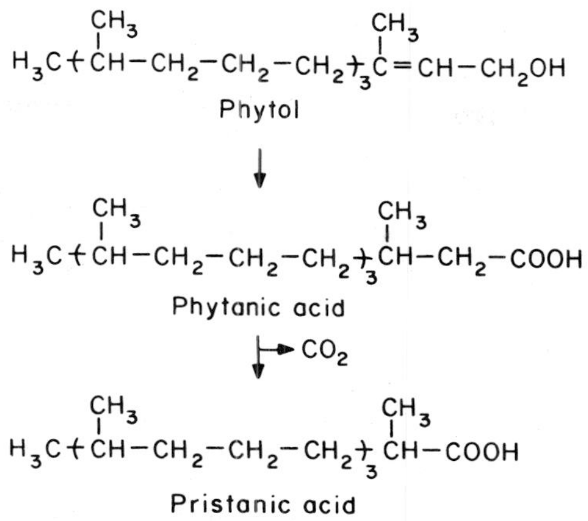

Figure 4.7. Metabolism of phytol.

makes fatty acids available for microsomal ω-oxidation, by which fatty acids are oxidized at their terminal (ω) methyl group or at their penultimate carbon. Molecular oxygen is required for this oxidation, and dicarboxylic acids are formed. Long-chain dicarboxylic acids can be β-oxidized to medium-chain dicarboxylic acids, which are excreted in the urine.

Some patients with nonketotic aciduria excrete 2-methyl-3-hydroxybutyric acid, a metabolite of isoleucine degradation. The possibility that this aciduria may be due to the deficiency of a β-oxidation enzyme was substantiated when the absence of thiolase active with 2-methylacetoacetyl-CoA as a substrate was demonstrated.

The importance of α-oxidation in humans has been established as a result of studying *Refsum's disease*, a rare and inherited neurological disorder. Patients afflicted with this disease accumulate large amounts of phytanic acid (Figure 4.7), which is derived from phytol, a component of chlorophyll. Because of a methyl substituent at its β-carbon, phytanic acid cannot be β-oxidized, but it can undergo α-oxidation to pristanic acid. This minor pathway of fatty acid oxidation involves the hydroxylation at the α-carbon followed by decarboxylation. Pristanic acid, in contrast with phytanic acid, can be degraded by β-oxidation. A deficiency of α-oxidation prevents the metabolism of phytanic acid and results in its accumulation in various body compartments.

FATTY ACID OXIDATION IN *E. COLI*

The presence of an active pathway of fatty acid oxidation in *E. coli* was first demonstrated in 1967 by Overath and co-workers (1967), who showed that growth of *E. coli* on long-chain fatty acids instead of on glucose resulted in a 200-fold induction of the enzymes of fatty acid oxidation. The isolation of *E. coli* mutants unable to grow on

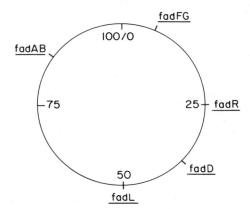

Figure 4.8. *E. coli* chromosome map with locations of genes required for fatty acid degradation (*fad*).

long-chain fatty acids permitted more-detailed genetic studies of this pathway (Overath, Pauli, and Schairer 1969). Genes essential for fatty acid oxidation have been mapped at five different locations on the *E. coli* chromosome, as shown in Figure 4.8. Position *fadAB* (fatty acid degradation AB) refers to an operon which contains the genetic information for 3-ketoacyl-CoA thiolase, enoyl-CoA hydratase (crotonase), 3-hydroxyacyl-CoA dehydrogenase, *cis*-Δ^3-*trans*-Δ^2-enoyl-CoA isomerase and 3-hydroxyacyl-CoA epimerase. Acyl-CoA synthetase is coded for by gene *fadD*. The region designated *fadFG* is believed to harbor the closely linked genes for at least one short-chain acyl-CoA dehydrogenase and one long-chain acyl-CoA dehydrogenase. Position *fadL* is that of a gene essential for fatty acid uptake. The product of the *fadR* gene appears to be a repressor protein involved in the expression of the enzymes of fatty acid oxidation.

The regulation of fatty acid oxidation in *E. coli* is mechanistically similar to the regulation of lactose degradation in the same organism. A difference between these two inducible pathways is that the *lac* genes are organized within one operon, whereas the *fad* genes form a regulon composed of several operons. Indirect evidence suggests that CoA derivatives of fatty acids with more than 10 carbon atoms induce the fatty acid oxidation system in *E. coli*. Induction of this system, possibly by the binding of long-chain acyl-CoA to the repressor protein, results in the coordinate induction of all enzymes required for fatty acid oxidation except for ETF, which is constitutively expressed. Synthesis of the enzymes of fatty acid oxidation is strongly repressed when glucose, in addition to long-chain fatty acids, is present in the growth medium. This catabolite repression of the *fad* system is a consequence of a decrease in the cellular concentration of cAMP caused by glucose. When the concentration of cAMP is low, the catabolite gene activator protein (CAP) does not bind to the *fad* promoters, thereby preventing the efficient transcription of the *fad* regulon.

The uptake of fatty acids with more than 10 carbon atoms by *E. coli* is dependent on the *fadL* gene product, which is apparently membrane-bound and is believed to function as a fatty acid permease (Nunn and Simons 1978). The uptake of these long-chain fatty acids is closely coupled to their activation by acyl-CoA synthetase. This conclusion is based on the observation that an *E. coli* mutant constitutive for the enzymes of fatty acid oxidation, but with a defective acyl-CoA synthetase, was unable to take up and metabolize fatty acids. Medium-chain fatty acids (fatty acids with 6 to 10 carbon atoms) can enter *E. coli* cells either via the long-chain uptake system or by a nonsaturable process characteristic of simple diffusion.

E. coli apparently contains only one acyl-CoA synthetase, the product of the *fadD* gene, which can activate both medium-chain and long-chain fatty acids. This ATP-dependent (AMP-forming) acyl-CoA synthetase has been purified to homogeneity. The molecular weights of the native enzyme and its subunits are estimated to be 130,000 and 47,000, respectively. This synthetase activates fatty acids with 6 to 18 carbon atoms, but is most active with dodecanoic acid.

Purification of the enzymes of β-oxidation from *E. coli* resulted in the isolation of a homogeneous protein which exhibited enoyl-CoA hydratase (crotonase), L-3-hydroxyacyl-CoA dehydrogenase, 3-ketoacyl-CoA thiolase, *cis*-Δ^3-*trans*-Δ^2-enoyl-CoA isomerase, and 3-hydroxyacyl-CoA epimerase activities (Figure 4.9) (Binstock, Pramanik, and Schulz 1977). This multienzyme complex of fatty acid oxidation contains all β-oxidation enzymes, with the exception of acyl-CoA dehydrogenase. The chain-length specificities of 3-ketoacyl-CoA thiolase, L-3-hydroxyacyl-CoA dehydrogenase, and enoyl-CoA hydratase of the *E. coli* complex are similar to those of the mammalian enzymes. All three enzymes act on substrates having between 4 and 16 carbon atoms. However, enoyl-CoA hydratase is most active with short-chain substrates, whereas 3-ketoacyl-CoA thiolase and L-3-hydroxyacyl-CoA dehydrogenase exhibit their optimal activities with medium-chain substrates. The complex has an estimated molecular weight of 260,000 and is composed of two types of subunits with molecular weights of 78,000 and 42,000. The quaternary structure of the complex is $\alpha_2\beta_2$, where α and β denote the 78,000-dalton and 42,000-dalton subunits, respectively. Phospholipids characteristic of the *E. coli* membrane are associated with the purified complex. They constitute approximately 4% of the protein mass. Immunological studies suggest that the total 3-ketoacyl-CoA thiolase, L-3-hydroxy-acyl-CoA dehydrogenase, and crotonase activities present in *E. coli* extracts are associated with the complex. However, a long-chain enoyl-CoA hydratase is apparently not part of the complex. The

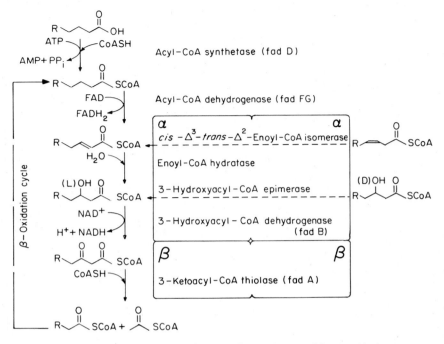

Figure 4.9. Pathway of fatty acid oxidation and organization of the β-oxidation enzymes in *E. coli*. The 78,000-dalton and 42,000-dalton subunits are marked α and β, respectively.

subunit locations of the five component enzymes were determined by chemical modifications (Yang and Schulz 1983). 3-Ketoacyl-CoA thiolase appears to be the only enzymatic activity associated with the 42,000-dalton subunit, whereas enoyl-CoA hydratase, L-3-hydroxyacyl-CoA dehydrogenase, *cis*-Δ^3-*trans*-Δ^2enoyl-CoA isomerase, and 3-hydroxyacyl-CoA epimerase are located on the 78,000-dalton subunit. Thus, the *fadAB* operon which codes for the fatty acid oxidation complex, contains two structural genes: gene *fadA*, which codes for the 42,000-dalton subunit, and gene *fadB*, which codes for the 78,000-dalton polyfunctional polypeptide. Kinetic studies performed with the fatty acid oxidation complex are indicative of the channeling of β-oxidation intermediates, like L-3-hydroxyacyl-CoA, from the active site of crotonase to that of L-3-hydroxyacyl-CoA dehydrogenase without equilibrating with the bulk medium. This process can explain why intermediates of β-oxidation do not accumulate within cells. The isolation of the *E. coli* fatty acid oxidation complex represents the first proof for the existence of such a multi-enzyme complex, which has been proposed, but not proven, to be present in mitochondria.

FUTURE DIRECTIONS Although fatty acid oxidation has been studied extensively, the regulation of this important metabolic pathway remains poorly understood. Following are some major questions that have been raised, but have not been fully answered. (a) What is the molecular basis for the effect of hormones like insulin on extrahepatic fatty acid oxidation? (b) Is fatty acid oxidation regulated by changes in coenzyme concentrations and if so, how? (c) Are enzymes of fatty acid oxidation regulated by covalent modifications? (d) Is fatty acid oxidation controlled at a transport step other than the carnitine-dependent uptake of fatty acids? (e) Do the enzymes of β-oxidation exist in mitochondria as a multienzyme complex? (f) Which of the β-oxidation steps are catalyzed by more than one enzyme? (g) What is the function of peroxisomal β-oxidation? Clearly, this well-established area of biochemistry is alive with opportunities for further research.

PROBLEMS

1. 2-Bromopalmitoyl-CoA inhibits the oxidation of palmitoyl-CoA by isolated mitochondria, but has no effect on the oxidation of palmitoylcarnitine. What is the most likely site at which 2-bromopalmitoyl-CoA inhibits fatty acid oxidation?

2. Explain why malonyl-CoA inhibits the oxidation of palmitic acid by isolated mitochondria, but has no effect on the oxidation of octanoic acid.

3. Palmitic acid can be oxidized by first passing five times through the peroxisomal β-oxidation cycle, followed by the mitochondrial oxidation of medium-chain acyl-CoA, acetyl-CoA, and NADH formed in peroxisomes. How many moles of ATP are generated when palmitic acid is completely oxidized by this cooperation of peroxisomes and mitochondria? How energy efficient is this pathway of palmitate oxidation compared with the oxidation of palmitate by mitochondria alone?

4. The growth of two *E. coli* mutants on various fatty acids was studied. Both mutants were constitutive for the enzymes of fatty acid oxidation. One mutant grew well on all even-numbered fatty acids from decanoate to oleate, whereas the other mutant grew only on decanoate. What is the most likely genetic difference between these two mutants?

5. When mitochondria are uncoupled, the oxidation of oleoylcarnitine is stimulated, whereas the oxidation of docosahexaenoylcarnitine, 22:6 (n − 3), is inhibited. The addition of glutamate restores the oxidation rate of docosahexaenoylcarnitine in uncoupled mitochondria to that observed with coupled mitochondria. Explain these observations.

BIBLIOGRAPHY

Entries preceded by an asterisk are suggested for further reading.

Billington, D., and Sherratt, H. S. A. 1981. Hypoglycin and metabolically related inhibitors. *Methods Enzymol.* 72:610–16.

Binstock, J. F.; Pramanik, A.; and Schulz, H. 1977. Isolation of a multi-enzyme complex of fatty acid oxidation from *Escherichia coli*. *Proc. Natl. Acad. Sci. USA* 74:492–95.

Cuebas, D., and Schulz, H. 1982. Evidence for a modified pathway of linoleate degradation. Metabolism of 2,4-decadienoyl coenzyme A. *J. Biol. Chem.* 257:14140–44.

Dakin, H. 1909. The mode of oxidation in the animal organism of phenyl derivatives of fatty acids.

Part IV, Further studies on the fate of phenylpropionic acid and some of its derivatives. *J. Biol. Chem.* 6:203–19.

de Duve, C. 1983. Microbodies in the living cell. *Scientific American,* 248:74–84.

Di Mauro, S.; Trevisan, C.; and Hays, A. 1980. Disorders of lipid metabolism in muscle. *Muscle and Nerve* 3:369–88.

*Groot, P. H. E.; Scholte, H. R.; and Hülsmann, W. C. 1976. "Fatty acid activation: specificity, localization, and function." In *Advances in lipid research,* ed. R. Paoletti and D. Krichevsky, vol. 14, 75–126. New York: Academic Press.

*Hashimoto, T. 1982. Individual peroxisomal β-oxidation enzymes. *Ann. N.Y. Acad. Sci.* 386:5–12.

*Hoppel, C. L. 1982. Carnitine and carnitine palmitoyltransferase in fatty acid oxidation and ketosis. *Fed. Proc.* 41:2853–57.

Knoop, F. 1904. *Der Abbau aromatischer Fettsäuren im Tierkörper.* Freiburg, Germany: Ernst Kuttruff.

Kunau, W. H., and Dommes, P. 1978. Degradation of unsaturated fatty acids. Identification of intermediates in the degradation of *cis*-4-decenoyl-CoA by extracts of beef liver mitochondria. *Eur. J. Biochem.* 91:533–44.

Lazarow, P. B., and de Duve, C. 1976. A fatty acyl-CoA oxidizing system in rat liver peroxisomes; Enhancement by clofibrate, a hypolipidemic drug. *Proc. Natl. Acad. Sci. USA* 73:2043–46.

*Lynen, F. 1952–53. Acetyl coenzyme A and the fatty acid cycle. *Harvey Lectures* ser. 48:210–44.

*McGarry, J. D., and Foster, D. W. 1980. Regulation of hepatic fatty acid oxidation and ketone body production. *Ann. Rev. Biochem.* 49:395–420.

Nunn, W. D., and Simons, R. W. 1978. Transport of long-chain fatty acids by *Escherichia coli.* Mapping and characterization of mutants in the *fadL* gene. *Proc. Natl. Acad. Sci. USA* 75:3377–81.

Ockner, R. K.; Manning, J. A.; Poppenhausen, R. B.; and Ho, W. K. L. 1972. A binding protein for fatty acids in cytosol of intestinal mucosa, liver, myocardium, and other tissues. *Science* 177:56–58.

Overath, P.; Pauli, G.; and Schairer, H. U. 1969. Fatty acid degradation in *Escherichia coli.* An inducible acyl-CoA synthetase, the mapping of old mutants, and the isolation of regulatory mutants. *Eur. J. Biochem.* 7:559–74.

Overath, P.; Raufuss, E. M.; Stoffel, W.; and Ecker, W. 1967. The induction of the enzymes of fatty acid degradation in *Escherichia coli. Biochem. Biophys. Res. Commun.* 29:28–33.

Pande, S. V. 1975. A mitochondrial carnitine acylcarnitine translocase system. *Proc. Natl. Acad. Sci. USA* 72:883–87.

Rhead, W. J.; Amendt, B. A.; Fritchman, K. S.; and Felts, S. J. 1983. Dicarboxylic aciduria: Deficient [1-^{14}C]octanoate oxidation and medium-chain acyl-CoA dehydrogenase in fibroblasts. *Science* 221:73–75.

Schulz, H., and Fong, J. C. 1981. 4-Pentenoic acid. *Methods Enzymol.* 72:604–10.

Stoffel, W., and Caesar, H. 1965. Der Stoffwechsel der ungesättigten Fettsäuren. V. Zur β-Oxidation der Mono- und Polyenfettsäuren. Der Mechanismus der enzymatischen Reaktionen an Δ^{2cis}-enoyl-CoA-Verbindungen. *Hoppe-Seyler's Z. Physiol. Chem.* 341:76–83.

Stoffel, W.; Ditzer, R.; and Caesar, H. 1964. Der Stoffwechsel der ungesättigten Fettsäuren. III. Zur β-Oxidation der Mono- und Polyenfettsäuren. Der Mechanismus der enzymatischen Reaktionen an Δ^{3cis}-enoyl-CoA-Verbindungen. *Hoppe-Seyler's Z. Physiol. Chem.* 339:167–81.

Tanaka, T.; Hosaka, K.; Hoshimaru, M.; and Numa, S. 1979. Purification and properties of long-chain acyl coenzyme A synthetase from rat liver. *Eur. J. Biochem.* 98:165–72.

*Yang, S., and Schulz, H. 1983. The large subunit of the fatty acid oxidation complex from *Escherichia coli* is a multifunctional polypeptide. *J. Biol. Chem.* 258:9780–85.

CHAPTER 5

Fatty Acid Synthesis in Eucaryotes

Alan G. Goodridge

Long-chain fatty acids serve two primary functions in animals. As parts of phospholipids and other complex lipids, fatty acids are critical structural components of cellular membranes. As parts of triacylglycerols, fatty acids represent stored energy. The latter function is the primary concern of this chapter. *Homeothermic* animals maintain a constant body temperature and, in adults, a constant body weight. To do so, birds and mammals balance the amount of energy consumed in their diets with the amount lost as heat. If food were available continuously, special mechanisms for storing energy would be unnecessary. In reality, however, the food supply is often erratic. Furthermore, cyclic patterns of foraging may be imposed on animals by their environments (for example, light, temperature). Energy must be stored during feeding periods so that body temperature and metabolic rate can be maintained during nonfeeding periods. Nature has solved the problem of erratic or cyclic food availability by evolving a complex, highly regulated system which ensures that energy can be stored when food is abundant; the stored energy can be utilized when food is scarce.

Triacylglycerol contains about twice as many calories per gram as protein or carbohydrate and is the form in which energy is stored. In addition, the energy is stored without the concomitant deposition of large quantities of water. In an average 70-kg human male, triacylglycerols constitute 85% of the total 166,000 cal stored in body tissues. Carbohydrate, the other easily mobilized form of energy, constitutes less than 1000 cal. By contrast, the diets of many animals contain a large amount of carbohydrate. Clearly, energy storage involves conversion of carbohydrates to fatty acids.

The rate of de novo synthesis of long-chain fatty acids is rapid in well-fed animals, especially when the diet has little or no fat, and slow in starved animals. This regulation is important because glucose, a substrate for lipogenesis, is *required* as an energy source for the brain and for erythrocytes, even during starvation. Thus, inhibition of the conversion of glucose to fatty acids during starvation preserves glucose for those tissues which require it. Inhibition of the terminal segment of the pathway for fatty acid synthesis during starvation also prevents futile cycling. Fatty acid oxidation is very active in the livers of starved animals. The product of the oxidation of fatty acids is acetyl-CoA, the same compound which is the substrate for the terminal segment of the fatty acid synthesis pathway. If inhibition did not occur during starvation, energy would be wasted in the futile cycling of acetyl-CoA to fatty acids and back to acetyl-CoA again.

The cells of most tissues synthesize fatty acids at very low rates. The liver, however, has a large capacity to synthesize fatty acids. Triacylglycerols are synthesized in the liver and transported to adipose tissue, where they are stored. A few species, especially rodents, convert dietary carbohydrate to triacylglycerol in both liver and adipose tissue. Certain other specialized tissues also synthesize large amounts of long-chain fatty acids but for other purposes. For example, the lactating mammary gland converts carbohydrate into triacylglycerol which is used to nourish the newborn. Fatty acids synthesized by sebaceous glands are secreted as wax esters and triacylglycerols, secretions which are used to condition skin, hair, and feathers and to lubricate external surfaces. Regulation of fatty acid synthesis in mammary gland or in sebaceous glands is different from that in liver or adipose tissue. This chapter will focus on regulation of fatty acid synthesis in the latter organs.

SIGNALS IN BLOOD THAT MEDIATE THE EFFECTS OF DIET ON LIPOGENESIS

Incorporation of [^{14}C]glucose or [^{14}C]acetate into long-chain fatty acids is inhibited by starvation, diabetes, and diets high in fat. Treatment of diabetic animals with insulin stimulates fatty acid synthesis, as does feeding starved animals, especially if the diet contains a high proportion of carbohydrate. These early findings suggest that insulin mediates the effects of diet on lipogenesis. Two other facts strengthen this postulate. Glucose stimulates secretion of insulin from the β-cells of the islets of Langerhans, and insulin stimulates the metabolism of glucose by liver and adipose tissue. However, the failure of insulin, in vivo or in vitro, to restore fatty acid synthesis in tissues from starved rats suggests the involvement of other factors.

Glucagon, another pancreatic hormone, is a second important factor in the regulation of fatty acid synthesis. If glucagon is added to incubations of isolated tissues, fatty acid synthesis is inhibited. In vivo, glucagon blocks the stimulation of hepatic fatty acid synthesis caused by the refeeding of starved rats. Thus, glucagon may mediate a major part of the inhibition of fatty acid synthesis caused by starvation. The inverse relationship between glucose concentration in the blood and secretion of glucagon is consistent with this hypothesis. Glucagon may be even more important than insulin in the regulation of fatty acid synthesis. In tissue preparations such as perfused livers or isolated liver cells, the effects of glucagon are usually greater in magnitude than those of insulin. Furthermore, the glucagon concentration in blood is elevated in diabetic animals and decreased when diabetic animals are treated with insulin. This is due to the inability of glucose to inhibit glucagon secretion in the absence of insulin. Therefore, the low rate of fatty acid synthesis in diabetic animals may be attributable to a combination of inhibition by glucagon and lack of stimulation by insulin rather than simply to a lack of stimulation by insulin.

Glucagon and insulin are polypeptide hormones that interact with specific, but different, receptors on the outer surface of the plasma membrane. Several intracellular compounds have been postulated to mediate the actions of insulin. Discussion of this controversial subject is beyond the scope of this chapter. In contrast, the intracellular mediator of the action of glucagon is well known. Binding of glucagon to its receptor activates adenylate cyclase, an enzyme bound to the cytoplasmic face of the plasma membrane. The resulting increase in the intracellular concentration of cAMP causes activation of the catalytic subunit of cAMP-dependent protein kinase. The rapid inhibition of fatty acid synthesis caused by glucagon is due primarily to the phosphorylation by this protein kinase of enzymes which are rate-controlling in the pathway or of enzymes which regulate the production of important regulatory intermediates. The addition of exogenous cAMP to isolated liver preparations mimics the inhibitory effect of glucagon on fatty acid synthesis.

The concentrations of unesterified fatty acids in plasma and long-chain fatty acyl-CoAs in liver are elevated during starvation. In addition, diets containing a high percentage of calories as fat cause inhibition of fatty acid synthesis and elevation of the hepatic concentration of long-chain fatty acyl-CoA. These observations suggest that unesterified fatty acids may be a third agent in the plasma which regulates fatty acid synthesis. The inhibition of fatty acid synthesis

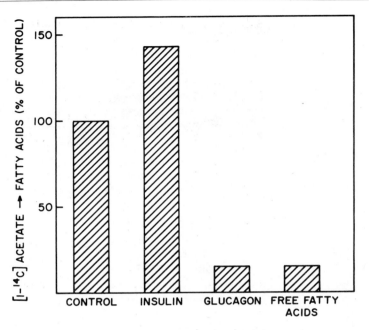

Figure 5.1. The effects of insulin, glucagon, and free fatty acids on incorporation of [1-^{14}C] acetate into total fatty acids in isolated hepatocytes from the newly hatched chick (Goodridge 1973).

caused by unesterified fatty acids in isolated hepatocytes and in perfused livers provides further evidence for this hypothesis. The effects of glucagon and insulin may be amplified via this mechanism because glucagon stimulates, and insulin inhibits, the release of fatty acids from adipose tissue. (Figure 5.1 demonstrates the effects of insulin, glucagon, cAMP, and unesterified fatty acids on fatty acid synthesis from labeled acetate in isolated hepatocytes.)

Insulin, glucagon, and unesterified fatty acids are, therefore, important regulators of the pathway of fatty acid synthesis during dietary manipulations. They are not, however, the only agents in blood which regulate the pathway. Fatty acid synthesis in liver is increased in the hyperthyroid state and decreased in the hypothyroid state. Furthermore, thyroid hormones stimulate fatty acid synthesis in liver cells in culture. The role of thyroid hormone in the regulation of fatty acid synthesis during cycles of starvation and feeding is not clear. Limited space prevents the cataloging of every compound which affects the rate of fatty acid synthesis in adipose tissue or liver. Insulin, thyroid hormone, glucagon, and unesterified fatty acids act directly on isolated hepatocytes or adipocytes and have physiologically relevant effects. We shall concentrate on the mechanisms by which these agents exert their effects on fatty acid synthesis.

ENZYMATIC BASIS
FOR THE SYNTHESIS
OF LONG-CHAIN
FATTY ACIDS FROM
TWO-CARBON
PRECURSORS

Efforts to identify extracellular regulators began in the early 1950s, when ^{14}C-labeled compounds became readily available. The contributions from Chaikoff's laboratory were particularly important during this period. Although it was known that acetate or a closely related compound was an intermediate in the conversion of carbohydrate to fat, early studies were executed without a knowledge of the enzymatic basis for the conversion of the two-carbon intermediate to long-chain fatty acids. In fact, prior to about 1958, fatty acid synthesis was thought to proceed via a reversal of the β-oxidation pathway. In the early 1950s, Gurin and co-workers and Popjak and Teitz successfully prepared soluble extracts of pigeon liver and mammalian mammary gland, respectively, which synthesized long-chain fatty acids from acetate. Wakil and colleagues then fractionated the soluble extract of pigeon liver into four partially purified components, each of which was required for synthesis of long-chain fatty acids from acetate in the presence of ATP, CoA, NADH, Mn^{2+}, isocitrate, $NADP^+$, and a sulfhydryl compound such as cysteine or glutathione. Subsequent work from the same group narrowed the requirements to two purified fractions when acetyl CoA, NADPH, and bicarbonate were present in the incubations. None of the activities of the fatty acid oxidation sequence was present in the two highly purified fractions, thus establishing that fatty acid synthesis was not the simple reversal of fatty acid oxidation.

Two observations were particularly important in the discovery that activation of acetyl-CoA to malonyl-CoA (acetyl-CoA carboxylase) was a key step in fatty acid synthesis. First, bicarbonate was required for the synthesis of long-chain fatty acids despite the fact that [^{14}C]CO_2 was not incorporated into the product. Second, malonyl-CoA was an intermediate in the synthesis of fatty acids from acetyl-CoA. Malonyl-CoA was formed from acetyl-CoA and bicarbonate in the presence of ATP and one of the two highly purified protein fractions prepared by Wakil and colleagues. Furthermore, malonyl-CoA could be incorporated into fatty acids in the absence of ATP.

The individual reactions involved in the incorporation of malonyl-CoA into fatty acids were largely worked out in bacterial systems (Chapter 3). Lynen and co-workers characterized the enzymatic properties of fatty acid synthase in yeast, while the laboratories of Wakil, Vagelos, and Porter did the same for the vertebrate enzyme.

At this stage, one additional feature of the enzymatic basis of fatty acid synthesis remained to be elucidated. Acetyl-CoA, the substrate for acetyl-CoA carboxylase, is formed in the mitochondrion. The enzyme itself, however, is present in the cytosol and not in the mitochondrion. Furthermore, the mitochondrial inner membrane is impermeable to acetyl-CoA. Lowenstein's laboratory was instrumental in the analysis of this dilemma. Citrate synthase in the mitochondrion catalyzes the conversion of oxaloacetate and acetyl-CoA to citrate. The citrate diffuses into the cytosol, where it is cleaved to acetyl-CoA and oxaloacetate in a reaction catalyzed by ATP:citrate lyase (originally called *citrate cleavage enzyme*). The four carbons of oxaloacetate can be recycled to the mitochondrion, either as malate or as pyruvate and CO_2.

WHICH ENZYMES REGULATE FATTY ACID SYNTHESIS?

Having identified extracellular regulators and established the enzymatic mechanisms for the terminal segment of fatty acid synthesis, our next step in the analysis is to identify those enzymes with the potential to regulate flux through the pathway for fatty acid biosynthesis. There are more than 25 enzymes involved in the conversion of glucose to long-chain fatty acids. These include 14 enzymes or multienzyme complexes in the pathway per se, 4 to 6 which catalyze potential futile cycles, 3 which produce essential cofactors, 2 or more transporter proteins, and several enzymes which synthesize or degrade regulatory intermediates. Regulation of a metabolic pathway through alteration of the catalytic activity of an enzyme in that pathway must be exerted at a reaction in which the concentrations of the reactants are far from thermodynamic equilibrium. If the concentrations of the reactants were close to equilibrium, the reverse reaction would proceed at almost the same rate as the forward reaction, despite net flux through the overall pathway in the forward direction. A change in the catalytic activity of such an enzyme would have little effect on unidirectional flux through the pathway because both forward and reverse reactions would be activated to the same extent. An enzyme is classified as *regulatory* if its mass action ratio (product of the concentrations of the products divided by the product of the concentrations of the substrates) is displaced 50-fold or more from thermodynamic equilibrium. (The reader is referred to a review by Rolleston [1972] for a more detailed analysis.) This classification applies to activity changes caused either by changes in catalytic efficiency of a constant number of enzyme molecules or by changes in the concentration of enzyme molecules of equivalent catalytic efficiency. Several essentially irreversible enzymes involved in the conversion of glucose to fatty acids meet this criterion.

In the third stage of our analysis, we shall examine the properties of regulatory enzymes. Physical and kinetic properties of purified enzymes will be examined to determine the potential for regulation of catalytic efficiency in intact cells. Enzyme concentrations will be analyzed to assess regulation of the number of enzyme molecules. For the terminal section of the pathway for fatty acid synthesis, conversion of citrate to long-chain fatty acids, all three enzymes, ATP-citrate lyase, acetyl-CoA carboxylase, and fatty acid synthase, are candidates for regulation. ATP-citrate lyase and fatty acid synthase do not exhibit physical or kinetic properties consistent with an ability to regulate catalytic efficiency in intact cells, whereas acetyl-CoA carboxylase does. Analysis of the physical and kinetic properties of this carboxylase is the subject of a subsequent section of this chapter.

Another way to regulate flux through a metabolic pathway is to control delivery of substrates to the pathway. Therefore, before discussing acetyl-CoA carboxylase, we shall briefly review regulation of the delivery of citrate, NADPH, ATP, and CoA to the terminal segment of the pathway for fatty acid synthesis.

REGULATION OF SUBSTRATE SUPPLY
Production of Pyruvate from Glucose

Most of the carbon destined for fatty acid synthesis flows through the pyruvate pool. In liver, glucose can be synthesized from pyruvate via gluconeogenesis, and pyruvate can be produced from glucose via glycolysis. Most of the enzymes in these pathways are freely reversible and "near equilibrium," functioning equally well in either direction. Complete reversibility at every step would make regulation impossible, and energy would be wasted in futile cycles. Different and essentially irreversible reactions catalyze the interconversion of glucose and glucose-6-phosphate, fructose-6-phosphate and fructose-1,6-bisphosphate, and pyruvate and phosphoenolpyruvate. Regulation occurs at each of these steps.

When animals are fed high-carbohydrate diets and/or when the ratio of insulin to glucagon in the blood is high, the activities of glucokinase, phosphofructokinase I, and pyruvate kinase are increased, and the activities of glucose-6-phosphatase, fructose-1,6-bisphosphatase, and phosphoenolpyruvate carboxykinase are decreased. These changes in activities are due to a combination of changes in enzyme concentration and changes in catalytic efficiency. Catalytic efficiency is regulated by a combination of covalent modification (phosphorylation) and allosteric mechanisms. (El-Maghrabi et al. [1982]; Spence [1983]; Hers and van Schaftingen [1982]; and Denton and Halestrap [1979] contain more detailed discussions of these mechanisms). These regulatory adjustments cause net pyruvate production from glucose in the livers of well-fed animals and net glucose production from pyruvate in livers of starved animals.

Production of Citrate from Pyruvate

Metabolism of pyruvate is the key to the disposition of carbon from carbohydrate or protein precursors (Figure 5.2). Inhibition of pyruvate kinase during active gluconeogenesis prevents phosphoenolpyruvate from being reconverted to pyruvate in the cytosol, thus inhibiting futile cycling. Other mechanisms direct incoming pyruvate

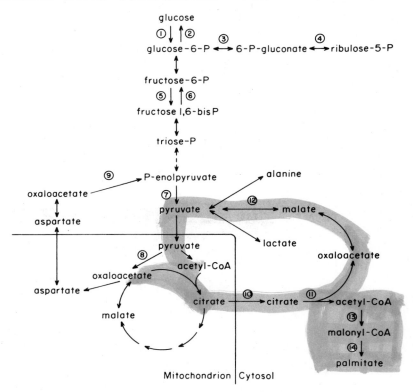

Figure 5.2. A schematic representation of the pathways involved in the conversion of glucose to fatty acids in liver. Key reactions are identified by numbers.
1, glucokinase; 2, glucose-6-phosphatase; 3, glucose-6-phosphate dehydrogenase;
4, 6-phosphogluconate dehydrogenase; 5, phosphofructokinase I; 6, fructose
1,6-bisphosphatase; 7, pyruvate kinase; 8, pyruvate carboxylase;
9, phosphoenolpyruvate carboxykinase; 10, mitochondrial tricarboxylate anion carrier;
11, ATP:citrate lyase; 12, malic enzyme; 13, acetyl-CoA carboxylase; 14, fatty acid synthase.

toward gluconeogenesis, oxidation, or lipogenesis depending on the fuel and hormonal milieu. Conditions favoring glucose synthesis (such as starvation) cause the inhibition of pyruvate dehydrogenase in the mitochondrion. As a consequence, pyruvate flux is directed toward oxaloacetate and thence to glucose via the pyruvate carboxylase reaction. The mechanisms by which the pyruvate dehydrogenase multienzyme complex are regulated are complex. Both covalent modification and allosteric regulation by substrates, products, and other mitochondrial metabolites are involved (Denton and Halestrap 1979). The net result of stimulation of this enzyme activity, however, is rapid formation of citrate in the mitochondria when environmental conditions favor lipogenesis. Citrate is transported from the mitochondrion to the cytosol, where fatty acid synthesis occurs. The mitochondrial inner membrane is impermeable to citrate; hence, citrate leaves the mitochondrion on the tricarboxylate anion carrier in exchange for another organic anion, probably malate. The rate of efflux of citrate is a function of the citrate concentration gradient, the malate concentration gradient, and the activity of the tricarboxylate anion carrier. Fatty acyl-CoA, a product of the fatty acid synthesis pathway, may inhibit this carrier. The amount of citrate in the cytosol is an important determinant of the rate of fatty acid synthesis, both as substrate for the pathway and as a positive regulator of acetyl-CoA carboxylase.

Production of NADPH

The process of fatty acid synthesis requires the utilization of two molecules of NADPH for each molecule of acetate incorporated into long-chain fatty acids. In liver, four cytosolic enzymes have the potential to supply NADPH. Glucose-6-phosphate dehydrogenase and 6-phosphogluconate dehydrogenase probably furnish about half of the NADPH used in fatty acid synthesis, with the other half coming from malic enzyme [L-malate:NADP$^+$ oxidoreductase (decarboxylating)]. The fourth enzyme, isocitrate dehydrogenase [$threo$-D$_s$-isocitrate:NADP$^+$ oxidoreductase (decarboxylating)], appears to function primarily in the utilization of NADPH rather than its production. The activities of the two dehydrogenases of the pentose phosphate pathway and of malic enzyme correlate positively with the rate of fatty acid synthesis under a wide variety of conditions. Thus, the rate of production of NADPH could regulate fatty acid synthesis. However, in liver, each of these enzymes is near equilibrium with respect to its substrates and products under most conditions. As a consequence, changes in the activities of these enzymes do not alter the rate of production of NADPH. Dietary and hormonal conditions which stimulate hepatic fatty acid synthesis cause a decrease in the intracellular concentration of NADPH, consistent with the foregoing interpretation. Under most conditions, therefore, the rate of production of NADPH is a function of the rate of utilization of NADPH, rather than the converse.

REGULATION OF THE CATALYTIC EFFICIENCY OF ACETYL-CoA CARBOXYLASE A Key Regulatory Reaction

Acetyl-CoA carboxylase catalyzes the carboxylation of acetyl-CoA to malonyl-CoA. Hydrolysis of ATP provides the energy to drive this essentially irreversible reaction. Acetyl-CoA carboxylase is considered to be the key regulatory enzyme in the conversion of citrate to long-chain fatty acids, for several reasons. First, the concentrations of its substrates and products are far from thermodynamic equilibrium, indicating that this reaction is catalyzed by a regulatory enzyme.

Second, the maximum velocity of the enzyme, as measured in cell extracts under optimal conditions, is usually the slowest of all enzymes in the pathway. Third, the concentration of the product of the enzyme, malonyl-CoA, increases when flux through the pathway increases. If the catalytic efficiency of the next enzyme in the pathway, fatty acid synthase, were regulated, malonyl-CoA concentration would decrease with increased flux. Fourth, acetyl-CoA carboxylase catalyzes the first committed step in the pathway, the most appropriate step at which to regulate a metabolic pathway. Finally, despite extensive analysis, there is no compelling evidence for physiologically relevant regulation of the catalytic efficiency of either ATP:citrate lyase or fatty acid synthase, the other two enzymes in this section of the pathway. Therefore, having concluded that acetyl-CoA carboxylase plays an important role in regulating the rate of fatty acid synthesis, we shall now examine the physical, kinetic, and regulatory properties of this enzyme.

Structure and Reaction Mechanism

Acetyl-CoA carboxylase has two distinct catalytic sites, each of which carries out one of the following partial reactions:

1. E-Biotin + HCO^-_3 + ATP $\longleftrightarrow$ E-biotin-CO_2 + ADP + P_i
2. E-Biotin-CO_2 + acetyl-CoA $\longrightarrow$ E-biotin + malonyl-CoA

   Net: ATP + HCO_3^- + acetyl-CoA $\longrightarrow$ malonyl-CoA + ADP + P_i

In *E. coli* these reactions require the participation of three different proteins, as described in Chapter 3. In both yeast and animals, the analogous reactions are catalyzed by multifunctional polypeptides. The subunit molecular weight of animal acetyl-CoA carboxylase is 230,000 to 260,000. The smallest form of the native enzyme is a dimer with a molecular weight of about 480,000 (protomer) which lacks enzyme activity. Citrate and certain related substances activate the enzyme. Concomitant with the increase in enzyme activity, the protomers assemble into filamentous polymers with molecular weights of up to 1×10^7 (polymer).

Regulation by Citrate

Conversion from the catalytically inactive protomer to the catalytically active polymer is stimulated by citrate. Several different tricarboxylate anions are as effective as citrate, but citrate is generally considered to be the physiologically relevant anion. Activation of acetyl-CoA carboxylase is unusual in that the purified enzyme has no activity in the absence of an activator such as citrate. Furthermore, citrate increases the rate of reaction of bound substrate, V_{max}, rather than affecting the apparent affinity, K_m, of the enzyme for substrates. Both reactions 1

mitochondrial citrate

↓

Cytosol citrate

↗   ↘

↑ [acetyl CoA]   ↓

activates

Acetyl CoA carboxylase

and 2 are stimulated by citrate. Thus, citrate appears to induce a conformational change which causes the biotin prosthetic group to be reoriented with respect to the biotin carboxylase and carboxyl transferase active sites, facilitating efficient catalysis (Lane, Moss, and Polakis 1974).

The ability of citrate to activate the pace-setting enzyme in the terminal segment of the pathway for fatty acid synthesis suggests a teleologically satisfying mechanism for regulation of flux from glucose to fatty acids in intact cells. Under conditions favoring fatty acid synthesis (for example, animals fed a high-carbohydrate diet), citrate production in the mitochondria should be high. Efflux of this citrate into the cytosol should also be high, leading to an increased concentration of citrate in that compartment. As noted earlier, this would supply substrate for fatty acid synthesis via ATP:citrate lyase and stimulate the activity of acetyl-CoA carboxylase. The results of experiments which were designed to test this hypothesis are conflicting and sometimes difficult to interpret. Some reports indicate that citrate content of liver from intact animals, perfused liver, and isolated liver cells did not correlate with the rate of fatty acid synthesis and, furthermore, that the intracellular concentration of citrate was lower than that required to activate acetyl-CoA carboxylase. However, many of these studies failed to distinguish between mitochondrial and cytosolic citrate or used indirect methods for calculating the subcellular distribution of citrate. In addition, they used rapid-stop techniques such as freeze-clamping to stop tissue metabolism. How fast is fast enough, especially with intact organs? In isolated avian hepatocytes, the rate of fatty acid synthesis correlates positively with the concentration of cytosolic citrate. In some instances, therefore, citrate may be a physiological regulator of acetyl-CoA carboxylase. The relative contribution of citrate to regulation of this enzyme may very greatly in different tissues and under different conditions (Lane, Watkins, and Meredith 1979).

There remains the question of the biological significance of the polymerization of acetyl-CoA carboxylase caused by citrate. Two different approaches have been used to investigate this question. The first approach makes use of the fact that avidin binds tightly to biotin or biotin-containing proteins. When added to purified acetyl-CoA carboxylase in the protomeric form, avidin inhibits subsequent activation of catalysis by citrate. However, treatment of the purified enzyme with avidin has little effect on enzyme activity if the enzyme is first activated by citrate, putting it into the polymerized form. In crude cell extracts prepared from cells with a low rate of fatty acid synthesis, avidin blocks the stimulation of enzyme activity caused by citrate. However, if the extracts are prepared from tissue with a high rate of fatty acid synthesis, avidin has only a small inhibitory effect on

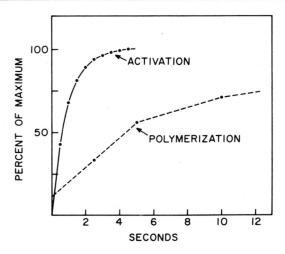

Figure 5.3. Kinetics of activation and polymerization of purified acetyl-CoA carboxylase caused by citrate. Polymerization (dashed line) was measured in a Durrum stopped-flow apparatus. Activation (solid line) was measured in a rapid-quench apparatus. These curves have been replotted from data presented in Beatty and Lane (1983a, 1983b), to which the reader should refer for more detail.

the stimulation of enzyme activity caused by citrate. This result has been interpreted to mean that the protomeric and polymeric forms of the enzyme are present in livers of animals with low and high rates of fatty acid synthesis, respectively.

When measured in minutes, the relative stimulations of both catalysis and polymerization are similar as the citrate concentration is increased. However, independent measurements of catalytic activity and rate of polymerization, using stopped-flow techniques, indicate that activation by citrate occurs much more rapidly than polymerization (Figure 5.3) (Beatty and Lane 1983a, 1983b). Thus, polymerization is not a prerequisite for increased catalytic efficiency. Furthermore, the rapid, citrate-induced conformational change also may lead to sequestration of the biotinyl moiety, restricting accessibility to added avidin. Thus, resistance to avidin does not necessarily imply that the polymeric form of the enzyme existed in the intact cell. In addition, aggregation and the acquisition of avidin resistance could occur during preparation of the cell extracts.

The second indirect approach to the question of whether intact cells with high rates of fatty acid synthesis contain polymeric acetyl-CoA carboxylase involves treating isolated hepatocytes with a low concentration of digitonin, a procedure which causes the selective loss of cytosolic enzymes. Under conditions which should cause acetyl-CoA carboxylase to be in the protomeric form (for instance, low rates of fatty acid synthesis), this enzyme is released from cells by digitonin with the same kinetics as other "typical" cytosolic enzymes. However, under conditions favoring high rates of fatty acid synthesis, acetyl-CoA carboxylase is retained by digitonin-treated hepatocytes. This result may be interpreted in two ways. First, cells with high rates of fatty acid synthesis may have the polymerized form of acetyl-CoA carboxylase, a form which is too large to diffuse out of cells made permeable with digitonin. Alternatively, the active form of acetyl-CoA carboxylase may be bound to cellular structures and restricted from leaving the cell. Filamentous structures with dimensions similar to those of polymeric acetyl-CoA carboxylase have been observed in situ in adipocytes. However, there is no convincing evidence that the structures observed with the electron microscope are polymeric forms of acetyl-CoA carboxylase. In summary, the evidence for polymeric forms of acetyl-CoA carboxylase in intact cells is not compelling.

Regulation by Long-Chain Fatty Acyl-CoA

Diets containing a high concentration of fat inhibit fatty acid synthesis, which suggests that long-chain fatty acids and/or long-chain acyl-CoA derivatives may regulate a key step in the pathway. Their ability to regulate acetyl-CoA carboxylase activity was tested with the earliest preparations of partially purified acetyl-CoA carboxylase. Unesterified fatty acids had no effect on enzyme activity when used at physiological concentrations. Fatty acyl-CoA derivatives, however, were potent inhibitors of the enzyme, effective at what appeared to be very low concentrations. Unfortunately, fatty acyl-CoAs inhibited the activity of almost every enzyme which was tested. In most instances, the inhibition was irreversible and occurred at concentrations higher than the critical micellar concentration of the acyl-CoA. Thus, it was concluded that all the inhibitory effects of long-chain fatty acyl-CoA derivatives were caused by their detergent properties and were therefore of no physiological significance. However, more detailed studies of the actions of fatty acyl-CoAs on acetyl-CoA carboxylase have led to quite a different conclusion. Inhibition of the activity of acetyl-CoA carboxylase by fatty acyl-CoA is competitive with citrate, is reversible by citrate or albumin, and has an apparent K_i of about $0.2 \mu M$, well below the critical micellar concentration (Goodridge 1972). When several different enzyme activities are challenged with submicromolar concentrations of long-chain acyl-CoA, only acetyl-CoA carboxylase shows significant inhibition. Furthermore, binding of 1 mol of palmitoyl-CoA to 1 mol of acetyl-CoA carboxylase completely inhibits enzyme activity. Thus, long-chain fatty acyl-CoAs are specific inhibitors of acetyl-CoA carboxylase. In additional studies, glucokinase and citrate synthase have been added to the list of enzymes which are inhibited by specific interactions with long-chain acyl-CoAs.

long chain F. A is an inhibitor to the acetyl CoA and glucokinase and citrate synthase

Table 5.1. The Correlation between Rate of Fatty Acid Synthesis and Concentration of Long-Chain Fatty Acyl-CoA in Livers of Rats

Treatment	Fatty acid synthesis*	Long-chain fatty acyl-CoA[†]
Control	1.0	83
Starved 3 d	0.2	131
Starved 3 d, high-fat diet 3 d	0.2	144
Starved 3 d high-carbohydrate diet 3 d	2.4	72
Diabetic	0.1	105
Diabetic treated with insulin	0.8	73

Source: Greenbaum, Gumaa, and McLean (1971).
* Fatty acid synthesis was estimated by measuring the incorporation of [U-^{14}C] glucose into total fatty acids in liver slices (micromoles per gram of liver per hour).
[†] Fatty acyl-CoA was measured in the insoluble fraction of perchloric acid extracts of freeze-clamped liver (nanomoles per gram of liver).

[handwritten margin note: most of F.A are bound to protein in cytoplasma so the role of long chain F.A is unknown]

The question of physiological relevance remains. In addition to the specific inhibition of enzyme activity by long-chain acyl-CoA, there is an inverse relationship between the rate of fatty acid synthesis and the total concentration of long-chain fatty acyl-CoA in liver (Greenbaum, Gumaa, and McLean 1971) (Table 5.1). However, virtually all the long-chain fatty acyl-CoA in a cell is bound to protein. The concentration and affinity constants for cellular proteins which bind fatty acyl-CoA are largely unknown. With the exception of the enzymes which metabolize fatty acyl-CoA and Z-protein, an abundant cytosolic protein which binds fatty acids and fatty acyl-CoAs, the identity of binding proteins is also unknown. Changes in the concentrations of specific binding proteins could profoundly alter the free concentrations of fatty acyl-CoAs. Further complicating the analysis is our lack of understanding of which species of long-chain fatty acyl-CoA are present under different conditions. The ability of different species of fatty acyl-CoA to inhibit acetyl-CoA carboxylase is quite variable. In summary, indirect evidence suggests that long-chain fatty acyl-CoA may participate in the regulatory process in intact cells; however, formal proof is lacking.

Regulation by Covalent Modification

The first indication that acetyl-CoA carboxylase might be regulated by phosphorylation was the finding of 2 mol of covalently bound phosphate per mole of subunit in highly purified enzyme from rat liver. Newer procedures for purifying acetyl-CoA carboxylase are rapid and minimize protease and phosphatase activity. Enzyme

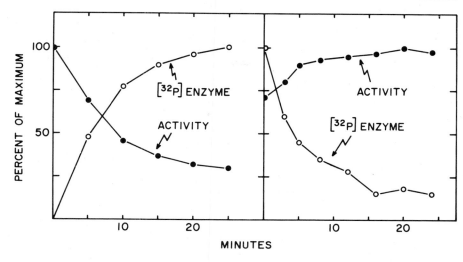

Figure 5.4. Kinetics of inactivation and phosphorylation of acetyl-CoA carboxylase activity (left panel) and of activation and dephosphorylation of the enzyme (right panel) in partially purified enzyme preparations. In the left panel the reactions contained [γ-^{32}P]ATP and endogenous protein kinase. In the right panel, ATP was removed after the phosphorylation was completed and fresh liver extract added as a source of protein phosphatase. At different time points, the reaction was terminated and acetyl-CoA carboxylase precipitated by a specific antiserum. The immunoprecipitates were assayed for [^{32}P] radioactivity. These curves have been replotted from the data presented in Carlson and Kim (1974), to which the reader should refer for more detail.

prepared by such procedures has at least 6 mol of alkali-labile phosphate per mole of enzyme subunit. Phosphorylation of partially purified acetyl-CoA carboxylase is accompanied by loss of enzyme activity, dephosphorylation by an increase in activity (Carlson and Kim 1974) (Figure 5.4). Purified acetyl-CoA carboxylase can be phosphorylated by purified cAMP-dependent protein kinase; loss of enzyme activity is concomitant with phosphorylation. Addition of purified protein phosphatase reverses both the phosphorylation and the inhibition of enzyme activity (Hardie and Guy 1980) (Figure 5.5). Phosphorylation by the cAMP-dependent protein kinase does not change the apparent K_m values for substrates, but does decrease V_{max} by about 50%. The decrease in V_{max} appears to be due to decreased affinity of the more phosphorylated enzyme for its allosteric activator, citrate. The more phosphorylated enzyme also is more sensitive to inhibition by long-chain fatty acyl-CoA. These results establish the potential for regulation of acetyl-CoA carboxylase by a phosphorylation/dephosphorylation mechanism. The allosteric regulators, citrate and long-chain fatty acyl-CoA, may act synergistically with covalent modification in regulation of the activity of this enzyme.

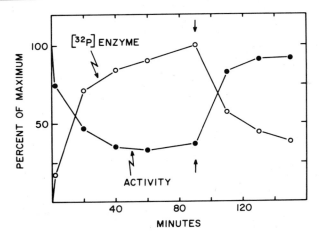

Figure 5.5. Effect of phosphorylation by cAMP-dependent protein kinase and dephosphorylation by protein phosphatase-1 on the activity of highly purified acetyl-CoA carboxylase. The initial reaction was carried out in the presence of the cAMP-dependent protein kinase. At the point indicated by the arrows, protein kinase inhibitor, $MnCl_2$, and protein phosphatase-1 were added to the incubations. These curves have been replotted from the data presented in Hardie and Guy (1980), to which the reader should refer for more detail.

Phosphorylation of acetyl-CoA carboxylase also has been demonstrated in intact adipocytes and hepatocytes. In adipocytes, phosphorylation is stimulated by epinephrine. Increased phosphorylation is correlated with a decreased acetyl-CoA carboxylase activity in adipose tissue extracts. Similarly, glucagon stimulates phosphorylation of the hepatocyte enzyme and decreases enzyme activity. Six or more different sites in acetyl-CoA carboxylase can be phosphorylated. The site or sites phosphorylated when intact cells are incubated with epinephrine are on the same tryptic peptide as the site or sites phosphorylated when the purified enzyme is incubated with cAMP-dependent protein kinase. Both epinephrine and glucagon cause an elevation of the intracellular concentration of cAMP, and the concentrations of both hormones are elevated during starvation or diabetes, conditions associated with marked inhibition of fatty acid synthesis. Collectively, these results suggest that phosphorylation of acetyl-CoA carboxylase plays a physiological role in the inhibition of fatty acid synthesis caused by agents which elevate the intracellular concentration of cAMP.

Insulin, a hormone which stimulates fatty acid synthesis and the activity of acetyl-CoA carboxylase, also stimulates phosphorylation of the enzyme. The site (or sites) for increased phosphorylation caused by insulin in intact cells have been mapped to the same tryptic peptide which is phosphorylated when purified acetyl-CoA carboxylase is incubated with a cAMP-independent protein kinase. Thus, phosphorylation on one peptide may be catalyzed by a cAMP-independent protein and may stimulate activity, while phosphorylation on another peptide is catalyzed by a cAMP-dependent protein kinase and inhibits activity. Only two kinds of protein kinases have been discussed; there may be more, each of which phosphorylates specific sites on the enzyme. Phosphorylation at other sites may serve regulatory functions yet undescribed.

Allosteric mechanisms and phosphorylation-dephosphorylation mechanisms probably play complementary roles in regulating the catalytic activity of acetyl-CoA carboxylase. Glucagon, for example, stimulates phosphorylation of the hepatic enzyme at an inhibitory site. Glucagon also stimulates the hormone-sensitive lipase in adipose tissue causing an increase in the rate of release of unesterified fatty acids to the plasma and, consequently, an increase in unesterified fatty acid concentration in plasma and other tissues including liver. In liver, the concentration of long-chain fatty-acyl CoA varies in equal proportion with that of unesterified long-chain fatty acids. The increased phosphorylation of acetyl-CoA carboxylase caused by glucagon increases sensitivity of the enzyme to inhibition by fatty acyl-CoA. As described previously, glucagon also inhibits the flux of carbon from glucose to pyruvate. Decreased production of extramitochondrial citrate is a consequence of the decreased production of pyruvate via glycolysis. Increased phosphorylation of acetyl-CoA carboxylase at its inhibitory site decreases sensitivity of this enzyme to its activator, citrate. The combined result of the three actions (direct inhibition of activity due to phosphorylation, elevation of the long-chain fatty acyl-CoA level, and diminution of the citrate level) results in an amplification of the inhibitory response which neither allosteric mechanisms nor covalent modification alone could have achieved.

Regulation by Proteolytic Removal of Specific Peptides

Acetyl-CoA carboxylase purified from rat liver by traditional procedures has a subunit molecular weight of about 230,000. Analysis of enzyme purified by new procedures, which are rapid and minimize protease activity, indicates that the acetyl-CoA carboxylase has a subunit molecular weight of about 260,000. The loss of a peptide(s) of about 30,000 daltons correlates with a fivefold increase in specific activity. A similar result is obtained when crude extracts of rat liver are incubated with trypsin: Acetyl-CoA carboxylase activity is stimulated. Interestingly, trypsin treatment also causes a decrease in the ability of long-chain fatty acyl-CoAs to inhibit enzyme activity. The physiological relevance of the stimulation of acetyl-CoA carboxylase activity caused by the proteolytic removal of a small peptide is unknown.

Other Regulatory Mechanisms

A variety of drugs and intracellular intermediates have been reported to regulate the activity of acetyl-CoA carboxylase. The relevance of these effects to the physiological regulation of enzyme activity is unknown, and they will not be discussed in this chapter.

Thus far, we have discussed regulation by allosteric factors and covalent modification, mechanisms which alter the catalytic efficiency of enzyme molecules. This type of regulation accounts for the prompt changes in enzyme activity which occur under different hormonal and nutritional conditions. The concentration of acetyl-CoA carboxylase molecules also is regulated by hormones and nutritional status. However, since the concentrations of the lipogenic enzymes are regulated coordinately, this kind of regulation will be analyzed after a discussion of the physical and enzymatic properties of fatty acid synthase.

FATTY ACID SYNTHASE

The synthesis of long-chain fatty acids from acetyl-CoA and malonyl-CoA involves a sequence of six reactions for each two-carbon addition; the sequence is repeated several times to produce a long-chain fatty acid. Although the structural organization of the process varies greatly, the enzymatic mechanisms are very similar in all organisms in which this process has been analyzed. In E. coli and in the plastids of certain green plants, the enzymes catalyzing the individual reactions are discrete monofunctional proteins, which can be separated and analyzed individually (Chapter 3). In yeast, synthesis of fatty acids from acetyl-CoA and malonyl-CoA is catalyzed by a fatty acid synthase complex which consists of two multifunctional polypeptides, each coded by a different gene. Mammalian and avian fatty acid synthases are also multifunctional polypeptides, but all enzyme activities are localized on a single polypeptide chain which is encoded by a single gene. This section will concentrate on the structural and functional organization of this fascinating multifunctional polypeptide.

Animal Fatty Acid Synthase: The Component Reactions

The enzymatic reactions for the synthesis of palmitate catalyzed by animal fatty acid synthase are listed in Figure 5.6. In addition to the 31 listed reactions, each cycle requires the transfer of malonyl-CoA to the 4'-phosphopantetheine residue of the enzyme, making a total of 37 enzymatic reactions for the synthesis of each molecule of palmitate. In the nonaggregated fatty acid synthase of E. coli, the growing acyl chain is attached to a small peptide (acyl carrier protein) via a 4'-phosphopantetheine residue. In addition to its role in fatty acid synthesis, the bacterial acyl carrier protein participates in several other acyl transfer reactions. The eucaryotic equivalent of the bacterial acyl carrier protein is part of the linear structure of the multifunctional fatty acid synthase polypeptide and functions only in the

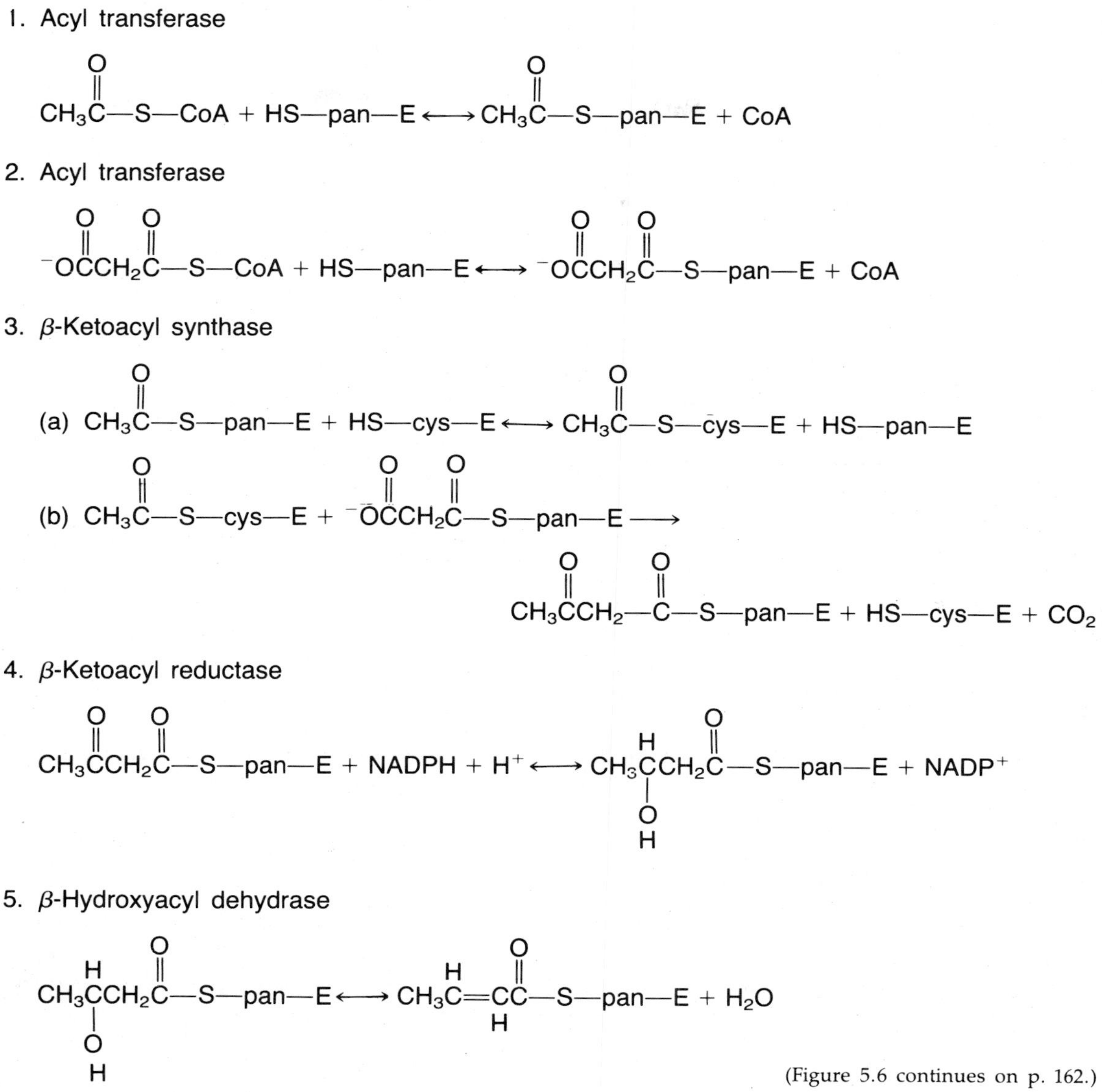

1. Acyl transferase

$$CH_3\overset{\displaystyle O}{\overset{\|}{C}}-S-CoA + HS-pan-E \longleftrightarrow CH_3\overset{\displaystyle O}{\overset{\|}{C}}-S-pan-E + CoA$$

2. Acyl transferase

$$^-O\overset{\displaystyle O}{\overset{\|}{C}}CH_2\overset{\displaystyle O}{\overset{\|}{C}}-S-CoA + HS-pan-E \longleftrightarrow {}^-O\overset{\displaystyle O}{\overset{\|}{C}}CH_2\overset{\displaystyle O}{\overset{\|}{C}}-S-pan-E + CoA$$

3. β-Ketoacyl synthase

(a) $CH_3\overset{\displaystyle O}{\overset{\|}{C}}-S-pan-E + HS-cys-E \longleftrightarrow CH_3\overset{\displaystyle O}{\overset{\|}{C}}-S-cys-E + HS-pan-E$

(b) $CH_3\overset{\displaystyle O}{\overset{\|}{C}}-S-cys-E + {}^-O\overset{\displaystyle O}{\overset{\|}{C}}CH_2\overset{\displaystyle O}{\overset{\|}{C}}-S-pan-E \longrightarrow$

$$CH_3\overset{\displaystyle O}{\overset{\|}{C}}CH_2-\overset{\displaystyle O}{\overset{\|}{C}}-S-pan-E + HS-cys-E + CO_2$$

4. β-Ketoacyl reductase

$$CH_3\overset{\displaystyle O}{\overset{\|}{C}}CH_2\overset{\displaystyle O}{\overset{\|}{C}}-S-pan-E + NADPH + H^+ \longleftrightarrow CH_3\underset{\underset{\displaystyle H}{\overset{\displaystyle |}{\underset{\displaystyle O}{|}}}}{\overset{\displaystyle H}{\overset{\displaystyle |}{C}}}CH_2\overset{\displaystyle O}{\overset{\|}{C}}-S-pan-E + NADP^+$$

5. β-Hydroxyacyl dehydrase

$$CH_3\underset{\underset{\displaystyle H}{\overset{\displaystyle |}{\underset{\displaystyle O}{|}}}}{\overset{\displaystyle H}{\overset{\displaystyle |}{C}}}CH_2\overset{\displaystyle O}{\overset{\|}{C}}-S-pan-E \longleftrightarrow CH_3\overset{\displaystyle H}{\overset{\displaystyle |}{C}}=\underset{\underset{\displaystyle H}{\displaystyle |}}{C}\overset{\displaystyle O}{\overset{\|}{C}}-S-pan-E + H_2O$$

(Figure 5.6 continues on p. 162.)

Figure 5.6. The component reactions of animal fatty acid synthase. The abbreviations HS-cys and HS-pan indicate cysteinyl residues and 4'-phosphopantetheine groups, respectively.

② mammalian

the eucaryotic equivalent of the bacterial acyl carrier protein is part of the linear structure of multifunctional F. A. synthetase and functions only in the reaction of the F. A synthetase.

6. Enoyl reductase

$$CH_3\overset{H}{\underset{H}{C}}{=}CC\overset{O}{\overset{\|}{C}}{-}S{-}pan{-}E + NADPH + H^+ \longrightarrow CH_3CH_2CH_2\overset{O}{\overset{\|}{C}}{-}S{-}pan{-}E + NADP^+$$

7. β-Ketoacyl synthase

(a) $CH_3CH_2CH_2\overset{O}{\overset{\|}{C}}{-}S{-}pan{-}E + HS{-}cys{-}E \longleftrightarrow CH_3CH_2CH_2\overset{O}{\overset{\|}{C}}{-}S{-}cys{-}E + HS{-}pan{-}E$

(b) $CH_3CH_2CH_2\overset{O}{\overset{\|}{C}}{-}S{-}cys{-}E + {}^-O\overset{O}{\overset{\|}{C}}CH_2\overset{O}{\overset{\|}{C}}{-}S{-}pan{-}E \longrightarrow CH_3(CH_2)_2\overset{O}{\overset{\|}{C}}CH_2\overset{O}{\overset{\|}{C}}{-}S{-}pan{-}E$
$$+ HS{-}cys{-}E + CO_2$$

8–10. Repeat reactions 4, 5, and 6, forming Hexanoyl—pan—E

11–30. Repeat reaction 3, 4, 5, and 6 five times, with the molecule growing by 2 carbons with each cycle to produce palmitoyl—pan—E

31. Thioesterase
Palmitoyl—pan—E + H_2O $\longrightarrow$ Palmitate + HS—pan—E

Figure 5.6. (Continued)

reactions of the fatty acid synthase complex. In other respects, the reactions of the animal fatty acid synthases are like those of *E. coli* (Chapter 3).

Animal Fatty Acid Synthase: The Subunits are Identical

Despite the successful purification and characterization of the component enzymes of *E. coli* fatty acid synthase, the animal enzyme proved highly resistant to dissociation. Since the bacterial complex comprised several independent gene products, the animal enzyme was generally assumed to be a multienzyme complex highly resistant to dissociation rather than a multifunctional polypeptide. Several groups reported the separation of multiple peptides and activities from purified animal fatty acid synthase. However, if protease activity was inhibited during purification and analysis, the totally denatured complex had only a single component, with a molecular weight of 250,000 to 270,000. The native enzyme has a molecular weight of about 500,000, indicating a dimeric structure.

The two subunits of avian and mammalian fatty acid synthases are almost certainly identical. Within the limits of discrimination of polyacrylamide gel electrophoresis in the presence of sodium dodecyl

prove the subunits are identical

sulfate, the two subunits are identical in size. Other evidence indicates that they have the same shape and charge. Free N-terminal amino acid residues cannot be detected in the purified enzyme. Thus, both subunit peptides must have blocked N-terminal residues. Each fatty acid synthase dimer contains 1.4 to 1.8 mol of 4'-phosphopantetheine per mole of dimer. Each dimer contains two sites each for the enzymatic activities of thioesterase, ketoreductase, enoyl reductase, and β-ketoacyl synthase. These stoichiometries are based on the measurement of the extent of reaction of inhibitors of the specific subreactions of fatty acid synthase. Thus, each of the two subunits of a fatty acid synthase molecule probably contains all of the component reactions of the enzyme. These results suggest, but do not prove, that the subunits are identical.

Further evidence for identity of the subunits comes from studies of proteolytic fragments of pure synthase. Peptides containing three of the enzyme's active sites have been purified from the homogeneous enzyme after it was cleaved by proteases. Thioesterase can be removed from the native enzyme by mild treatment with several different proteases (Figure 5.7). The isolated thioesterase, with a molecular weight of approximately 35,000, has been purified from several different fatty acid synthases with retention of enzymatic activity. The thioesterase purified from chicken liver fatty acid synthase has a unique N-terminal amino acid sequence of at least eight amino acids. A chymotryptic peptide containing the active serine of the thioesterase of fatty acid synthase purified from goose uropygial gland contains a unique sequence of eight amino acids. If these subunits were not identical, one would expect heterogeneity of these sequences. However, none was found. Similarly, peptides containing the 4'-phosphopantetheine prosthetic group of the acyl carrier peptide region and the NADPH binding site of the enoyl reductase region have unique sequences of 64 and 11 amino acids, respectively. Despite their purification from a mixture of the two subunits, each of the foregoing activities or active sites contained unique sequences. The probability that each of four such sequences would be identical in peptides from two different genes is very small. The two subunits must therefore have identical sequences throughout their entire lengths.

Animal Fatty Acid Synthase: Structural Organization

Multifunctional proteins such as immunoglobulins, the CAD protein (catalyzes the first three steps of UMP biosynthesis in animal cells), and DNA polymerase I are organized into globular domains. The component catalytic activities and regulatory sites for these proteins are located on different domains. These domains are connected to one another by polypeptide bridges which are susceptible to proteolytic

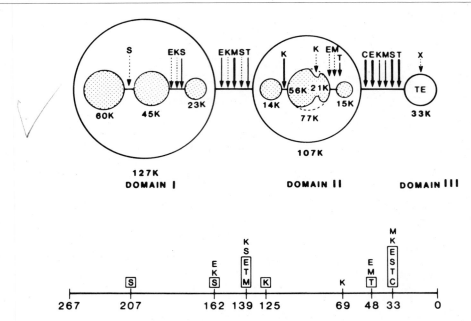

Figure 5.7. Proteolytic peptide map of chicken fatty acid synthase. TE = thioesterase. The other letters indicate the different proteases which were used to construct this map (T, trypsin; E, elastase; M, Myxobacter protease; S, subtilisin A or B; C, α-chymotrypsin; K, kallikrein; X, all of these proteases). The heavy arrows indicate primary cleavage sites; the light arrows indicate secondary cleavage sites; the dotted arrows indicate tertiary cleavage sites after longer incubation times. The bottom figure shows the relative distances (in daltons $\times 10^{-3}$) of the protease cleavage sites from the thioesterase terminus of the monomer. Taken from Mattick et al. (1983) with permission of the authors.

attack. Wakil and co-workers analyzed the fragmentation pattern of chicken liver fatty acid synthase using several different proteases (Mattick et al. 1983) and identified three principal domains (Figure 5.7). The locations of the various functional centers of the enzyme were determined by analyzing enzyme activity of fragments and by localizing labeled, site-specific reagents to specific fragments (Tsukamoto et al. 1983) (Figure 5.8). A similar map has been developed for the fatty acid synthase purified from rabbit mammary gland (McCarthy, Goldring, and Hardie 1983). The following description summarizes the analysis of chicken liver fatty acid synthase (shown diagrammatically in Figures 5.7 and 5.8).

Thioesterase (TE), which catalyzes the last reaction in the synthesis of fatty acids, is located in the COOH-terminal part of the enzyme in domain III. The peptide bridge connecting the thioesterase component to the rest of the enzyme is particularly sensitive to proteolytic attack. Brief treatment of the intact fatty acid synthase enzyme with

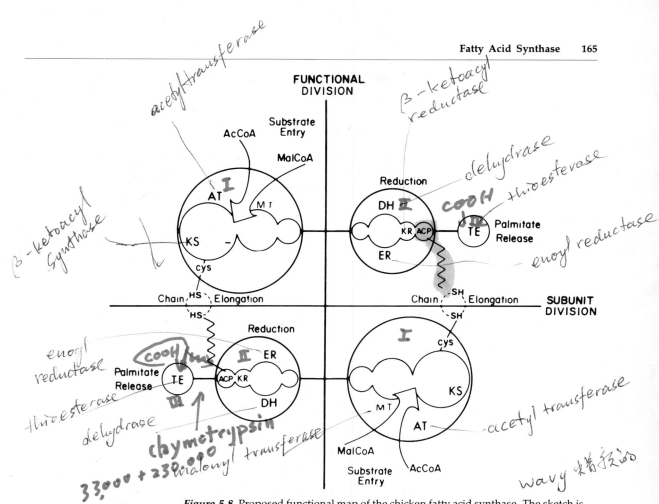

Figure 5.8. Proposed functional map of the chicken fatty acid synthase. The sketch is based on the proteolytic map (Figure 5.7) and the results discussed in the text. The abbreviations for partial activities are AT, acetyl transferase; MT, malonyl transferase; KS, β-ketoacyl synthase; KR, β-ketoacyl reductase; DH, dehydrase; ER, enoyl reductase; TE, thioesterase; and ACP, acyl carrier peptide. The wavy line represents the 4'-phosphopantetheine prosthetic group. Taken from Tsukamoto et al. (1983) with permission of the authors.

chymotrypsin produces a 33,000-dalton peptide containing the thioesterase activity and a 230,000-dalton peptide containing all the other activities. The COOH-terminal location of thioesterase was deduced from the identity of the COOH-terminal amino acids of the intact fatty acid synthase and the purified thioesterase domain. In addition, the NH_2-terminal amino acids of both the intact enzyme and the 230,000-dalton chymotryptic fragment were blocked, whereas that of the purified thioesterase fragment was not.

Localization of the region of fatty acid synthase which contained the 4'-phosphopantetheine prosthetic group was determined after labeling the intact enzyme with [^{14}C]pantetheine. After proteolytic cleavage, the smallest fragment containing radioactivity had a molecular weight of about 15,000. When the intact fatty acid synthase was treated with chymotrypsin the radioactivity was associated with the 230,000-dalton peptide. On the other hand, [^{14}C] was associated with the thioesterase activity in one of the intermediates formed during digestion with kallikrein. Therefore, the peptide fragment analogous to acyl carrier peptide was localized to the COOH-terminal part of domain II (15-K peptide, Figure 5.7), adjacent to the thioesterase.

The two reductase activities of fatty acid synthase (77-K peptide, Figure 5.7) were located in domain II, close to the peptide which contained the 4'-phosphopantetheine. β-Ketoreductase was assigned to domain II on the basis of the β-ketoreductase activity of this fragment. It was placed adjacent to the acyl carrier peptide because some peptide intermediates contained both the phosphopantetheine and ketoreductase activity. Enoyl reductase activity was lost rapidly during proteolysis, so a different approach was used. Pyridoxal phosphate is a specific inhibitor of this reductase. Thus, enoyl reductase was localized by labeling the enzyme with radioactive pyridoxal phosphate in the presence and absence of the (protective) substrate, NADPH. The location was confirmed by labeling the synthase with a ^{14}C-labeled photoaffinity analogue of NADP$^+$ in the presence or absence of pyridoxal phosphate.

The hydratase activity (77-K peptide, Figure 5.7) also was sensitive to proteolysis. Since there are no known site-specific reagents, this site has not yet been mapped by experiment. It was placed tentatively in the reductase domain because of its position between the reductase steps in the overall reaction sequence.

The acyl transferase activity of the fatty acid synthase was mapped to domain I by determining the binding of labeled acetyl-CoA and malonyl-CoA. These studies could not distinguish between one transferase for both substrates or one transferase for each of the two acyl-CoAs. However, isolation and sequencing of peptides which bind both malonyl-CoA and acetyl-CoA indicate that there is only one acyl transferase for both substrates (McCarthy et al. 1983).

The β-ketoacyl synthase active site (domain I) contains a highly reactive cysteine-SH. This cysteine is the target for attack by inhibitory sulfhydryl reagents. One such reagent, dibromopropanone, cross-links the fatty acid synthase subunits by reacting with a cysteine-SH of one subunit (domain I) and a cysteamine-SH of the adjacent subunit (domain II). The cysteamine-SH is derived from the 4'-phosphopantetheine prosthetic group attached to the acyl carrier region of domain II. Addition of malonyl-CoA prevented cross-linking but did not prevent inhibition of β-ketoacyl synthase activity by dibromopropanone. Acetyl-CoA prevented both cross-linking and inhibition of β-ketoacyl synthase by dibromopropanone. These results suggest that dibromopropanone and acetyl-CoA compete for the same thiol at the active site of the β-ketoacyl synthase. Dibromopropanone increased the susceptibility of fatty acid synthase to proteolysis, making it difficult to map the cross-linked peptides. However, when the dibromopropanone was added after production of the 127,000-dalton tryptic peptide (domain I), that peptide was cross-linked to a peptide containing cysteamine-SH. The cross-linking of these tryptic peptides was inhibited by acetyl-CoA and malonyl-CoA. Further evidence for this location for the β-ketoacyl synthase activity was obtained by labeling the enzyme with [14C]cerulenin before treatment with proteases. The antibiotic cerulenin selectively inhibits the β-ketoacyl synthase activity of fatty acid synthase. Essentially complete inhibition of fatty acid synthesis is associated with the binding of 2 mol of cerulenin per mole of synthase dimer. The only tryptic peptide containing significant radioactivity was the 127,000-dalton domain I. Taken together, the cerulenin and dibromopropanone experiments provide strong evidence that the β-ketoacyl synthase activity is located in domain I.

The cross-linking of the cysteine-SH in the active site of the β-ketoacyl synthase with the cysteamine-SH of the 4'-phosphopantetheine prosthetic group of the acyl carrier peptide provides part of the rationale for organization of the subunits in the head-to-tail fashion shown in Figure 5.8. This organization also explains why the synthesis of palmitate is blocked when the identical subunits of fatty acid synthase are dissociated. Upon dissociation, the β-ketoacyl synthase reaction is disrupted because this reaction requires the participation of the 4'-phosphopantetheine prosthetic group of the opposite subunit. The other activities which use the 4'-phosphopantetheine prosthetic group are in the same subunit as the pantetheine group which they use. Although further work is required to substantiate the model described here, this hypothesis does provide a satisfying synthesis of the known experimental findings.

Comparison of Yeast and Animal Fatty Acid Synthases

The structural organization of the yeast enzyme is intermediate between that of E. coli and animals. The six catalytic sites and the acyl carrier function are present on two different multifunctional polypeptides, named the alpha and beta peptides. Elegant genetic studies carried out in Schweizer's laboratory (Knobling and Schweizer 1975) established that the acyl carrier function, ketoacyl synthase activity, and ketoacyl reductase are on the alpha subunit and the acetyl transferase, acyl transferase, dehydrase, and enoyl reductase activities are on the beta subunit. The native molecular weight of the yeast enzyme is about 2.3×10^6. Since the α- and β-subunits have

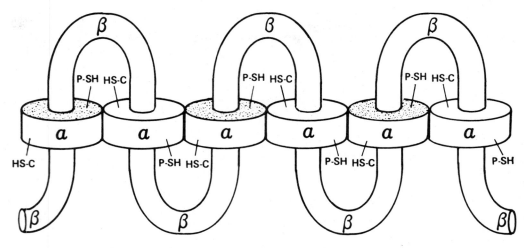

Figure 5.9. A linear drawing of the model for the subunit arrangement of yeast fatty acid synthase depicting six sites of fatty acid synthesis and the complementary arrangement of the 4'-phosphopantetheine-SH (P-SH) and the active cysteine-SH (HS-C) at the β-ketoacyl synthase centers in each of these sites. Taken from Wakil, Stoops, and Joshi (1983) with permission of the authors.

organization *difference* {

difference {

molecular weights of about 200,000 each, the structure of the native enzyme must be $\alpha_6\beta_6$. The model shown in Figure 5.9 is based on electron microscope studies of negatively stained yeast fatty acid synthase (Wakil, Stoops, and Joshi 1983).

In addition to the organizational differences between the yeast and animal enzymes, there are several functional differences. The yeast enzyme has separate acetyl and malonyl transferases, whereas the animal enzyme has a single acyl transferase for both substrates. The product of the yeast enzyme is palmitoyl-CoA, whereas that of the animal enzyme is free palmitate. Finally, the yeast enoyl reductase component requires FMN in addition to NADPH as a cofactor, while the animal enzyme does not.

The multifunctional character and the domain structure of animal fatty acid synthases suggest that the animal enzymes have evolved from independently coded monofunctional enzymes, in part at least, through a series of gene fusion events. If so, does the yeast enzyme represent a step in the evolution of the animal enzyme, or does it represent the result of an independent series of gene fusion events? According to the protease mapping studies, the linear order of the individual enzymatic activities is different for the animal and yeast enzymes. This observation, coupled with the other functional and structural differences, suggests that these two fatty acid synthases have evolved by independent gene fusion events.

Animal Fatty Acid Synthase: Regulation

The complexity of the multifunctional fatty acid synthase polypeptide suggests a considerable potential for regulatory mechanisms based on modulation of the efficiency of catalysis. There is, however, no convincing evidence for regulation of activity by covalent modification or by allosteric effectors at physiologically relevant concentrations. In the next section, we shall discuss regulation of the concentration of animal fatty acid synthase, acetyl-CoA carboxylase, and other lipogenic enzymes.

REGULATION OF ENZYME CONCENTRATION

The enzyme of carbohydrate metabolism and F.A synthesis have high activity in well-fed animal. and low activity in the starvation animal.

Rapid changes in the flux of carbon from glucose to long-chain fatty acids are initiated by a combination of changes in delivery of substrates, concentration of allosteric effectors, and degree of phosphorylation of enzymes, as discussed previously. In contrast, increases or decreases in fatty acid synthesis caused by changes in the total activities (that is, amounts) of the lipogenic enzymes occur over periods of hours or days rather than the seconds or minutes typical of changes in catalytic efficiency caused by allosteric or phosphorylation mechanisms. Total activity is defined here as the maximum activity which can be demonstrated in cell-free extracts under optimal assay conditions with respect to substrates, cofactors, and effectors. Thus, the total activities of glucose-6-phosphate dehydrogenase, 6-phosphogluconate dehydrogenase, malic enzyme, ATP-citrate lyase, acetyl-CoA carboxylase, and fatty acid synthase are high in the livers of well-fed animals, especially if the diet is high in carbohydrate and low in fat. These activities are decreased by starvation, high-fat/low-carbohydrate diets, or diabetes. The slowness of the changes and their manifestation under optimal assay conditions suggests that they are due to changes in enzyme concentration.

Regulation of the catalytic efficiency of a constant quantity of enzyme molecules can be distinguished experimentally from regulation of the number of enzyme molecules per cell by using immunological techniques. Each of the lipogenic enzymes just listed has been purified to homogeneity, and the homogeneous enzyme has been used to raise monospecific antisera in rabbits or goats. Immunological analyses have been done for each of these lipogenic enzymes under the dietary and hormonal conditions described in the previous paragraph. With a few quantitatively minor exceptions, dietary and hormonal control of the total activities of all the lipogenic enzymes involves regulation of enzyme protein concentration.

Enzyme Synthesis is Regulated

The concentration of an enzyme protein is a function of both its rate of synthesis and its rate of degradation. Therefore, during an increase in enzyme concentration, the rate of accumulation of that protein is not the same as its rate of synthesis. Although there are complex methods for estimating absolute synthesis rates for enzymes, studies of the lipogenic enzymes have invariably measured relative synthesis. The rate of incorporation of a radioactive amino acid into the enzyme in question is expressed as a fraction of the rate of incorporation into total protein. Loss of newly synthesized radioactive enzyme due to degradation is minimized by keeping the length of the labeling period very short with respect to the half-life of the enzyme.

The validity of this procedure for measuring enzyme synthesis is critically dependent on the availability of a quantitative method for the isolation of pure enzyme. Contamination by even a small mass of other proteins may be unacceptable if the contaminating species are synthesized much more rapidly than the enzyme in question. Few standard protein purification procedures are quantitative. Antibodies, on the other hand, can precipitate their antigens quantitatively and with great selectivity. The degree of selectivity of an antiserum or immunoglobulin preparation depends on the specificity of the antibody molecules therein and on the reaction conditions during precipitation. Detergents and high ionic strength minimize nonspecific adsorption of contaminating proteins to antibodies. A useful test of the specificity of the antisera is to dissociate the immunoprecipitate with sodium dodecyl sulfate (SDS) and separate the products by electrophoresis in polyacrylamide gels containing SDS (Figure 5.10) (Fischer and Goodridge 1978). To ensure specificity in

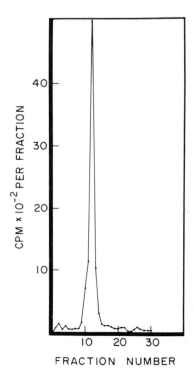

Figure 5.10.
SDS–polyacrylamide gel electrophoresis of fatty acid synthase immunoprecipitates from the livers of [^{3}H]leucine-injected, fed, 7-d-old chicks. Migration was from left to right. Only the upper half of the gel was sliced and counted. Taken from Fischer and Goodridge (1978) with permission of Academic Press, Inc.

Table 5.2. Enzyme Activity and Relative Synthesis of
Fatty Acid Synthase in Fed and Starved Chicks

Age (d)	Treatment	Activity*	Relative synthesis[†]
0	Starved	6	0.09
4	Starved	18	0.19
4	Fed	280	0.86

Source: Fischer and Goodridge (1978).
* Activity is expressed as nanomoles of palmitate formed per minute per gram of liver.
† Synthesis is expressed as cpm of [^{3}H]leucine incorporated into fatty acid synthase per 100 cpm incorporated into total protein.

the measurement of enzyme synthesis, many investigators routinely purify enzymes by a combination of immunoprecipitation and electrophoresis before measuring radioactivity in the enzyme.

Procedures such as these have been used to estimate the relative rates of synthesis of the lipogenic enzymes under a variety of dietary and hormonal conditions. An example of regulation of fatty acid synthase by diet in newly hatched chicks (Fischer and Goodridge 1978) is illustrated in Table 5.2. Without exception, changes in the concentrations of lipogenic enzymes have been associated with quantitatively comparable changes in the relative synthesis rates for those enzymes.

Enzyme Degradation is Not Regulated Selectively

Even though regulation of synthesis rates appears to account for control of the concentrations of the lipogenic enzymes, significant contributions from changes in degradation rate constants remain possible. Furthermore, one must have an estimate of an enzyme's half-life (the degradation rate constant, k_d, is equal to ln 2 divided by the half-life, $t_{\frac{1}{2}}$) before deciding on the length of the labeling period in the synthesis measurements. The half-life is the time required to degrade one-half of the enzyme molecules. Hence, the k_d of an enzyme can be estimated with a pulse-chase protocol. After sufficient incorporation of label into the enzyme has been achieved, periodic samples are taken during a chase with nonradioactive precursor. Radioactivity in the enzyme is assessed by the combination of immunoprecipitation and gel electrophoresis described in the preceding section. Since degradation of cellular protein is a first-order process, the half-life of the enzyme can be determined from a plot of the logarithm of radioactivity in the enzyme versus time. (Schimke [1973] contains a more complete discussion of methods used to measure degradation of animal enzymes.)

The rates of degradation of the lipogenic enzymes are stimulated to a small extent by starvation at the same time that the relative rates of synthesis of these enzymes are inhibited. However, starvation also stimulates the rate of degradation of total protein, whereas it has little effect on rate of synthesis of total protein. Since degradation of all proteins is stimulated to about the same extent, the changes in degradation do not contribute to the decreases in concentrations of the lipogenic enzymes caused by starvation.

Half-lives of the lipogenic enzymes vary from as short as 20 h to as long as 70 h. The slow turnover rates for these enzymes provides an explanation for the slow accumulation of enzyme protein which occurs after the application of a stimulus which increases enzyme synthesis, for example, feeding a starved animal. The magnitude of the increase in enzyme concentration is a function of the increase in enzyme synthesis. The time required to achieve the new steady-state level of enzyme protein, however, is the exclusive function of the degradation rate constant. This conclusion is derived from the kinetic equations which describe the changes in enzyme concentration as a function of the rates of synthesis and degradation of that enzyme. Thus, enzyme synthesis is a zero-order process, and enzyme degradation a first-order process, such that

$$\frac{d[E]}{dt} = k_s - k_d[E]$$

where t is the unit of time, $[E]$ is the enzyme concentration expressed per unit of tissue mass, k_s is the rate constant for enzyme synthesis expressed per unit of time, and k_d is the first-order rate constant for enzyme degradation expressed as the reciprocal of the time. At steady state,

$$\frac{d[E]}{dt} = 0$$

Therefore,

$$k_s = k_d[E]$$

and

$$[E] = \frac{k_s}{k_d}$$

We can now derive an equation that describes the time course of an increase in the concentration of a lipogenic enzyme caused by feeding starved animals, where k_s is changed to k'_s, and there is no change in k_d:

$$\frac{[E]_t}{[E]_0} = \frac{k'_s}{k_d[E]_0} - \left\{\frac{k'_s}{k_d[E]_0} - 1\right\}e^{-k_d t}$$

where $[E]_t$ is the concentration of the lipogenic enzyme at any time t and where $[E]_0$ is the concentration under the steady-state conditions defined by k_s and k_d. This equation shows that although a new steady state is defined by the new value for k'_s, the time required to achieve the new steady state is determined only by the rate constant for degradation, k_d. Berlin and Schimke (1965) solved these equations for the amount of time t required to go halfway from one steady state to a new, higher steady state and found that

$$t = \frac{\ln 2}{k_d}$$

where $\ln 2$ is the natural logarithm of 2, and k is the degradation rate constant characteristic of the new, higher steady state. The half-life of an enzyme is defined by the equation

$$t_{\frac{1}{2}} = \frac{\ln 2}{k_d}$$

It follows, therefore, that it takes one half-life to achieve 50% of the enzyme concentration at the final steady state. The same considerations apply when enzyme concentration decreases to a lower steady state. Since half-lives of the lipogenic enzymes are long, changes in the total activities of the lipogenic enzymes occur slowly. (Space limitations preclude a complete mathematical derivation of these relationships. The reader is referred to Schimke [1973] and Berlin and Schimke [1965] for more detail.)

Physiological Significance of Changes in Enzyme Concentration

Rapid inhibition of flux through the fatty acid synthesis pathway is achieved by inhibiting the catalytic efficiency of key enzymes in the pathway. For regulatory enzymes (reactants far from thermodynamic equilibrium, such as acetyl-CoA carboxylase), a slow reduction in enzyme concentration will cause a similarly slow reduction in the rate of fatty acid synthesis at any given state of phosphorylation or set of concentrations for allosteric effectors. However, for enzymes which catalyze reactions which are near equilibrium, malic enzyme or the dehydrogenases of the pentose phosphate pathway, for example, a slow decrease in enzyme concentration will have no effect on flux through the pathway.

Why is it that lipogenic tissues do not express malic enzyme and the dehydrogenases of the pentose pathway constitutively, at levels sufficient to maintain catalytic capacity well in excess of maximum flux rates? The concentrations of the cytosolic malate (NAD^+), lactate (NAD^+), and isocitrate ($NADP^+$) dehydrogenases appear to be expressed in just such constitutive fashions. Teleologically, the only obvious reason for nonconstitutive synthesis of malic enzyme, glucose-6-phosphate dehydrogenase, and 6-phosphogluconate dehydrogenase is elimination of the energy cost associated with high rates of synthesis of proteins which are not necessary under many environmental conditions. Alternatively, there may be regulatory features we do not yet understand which would explain why hepatocytes and adipocytes modulate the concentrations of malic enzyme and the dehydrogenases of the pentose pathway over a 20- to 100-fold range. Even the regulatory mechanisms themselves must have significant energy costs.

Enzyme Synthesis Rates Correlate with Messenger RNA Concentration

Synthesis of a specific enzyme could be regulated by two alternative mechanisms. First, the efficiency with which a constant quantity of its mRNA is translated into protein could be controlled. This would be analogous to regulating enzyme activity by controlling the catalytic efficiency of a constant amount of enzyme protein. Alternatively, translation of the specific protein could be regulated by controlling the abundance of its mRNA, analogous to regulation of enzyme activity by controlling enzyme concentration. Although there are techniques for measuring the translational activity of specific mRNAs, the results of such experiments are subject to the same interpretive difficulties inherent in measuring enzyme activities. Therefore, analysis of the regulation of expression of specific mRNAs required the development of techniques and the isolation of reagents which would be as useful for measuring mRNA concentration as antibodies are for measuring enzyme concentration.

Cloned complementary DNAs (cDNAs) for ATP:citrate lyase, malic enzyme, 6-phosphogluconate dehydrogenase, and fatty acid synthase have been isolated. These cloned cDNAs have been used in hybridization assays to determine the abundance of the specific mRNAs in crude mixtures of total cellular RNA. The hepatic concentrations of the mRNAs for ATP:citrate lyase, malic enzyme, 6-phosphogluconate dehydrogenase, and fatty acid synthase correlate with enzyme synthesis rates in starved versus fed animals and in animals treated with thyroid hormone. Thus, synthesis rates of the lipogenic enzymes are controlled by regulating the concentrations of their respective mRNAs. The next step in these analyses is to determine whether accumulation of the specific mRNAs is regulated at the level of gene transcription, processing of nuclear transcripts, or stability of nuclear or mature mRNA.

Regulation of the Synthesis of Lipogenic Enzymes in Cells in Culture

Nutritional regulation of lipogenesis and its associated enzymes occurs in two organs, liver and adipose tissue. There are complex interactions of liver and adipose tissue with other organs in the body, partly via the nervous system, but more importantly via numerous hormones and fuels in the blood. These interactions make it difficult to identify extracellular regulatory molecules and to analyze the intracellular molecular mechanisms which regulate fatty acid synthesis. Regulation of catalytic efficiency of pace-setting enzymes has been studied in isolated preparations of liver and adipose tissue for 30 years. Studies of short-term regulatory mechanisms require incubation of the isolated tissue for a few hours or less. By contrast, regulation of enzyme concentration occurs over periods of hours or days. Hepatocytes or adipocytes which retain functions characteristic of the differentiated state are difficult to maintain in culture for the required lengths of time. There are many cell lines derived from minimum-deviation hepatomas. However, neither the tumors in vivo nor the derived cell lines have the capacity to modulate lipogenesis or the activities of the lipogenic enzymes in a manner which is typical of normal liver. However, two suitable model systems have been developed over the past 10 years; one is the result of a serendipitous observation in a permanent fibroblast cell line and the other is due to improvements in culture conditions, which allow long-term maintenance of hepatocytes retaining characteristic differentiated functions.

3T3-L1 Cells—A Pre-adipocyte Cell Line

The 3T3 cell line is derived from mouse embryo fibroblasts which survived senescence in culture. Foci of cells containing fat droplets appear when 3T3 cells are held at confluence for more than a week. The 3T3-L1 subline was derived from cells in one of these foci of adipocyte-like cells. In the 3T3-L1 cell line, almost all cells are converted into adipocytes if the cells are held at confluence for a sufficiently long time. Treatment of confluent 3T3-L1 cells with dexamethasone (a synthetic glucocorticoid hormone) and isobutyl-methyl xanthine (an inhibitor of phosphodiesterase) causes a rapid and synchronous differentiation of the pre-adipocytes into adipocytes. Based on both morphological and biochemical criteria, the differentiated 3T3-L1 cells are remarkably similar to normal adipocytes. In the course of differentiation, many proteins, including all of the lipogenic enzymes, accumulate to levels characteristic of rodent adipose tissue. The increases in the levels of malic enzyme (Figure 5.11), ATP:citrate lyase, and fatty acid synthase which occur when 3T3-L1 pre-adipocytes are converted to adipocytes are due to increased rates of synthesis of these enzymes in the differentiating cells. The process of conversion of fibroblast-like pre-adipocytes into adipocytes probably mimics the terminal differentiation of adipocytes which occurs in vivo. Both insulin and growth hormone stimulate conversion of pre-adipocytes to adipocytes. At a molecular level, however, little is known about the intracellular events which initiate and maintain this differentiation process.

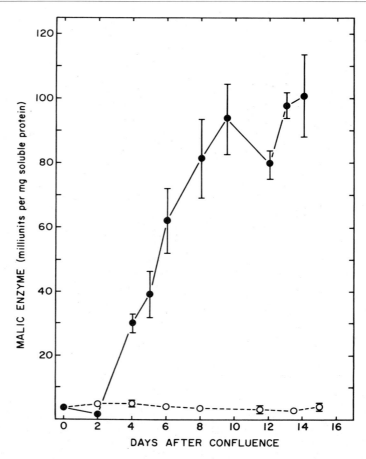

Figure 5.11. Malic enzyme activity in 3T3-L1 cells as a function of time after confluence in cells treated with (solid line) or without (dashed line) dexamethasone and methylisobutyl xanthine. These initiating drugs were present for the first 2 d of the experiment. The bars represent standard errors. Replotted from data presented in Goodridge, Fisch, and Glynias (1984).

Mechanisms involved in the commitment of cells to a particular differentiation path are unlikely to be the same as those which regulate metabolic function in terminally differentiated cells. 3T3-L1 adipocytes do not exhibit all of the regulatory properties exhibited by intact adipose tissue. For example, the activities of the lipogenic enzymes in mature adipocytes of intact rats are stimulated by insulin and thyroid hormone and inhibited by treatments which increase the intracellular concentration of cAMP. In mature 3T3-L1 adipocytes, this type of regulation is either lacking entirely or small in magnitude compared with the in vivo phenomena. Thus, insulin and thyroid hormone, potent positive regulators of malic enzyme in intact rat adipose tissue, have little effect on the enzyme in 3T3-L1 adipocytes. Cyclic AMP, a putative mediator of the inhibition of the synthesis of lipogenic enzymes in vivo, does inhibit synthesis of fatty acid synthase in mature 3T3-L1 adipocytes but has only a small effect on synthesis of malic enzyme. The basis for these differences between 3T3-L1 adipocytes and intact adipose tissue is unknown.

Hepatocytes in Maintenance Culture

Mass maintenance cultures of hepatocytes from both rats and chickens are used to analyze regulation of the concentration of the lipogenic enzymes. In avian hepatocytes maintained in a chemically defined medium for 3 d, insulin plus thyroid hormone causes 33- and 8-fold increases in the activities of malic enzyme and fatty acid synthase, respectively (Table 5.3). The addition of glucagon blocks the increases caused by insulin and thyroid hormone. The intracellular mediator of the inhibition caused by glucagon is cAMP. The immunological and nucleic acid hybridization procedures outlined in the previous section were used to establish that enzyme activity was regulated by controlling enzyme concentration, that enzyme concentration was regulated by controlling enzyme synthesis, and that enzyme synthesis was regulated by controlling mRNA concentration (Table 5.3). Gene transcription, processing of nuclear precursor RNA, transport of mRNA from the nucleus to the cytoplasm, and degradation of either precursor or mature mRNA are being studied as potential points at which abundance of the cytoplasmic mRNAs for malic enzyme and fatty acid synthase may be regulated.

Table 5.3. Effects of Triiodothyronine and Glucagon on the Activities, Synthesis Rates, and mRNA Abundances of Malic Enzyme and Fatty Acid Synthase in Avian Hepatocytes in Maintenance Culture

Measurement	Control	Triiodothyronine	Triiodothyronine plus glucagon
Malic enzyme			
Activity	3	100	3
Relative synthesis	1	100	3
mRNA abundance	3	100	3
Fatty acid synthase			
Activity	12	100	21
Relative synthesis	6	100	16
mRNA abundance	8	100	35

Sources: Fischer and Goodridge (1978); Goodridge and Adelman (1976).
Note: The results are expressed as a percentage of the values in cells incubated with triiodothyronine. All incubations contained insulin.

FUTURE DIRECTIONS

There are four stages in the analysis of the regulation of a metabolic pathway. First, regulatory enzymes are identified; second, their physical, kinetic, and regulatory properties are investigated in detail; third, the physical, kinetic, and regulatory properties are woven into a hypothesis to explain observed regulation of the pathway; and finally, the hypothesis is tested and modified as necessary. For eucaryotic fatty acid synthesis, stages one to three are rather complete. Thus, the hypotheses outlined in this chapter must now be tested and refined. In addition, there is a fifth stage: analysis of the mechanisms by which alterations in the structure of regulatory enzymes result in functional regulation. The new technologies of genetic engineering will play crucial roles in the analyses at stages four and five and in analyzing aspects of the regulation of enzyme activity which remain poorly understood (stage two). Areas which will be actively studied include (1) regulation of gene expression, (2) regulation of the catalytic efficiency of acetyl-CoA carboxylase, and (3) structure-function relationships in the multifunctional fatty acid synthase polypeptide.

The molecular mechanisms by which hormones regulate gene expression is an area of intense research. The genes for the lipogenic enzymes are probably not unusual, but several features of their expression make them especially interesting. Regulation of these genes by thyroid hormone, insulin, and cAMP make them part of a small set of identified proteins regulated by these agents. Other features which make the lipogenic enzymes interesting include coordinate regulation of their synthesis, tissue-specific expression and regulation, availability of cloned cDNAs, and availability of tissue culture models.

The physiological significance of phosphorylation and allosteric effectors in the regulation of acetyl-CoA carboxylase activity will continue to receive considerable attention. So too will the molecular mechanisms whereby phosphorylation and allosteric effectors regulate catalytic efficiency of the enzyme. Despite its potential importance to such studies, and to studies of gene expression, the cDNA for acetyl-CoA carboxylase has not been cloned. The difficulty in isolating cDNA clones for acetyl-CoA carboxylase mRNA is related to its large size, which increases susceptibility to degradation by RNAse, and low concentration. New approaches to the cloning of mRNAs of low abundance should result in the isolation of cDNA clones for this important enzyme in the near future. Availability of this clone should lead to a complete sequence of the acetyl-CoA carboxylase protein and to physical, kinetic, and functional studies of the protein altered by site-directed mutagenesis. This will be an exceptionally powerful approach to understanding physiological relevance and structure-function relationships in the regulation of the catalytic efficiency of acetyl-CoA carboxylase.

Genomic clones for yeast fatty acid synthase and cDNA clones for avian fatty acid synthase have been isolated. The large size of the avian fatty acid synthase mRNA, about 12 to 16 kilobases, may delay its complete sequencing. When the sequence is available, analyses aimed at understanding the relationship between enzymatic function and domain structure of fatty acid synthase should be possible.

PROBLEMS

1. Describe the evidence which indicates that acetyl-CoA carboxylase is pace-setting for the conversion of cytosolic citrate to fatty acids.
2. How do stearoyl-CoA and isocitrate regulate the catalytic efficiency of purified acetyl-CoA carboxylase?
3. In liver, the concentrations of the reactants in the reaction catalyzed by malic enzyme are near thermodynamic equilibrium. Can changes in the activity of malic enzyme regulate production of NADPH? Give reasons for your answer.
4. Why are the overall synthesis of fatty acids and the β-ketoacyl synthase reactions the only activities to be lost when the identical subunits of avian fatty acid synthase are dissociated in vitro?
5. Describe four independent routes by which glucagon may inhibit the activity of acetyl-CoA carboxylase in the liver.
6. Insulin stimulates, and epinephrine inhibits, acetyl-CoA carboxylase activity. If the catalytic efficiency of acetyl-CoA carboxylase is regulated by phosphorylation-dephosphorylation mechanisms, how do you explain that both hormones cause a stimulation of phosphorylation?

BIBLIOGRAPHY

Entries preceded by an asterisk are suggested for further reading.

The following reviews emphasize control mechanisms: Rolleston (1972); Lane, Watkins, and Meredith (1979); Wakil, Stoops, and Joshi (1983); Schimke (1973); Volpe and Vagelos (1973); and Bloch and Vance (1977).

Beatty, N. B., and Lane, M. D. 1983a. Kinetics of activation of acetyl-CoA carboxylase by citrate. Relationship to the rate of polymerization of the enzyme. *J. Biol. Chem.* 258:13043–50.

———. 1983b. The polymerization of acetyl-CoA carboxylase. *J. Biol. Chem.* 258:13051–55.

Berlin, C. M., and Schimke, R. T. 1965. Influence of turnover rates on the responses of enzymes to cortisone. *Mol. Pharmacol.* 1:149–56.

*Bloch, K., and Vance, D. 1977. Control mechanisms in the synthesis of saturated fatty acids. *Ann. Rev. Biochem.* 46:263–98.

Carlson, C. A., and Kim, K.-H. 1974. Regulation of hepatic acetyl coenzyme A carboxylase by phosphorylation and dephosphorylation. *Arch. Biochem. Biophys.* 164:478–89.

*Denton, R. M., and Halestrap, A. P. 1979. Regulation of pyruvate metabolism in mammalian tissues. *Essays Biochem.* 15:37–77.

El-Maghrabi, M. R.; Claus, T. H.; McGrane, M. M.; and Pilkis, S. J. 1982. Influence of phosphorylation on the interaction of effectors with liver pyruvate kinase. *J. Biol. Chem.* 257:233–40.

Fischer, P. W. F., and Goodridge, A. G. 1978. Coordinate regulation of acetyl coenzyme A carboxylase and fatty acid synthetase in liver cells of the developing chick in vivo and in culture. *Arch. Biochem. Biophys.* 190:332–44.

Goodridge, A. G. 1972. Regulation of the activity of acetyl coenzyme A carboxylase by palmitoyl coenzyme A and citrate. *J. Biol. Chem.* 247:6946–52.

———. 1973. Regulation of fatty acid synthesis in

isolated hepatocytes prepared from the livers of neonatal chicks. *J. Biol. Chem.* 248:1924–31.

Goodridge, A. G., and Adelman, T. G. 1976. Regulation of malic enzyme synthesis by insulin, triiodothyronine, and glucagon in liver cells in culture. *J. Biol. Chem.* 251:3027–32.

Goodridge, A. G.; Fisch, J. E.; and Glynias, M. J. 1984. Regulation of the activity and synthesis of malic enzyme in 3T3-L1 cells. *Arch. Biochem. Biophys.* 228:54–63.

Greenbaum, A. L.; Gumaa, K. A.; and McLean, P. 1971. The distribution of hepatic metabolites and the control of the pathways of carbohydrate metabolism in animals of different dietary and hormonal status. *Arch. Biochem. Biophys.* 143:617–63.

Hardie, D. G., and Guy, P. S. 1980. Reversible phosphorylation and inactivation of acetyl-CoA carboxylase from lactating rat mammary gland by cyclic AMP–dependent protein kinase. *Eur. J. Biochem.* 110:167–77.

Hers, H.-G. and van Schaftingen, E. 1982. Fructose 2,6-bisphosphate 2 years after its discovery. *Biochem. J.* 206:1–12.

Knobling, A., and Schweizer, E. 1975. Temperature-sensitive mutants of the yeast fatty-acid-synthase complex. *Eur. J. Biochem.* 59:415–21.

Lane, M. D.; Moss, J.; and Polakis, S. E. 1974. Acetyl coenzyme A carboxylase. *Curr. Top. Cell. Regul.* 8:139–95.

Lane, M. D.; Watkins, P. A.; and Meredith, M. J. 1979. Hormonal regulation of acetyl-CoA carboxylase activity in the liver cell. *CRC Crit. Rev. Biochem.* 7:121–41.

Mattick, J. S.; Tsukamoto, Y.; Nickless, J.; and Wakil, S. J. 1983. The architecture of the animal fatty acid synthetase. I. Proteolytic dissection and peptide mapping. *J. Biol. Chem.* 258:15291–99.

McCarthy, A. D.; Aitken, A.; Hardie, D. G.; Santikarn, S.; and Williams, D. H. 1983. Amino acid sequence around the active serine in the acyl transferase domain of rabbit mammary fatty acid synthase. *FEBS Lett.* 160:296–300.

McCarthy, A. D.; Goldring, D.; and Hardie, D. G. 1983. Evidence that the multifunctional polypeptides of vertebrate and fungal fatty acid synthases have arisen by independent gene fusion events. *FEBS Lett.* 162:300–304.

*Rolleston, F. S. 1972. A theoretical background to the use of measured concentrations of intermediates in study of the control of intermediary metabolism. *Curr. Top. Cell. Regul.* 5:47–75.

*Schimke, R. T. 1973. Control of enzyme levels in mammalian tissues. *Adv. Enzymol.* 37:135–87.

Spence, J. T. 1983. Levels of translatable mRNA coding for rat liver glucokinase. *J. Biol. Chem.* 258:9143–46.

Tsukamoto, Y.; Wong, H.; Mattick, J. S.; and Wakil, S. J. 1983. The architecture of the animal fatty acid synthetase complex. IV. Mapping of active centers and model for the mechanism of action. *J. Biol. Chem.* 258:15312–22.

*Volpe, J. J., and Vagelos, P. R. 1973. Saturated fatty acid biosynthesis and its regulation. *Ann. Rev. Biochem.* 42:21–60.

*Wakil, S. J.; Stoops, J. K.; and Joshi, V. C. 1983. Fatty acid synthesis and its regulation. *Ann. Rev. Biochem.* 52:537–79.

CHAPTER 6

Fatty Acid Desaturation and Chain Elongation in Eucaryotes

Harold W. Cook

Fatty acyl chains, esterified to the complex lipids of biological membranes, are major components of these membranes, both quantitatively and qualitatively. Fatty acyl chains account for more than half the mass of most major phospholipids and are primarily responsible for the apolar nature of the membrane bilayer. Depending on their chain length and degree of unsaturation, they contribute to fluidity and other physical and chemical properties of the membrane.

As major membrane components, acyl chains influence a variety of membrane functions, such as transport, endocytosis and exocytosis, and the activities of membrane-associated enzymes. Furthermore, polyunsaturated fatty acids derived from essential fatty acids also serve as precursors of biologically active molecules, such as prostaglandins, leukotrienes, and hydroxy fatty acids (Chapter 11).

Why is the variety of acyl chains in the lipids of biological membranes of eucaryotic organisms necessary? How are these chains derived and modified, and what regulates metabolism of acyl chains prior to their esterification to membrane lipids?

Consider the apolar character contributed to membrane lipid by fatty acyl chains. Primarily three major forces hold lipid molecules within the membrane: (1) electrostatic interactions between polar groups of lipids and oppositely charged groups in adjacent proteins, (2) hydrogen bonding between oxygen and nitrogen atoms in lipids and adjacent proteins, and (3) London–van der Waals dispersion forces between CH_2 pairs in hydrocarbon tails of adjacent lipid molecules. Of these, only London–van der Waals forces are involved along the acyl chains and may be the major force holding membrane lipid molecules together.

The carbon-carbon bonds of a fatty acyl chain are arranged to form a linear chain. Since there is a symmetrical electron distribution along the acyl chain, no net charge results, but there are gravitational forces between adjacent acyl chains and with other closely aligned molecules. These London–van der Waals forces are relatively weak; nevertheless, they are additive and proportional to the number of overlapping methylene groups and the distances between them. For a long acyl chain, the total bonding strength becomes significant and, in most cases, is greater than the electrostatic and hydrogen bonding of polar head groups.

Thus, length is a crucial parameter in acyl chain contribution to membrane structure and stability. The weak apolar interactions determine how the membrane will interact with its environment. For example, restricted solubility of lipids in the surrounding aqueous milieu is governed by the tendency of acyl chains to remain in association with one another rather than to associate with the aqueous environment. Similarly, response to temperature fluctuations is modulated by the extent to which thermal influences cause acyl chains to dissociate from one another and assume more ran-

FATTY ACID	ABBREVIATION	MELTING POINT	SPATIAL WIDTH (nm)	POSSIBLE CONFIGURATIONS
Stearic Acid	18:0	70°	0.25	
Oleic Acid	c-18:1$(n-9)$	16°	0.72	
Elaidic Acid	t-18:1$(n-9)$	43°	0.31	
Linoleic Acid	c,c-18:2$(n-6)$	−5°	1.13	

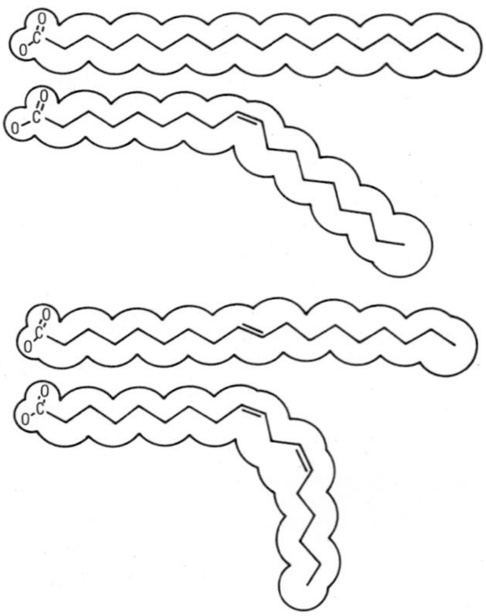

Figure 6.1. Some physical characteristics of fatty acids. See Table 6.1 for nomenclature of the fatty acids. Single bonds have a length of 0.154 nm and an angle of approximately 111°; double bonds have a length of 0.133 nm and an angle of approximately 123°.

domized structure. Accordingly, solubility is decreased and melting point increased as fatty acyl chain length is increased.

Where a more loosely packed membrane structure is advantageous, the rigidity of lengthy saturated acyl chains can be countered by acyl chains with double bonds. Introducing a double bond of cis geometric configuration results in a bending of the chain with a change of approximately 30° from the linearity of the saturated chain (Figure 6.1). As London–van der Waals forces vary inversely as the sixth power of the distance between the acyl chains, factors that increase that distance markedly decrease chain-chain interaction. Double bonds also are nonrotating and restrict acyl chain movement. Furthermore, some charge concentration around the double bond increases polarity in the acyl chain.

The extent to which double bonds actually cause bends or curved shapes (potentially, a hexaenoic acyl chain, with six double bonds, could assume a U-shape or be nearly circular) within the biological membrane is not well understood. Physical and biochemical data clearly show that unsaturated acids caused a marked decrease in membrane rigidity. Accordingly, within membrane phospholipids, acyl chain length and the number and position of double bonds in it markedly influence fluidity, permeability, and stability of biological membranes and interaction with their environment.

HISTORICAL BACKGROUND

The vital contribution of lipid molecules to the hydrophobic character of membranes was recognized late in the nineteenth century. However, the nutritional importance of specific lipid molecules was first revealed through the pioneering work of Burr and Burr (1929). They fed rats a fat-free diet and observed retarded growth, scaly skin, tail necrosis, and eventual death. This disorder was reversed by feeding a specific fatty acid fraction. Linoleic acid was recognized as the active agent, and the term *essential fatty acid* was coined.

During the following two decades, progress toward understanding the metabolism of unsaturated fatty acids was limited by available analytical techniques. However, the number of double bonds in fatty acids could be determined by iodination, and progress was made toward describing the process of fatty acid desaturation.

Two factors emerged during the 1950s that led to our current understanding of fatty acid metabolism. The advent of chromatographic techniques (gas-liquid and thin-layer chromatography were prime contributors) and the greater availability of appropriate substrates and precursors labeled with isotopes such as ^{14}C and ^{3}H greatly enhanced capabilities for studying single species of fatty acids and monitoring their conversions by analyzing relatively small amounts of sample.

During the last three decades, considerable detail has been revealed about metabolic pathways of elongation and desaturation of saturated, monounsaturated, and several families of polyunsaturated fatty acids. In the 1950s and 1960s, in vivo evaluation of fatty acid metabolism, primarily by Holman (1981) and by Mead (1981), was supplemented by in vitro assays of specific enzymatic steps. Activity of a Δ^9 desaturase was measured in yeast by Bloomfield and Bloch (1960), in rat liver microsomes by Bernhard, von Bulow-Koster, and Wagher (1959), and in plants by Nagai and Bloch (1966). Nugteren (1962) and Marcel, Christiansen, and Holman (1968) reported polyunsaturated fatty acid formation in animal tissues involving Δ^6 and Δ^5 desaturation. Further desaturation at the Δ^{12} and Δ^{15} positions was described by Meyer and Bloch (1963), MacMahon and Stumpf (1964), and Harris and James (1965). Initial studies of elongation of long chain fatty acids in mitochondria were reported by Wakil (1961) and in microsomes by Nugteren (1965). During this period a direct relationship between essential fatty acids and prostaglandins was elucidated. Through the subsequent work of many others, we now have a firm basis for understanding primary and alternate pathways in fatty acid desaturation and chain elongation, interactions of fatty acids in competitive reactions, and factors that influence the regulation of long chain fatty acid metabolism.

CHAIN ELONGATION OF LONG-CHAIN FATTY ACIDS

De novo synthesis of fatty acids by the soluble, cytosolic enzymes of the acetyl-CoA carboxylase and fatty acid synthase complexes (Chapter 5) produces mainly palmitate, with minor amounts of stearate. Quantitatively, these chain lengths are major components of many membrane lipids and qualitatively appear to be related to the optimum width of the membrane lipid bilayer. On the other hand, there are in membranes many major acyl chains longer than 16 carbons. For example, in the myelin surrounding processes of neuronal cells, fatty acyl chains of 18 carbons or greater make up more than 60% of the total, and in sphingolipids in particular, acyl chains of 24 carbons are especially prominent.

Many eucaryotic cells have the capacity for two-carbon chain elongation, both of endogenously synthesized acids and of exogenous, dietary acids. In liver, brain, and other tissues there are two primary systems for elongation, one in the endoplasmic reticulum, and the other in mitochondria.

The Microsomal Elongation System

The more active fatty acyl chain elongation system is associated with the endoplasmic reticulum. The two-carbon condensing unit is malonyl-CoA; very limited activity has been seen with acetyl-CoA.

micellar
脂中沉.

Avidin, a protein that binds biotin and inhibits biotin-dependent carboxylases, does not alter microsomal elongation activities in vitro, indicating that a microsomal acetyl-CoA carboxylase is not involved. CoA derivatives appear to be the active form of the malonyl group and of the fatty acyl acceptors. The latter can be formed from free fatty acids by fatty acyl-CoA ligase in the presence of ATP, Mg^{2+}, and CoA. Some evidence suggests that microsomal elongation of polyunsaturated acids could occur in the absence of CoA; however, a soluble acyl carrier protein is not involved. Microsomal elongation is active with both saturated and unsaturated fatty acids, the latter having higher activity; γ-linolenate is the most effective of the unsaturated substrates. Saturated or polyunsaturated acids that are not to be elongated further generally inhibit elongation. Both NADPH and NADH are utilized as electron donors, and although results vary somewhat in different studies, NADPH usually has been found to be more active.

Details of the component reactions of chain elongation have come from the laboratories of Seubert and Sprecher, among others. In a manner analogous to that of fatty acid synthase, four component reactions occur in the two-carbon elongation process (Figure 6.2). The first, a condensation of the fatty acyl-CoA and malonyl-CoA to form the β-ketoacyl CoA derivative, is the rate-limiting reaction and can be measured in vitro in the absence of NADPH. When several saturated and unsaturated acyl-CoAs are used, the rate of condensation is equivalent to the overall rate of elongation and is dependent on the chain length and number and position of double bonds in the primer. At least two condensation enzymes, one for saturated and one for unsaturated primers, are indicated from differences in rates with various labeled substrates and response to inhibitors such as N-ethylmaleimide. In vitro, the condensation reaction is enhanced by bovine serum albumin, which suggests that reduction of micellar inhibition supports interaction of a nonmicellar monomeric form. Furthermore, albumin may also inhibit thiolases and prevent hydrolysis of the acyl-CoA substrate. On the other hand, elongation is not altered by albumin when microsomes are prelabeled with fatty acid, suggesting that acyl intermediates do not readily exchange with acyl pools in their aqueous environment during elongation.

The condensing enzymes funnel the β-keto acyl-CoAs to a common set of enzymes for completion of elongation. The second reaction in elongation, catalyzed by a reductase which utilizes NADPH in the formation of β-hydroxy acyl-CoA, cannot readily be assayed as a single reaction using microsomal preparations, since the fully elongated end product is formed when NADPH is present. The third reaction is catalyzed by a dehydrase, which can be monitored by measuring 2-trans-enoyl-CoA formation from the β-hydroxyacyl-CoA

1. Condensation

a. Microsomes

$$R-\overset{O}{\overset{\|}{C}}-S-CoA + {}^-O-\overset{O}{\overset{\|}{C}}-CH_2-\overset{O}{\overset{\|}{C}}-S-CoA \rightleftharpoons R-\overset{O}{\overset{\|}{C}}-CH_2-\overset{O}{\overset{\|}{C}}-S-CoA + H-S-CoA$$
$$+ CO_2$$

b. Mitochondria

$$R-\overset{O}{\overset{\|}{C}}-S-CoA + CH_3-\overset{O}{\overset{\|}{C}}-S-CoA \rightleftharpoons R-\overset{O}{\overset{\|}{C}}-CH_2-\overset{O}{\overset{\|}{C}}-S-CoA + H-S-CoA$$

2. Reduction (β-keto acyl-CoA reductase)

$$R-\overset{O}{\overset{\|}{C}}-CH_2-\overset{O}{\overset{\|}{C}}-S-CoA + NAD(P)H + H^+ \rightleftharpoons R-CHOH-CH_2-\overset{O}{\overset{\|}{C}}-S-CoA$$
$$+ NAD(P)^+$$

3. Dehydration (β-hydroxyacyl-CoA dehydrase)

$$R-CHOH-CH_2-\overset{O}{\overset{\|}{C}}-S-CoA \rightleftharpoons R-CH=CH-\overset{O}{\overset{\|}{C}}-S-CoA + H_2O$$

4. Reduction (2-*trans*-enoyl-CoA reductase)

$$R-CH=CH-\overset{O}{\overset{\|}{C}}-S-CoA + NAD(P)H + H^+ \rightleftharpoons R-CH_2-CH_2-\overset{O}{\overset{\|}{C}}-S-CoA$$
$$+ NAD(P)^+$$

Figure 6.2. Reactions in two-carbon chain elongation of long-chain fatty acids.

precursor in the absence of reduced pyridine nucleotide. This dehydrase has been solubilized with detergent and purified nearly 100-fold. In vitro, a micellar substrate for the dehydrase has been implicated from the close agreement between optimal substrate concentrations for various β-hydroxy acyl derivatives and their critical micellar concentrations. The final reaction, catalyzed by 2-*trans*-enoyl-CoA reductase in the presence of NADPH, can be measured separately. In contrast with the condensation enzyme, the reductases and dehydrase are not influenced by diet or substrate modifications.

Cytochrome b_5 participation in transfer of electrons from the pyridine nucleotides to the reductases has been proposed based on increased reoxidation of microsomal cytochrome b_5 in the presence of NADPH, ATP, and malonyl-CoA and on marked reduction of malonyl-CoA incorporation in the presence of antibodies to cytochrome b_5. Developmental profiles of microsomal cytochrome reductases closely parallel those of fatty acid elongation in both liver and brain.

Relationships between component enzymes of the elongation process within endoplasmic reticulum membranes are not clear. Sprecher has argued that each reaction is catalyzed by a discrete enzyme, rather than by a multifunctional complex, since CoA derivatives of all the intermediates have been isolated. As additional support, he cites evidence for two or more condensation enzymes and observations that partially purified β-hydroxyacyl-CoA dehydrase neither contains nor requires the presence of the other activities. Others have suggested a covalent linkage of the acyl-CoA to a multifunctional enzyme after condensation.

Most studies have utilized liver as an enzyme source because of high activity in this tissue; however, in preweanling rats, brain activity also is high and generally exceeds that in liver. Studies with brain, using various radioactively labeled fatty acyl precursors, such as palmitate, stearate, arachidate, and behenate, present strong evidence for the existence of different enzymes reacting with dissimilar acyl chain lengths. For example, differences in pH optima, relative rates of substrate utilization, and activity profiles during development have been documented. Potentially most convincing are reports that in the Quaking strain of mice (a mutant with defective myelination), chain elongation of 20:0 to 22:0 and 24:0 was reduced by about 70%, whereas elongation of 16:0 and 18:0 was unaltered relative to control mice.

In studies of the elongation of saturated, dienoic, and trienoic substrates by rat liver microsomes, fasting depressed all three activities similarly; in contrast, the increase in response to refeeding was much greater with the saturated substrate than with the unsaturated fatty acids. This suggestion of more than one elongation system in liver was further supported by competitive substrate experiments. Thus, while most evidence is indirect, and complexities in isolating such membrane-associated enzymes remain, it appears that several elongation systems with specificities based on chain length and degree of unsaturation may be operative in the endoplasmic reticulum.

The Mitochondrial Elongation System

Although less active than the microsomal system, the mitochondrial chain elongation system has been extensively investigated in many tissues, particularly in liver and brain. In contrast with the microsomal system, the two-carbon condensing donor in mitochondria is acetyl-CoA (Figure 6.2). Saturated and monoenoic fatty acyl-CoA derivatives are major acceptors, supporting higher activity than polyunsaturates; generally, monoenes are more active than saturates, particularly in brain. While there is no uniform agreement about nucleotide requirements, maximal mitochondrial elongation in most tissues, including liver, brain, kidney, and adipose tissue, seems to require both NADPH and NADH; heart, aorta, and muscle, however, may require only NADH. The effectiveness of mixtures of NADPH and NADH can probably be explained by the fact that NADH is optimal in the first reduction, and NADPH in the second.

During the early 1970s, Seubert and co-workers elucidated the mechanism of mitochondrial chain elongation. Subcellular location of the enzymes and other similarities between β-oxidation (Chapter 4) and chain elongation led them to propose that reversal of β-oxidation would not be thermodynamically feasible; the FAD-dependent acyl-CoA dehydrogenase of β-oxidation would have to be substituted with a more thermodynamically favorable enzyme, such as enoyl-CoA reductase, to produce an overall negative free energy for the sequence. Thus, chain elongation and β-oxidation did not utilize all of the same enzymes in reverse.

An enoyl-CoA reductase enzyme was isolated from liver mitochondria by centrifugation and chromatographic techniques. The mitochondrial reductase was distinct from the microsomal reductase, based on pH optima and specificities for saturated and unsaturated acyl-CoA derivatives. The mitochondrial reductase showed a specificity corresponding to overall mitochondrial elongation. Kinetic studies suggested that enoyl-CoA reductase was rate limiting in mitochondrial chain elongation.

Functions of the Two Elongation Systems

Microsomal chain elongation appears to be the most important source of acyl chains greater than 16 carbons in many tissues. During tissue growth and maturation, the required long-chain acids may not be supplied adequately in the diet. For example, during development of the central nervous system, a continuous, reliable supply of 18- to 24-carbon saturates and monoenes, and 20- and 22-carbon polyunsaturates, is required for normal brain myelination, regardless of dietary fluctuations.

The function of the mitochondrial elongation system, including its involvement in biogenesis of mitochondrial membranes per se, is less clear. Enzyme systems of different molecular weights have been isolated from the inner and outer membranes of mitochondria. In view of the relatively low activity toward 16- and 18-carbon acyl chains, a primary role for the mitochondrial system in the formation of long acyl chains for membrane synthesis is questionable. It has been proposed that mitochondrial elongation in liver may serve a transhydrogenase function, moving electron equivalents from NADPH-generating substances to the respiratory chain. During cellular anoxia, chain elongation could conserve reducing equivalents or acetate units through the formation of acyl chains. Indeed, glutamate and isocitrate do stimulate liver mitochondrial chain elongation. Considering the low capacity for chain elongation in mitochondria relative to other reactions involving NADH or NADPH generation and utilization, a quantitatively significant role in the reduced-to-oxidized nucleotide balance must be viewed cautiously.

FORMATION OF MONOUNSATURATED FATTY ACIDS BY OXIDATIVE DESATURATION

The spectrum of fatty acyl chains required to meet the physical and biochemical requirements of lipid storage and of membrane synthesis and maintenance cannot be supplied by diet, de novo synthesis, and chain elongation alone. Unsaturated fatty acids must be supplied by oxidative desaturation of saturated chains and by further modification of a wide variety of dietary acids of plant and animal origin. This ensures that the fatty acyl compositions of phospholipids and neutral glycerides are in a physical state that can be utilized at physiological temperatures.

Nomenclature to Describe Double Bonds

Before discussing desaturation enzymes, we will outline, using oleic acid as an example, the abbreviations used to describe the number and position of double bonds in acyl chains (Table 6.1).

1. To indicate that oleic acid is an 18-carbon fatty acid with one double bond, the shorthand 18:1 will be used. The number before the colon denotes the number of carbon atoms, and the number following refers to the number of double bonds. This form is adequate only where the position or nature of the double bond is not known, for example, in chromatographic analysis of an acyl chain mixture where positional isomers are not separated.

2. To assign the position of an individual double bond or the specificity of an enzyme inserting it, the delta (Δ) nomenclature will be used. This describes a bond position relative to the carboxyl carbon of the acyl chain. For oleic acid, the double bond is in the Δ^9 position, between carbons 9 and 10, and would be introduced into the 18-carbon chain by a Δ^9 desaturase enzyme.

3. To designate an individual fatty acid within a family of structurally related acids, the $(n -)$ nomenclature, promoted by the IUPAC-IUB Commission on Nomenclature, will be used. Here, the position of the first double bond from the methyl end is described. Thus, 18:1$(n - 9)$ indicates that the double bond closest to the methyl end (the only double bond in this example) is in the Δ^9 position, that is, since $n = 18$ and $n - 9 = 9$, it is 9 carbons from both the carboxyl and the methyl ends. In

Table 6.1. Nomenclature and Bond Positions of Major Long-Chain Fatty Acids

Common name	Systematic name*	Abbreviation	nd positions
Palmitic acid	hexadecanoic acid	16:0	
Palmitoleic acid	9-hexadecenoic acid	16:1 $(n-7)$	Δ^9
	6-hexadecenoic acid	16:1 $(n-10)$	Δ^6
Stearic acid	octadecanoic acid	18:0	
Oleic acid	9-octadecenoic acid	18:1 $(n-9)$	Δ^9
Vaccenic acid	11-octadecenoic acid	18:1 $(n-7)$	Δ^{11}
Petroselenic acid	6-octadecenoic acid	18:1 $(n-12)$	Δ^6
Elaidic acid	t-9-octadecenoic acid	t-18:1 $(n-9)$	$t\text{-}\Delta^9$
Linoleic acid	9,12-octadecadienoic acid	18:2 $(n-6)$	$\Delta^{9,12}$
Linoelaidic acid	t,t-9,12-octadecadienoic acid	t,t-18:2 $(n-6)$	$t,t\text{-}\Delta^{9,12}$
α-Linolenic acid	9,12,15-octadecatrienoic acid	18:3 $(n-3)$	$\Delta^{9,12,15}$
γ-Linolenic acid	6,9,12-octadecatrienoic acid	18:3 $(n-6)$	$\Delta^{6,9,12}$
Arachidic acid	eicosanoic acid	20:0	
Gadoleic acid	9-eicosenoic acid	20:1 $(n-11)$	Δ^9
Gondoic acid	11-eicosenoic acid	20:1 $(n-9)$	Δ^{11}
Dihomo-γ-linolenic acid	8,11,14-eicosatrienoic acid	20:3 $(n-6)$	$\Delta^{8,11,14}$
Mead acid	5,8,11-eicosatrienoic acid	20:3 $(n-9)$	$\Delta^{5,8,11}$
Arachidonic acid	5,8,11,14-eicosatetraenoic acid	20:4 $(n-6)$	$\Delta^{5,8,11,14}$
Timnodonic acid	5,8,11,14,17-eicosapentaenoic acid	20:5 $(n-3)$	$\Delta^{5,8,11,14,17}$
Behenic acid	docosanoic acid	22:0	
Cetoleic acid	11-docosenoic acid	22:1 $(n-11)$	Δ^{11}
Erucic acid	13-docosenoic acid	22:1 $(n-9)$	Δ^{13}
Adrenic acid	7,10,13,16-docosatetraenoic acid	22:4 $(n-6)$	$\Delta^{7,10,13,16}$
Docosapentaenoic acid	4,7,10,13,16-docosapentaenoic acid	22:5 $(n-6)$	$\Delta^{4,7,10,13,16}$
Clupanodonic acid	7,10,13,16,19-docosapentaenoic acid	22:5 $(n-3)$	$\Delta^{7,10,13,16,19}$
Cervonic acid	4,7,10,13,16,19-docosahexaenoic acid	22:6 $(n-3)$	$\Delta^{4,7,10,13,16,19}$
Lignoceric acid	tetracosanoic acid	24:0	
Nervonic acid	15-tetracosenoic acid	24:1 $(n-9)$	Δ^{15}

* The full designation for double bonds would be all-*cis*, for example, all-*cis*-9,12,15-octadecatrienoic acid; unless otherwise indicated, all bonds are of the cis geometric configuration.

$18:1 (n-7)$

$18-7 = 11 \cdot \Delta^{11}$

= 7 carbon from the methyl end.

contrast, 18:1$(n-7)$ specifies the positional isomer, vaccenic acid, with its double bond in the Δ^{11} position, that is, 7 carbons from the methyl end. This convention is particularly useful in designating groups of fatty acids that are derived from the same parent compound and in which metabolic reactions do not occur on the methyl side of an existing double bond.

4. To indicate the geometric configuration of a double bond, the designation will be preceded by a *c-* for cis or *t-* for trans. Thus, *c*-18:1$(n-9)$ distinguishes oleic acid from its trans isomer,

elaidic acid. Generally, double-bond configuration will be understood to be cis unless noted.

Among other conventions, the ω-designation is still widely used to designate the position of a double bond from the methyl end (ω-carbon) and is similar to the $(n -)$ nomenclature.

Characteristics of the Monoene-forming Desaturation Enzymes

Monounsaturated fatty acids are formed in mammalian systems by direct oxidative desaturation (a removal of two hydrogens) of a preformed long-chain saturated fatty acid. The oxygenase type of enzyme is associated with the endoplasmic reticulum and can be isolated in microsomes of many tissues, including liver, mammary gland, brain, testes, and adipose tissue. The Δ^9 desaturase usually is the predominant, if not exclusive, desaturation enzyme for saturated acids in these tissues.

The Δ^9 desaturase acts on substrates in the form of fatty acyl-CoA. With crude microsomal preparations in vitro, free fatty acids can be used if ATP, Mg^{2+}, and CoA are supplied to supplement activation by fatty acyl-CoA ligase, which is usually not rate limiting. For most tissues, 14- to 18-carbon saturated fatty acyl chains are good substrates, with stearoyl-CoA being most active. A reduced pyridine nucleotide is required, and generally NADH is more active than NADPH. The Δ^9 desaturase has an absolute requirement for molecular oxygen—which acts as an electron acceptor for two pairs of hydrogens, one from NADH and the other from the fatty acyl-CoA— and is highly sensitive to inhibition by cyanide.

Most assays of Δ^9 desaturase use stearoyl-CoA (or stearic acid and cofactors) labeled with ^{14}C in the carboxyl carbon or ^3H in the acyl chain. Following in vitro incubation, methyl esters can be formed from extracted lipids by transesterification. Saturated substrate and monounsaturated products can be separated by gas-liquid chromatography or argentation thin-layer chromatography. In the latter technique, silica gel impregnated with silver ions effectively separates acyl chains according to their degree of unsaturation. Monitoring radioactivity in separated fractions is reliable for quantitating desaturase activities.

In the early 1970s, it was established that the Δ^9 desaturation system consists of three major proteins: (1) NADH-cytochrome b_5 reductase, (2) cytochrome b_5, and (3) a terminal desaturase component or cyanide-sensitive protein (Figure 6.3). Initial observations using various electron donors and inhibitors were later supplemented by immunochemical techniques to demonstrate involvement of cytochrome b_5 and its reductase. Under most circumstances, the capacity for electron transport greatly exceeds the activity of the rate-limiting desaturase component. Functional association of the three

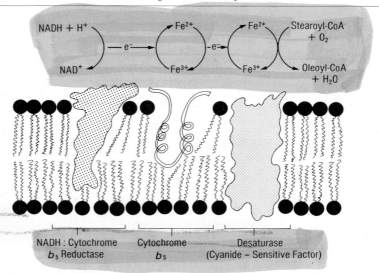

Figure 6.3. Diagrammatic representation of the Δ^9 desaturase complex, including the electron transport proteins.

components is dependent on the lipid milieu of the endoplasmic reticulum membrane. Most of the neutral lipid and some phospholipids can be extracted from microsomes without significant loss of Δ^9 desaturase activity; however, a specific phospholipid fraction, including phosphatidylcholine, is required for desaturation.

The membrane-bound nature of the Δ^9 desaturase system and its firm association with the endoplasmic reticulum retarded early characterization of the complex, but much has been learned from Holloway, James, Jeffcoat, Stritmatter, Wakil, and others whose laboratories have contributed to solubilization, purification, and reconstitution of the proteins. Initial loss of activity during solubilization was largely overcome through carefully controlled ratios of detergents to protein and combinations of mild extraction solvents. Activity in a purified complex was achieved by recombination of the three major protein components and a lipid fraction such as egg lecithin or synthetic phosphatidylcholine.

The desaturase component, highly sensitive to cyanide, has a molecular weight of 53,000, one atom of nonheme iron per molecule as the prosthetic group, and 62% nonpolar amino acid residues. Evidence suggests that the desaturase component is largely within the microsomal membrane, with the portion containing the active center exposed to the cytosol. Interaction of the desaturase with specific reagents suggests that arginyl residues may play a role at the binding site for the negatively charged CoA moiety of the substrate, and tyrosyl residues may be involved in chelation of the iron prosthetic group. Stearoyl-CoA desaturase, purified 150-fold, appears homogeneous as judged by SDS-polyacrylamide gel electrophoresis. Based on this observation, the level of Δ^9 desaturase in rat liver at maximal induction could account for approximately 0.5–0.8% of the microsomal protein.

NADH-cytochrome b_5 reductase (a flavoprotein of molecular weight 43,000) and cytochrome b_5 (a heme-containing protein of molecular weight 16,700) are more readily solubilized than the desaturase, and each has two major interactive domains in its peptide structure. For cytochrome b_5 there is a catalytic hydrophilic region of 85 residues (including the NH_2-terminal), and the protein terminates in a hydrophobic COOH-terminal tail of approximately 40 amino acids. The latter attaches the protein to the membrane in a favorable spatial orientation, looping back to the cytosolic side of the bilayer.

The mechanism of hydrogen removal from the saturated acyl chain is not fully understood. Stereochemical studies have shown that only the D-hydrogens at positions 9 and 10 are removed to give a cis double bond. This appears to occur by concerted removal of the hydrogens without involving an oxygen-containing intermediate, but the active form of the oxygen in the enzyme-substrate complex is unknown. Attempts to demonstrate involvement of a hydroxyacyl intermediate have been negative, and hydroxyacyl-CoAs are not readily desaturated. Furthermore, it remains uncertain whether one or two cytochrome b_5 molecules per complex are required to transfer two electrons from NADH to molecular oxygen.

Modification of Δ^9 Desaturase Activities In Vitro

To understand the mechanisms, control, and regulation of the Δ^9 desaturase system, it is important to have probes that alter its activity. One potent inhibitor is cyanide. Since $1mM$ KCN completely inhibits Δ^9 desaturation in rat liver by acting on the terminal desaturase component, the latter is frequently referred to as the *cyanide-sensitive factor*. Cyanide inhibition appears to be related to accessibility of the nonheme iron in the desaturase. Tissues with relatively low Δ^9 desaturase activity are less inhibited by cyanide. Some Δ^9 desaturases (for example, yeast) are not inhibited by cyanide.

Iron chelators, such as bathophenanthroline sulfonate in the presence of ascorbate, are inhibitory in vitro, which could be anticipated in view of the nonheme iron in the desaturase component. Less readily explained is the observation that copper supplementation increases the 18:1/18:0 ratio in adipose tissue, and copper deficiency reduces Δ^9 desaturase activity of liver microsomes. Neither copper chelators nor $CuSO_4$ added in vitro alters desaturase activity. Considering the absence of copper in purified proteins, direct involvement of copper as an active component of the desaturase seems unlikely.

Cyclopropenoid fatty acids found in stercula and cotton seeds are potent inhibitors of the Δ^9 desaturase. Sterculoyl- and malvaloyl-CoA

(18- and 16-carbon derivatives, respectively, with cyclopropene rings in the Δ^9 position) specifically inhibit Δ^9 desaturation in vitro. Their in vivo action can be seen when hens are fed meal containing these fatty acids and decreased 18:1/18:0 ratios in their egg yolks are monitored. Cyclopropene acids have been used in vitro to alter differentially Δ^9 and Δ^6 activities of developing brian and thus to distinguish between the relative contributions of these two enzymes in perinatal brain development.

Acceptors for the products of desaturation generally stimulate the Δ^9 desaturase system. Both glycerol-3-phosphate and lysolecithin remove monoenoic products as they are desaturated and probably stimulate overall activity by reducing inhibitory feedback. The monounsaturated products are found predominantly in phospholipids in in vitro studies with crude microsomal preparations.

Dietary and Hormonal Regulation of Δ^9 Desaturase

A remarkable feature of Δ^9 desaturase is the extreme response to dietary deprivation and alterations. When rats are not fed for 12 to 72 h, liver Δ^9 desaturase activity declines markedly to levels less than 5% of control values, as stored energy reserves are mobilized from adipose triacylglycerols (Figure 6.4). When the rats are refed, Δ^9 desaturase activity increases dramatically to levels of 2- to 4-fold above normal. The restoration process has been termed *super induction*, as levels of enzyme activity can rise more than 50-fold above the fasted state, particularly when the rats are refed a fat-free diet enriched in carbohydrate or protein. With protein synthesis inhibitors and immunological techniques it has been shown that synthesis of the desaturase component is altered quantitatively. Such responses of the liver enzyme to dietary intake probably explain the so-called circadian changes in Δ^9 desaturase activity, where activities can fluctuate 4-fold over a 24-h period; highest liver activity (around midnight) corresponds to maximal food intake in the nocturnal rat.

In contrast with the liver enzyme, brain Δ^9 desaturase is little altered by dietary restrictions or modification, which ensures a constant level of activity during crucial stages of brain development. Brain Δ^9 desaturase activity is greatest during the perinatal and suckling period in rats and is generally higher than in liver. However, when the rats are weaned, the liver activity rapidly increases in response to solid-food intake to achieve adult levels 100- to 200-fold greater than in neonates; in contrast, brain Δ^9 desaturase activity slowly declines.

In addition to the role that diets low in total or polyunsaturated fats play in increasing liver Δ^9 desaturase activity, there is also evidence that dietary polyunsaturated acids, particularly linoleic acid, selectively inhibit monoene formation. While the mechanism is not

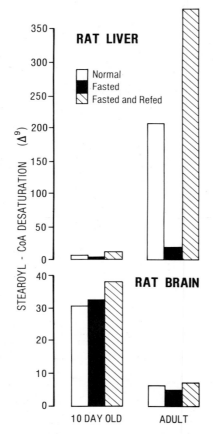

Figure 6.4. Effects of fasting for 48 h and subsequent refeeding of a normal chow diet for 24 h on the in vitro Δ^9 desaturase activities of brain and liver from 10-day-old and adult rats. Adapted from Cook and Spence (1973).

known, the effect of polyunsaturated acids on Δ^9 desaturase is much more rapid than on de novo fatty acid synthesis.

Hormonal regulation of Δ^9 desaturase is not completely defined, despite close association between the activities of lipogenic enzymes and carbohydrate availability and metabolism. Insulin appears to be an inducer of Δ^9 desaturase based on in vivo experiments. Rats made diabetic by streptozotocin-induced destruction of β-cells of the pancreas have depressed Δ^9 desaturase activity in liver, mammary gland, and adipose tissue, and insulin restores the activity if administered in vivo; insulin added to an in vitro assay has no effect on desaturase activity. The in vivo insulin effect is blocked by actinomycin D or puromycin, indicating that activity changes involve synthesis of new protein. Significant changes in the content of cytochrome b_5 and the reductase do not occur. Despite limited use of immunochemical techniques to assess amounts of desaturase protein, time-course studies with transcription and translation inhibitors indicate that the mRNA for the desaturase is rather stable. Other hormones and effectors,

insulin is an inducer of Δ^9 desaturase

glucagon . cAMP do not alter Δ^9 desaturase

such as glucagon and cAMP, do not appear to alter Δ^9 desaturase activity, whereas epinephrine and thyroxine may enhance monoene formation.

What remains to be demonstrated is whether the effects of insulin and other hormones are attributable to more than alteration of the flux of carbohydrate metabolites or if, as with de novo fatty acid synthesis, the desaturase is covalently modified.

Formation of Monounsaturated Fatty Acids in Plants

In plants, desaturation occurs in both the cytosol and chloroplasts. Most plants utilize stearoyl-ACP as substrate, with free fatty acid or fatty acyl-CoA being inactive. *Euglena* is an interesting case. When grown in the dark, its chloroplasts become nonfunctional and 18:0-CoA is the required form of the saturated substrate, whereas cells grown in light require 18:0-ACP. Somewhat analogous to animal systems, desaturation in plants requires molecular oxygen, NADPH, a flavoprotein reductase (ferrodoxin:NADPH reductase), and ferrodoxin as an intermediate electron receptor. The desaturase component is inhibited by cyanide but not by carbon monoxide.

FORMATION OF POLYUNSATURATED FATTY ACIDS
Characteristics and Restrictions in Animal Systems

All eucaryotic organisms contain polyenoic fatty acyl chains in the complex lipids of their membranes, and most mammalian tissues can modify acyl chain composition by introducing more than one double bond. To understand the limitations of polyunsaturated fatty acid formation in eucaryotes, several characteristics or restrictions must be considered:

1. The first double bond introduced into a saturated acyl chain is generally in the Δ^9 position, so substrates for further desaturation contain either a Δ^9 double bond or one derived from the Δ^9 position by chain elongation. An exception is the relatively large amount of $(n - 10)$ monounsaturated fatty acyl chains in neonatal rat brain ($16:1(n - 10)$ and $18:1(n - 10)$ comprise up to 35% of the monoene fraction). In vitro evidence indicates that these acids are formed from 16:0 and 18:0 by Δ^6 desaturation by an enzyme distinct from the Δ^9 desaturase; however, the qualitative significance of this unusual isomer composition during brain development is unclear.
2. Like the Δ^9 desaturation that inserts the first double bond, further desaturation is an oxidative process requiring molecular oxygen, reduced pyridine nucleotide, and an electron transfer system consisting of a cytochrome and related reductase enzyme.
3. Animal systems cannot introduce double bonds beyond the Δ^9 position. Thus, second and subsequent double bonds are always

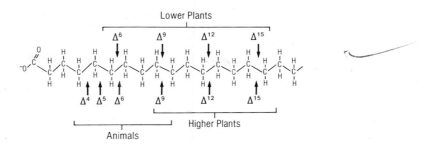

Figure 6.5. Positions of fatty acyl chain desaturation by enzymes in animals, plants, and lower plants such as *Euglena*.

inserted between an existing bond and the carboxyl end of the acyl chain, never on the methyl side of an existing bond (Figure 6.5). Plants, on the other hand, generally introduce second and third double bonds between the existing double bond and the terminal methyl group. The ability displayed by diatoms and *Euglena* to desaturate on either side of an existing bond has not been demonstrated in higher organisms. Consequently, in animals, double bonds are inserted at the Δ^9, Δ^6, Δ^5, and Δ^4 positions and in plants, at the Δ^9, Δ^{12}, and Δ^{15} positions. Well-established evidence confirms Δ^9, Δ^6 and Δ^5 desaturases in a variety of animal tissues. Although Δ^4 desaturase has not been measured directly in vitro, the abundance of long chain polyenoic acids containing Δ^4 double bonds in tissues such as brain, retina, and testes, coupled with the other general restrictions on polyene formation, argues strongly for a distinct Δ^4 desaturase. Comparative studies with tissue explants and neoplastic cells in culture suggest that Δ^4 desaturase is characteristic of differentiated tissue and may not be expressed in undifferentiated cells.

4. In most organisms, and certainly in higher animals, methylene interruption between double bonds must be maintained, and conjugated double bonds are extremely rare. Accordingly, given the limitations of mammalian desaturases, chain elongation usually alternates with desaturation to maintain methylene interruption in polyunsaturated chains.

5. All bonds introduced by oxidative desaturation are in the cis geometric configuration. When acyl chains containing trans double bonds are introduced to animal systems through diet or intestinal bacteria, the trans bonds are recognized as distinct from cis acids.

Essential Fatty Acids: A Contribution of Plant Systems

An important consequence of the restrictions to unsaturated fatty acid formation in animals is that requirements for polyunsaturated acyl chains cannot be met solely by de novo metabolic processes within animal tissues. Animals are absolutely dependent on plants for the two major precursors of the $(n - 6)$ and $(n - 3)$ fatty acids. Through dietary linoleic and linolenic acids, plants provide double bonds in the Δ^{12} and Δ^{15} positions. This permits ultimate formation of acyl chains containing four to six double bonds, using available desaturase and chain elongation enzymes of animal tissues.

Severe effects are observed in experimental animals and humans in the absence of these dietary acids. These include a dramatic decrease in weight; dermatosis and increased skin permeability to water; enlarged kidneys and reduced adrenal and thyroid glands; cholesterol accumulation and altered fatty acyl composition in many tissues such as liver, lung, adrenals, and skin; impaired reproduction; and ultimate death. The four $(n - 6)$ acids in the sequence from $18:2(n - 6)$ to $20:4(n - 6)$ (Figure 6.6) individually have similar potency in reversing these effects of deficiency, whereas the activity of $18:3(n - 3)$ alone is much lower. Thus, the term essential fatty acid clearly applies at least to the major $(n - 6)$ acids, although the essential role of $(n - 3)$ acids remains unclear.

A function for $18:2(n - 6)$, in addition to its role as precursor of $20:4(n - 6)$ seems likely. Recent studies have shown that cats, who are unable to synthesize $20:4(n - 6)$ from $18:2(n - 6)$, require both $18:2(n - 6)$ and $20:4(n - 6)$ in their diets. Hence, as carnivores, they rely on other animals to make $20:4(n - 6)$ for them. Also, some other fatty acids that cannot serve as prostaglandin precursors fulfill some requirements of essential fatty acids, which further suggests a role for essential fatty acids in their own right.

Families of Fatty Acids and Their Metabolism

Relationships between fatty acids in the pathways of metabolic conversions can be evaluated by considering groups or families of fatty acids based on the primary parent or initial unsaturated acid in the sequence. The predominant fatty acid families are the $(n - 6)$ acids derived from $18:2(n - 6)$, the $(n - 3)$ acids derived from $18:3(n - 3)$, the $(n - 9)$ acids derived from $18:1(n - 9)$, and the $(n - 7)$ acids derived from $16:1(n - 7)$ (Figure 6.6).

The (n − 6) family. Arachidonate [$20:4(n - 6)$], an abundant polyenoic acyl chain found in most animal tissues, can be formed from $18:2(n - 6)$ by the alternating sequence of Δ^6 desaturation, chain elongation of the $18:3(n - 6)$ intermediate, and Δ^5 desaturation of $20:3(n - 6)$. Not only is $20:4(n - 6)$ a component of phospholipids contributing to the structural integrity of membranes, but it also is the primary precursor of several classes of oxygenated derivatives (for example, prostaglandins) with a variety of biological activities (Chapter 11). Frequently, $20:4(n - 6)$ is referred to as an essential fatty acid.

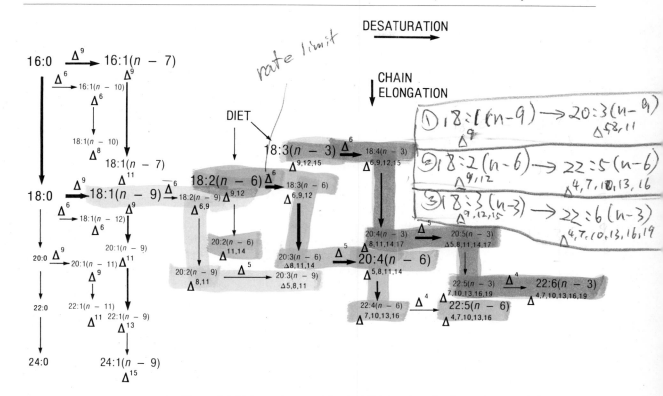

Figure 6.6. Major pathways of fatty acid biosynthesis by desaturation and chain elongation in animal tissues. Note the alternating sequence of desaturation in the horizontal direction and chain elongation in the vertical direction in the formation of polyunsaturated fatty acids from dietary essential fatty acids.

Indeed, it is absolutely required in many tissues, but an adequate dietary supply of 18:2(n − 6) can be converted to 20:4(n − 6) under most circumstances, so 20:4(n − 6) per se need not be an essential dietary component for most mammals. There are exceptions in specific tissues or cells. For example, neutrophils require 20:4(n − 6) for leukotriene production but cannot synthesize it from 18:2(n − 6); thus, these cells are dependent on extracellular supplies.

In liver and most other tissues of animals in a normal, balanced state, the only members of the (n − 6) family to accumulate in relatively large quantities are 18:2(n − 6) and 20:4(n − 6); much lower levels of the intermediates 18:3(n − 6) and 20:3(n − 6) are detected. Such observations support a rate-limiting role for the Δ^6 desaturase in the sequence, although Δ^5 desaturase activity measured in vitro with 20:3(n − 6) as substrate is approximately equivalent or lower (Table 6.2). Thus, a possible regulatory function for the Δ^5 desaturase in some circumstances should not be entirely dismissed.

Table 6.2. Relative Rates of Desaturation by Rat Liver and Brain

Desaturation	Substrate	Liver		Brain	
		Adult	10-day-old	Adult	10-day-old
Δ^9	18:0	100	2	4	12
Δ^6	18:2 $(n - 6)$	30	9	1	8
	18:3 $(n - 3)$	47	17	1	13
Δ^5	20:3 $(n - 6)$	17	4	1	1

Source: Adapted from Sprecher (1981) and Cook and Spence (1973).

[handwritten margin note: Δ⁶ and Δ⁵ desaturase exist separately]

Several lines of indirect evidence support the existence of distinct Δ^6 and Δ^5 desaturase enzymes. The enhancement of desaturase activity observed upon refeeding after fasting can be suppressed by glucagon or dibutyryl cAMP in the case of Δ^6 desaturase activity but not for Δ^5 desaturase activity. Further support for distinct enzymes comes from cell lines that have lost Δ^6 but retain Δ^5 desaturase activity. Also convincing is evidence that the Δ^5 desaturase of rat liver microsomes can act directly on a phospholipid substrate to form arachidonyl phosphatidylcholine from the eicosatrienoyl phospholipids precursor; no such activity with acyl chains esterified to phospholipid has been demonstrated for the Δ^6 desaturase.

Desaturation by Δ^9, Δ^6, and Δ^5 desaturases of liver and brain can be stimulated by cytosolic proteins, bovine serum albumin, or catalase, particularly when microsomal desaturase activity is reduced by repeated washings with buffer. At least part of the activation may be related to the acyl chain binding properties of these proteins, which may regulate availability of fatty acyl substrates for the reaction or removal of products. Catalase may also influence oxidation-reduction reactions of the electron transport sequence, particularly for the Δ^6 and Δ^5 desaturases. The physiological significance of a direct action by the soluble proteins on desaturation should be considered cautiously, however, as partially purified desaturase is not stimulated by either albumin or cytosolic protein.

The physical relationship of the desaturases and chain-elongation enzymes within the membranes of the endoplasmic reticulum remains to be demonstrated. There is indirect evidence that the sequence of reactions, including esterification of 20:4$(n - 6)$ to phospholipids, proceeds in a concerted manner without release of free fatty acyl intermediates to cellular pools. Dilution by intermediates of the desaturation-elongation sequence does not occur.

An alternate method for 20:3$(n - 6)$ and 20:4$(n - 6)$ formation from 18:2$(n - 6)$ would be possible if mammalian tissues had a Δ^8 desaturase whereby 20:2$(n - 6)$, formed by chain elongation of 18:2$(n - 6)$, could be desaturated to 20:3$(n - 6)$. However, rigorous testing of liver, brain, and mammary tissue in vivo and in vitro using substrates with the first double bond in the Δ^{11} position [for example, $20:1(n - 9)$ and $20:2(n - 6)$] has shown that addition of a double bond is always in the Δ^5 and not the Δ^8 position. Thus, these tissues apparently do not have a Δ^8 desaturase. The relatively low Δ^8 desaturation reported in testes and tumors may be due to minor lack of specificity of the Δ^9 desaturase, although existence of a distinct Δ^8 enzyme in these tissues has not been eliminated.

The polyunsaturated acyl chains, $22:4(n - 6)$ and $22:5(n - 6)$, are quantitatively significant components of some tissues, particularly in the central nervous system, adrenals, kidneys, and testes. Involvement of a Δ^4 desaturase in the formation of $22:5(n - 6)$ has been questioned based on feeding studies with rats in which $22:4(n - 6)$ did not give rise to $22:5(n - 6)$ in tissue lipids. On the other hand, Δ^4 desaturation of dienoic, trienoic, and tetraenoic acids of 20-carbons or fewer has been reported by several workers, confirming the existence of a Δ^4 desaturase. Exogenous $22:4(n - 6)$ may be a preferred substrate for *retroconversion* (that is, chain shortening). This mitochondrial process of partial degradation involves loss of either a single two-carbon fragment or a double bond and either two or four carbon atoms by reduction, isomerization, and β-oxidation reactions. In general, retroconversion utilizes fatty acids of 20 carbons or greater. Since only double bonds in the Δ^4 position are lost, this process could provide an alternate route for the production of acids with the first double bond in the Δ^5 position. The quantitative significance of this process is not well established.

The (n − 3) family. Generally, the most abundant $(n - 3)$ acyl chain in animal tissues is $22:6(n - 3)$, found in the membrane phospholipids of cerebral cortex, retina, testes, muscle, and several other tissues. In retinal rod outer segments, for example, the major phospholipids may contain 40–60% $22:6(n - 3)$. Clearly, the presence of six double bonds, which is not possible in $(n - 6)$ acids, markedly influences membrane fluidity. However, the full significance of the $(n - 3)$ family of acyl chains derived from $18:3(n - 3)$ is not well understood. Evidence continues to accumulate for a specific role for the $(n - 3)$ acids, and $18:3(n - 3)$ is widely considered an essential fatty acid. A vital role for the long chain $(n - 3)$ acids in membranes is supported by their tenacious retention during dietary deprivation. A major obstacle to defining precise roles and the essential nature of $(n - 3)$ acids is the difficulty in providing diets so balanced with $(n - 6)$ acids that one can distinguish signs of deficiency and effects of supplements attributable specifically to $(n - 3)$ acids in the absence of a deficiency of $(n - 6)$ acids. A reported case of human $18:3(n - 3)$ deficiency involving neurological abnormalities supports the dietary essentiality of this family.

Competition between the $(n - 3)$ and $(n - 6)$ fatty acids is a significant consideration at the level of desaturation and chain elongation and of oxygenation (Chapter 11). For example, with the Δ^6 desaturase enzyme, $18:3(n - 3)$ is a better substrate than $18:2(n - 6)$, which in turn is better than $18:1(n - 9)$. Thus, an abundance of $18:3(n - 3)$ can effectively decrease formation of $20:4(n - 6)$ from $18:2(n - 6)$. Incorporation of dietary long-chain $(n - 3)$ acids depresses the content of $20:4(n - 6)$ in practically all tissues examined. Claims that $20:5(n - 3)$ in the diet may reduce the tendency for blood platelet aggregation and arterial thrombosis probably relate to competition of the $(n - 3)$ and $(n - 6)$ acids for oxygenase enzymes (Chapter 11).

The $(n - 9)$ family. The prominent acyl chain in the $(n - 9)$ family is $18:1(n - 9)$. Generally, competition from $18:2(n - 6)$ and $18:3(n - 3)$ for the Δ^6 desaturase prevents formation and accumulation of more unsaturated $(n - 9)$ acids. However, in animals on a diet deficient in essential fatty acids, competition is removed and $18:1(n - 9)$ is utilized as a substrate for the rate-limiting Δ^6 desaturase (Figure 6.6). Further chain elongation of $18:2(n - 9)$ to $20:2(n - 9)$ and Δ^5 desaturation results in accumulation of $20:3(n - 9)$. While $20:3(n - 9)$ may partially substitute for some physical functions of the essential fatty acids within membranes, it is not a precursor of prostaglandins. Since the deficiency of essential fatty acids markedly reduces $20:4(n - 6)$ while increasing $20:3(n - 9)$, an increase in the ratio of triene to tetraene in tissues and serum has often been used as an index of essential fatty acid deficiency. With capillary and high-efficiency gas chromatography, isomer separations can be achieved in analyses of tissue and fluid lipids, so the ratio of $20:3(n - 9)$ to $20:4(n - 6)$ (normally less than 0.2) becomes an even more precise indication of a deficiency. Even the use of this ratio has limitations; for example, inhibition of Δ^6 desaturase would reduce formation of both $20:3(n - 9)$ and $20:4(n - 6)$, resulting in a deficiency state without a marked alteration of the ratio. The total amount of $(n - 6)$ acids has been proposed as a better reflection of essential fatty acid deficiency.

The $(n - 7)$ family. The fatty acid $16:1(n - 7)$ is the primary $(n - 7)$ acid in membranes and circulating lipids. However, as most analyses do not distinguish specific 18:1 isomers, the contribution of $18:1(n - 7)$ to the 18:1 fraction is seldom appreciated. In developing brain, for example, $18:1(n - 7)$ formed by chain elongation of $16:1(n - 7)$ comprises up to 25% of the total 18-carbon monoene pool. On the other hand, high levels of polyunsaturated fatty acids derived from $16:1(n - 7)$ are not detected even on a fat-free diet, although increased levels of $16:1(n-7)$ frequently accompany deficiencies of essential fatty acids. Potentially, $20:4(n-7)$ could be formed from $16:1(n-7)$. This 20:4 isomer has only a single carbon shift of the double bonds compared with $20:4(n-6)$. Thus, lack of $20:4(n-7)$ formation in deficiency states supports a high degree of specificity in the formation and utilization of 20:4.

Dietary and Hormonal Alterations of Polyunsaturated Acid Synthesis

Regulation of the desaturase and chain-elongation enzyme activities involved in polyunsaturated fatty acid synthesis is not well defined, possibly reflecting the complexity of the mechanisms involved. Since the capacity of the cytochrome system greatly exceeds the requirements of the desaturases, electron transfer is not considered a point of control. Responses of the desaturases to dietary alterations observed in different laboratories have sometimes been conflicting, but the Δ^6 and Δ^5 desaturases are less dramatically influenced by fluctuations in dietary intake than is the Δ^9 desaturase (Table 6.3).

Table 6.3. Effects of Dietary, Hormonal, and Other Manipulations on Δ^9, Δ^6, and Δ^5 Desaturation Activities in Experimental Animals

Alterations of desaturation activities			
	Effect on desaturation		
Treatment	Δ^9	Δ^6	Δ^5
Dietary:			
High Glucose-short term	↑	↑	↑
long term	↑	↓	
High Protein	↑	↑ ↑	↑
Fasting	↓ ↓	↓	↓
Refeeding	↑ ↑	↑	↑
Hormonal:			
Insulin	↑	↑	↑
Glucagon	−	↓	↓
Epinephrine	↑	↓	↓
cAMP	−	↓	↓
Glucocorticoids		↓	↓
Thyroxine	↑	↓	
Hypothyroidism	↓ ↓	↓	
Others:			
Sterculic Acid	↓	−	
Cytosolic Proteins	↑	↑ ↑	↑
Retinoic Acid	↓	↑	

The − indicates no significant change and blank spaces indicate an absence of definitive information for that particular treatment. Adapted primarily from Jeffcoat (1979) and Brenner (1981).

Generally, Δ^6 and Δ^5 desaturase activities are reduced during fasting, but a major increase above normal levels—the "superinduction" seen with the Δ^9 desaturase—is not observed upon refeeding (Figure 6.4 and Table 6.3). Protein-enriched diets induce an increase in Δ^6 desaturase activity, which is blocked by inhibitors of protein synthesis. The observations that glucose and fructose seem to increase desaturase activities in fasted animals in the short term, but repress activity with prolonged feeding, may reflect an initial positive response of the desaturases to insulin. Insulin does increase the activities of all three desaturases, particularly when they have been suppressed by induced experimental diabetes. Other hormones induce a differential effect on the Δ^9, Δ^6, and Δ^5 desaturases. For example, glucagon and dibutyryl cAMP block the response of Δ^6 desaturase to refeeding after starvation but have little or no effect on Δ^9 desaturase. Epinephrine also suppresses Δ^6 desaturase but enhances Δ^9 desaturase. Generally, Δ^6 and Δ^5 desaturases respond similarly. Cyclic nucleotides and protein kinases as mediators in the action of hormones require further investigation, and more research is necessary to define hormonal control of polyunsaturated fatty acid biosynthesis.

Two or More Double Bonds in Plants   Studies by James and co-workers of specificities of plant desaturases that insert second and third double bonds into monounsaturated acyl chains suggested two operative systems: (1) an enzyme with an active site that interacts with an existing double bond at a fixed distance (optimally nine carbons) from the carboxyl end and (2) another enzyme with an active center that interacts with an existing double bond at a fixed distance (optimally nine carbons) from the methyl end. It has been speculated that the former accepts thioesters as substrates, whereas the latter may accept an acyl chain esterified to a complex lipid substrate. Since 18:1$(n - 9)$ meets both criteria, it could be acted on by both systems and, therefore, would be the most actively desaturated monoenoic acid in plants. However, recent studies indicate that 18:1$(n - 9)$ desaturation to 18:2$(n - 6)$ occurs primarily on phosphatidylcholine in plant microsomes, not by acyl-CoA desaturation. Further desaturation of 18:2$(n - 6)$ to 18:3$(n - 3)$ also occurs on microsomal phosphatidylcholine.

UNSATURATED FATTY ACIDS WITH TRANS DOUBLE BONDS   In recent years lipid biochemists and nutritionists have focused intently on the influence and metabolism in various animal tissues and cells of unsaturated fatty acids containing trans double bonds. Trans unsaturated fatty acids, the geometric isomers of naturally occurring cis acids, are not produced by mammalian enzymes but are formed enzymatically by microorganisms in the gastrointestinal tract of ruminant mammals and chemically during commercial partial hydrogenation of fats and oils. Trans acids have been described as unnatural, foreign, or nonphysiological, but these terms are not completely accurate. Through diets containing beef fat, milk fat, margarines, and partially hydrogenated vegetable oils, trans acids are ingested and subsequently incorporated and modified in animal tissues.

Early studies suggested selective exclusion of trans acids from certain metabolic processes and from incorporation into membrane lipids, particularly in the central nervous system. Recent investigations from many laboratories support various levels of selectivity in utilization of trans acids. Some studies have indicated no appreciable selectivity for absorption, esterification, or oxidation of trans isomers compared with cis isomers, whereas others have indicated a significant degree of discrimination of specific positional isomers of trans acids (for example, Δ^{13} trans acids). In general, trans acids appear to be recognized as a distinct class of acyl chains with properties intermediate between those of saturated and cis-monounsaturated acids, particularly in specificity for the position of esterification to phospholipids. Exceptions can be found, but short-term accumulation of trans acids in tissues generally is proportional to dietary levels. Lack of preferential accumulation of positional isomers in tissues over the long term suggests that all isomers turn over similarly. For example, in human body tissues and fluids, t-18:1 (a mixture of positional isomers with the trans bonds distributed between the Δ^6 and Δ^{14} positions) is 2–14% of the total fatty acyl composition of the lipids, and patterns of positional isomer distribution do not appreciably differ from dietary lipids.

Many broader implications of the influence of trans-unsaturated fatty acids in biological systems continue to be controversial. Studies of the effects of specific trans isomers in comparison with their cis analogues are clarifying the influence that trans acids may have on normal lipid metabolism. Specific interactions of trans acids with the desaturation and chain-elongation enzymes of animal tissues have been reported. Positional isomers of t-18:1 (except for the Δ^9 isomer) are desaturated by the Δ^9 desaturase of rat liver microsomes, resulting in a series of cis- and trans-dienoic isomers that, in some cases, are desaturated again by the Δ^6 desaturase to unusual polyunsaturated structures. Chain elongation of some positional t-18:1 isomers also occurs, although at a much slower rate than Δ^9 desaturation. Thus, certain monoenoic trans isomers found in partially hydrogenated oils can both inhibit desaturase activities and form products with as yet undetermined biological activities.

Several trans,trans-dienoic fatty acid isomers, including t,t-18:2$(n - 6)$, act as substrates for Δ^6 desaturation in liver and brain, albeit at a much lower rate than for c,c-18:2$(n - 6)$. Dienoic isomers of 18:2 containing trans bonds clearly lack the properties of essential fatty acids and have been shown to interfere with normal conversion of 18:2$(n - 6)$ to 20:4$(n - 6)$ in a variety of in vitro and in vivo experimental models. Inhibition is primarily at the Δ^6 desaturase, although Δ^5 desaturase activity actually appears to increase in some tissues. The competitive effect of trans acids is accentuated in mild deficiency of essential fatty acids, where dietary supplements containing trans acids greatly intensify the signs of deficiency. Since the Δ^6 desaturase is inhibited, the conversion of available 18:2$(n - 6)$ to 20:4$(n - 6)$ is reduced, and accumulation of 20:3$(n - 9)$ from conversion of 18:1$(n - 9)$ also is blocked. While it should be borne in mind that trans-dienes are minor components (usually less than 1%) of hydrogenated vegetable oils, complex interactions and interconversions of a variety of trans isomers are possible.

Even though many suggestions of adverse effects of trans acids in biological systems were based on evidence not directly or uniquely attributable to trans acids themselves, questions about long-term effects of increased trans fatty acid intake still remain unresolved. Clearly, continued investigations of trans acids as a distinct, but certainly not insignificant, class of fatty acids are necessary to determine their influence on such aspects of lipid metabolism as normal tissue development and function, the development of atherosclerosis, and altered cell metabolism as seen in tumor tissue.

ABNORMAL PATTERNS OF DISTRIBUTION AND METABOLISM OF LONG-CHAIN SATURATED AND UNSATURATED FATTY ACIDS

Despite the diversity of synthetic enzymes involved in fatty acyl chain formation, documented clinical defects in acyl chain metabolism are relatively few. Most alterations of acyl chain patterns reflect deficiency states, not absolute absence of specific enzymes. This probably indicates that all the major desaturation and elongation enzymes are essential in supporting life-sustaining cellular processes. However, defects or deficiencies resulting in abnormal patterns of unsaturated fatty acid distribution have been documented and warrant consideration.

Essential Fatty Acid Deficiency

An inadequate supply of essential fatty acids resulting in the deficiency signs described previously for rats is very rare in humans. Most normal diets contain enough 18:2$(n - 6)$ and 18:3$(n - 3)$ or their metabolic products to meet tissue demands, and adipose stores provide a protective buffer for extended periods of limited intake. However, severe deficiency states have been observed in humans (especially in premature infants with restricted adipose stores) on

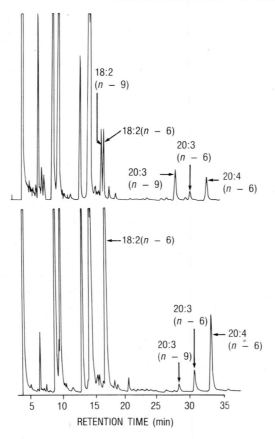

Figure 6.7. Gas-liquid chromatographic pattern of acyl chain distribution in the serum of an infant on total parenteral nutrition (intravenous feeding) for 5 weeks without appropriate lipid supplementation (upper panel) and after 1 week on lipid and zinc supplementation (lower panel). Fatty acyl chain methyl esters from the serum phospholipids were analyzed by gas-liquid chromatography at 180°C using a 50-m capillary column coated with Silar 9CP.

prolonged intravenous feeding or artificial milk formulations without adequate lipid supplements (Figure 6.7). Marked alterations of serum fatty acid patterns, which are characterized by depletion of $(n - 6)$ acids and a major increase in the $20:3(n - 9)$ to $20:4(n - 6)$ ratio, are accompanied by severe skin rash, loss of hair, and irritability. These signs are reversed rapidly by supplementing with lipid emulsions containing esterified $18:2(n - 6)$. A deficiency of $18:2(n - 6)$ and $18:3(n - 3)$ in the plasma of patients on fat-free intravenous feeding reflects absence of dietary essential fatty acids and also may be a result of inhibition of release of free fatty acids from adipose tissues due to the high insulin levels secondary to constant hypertonic glucose infusion.

In experimental animals, a dietary deficiency of essential fatty acids is accompanied by changes in fatty acyl composition of tissue and circulating lipids. Brain tissue shows exceptional resistance to loss of essential fatty acids during restrictions, but modification of acyl patterns can be achieved if the diet is started at an early age and continued for several generations.

Zinc Deficiency

Many gross signs of zinc deficiency in experimental animals are similar to those observed in essential fatty acid deficiency. Possible relationships between zinc and essential fatty acids and prostaglandins have been proposed often, but direct connections at the metabolic level have not been shown. Some studies, including those with humans, have shown a positive correlation between plasma zinc and $20:4(n - 6)$ levels. In chicks and rats, however, zinc deficiency increases accumulation of $20:4(n - 6)$ and apparently interferes with its normal metabolism, possibly causing prostaglandin deficiency. Direct involvement of zinc in desaturation and chain elongation and specific impairment of these processes in deficiency states has been proposed but not demonstrated. Other potential sites for zinc involvement include intestinal absorption of fatty acids, release and mobilization of acyl chains from complex lipids, and conversion of fatty acids to oxygenated derivatives.

Acrodermatitis enteropathica is a rare and serious disorder characterized by skin lesions, gastrointestinal disturbances, and retarded growth. It is accompanied by low polyunsaturated fatty acid content in serum lipids, yet without accumulation of $20:3(n - 9)$. Administration of zinc elicits a dramatic reversal of the symptoms, with an apparent correlation between zinc and $18:2(n - 6)$ levels. Clearly, the relationships between zinc and essential fatty acid metabolism in this disorder, and in general, are complex.

Other Clinical Disorders

In a collective assessment of several human diseases or disorders, Holman and Johnson (1981) found that many involved abnormal patterns of polyunsaturated fatty acids, attributable to insufficient dietary $18:2(n - 6)$ or to abnormal metabolism of the essential fatty acids (Figure 6.8). Statistical comparisons of total families of fatty acids, individual fatty acid components, and total products of chain elongation and individual desaturation steps permitted broad groupings. Some disorders, including cystic fibrosis, Crohn's disease, Sjögren-Larsson syndrome, peripheral neuropathy, and congenital liver disease of unknown etiology were assessed to have diminished metabolic capabilities for desaturation or chain elongation of polyunsaturated fatty acids. Essential fatty acid deficiency, alcoholism,

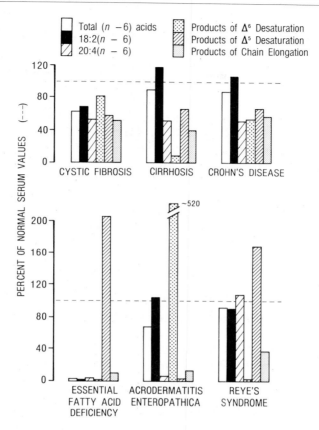

Figure 6.8. Alterations in the patterns of long chain fatty acids in the serum lipids of patients with various diseases or disorders compared with control individuals. Note that classifications include only accumulating end products and that other reactions may be involved in prior steps leading to their formation. Adapted from Holman and Johnson (1981).

cirrhosis, Reye's syndrome, and chronic malnutrition were accompanied by significantly abnormal patterns of essential fatty acids in serum phospholipids. While such analyses point to potential defects in metabolic steps of polyunsaturated fatty acid metabolism, they also underline our limited knowledge of specific details and the potential for secondary deficits in essential fatty acid metabolism in many human disorders.

Relationships to Plasma Cholesterol   In the past three decades, considerable evidence has emerged to support a correlation between a high dietary intake of saturated fats relative to polyunsaturated fats and the occurrence of atherosclerosis and coronary disease. Although some earlier epidemiological data have been the subject of controversy, there is reliable evidence that the risk of coronary disease is proportional to serum cholesterol levels. Furthermore, total serum cholesterol (particularly in low-density lipoproteins; see Chapter 13) can be decreased by dietary intake of lipids enriched in polyunsaturated fatty acyl chains. In addition, factors such as platelet aggregation, blood pressure, and vascular obstruction may be influenced through some of the potent oxygenated derivatives of polyunsaturated fatty acids (Chapter 11). Although the appropriate balance of polyunsaturated fatty acid intake cannot be defined precisely from available evidence, and indeed it may be impossible to generalize in a manner applicable to all individuals, greater attention to the overall composition of dietary fats would seem to be individually beneficial.

FUTURE DIRECTIONS

Progress in understanding the desaturation and chain elongation of fatty acyl chains and the influence of altered fatty acyl composition on a range of lipid-related functions has been steady and exciting over the last half century. Yet each progressive step reveals a need to expand our knowledge of details of the mechanism, regulation, and functions of these processes in higher organisms and the extent to which available information applies to different mammalian tissues. New and innovative approaches must be added to the cumulative efforts of many disciplines.

A more specific definition of the essential role of $18:3(n-3)$ and the long-chain polyunsaturated fatty acids of the $(n-3)$ family relative to $18:2(n-6)$ and its derivatives is needed. Regulation of fatty acid modification at the cellular level and intermembrane associations of desaturation and chain-elongation enzymes within membrane domains can be evaluated only through improved isolation, purification, and reconstitution techniques coupled with perseverance and ingenuity, not to mention serendipidity.

Several promising developments are beginning to be applied to the study of acyl chain modification. Studies based on mutant cell lines deficient in one or more components of the desaturation–chain-elongation sequences, in combination with cloning and recombinant DNA techniques, will be used to assess regulation at the level of gene expression. A broader application of cultured normal, neoplastic, and hybrid cell lines also will facilitate regulation and manipulation of environmental factors. Highly specialized biochemical approaches must cautiously be extrapolated to the complexities of a continuously fluctuating cellular milieu. Despite the wealth of information gained over the last half century, it is safe to say we have only just begun.

PROBLEMS

1. In some biological reactions, the trans-monoenoic fatty acids act as a class of acyl chains with properties that are intermediate between saturated and cis-monoenoic acyl chains. Using physical considerations, explain why this intermediate character might be predicted.
2. Explain why the rate of the condensation step of 16:0-CoA chain elongation by liver microsomes is increased fourfold and occurs at $60–100\,\mu M$ acyl-CoA in the presence of bovine serum albumin, versus at $10–20\,\mu M$ acyl-CoA without albumin. Why might the rate of condensation be lower than the rate of overall chain elongation in the absence of albumin?
3. In neonatal rat brain, $16:1(n - 10)$ accounts for a significant portion of the 16-carbon monoene. How could this positional isomer of $16:1(n - 7)$ be quantitated, and how might the means of its formation be assessed?
4. The triene:tetraene ratio is a valuable index of essential fatty acid deficiency in many cases of restricted intake of dietary $18:2(n - 6)$. When might this ratio inadequately reflect a deficiency state?
5. An infant on intravenous feeding for 5 weeks due to a bowel obstruction had the following pattern of polyunsaturated acyl chains in serum phospholipids. Explain the relationship among these acids and why the patterns differ in the patient and control. How might the patient's situation be corrected? How might a return to normal be monitored?

Fatty acid	Patient	Control
	(% of total)	
$18:2(n - 6)$	2.2	38.1
$18:2(n - 9)$	1.9	0.0
$20:3(n - 6)$	0.2	0.6
$20:3(n - 9)$	1.4	0.1
$20:4(n - 6)$	1.2	3.6

BIBLIOGRAPHY

Entries preceded by an asterisk are suggested for further reading.

Bernhard, K.; von Bulow-Koster, J.; and Wagher, H. 1959. Die enzymatische Dehydrierung des Stearinsaure zu Olsaure. *Helv. Chim. Acta* 42:152–55.

Bloomfield, D. K., and Bloch, K. 1960. The formation of Δ^9-unsaturated fatty acids. *J. Biol. Chem.* 235:337–45.

Brenner, R. R. 1981. Nutritional and hormonal factors influencing desaturation of essential fatty acids. *Prog. Lipid Res.* 20:41–47.

Burr, G. O., and Burr, M. M. 1929. On the nature and role of the fatty acids essential in nutrition. *J. Biol. Chem.* 82:345–67.

Cook, H. W. 1979. Differential inhibition of Δ^9 and Δ^6-desaturation in rat brain preparations *in vitro*. *Lipids* 14:763–67.

Cook, H. W., and Spence, M. W. 1973. Formation of monoenoic fatty acids by desaturation in rat brain homogenate II. Effects of age, fasting and refeeding, and comparisons with liver enzyme. *J. Biol. Chem.* 248:1793–1796.

*Emken, E. A. 1983. Biochemistry of unsaturated fatty acid isomers. *J. Amer. Oil Chem. Soc.* 60:995–1004.

*Emken, E. A., and Dutton, H. J. 1979. *Geometrical and positional fatty acid isomers*. Champaign, Ill.: The American Oil Chemists' Society.

Harris, R. V., and James, A. T. 1965. Linoleic and α-linolenic acid biosynthesis in plant leaves and a green alga. *Biochim. Biophys. Acta* 106:456–64.

*Holman, R. T., ed. 1981. "Essential fatty acids and prostaglandins." In *Progress in lipid research*, vol. 20, particularly 13–22, 41–48, 123–28, 157–60. New York: Pergamon.

Holman, R. T., and Johnson, S. 1981. Changes in essential fatty acid profile of serum phospholipids in human disease. *Prog. Lipid Res.* 20:67–73.

*Jeffcoat, R. 1979. The biosynthesis of unsaturated fatty acids and its control in mammalian liver. *Essays Biochem.* 15:1–36.

*Kunau, W. H., and Holman, R. T. 1977. *Polyunsaturated fatty acids.* Champaign, Ill.: The American Oil Chemists' Society.

MacMahon, V., and Stumpf, P. K. 1964. Synthesis of linoleic acid by particular system from safflower seeds. *Biochim. Biophys. Acta* 84:359–61.

Marcel, Y. L.; Christiansen, K.; and Holman, R. T. 1968. The preferred metabolic pathway from linoleic acid to arachidonic acid *in vitro*. *Biochim. Biophys. Acta* 77:671–73.

Mead, J. F. 1981. The essential fatty acids: Past, present, and future. *Prog. Lipid Res.* 20:1–6.

*Mead, J. F., and Fulco, A. 1976. *The unsaturated and polyunsaturated fatty acids in health and disease.* Springfield, Ill.: Charles C. Thomas.

Meyer, F., and Bloch, K. 1963. Effect of temperature on the enzymatic synthesis of unsaturated fatty acids in *Torulopsis utilis*. *Biochim. Biophys. Acta* 77:671–73.

Nagai, J., and Bloch, K. 1966. Enzymatic desaturation of stearyl acyl carrier protein. *J. Biol. Chem.* 241:1925–27.

Nugteren, D. H. 1962. Conversion in vitro of linoleic acid into γ-linolenic acid by rat liver enzymes. *Biochim. Biophys. Acta* 60:656–57.

Nugteren, D. H. 1965. The enzymatic chain elongation of fatty acids by rat-liver microsomes. *Biochim. Biophys. Acta* 106:280–90.

*Seubert, W., and Podack, E. R. 1973. Mechanisms and physiological roles of fatty acid chain elongation in microsomes and mitochondria. *Mol. Cell Biochem.* 1:29–40.

*Sprecher, H. 1981. Biochemistry of essential fatty acids. *Prog. Lipid Res.* 20:13–22.

*Stubbs, C. D., and Smith, A. D. 1984. The modification of mammalian polyunsaturated fatty acid composition in relation to membrane fluidity and function. *Biochim. Biophys. Acta* 779:89–137.

*Thompson, G. A., Jr. 1980. *The regulation of membrane lipid metabolism.* Boca Raton, Fl.: CRC Press.

*Tinoco, J. 1982. Dietary requirements and functions of α-linolenic acid in animals. *Prog. Lipid Res.* 21:1–45.

*Vandenheuvel, F. A. 1965. Study of biological structure at the molecular level with stereomodel projections II. The structure of myelin in relation to other membrane systems. *J. Amer. Oil Chem. Soc.* 42:481–92.

*Wakil, S. J. 1961. Mechanism of fatty acid synthesis. *J. Lipid Res.* 2:1–24.

CHAPTER 7

Metabolism of Triacylglycerols

David N. Brindley

Triacylglycerols play a major role in energy storage in animals, where they are deposited in adipose tissue. When this storage is excessive it is manifested as obesity, and there is considerable medical interest in trying to understand why some people are so prone to this condition, whereas others find it equally difficult to gain weight. In plants, the storage of triacylglycerols is best illustrated by the oil seeds, in which the triacylglycerols provide energy for growth. These seeds constitute very important commercial crops. Triacylglycerols are ideally suited to this storage function because of the highly reduced state of their fatty acids. Thus, they have high energy contents—about 37 kJ/g, compared with 17 kJ/g for protein and 16 kJ/g for carbohydrate, including glycogen, which is also used to store energy. The other advantage of the triacylglycerols is their insolubility in water, which means that they do not alter the osmotic pressure of the cell.

Most of the triacylglycerols in animals are stored in adipose tissue. However, triacylglycerols can be deposited in liver, heart, and skeletal muscle under several conditions of metabolic stress when the supply of fatty acids from adipose tissue exceeds the need or capacity of the cells to oxidize them. This formation of triacylglycerols removes the potentially toxic effects of the fatty acids and of their acyl-CoA esters, which could damage membranes and inhibit enzymes. The process of triacylglycerol synthesis also regenerates CoA. When required, the triacylglycerols are hydrolyzed by intracellular lipases so that the fatty acids can then be oxidized.

The transport of fatty acids is a major function of the triacylglycerols. Since these compounds are insoluble in the aqueous environment of the cells and of the blood and lymph, the triacylglycerol must be stabilized by association with other lipids and proteins. Such aggregates are called *lipoproteins*. There are two major classes of triacylglycerol-rich lipoproteins, namely, *chylomicrons* and *very low density lipoproteins* (VLDL), as shown in Table 7.1. Chylomicrons carry absorbed dietary fat from the small intestine to other organs, whereas VLDL mainly carry triacylglycerol from the liver to other organs.

Lipoprotein metabolism is of considerable clinical relevance. A high rate of VLDL secretion from the liver, if coupled with a relatively low rate of removal from the circulation, leads to hypertriglyceridemia. This condition in combination with a low circulating concentration of high-density lipoprotein (HDL) is considered to be a major risk factor in the development of premature atherosclerosis (see also Chapter 13). VLDL are eventually converted to low-density lipoproteins (LDL) after the triacylglycerol is removed. These LDL particles still contain much of the cholesterol that was originally used to package and stabilize the VLDL. Thus, if VLDL secretion increases, then the flux into LDL is also increased, resulting in hypercholesterolemia if this flux is not balanced by an increased rate of removal of LDL and VLDL from the circulation. An increase in the concentration of LDL in the serum, especially when accompanied by a low concentration of HDL, is thought to provide one of the best indications of an increased risk for atherosclerosis (Chapter 13).

The initial understanding of the pathways of triacylglycerol synthesis began in the early 1950s with the discovery by Kornberg and Pricer that fatty acids are activated to acyl-CoA esters before they are esterified to phosphatidate. This discovery was followed by descriptions by the groups of Kennedy and Shapiro (Brindley and Sturton

Table 7.1. Composition and Properties of Human Triacylglycerol-rich Lipoproteins

Property	Chylomicrons	Very low density lipoprotein
Density (g/mL)	0.92–0.96	0.95–1.006
Diameter (nm)	75–1000	30–75
Composition (dry wt %)		
Triacylglycerol	80–95	55–65
Phospholipids	3–8	15–20
Free cholesterol	1–3	10
Esterified cholesterol	2–4	5
Protein	1–2	9–10
Major apoproteins	B, C	B, C, E
Minor apoproteins	A	A

1982) of how phosphatidate is converted to various phospholipids and triacylglycerol. In the early 1960s Hübscher and Clark showed that the small intestine can synthesize triacylglycerols from monoacylglycerols and that this conversion is involved in the transport of dietary fat across the *enterocytes,* which are the absorptive cells of the small intestine (Brindley 1984). Later in that decade Hajra and Agranoff provided evidence that dihydroxyacetone phosphate can act as an alternative acyl-acceptor to glycerolphosphate for initiating the de novo synthesis of glycerolipids (Brindley and Sturton 1982).

Most of the work in this area of biochemistry is now directed to understanding how the metabolism and transport of triacylglycerols is controlled. These investigations have generally not advanced as quickly as those dealing with the metabolism of water-soluble compounds. Lipids are difficult to use as enzyme substrates, and the kinetics are very complicated. Most of the enzymes are bound tightly to membranes, which makes them relatively more difficult to purify and characterize. The rest of this chapter will attempt to provide an overview of the routes of triacylglycerol metabolism in mammals and its control as we understand it at present.

BIOSYNTHESIS OF PHOSPHATIDATE

The synthesis of triacylglycerols begins with the activation of fatty acids to acyl-CoA esters (Figure 7.1), which involves the input of energy from ATP. Acyl-CoA is then used by the various acyltransferases in the synthesis of carboxylic ester bonds. Triacylglycerols generally contain long-chain fatty acids (C_{16} or greater). Consequently, the most important activating enzyme is the long-chain acyl-CoA synthetase, which is found mainly in the endoplasmic reticulum and mitochondria of mammalian cells. However, cells also contain distinct acyl-CoA synthetases that can activate short- and medium-chain acids. These acids can then be esterified in some situations. For example, the milk of many species contains short- and medium-chain acids, and coconut fat also has a high proportion of lauric acid (C_{12}).

The main acceptor for acyl-CoA in most tissues is thought to be the sn-glycerol-3-phosphate that is formed by glycolysis or by the phosphorylation of glycerol. Glycerol-3-phosphate is first acylated to 1-monoacylglycerol-3-phosphate (lysophosphatidate) by glycerolphosphate acyltransferase. A different acyltransferase is responsible for esterifying the hydroxyl at the 2-position of lysophosphatidate (Figure 7.2).

Figure 7.1. Reaction mechanism of acyl-CoA synthetase.

$$\text{Fatty acid} + \text{ATP} \xrightarrow{\text{Mg}^{2+}} \text{Fatty acyl-AMP} + \text{PP}_i$$

$$\text{Fatty acyl-AMP} + \text{CoA} \longrightarrow \text{Fatty acyl-CoA} + \text{AMP}$$

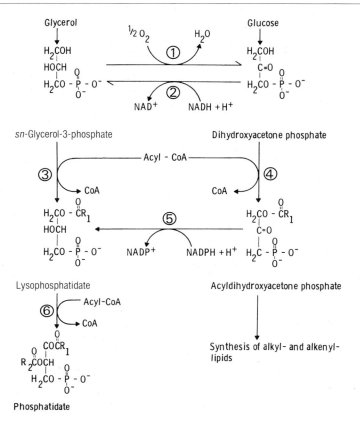

Figure 7.2. Synthesis of phosphatidate. Enzyme activities are indicated by (1) glycerol-3-phosphate dehydrogenase, (2) glycerol-3-phosphate dehydrogenase (NAD$^+$), (3) glycerolphosphate acyltransferase, (4) dihydroxyacetone phosphate acyltransferase, (5) acyldihydroxyacetone phosphate reductase, (6) monoacylglycerol phosphate (lysophosphatidate) acyltransferase.

In the liver, the glycerolphosphate acyltransferase is divided almost equally between the mitochondrial and microsomal fractions. By contrast, in heart, kidney, adrenal glands, and adipose tissue, the mitochondrial activity is only about 10% of the total activity, and the microsomal activity predominates (Brindley and Sturton 1982). The microsomal activity is found on the cytoplasmic side of the rough and smooth endoplasmic reticulum, and the mitochondrial enzyme is on the inner surface of the outer mitochondrial membrane. These acyltransferases are different enzymes, as is shown by the relative resistance of the mitochondrial acyltransferase to inhibition by heat, proteolytic enzymes, and N-ethylmaleimide (Table 7.2). The mitochondrial enzyme also has a lower K_m than the microsomal acyltransferase for acyl-CoA esters and for glycerolphosphate, and it

Table 7.2. Acyltransferases Involved in Phosphatidate Synthesis

Acyltransferase location	Substrates	Effect of N-ethylmaleimide
Endoplasmic reticulum	Glycerolphosphate or dihydroxyacetone phosphate Saturated and unsaturated acyl-CoA	Inhibition
Mitochondrial outer membrane	Glycerolphosphate Saturated acyl-CoA $\Big]$ low K_m	None
Peroxisomes	Dihydroxyacetone phosphate Saturated acyl-CoA	Slight stimulation

prefers saturated acyl-CoA esters, for example, palmitoyl-CoA. By contrast, the acyltransferase from the endoplasmic reticulum is able to use a variety of saturated and unsaturated acyl-CoA esters.

The esterification at the 2-position by monoacylglycerol phosphate (lysophosphatidate) acyltransferase is relatively selective for unsaturated fatty acids in both mitochondrial and microsomal fractions. However, this specificity is by no means absolute, and the fatty acid composition of the newly synthesized phosphatidate will depend upon the fatty acids available in the cell. The activity of the monoacylglycerol phosphate acyltransferase is relatively low in mitochondrial fractions, and lysophosphatidate is often the major product of esterification from glycerolphosphate. By contrast, phosphatidate is the main product of esterification with microsomal fractions.

The function of the relatively high mitochondrial glycerolphosphate acyltransferase activity, particularly in liver, is unclear. Phosphatidate is an intermediate in the synthesis of diphosphatidylglycerol (cardiolipin) (Chapter 8), which is characteristic of the inner mitochondrial membrane. However, the flux of phosphatidate into diphosphatidylglycerol is relatively low when compared with overall glycerolipid synthesis. The conversion of phosphatidate to triacylglycerol and to phosphatidylcholine is normally considered not to take place in mitochondria or to occur at very low rates.

The lower K_m values of the mitochondrial acyltransferase for glycerolphosphate and acyl-CoA esters (Table 7.2) should mean that the mitochondrial system is favored compared with the enzyme in the endoplasmic reticulum, particularly at low substrate supply. The physiological function of the mitochondrial acyltransferase is therefore uncertain and remains a major unanswered question in the control of glycerolipid synthesis. Phosphatidate can be hydrolyzed back to glycerolphosphate in a substrate cycle (Figure 7.3), which

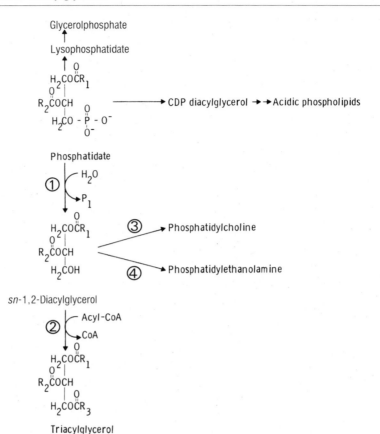

Figure 7.3. Metabolism of phosphatidate. Enzyme activities are indicated by (1) phosphatidate phosphohydrolase, (2) diacylglycerol acyltransferase, (3) choline phosphotransferase, (4) ethanolamine phosphotransferase.

might help in regulating the balance between β-oxidation and esterification (see later discussion). Alternatively, the phosphatidate might be transferred to the endoplasmic reticulum for the synthesis of triacylglycerol and phospholipids.

The other route for the de novo biosynthesis of phosphatidate is by the esterification of dihydroxyacetone phosphate (Figure 7.2). This reaction can be catalyzed in the endoplasmic reticulum by glycerolphosphate acyltransferase, which can use glycerolphosphate and dihydroxyacetone phosphate, which compete for the enzyme (Table 7.2). In addition, there is a dihydroxyacetone phosphate acyltransferase that neither uses glycerolphosphate as a substrate, nor is inhibited by it. The specific enzyme is not inhibited by N-ethylmaleimide, which further indicates that it is a different enzyme from the microsomal glycerolphosphate acyltransferase. The specific dihydroxyacetone phosphate acyltransferase is located mainly in peroxisomes. It has a higher reaction rate with the CoA esters of saturated than with unsaturated fatty acids; this specificity therefore favors the incorporation of saturated fatty acids into the 1-position of glycerolipids.

Acyldihydroxyacetone phosphate can undergo two different biosynthetic modifications (Figure 7.2). First, it can be converted into alkyldihydroxyacetone phosphate, which can then serve as the precursor for the synthesis of the alkyl- and alkenyl-lipids (Chapter 9). Alternatively, the acyldihydroxyacetone phosphate can be converted to 1-monoacylglycerol-3-phosphate (lysophosphatidate) by a reductase that uses NADPH. This reductase is found in both peroxisomes and the endoplasmic reticulum, and the same reductase probably acts on alkyldihydroxyacetone phosphate.

It is generally accepted that acyldihydroxyacetone phosphate is an obligatory intermediate in the synthesis of alkyl- and alkenyl-lipids. However, the quantitative importance of the esterification of dihydroxyacetone phosphate relative to glycerolphosphate for the synthesis of glycerolipids is very much in dispute. On the one hand it has been argued that the esterification of glycerolphosphate must predominate. This conclusion takes into account the relatively high concentration of glycerolphosphate in the cell and the K_m, K_i, and $V_{\max}$ values for glycerolphosphate and dihydroxyacetone phosphate for the glycerolphosphate acyltransferase that is located in the endoplasmic reticulum (Brindley and Sturton 1982). These calculations do not consider the specific acyltransferases that exist in the mitochondria and peroxisomes. However, it is not clear whether these organelles can efficiently donate phosphatidate for the synthesis of triacylglycerols and the zwitterionic phospholipids.

Other arguments are based upon the use of glycerol labeled either with ^{3}H at the 2-position or with ^{14}C. Conversion of glycerolphosphate to dihydroxyacetone phosphate before its incorporation causes the loss of ^{3}H, whereas ^{14}C is retained. The relative activities of the two pathways can be estimated by comparing the ^{3}H/^{14}C ratio in lipid with that in the glycerolphosphate. When this was done, 50–75% of the glycerolipid synthesized by rat liver slices appeared to be derived by acylation of dihydroxyacetone phosphate. A similar technique showed that about 56% of the phosphatidylglycerol and 64% of the phosphatidylcholine were formed from dihydroxyacetone phosphate in type II cells of lung (Brindley and Sturton 1982).

An alternative approach was to incubate cells with a mixture of D-[U-^{14}C]- and D-[3-^{3}H]-glucose and to measure the labeling of the lipids. This procedure generated [4-^{3}H]NADPH, which was incorporated into the 2-position of glycerolipids by the dihydroxyacetone phosphate pathway (Figure 7.2). It was then possible to calculate the relative contribution of the two pathways to the synthesis of glycerolipids on the assumption that the ^{3}H/^{14}C ratio in alkyl- and alkenyl-lipids results entirely from synthesis by the dihydroxyacetone phosphate pathway. A correction factor also had to be applied to compensate for some labeling of glycerolphosphate at the 2-position. From these results it was estimated that 49–61% of the glycerolipid synthesised by BHK-21 and BHK-Ts-a/1b-2 cells occurs by acylation of dihydroxyacetone phosphate.

It is also possible to postulate that the dihydroxyacetone phosphate pathway should be important in glycerolipid synthesis from theoretical considerations of cofactor requirements. Most anabolic pathways use NADPH as a cofactor in reductive synthesis. The formation of glycerolipids from glucose via the dihydroxyacetone phosphate pathway fulfills this requirement (Figure 7.2). NADPH production increases when there is active fatty acid synthesis. This increase could also favor the incorporation of the fatty acids into triacylglycerols by the dihydroxyacetone phosphate pathway. The conversion of dihydroxyacetone phosphate to glycerolphosphate requires NADH, which is normally involved in reductive degradation. In the liver, glycerolphosphate dehydrogenase, which catalyzes this reaction, is also involved in gluconeogenesis from glycerol and in maintaining the redox state.

We do not know the relative importance of the three acyltransferases that are described in Figure 7.2 in initiating glycerolipid synthesis or whether their relative contributions change in different physiological conditions. However, it is normally assumed that the synthesis of triacylglycerols takes place mainly in the endoplasmic reticulum by the esterification of glycerolphosphate.

CONVERSION OF PHOSPHATIDATE TO TRIACYLGLYCEROL

Phosphatidate lies at a branch point in glycerolipid synthesis which provides three routes for further metabolism (Figure 7.3). First, it can be hydrolyzed to glycerolphosphate in a substrate cycle by phospholipases of the A-type. This hydrolysis probably prevents the excessive accumulation of phosphatidate in the membranes. Second, it can be converted to CDP-diacylglycerol, which provides a precursor for the synthesis of the acidic phospholipids, phosphatidylinositol, phosphatidylglycerol, and diphosphatidylglycerol (Chapter 8). Third, it can be hydrolyzed to diacylglycerol, which in turn can serve as a common precursor in the synthesis of the zwitterionic phospholipids, phosphatidylcholine, and phosphatidylethanolamine (Chapter 8) and also for the synthesis of triacylglycerol. We shall now discuss the last of these routes of metabolism.

The conversion of phosphatidate to diacylglycerol is catalyzed by phosphatidate phosphohydrolase. This enzyme activity has been reported to occur in plasma membranes, lysosomes, mitochondria, the endoplasmic reticulum, and the cytosol. However, it is difficult to be certain about the distribution of the phosphohydrolase, since the assays used in many of these studies measured the release of inorganic phosphate or water-soluble phosphates from phosphatidate (Brindley and Sturton 1982). These phosphates can also arise by deacylation of the phosphatidate by phospholipase A activities to form glycerolphosphate, which can be further hydrolyzed by acid or alkaline phosphatases. The occurrence of a phosphatidate phosphohydrolase has been confirmed in the cytosol and endoplasmic reticulum in several cell types by specifically measuring the production of diacylglycerol. This activity is the one that is thought to be important for the synthesis of triacylglycerol, phosphatidylcholine, and phosphatidylethanolamine. Recent evidence shows that phosphatidate phosphohydrolase can be translocated between the cytosol and the endoplasmic reticulum (Brindley 1985). This phenomenon will be discussed later in terms of the control of glycerolipid synthesis.

This phosphatidate phosphohydrolase activity can be stimulated by Mg^{2+}. However, this is probably not a cation requirement in the normal sense of being essential for the binding of the phosphate group to the enzyme, since amphiphilic (detergentlike) amines such as chlorpromazine can substitute for Mg^{2+} in the presence of high concentrations of EDTA. The function of the Mg^{2+} is probably to adjust the surface charge and packing arrangement of the membrane that contains the phosphatidate, thus facilitating the interaction of substrate with the phosphohydrolase.

The final stage of triacylglycerol synthesis is completed by the action of diacylglycerol acyltransferase, which forms the final ester bond at the 3-position (Figure 7.3). This enzyme is located in the endoplasmic reticulum, and its activity can be stimulated by Mg^{2+}. As far as we know, this enzyme competes for a common pool of diacylglycerol with the choline- and ethanolamine-phosphotransferases that are responsible for the synthesis of phosphatidylcholine and phosphatidylethanolamine.

The pathways that have been described so far are the possible routes for the de novo synthesis of triacylglycerols that occurs in a wide variety of cell types. However, there is another route for triacylglycerol synthesis that operates in the enterocytes of the small intestine. This pathway will be discussed in relation to its role in lipid absorption and transport.

DIGESTION, ABSORPTION, AND TRANSPORT OF LIPIDS

Lipids constitute an important part of the diet, and in the more developed countries contribute 40–45% of energy intake. The largest proportion of this energy comes from triacylglycerols. This means that a person eats 60–130 g of fat each day and that the body has to digest, absorb, resynthesize and transport this quantity of lipid. Approximately 90–95% of the triacylglycerols from the diet are absorbed from the intestinal lumen, whereas only about 50% of the cholesterol is absorbed. Since some of the cholesterol in the intestinal lumen comes from the bile, the incomplete absorption of cholesterol enables the body to excrete it.

Digestion of Lipids

Relatively little digestion of triacylglycerols takes place in the stomach, but the action of the proteolytic enzymes releases lipid from food particles, and a coarse emulsion is formed by the churning action of the stomach. A gastric lipase that is distinct from pancreatic lipase has been identified in the stomach contents of several mammals, including man. This lipase is active at pH 3–4, and it preferentially releases short- and medium-chain-length fatty acids from acylglycerols. These fatty acids are found in the triacylglycerols of the milk of several species, and they are preferentially located at the 3-position (Figure 7.4). The partial hydrolysis of triacylglycerols to 1,2-diacylglycerols makes the milk fat micelles more susceptible to the subsequent action of pancreatic lipase in the small intestine. The short- and medium-chain fatty acids can be absorbed directly by the gastric mucosa. They are bound to albumin in the blood and carried to the liver, where they can be readily oxidized to produce energy.

The remaining lipid enters the proximal part of the small intestine, where the pH rises to 5.8–6.5. The coarse droplets of oil become

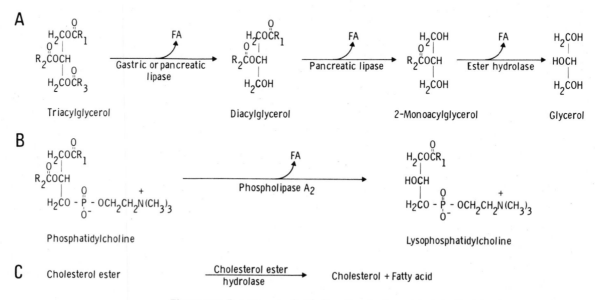

Figure 7.4. Reactions in lipid digestion in the gastrointestinal tract.

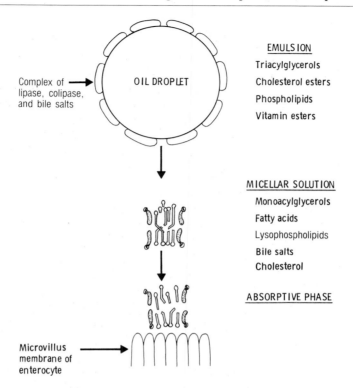

EMULSION
Triacylglycerols
Cholesterol esters
Phospholipids
Vitamin esters

Complex of →
lipase, colipase,
and bile salts

OIL DROPLET

MICELLAR SOLUTION
Monoacylglycerols
Fatty acids
Lysophospholipids
Bile salts
Cholesterol

ABSORPTIVE PHASE

Microvillus
membrane of
enterocyte

Figure 7.5. Schematic representation of lipid digestion and absorption in the small intestine.

coated with bile salts that are secreted by the liver. These compounds are amphiphilic, and they dissolve at the oil-water interface with their hydrophobic faces pointing into the oil and the hydrophilic surfaces pointing into the aqueous phase of the luminal contents. The bile salts help to disperse the oil droplets, and they donate a negative charge to their surfaces. A protein called colipase (molecular weight of about 10,000), which is secreted by the pancreas, is then adsorbed onto the surface of the oil droplets (Figure 7.5). This protein acts as an anchor for the attachment of pancreatic lipase at the oil-water interface. The bile salts, colipase, and lipase are thought to interact in a ternary complex which in the presence of Ca^{2+} is able to hydrolyze the 1- and 3-ester bonds of triacylglycerol (Figure 7.4A). About 85% of the digestion of dietary triacylglycerol in nonruminant animals ends with the formation of 2-monoacylglycerol, since pancreatic lipase is unable to hydrolyze the ester bond at the 2-position.

Limited hydrolysis of 2-monoacylglycerols does take place in nonruminant animals, but it is catalyzed by a nonspecific ester hydrolase that is secreted by the pancreas. The same enzyme probably also hydrolyzes cholesterol esters (Figure 7.4C) and the esters of fat-soluble vitamins.

In ruminants the complete digestion of triacylglycerols is effected by lipases that are produced by bacteria in the rumen. Bacteria are also responsible for hydrogenating unsaturated fatty acids that are present in the diets of ruminant animals, which accounts for the highly saturated nature of their body fats.

Pancreatic juice also contains phospholipases of the A_1- and A_2-types, which remove fatty acids from the 1- and 2-positions of phospholipids, respectively. These phospholipids can be derived either from the diet or from bile delivered to the small intestine. The digestion of phospholipids need not be complete: 30–40% of the dietary phosphatidylcholine can be absorbed as lysophosphatidylcholine (Figures 7.4B and 7.5).

The process of digestion converts lipids that have limited abilities to interact with water into more polar compounds with amphiphilic properties. Thus, triacylglycerols are transformed into 2-monoacylglycerols; cholesterol esters into cholesterol; and phospholipids into their lyso-derivatives, which are strong detergents. The fatty acids that are released are ionized at the pH that prevails in the intestinal lumen, so these compounds are also surface active, that is, they can interact with both lipid and aqueous environments (Brindley 1984).

As digestion proceeds, the monoacylglycerols, fatty acids, lysophospholipids, cholesterol, and bile salts dissociate from the surface of the oil droplets to form a micellar solution (Figure 7.5). The micelles are aggregates of amphiphilic lipids that orient themselves with the hydrophobic regions on the inside of the micelles and the polar groups exposed to the aqueous environment (Chapter 2). Further digestion of phospholipids and cholesterol esters can take place at the microvillus membrane of the enterocytes, which contains phospholipases and a cholesterol ester hydrolase.

Absorption of Lipids from the Intestines

Most of the lipids are absorbed from the small intestine (in the jejunum) with the exception of the bile salts, which remain in the lumen of the small intestine to facilitate further digestion. The bile salts are finally absorbed in the distal ileum and are transported back to the liver by the portal blood. This cycle constitutes the *enterohepatic circulation*.

Lipid absorption occurs when the micellar solution of lipids comes into contact with the microvillus membrane of the enterocytes (Figure 7.5). The lipids are transported across the membrane by an energy-independent process which relies on the maintenance of an inward diffusion gradient. This gradient can partly be achieved by the attachment of the fatty acids to a binding protein (Z-protein) with a

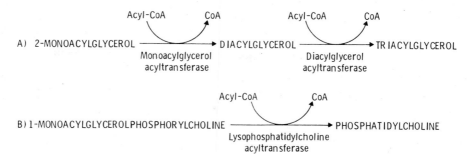

Figure 7.6. Major routes for the resynthesis of triacylglycerol and phosphatidylcholine in the enterocytes of the small intestine.

molecular weight of about 12,000. However, the main driving force for absorption comes from the rapid re-esterification of the lipids, which is an ATP-dependent process due to activation of fatty acids (Figure 7.1).

In the nonruminant animal the major acyl-acceptor is the 2-monoacylglycerol that is formed in the lumen of the intestine during the partial hydrolysis of triacylglycerol. Triacylglycerol is resynthesized by the sequential actions of monoacylglycerol acyltransferase and diacylglycerol acyltransferase (Figure 7.6A). This pathway is estimated to account for 75–85% of the total synthesis of triacylglycerol in the enterocytes of nonruminant animals. The remainder comes from the esterification of glycerolphosphate (and perhaps of dihydroxyacetone phosphate), as described earlier (Figures 7.2 and 7.3). Synthesis of triacylglycerol from glycerolphosphate is necessary, since some of the dietary triacylglycerol is completely hydrolyzed to glycerol. In ruminant animals, bacterial lipases are responsible for the hydrolysis of 2-monoacylglycerols; therefore the pathways for de novo synthesis of the triacylglycerol are required.

The diacylglycerol that is formed as an intermediate in triacylglycerol synthesis can also be used to synthesize phosphatidylcholine. However, the most prominent source of phosphatidylcholine in nonruminant animals is the esterification of the 1-monoacylglycerophosphorylcholine (lysophosphatidylcholine) (Figure 7.6B) that is formed by the partial digestion of phosphatidylcholine. The reactions of glycerolipid synthesis are very rapid—they take place within seconds after the precursors are absorbed.

The transport and metabolism of the absorbed cholesterol is much slower than that of triacylglycerols, and it is estimated that the $t_{\frac{1}{2}}$ for cholesterol in the enterocyte is about 12 h. During absorption the cholesterol becomes incorporated into the membranes of the enterocytes and diluted with existing cholesterol. A large proportion of the cholesterol that is transported from the enterocyte is esterified, mainly with oleic acid.

Formation of Chylomicrons and VLDL

The transport of lipids from the small intestine requires that they be packaged into a physically stable form that can exist in an aqueous environment. Thus, the most hydrophobic of the lipids (cholesterol esters and triacylglycerols) are coated with a layer of amphiphilic compounds including phosphatidylcholine, cholesterol, and various apoproteins. Apart from stabilizing the surface of the lipoproteins, the apoproteins also provide "address labels" that govern which cells in the body receive and metabolize these lipoproteins (Chapter 13). The main lipoproteins responsible for transporting dietary fat from the intestine are the chylomicrons. However, the intestine is also able to secrete VLDL, and it is responsible for about 10% of this lipoprotein that appears in the blood. The chylomicrons are bigger than the VLDL, and they have a higher content of triacylglycerol (Table 7.1). However, the size and composition of the lipoproteins can vary according to the composition of the diet, the rates of lipid absorption, and apoprotein synthesis. Thus, with a high intake of dietary fat and at the peak of absorption, the chylomicrons tend to be larger and contain more triacylglycerol than when the rate of lipid transport is lower.

The assembly of the chylomicrons and VLDL begins with the resynthesis of triacylglycerols, which takes place in the smooth endoplasmic reticulum in the apical region of the enterocyte. The triacylglycerols appear in the form of lipid droplets in the cisternae of the smooth endoplasmic reticulum within minutes of exposure of the cells to micellar lipid in the lumen (Figure 7.7). The droplets are stabilized with a coat of phospholipid and with proteins that are synthesized in the rough endoplasmic reticulum. As time progresses, the droplets increase in number as the endoplasmic reticulum extends and pinches off to form vesicles. It is thought that these lipid-filled vesicles fuse with the Golgi apparatus. The nascent chylomicrons and VLDL are then carried to the lateral surfaces of the enterocyte by the process of *exocytosis*. Fusion of the Golgi and surface membranes takes place, and the chylomicrons and VLDL are secreted into the intercellular spaces which drain into the lymph vessels (Figure 7.7). The lipoproteins in these vessels pass via the thoracic duct and enter the circulation at the level of the jugular vein.

The hydrolysis of the triacylglycerols of chylomicrons and VLDL involves their binding through the apo C-2 on their surfaces to an enzyme called lipoprotein lipase, which is found in the capillary beds of various extrahepatic tissues including skeletal muscle, cardiac muscle, and adipose tissue. The enzyme is anchored to polysaccharide chains on the endothelial wall of the capillaries (Figure 7.8). Lipoprotein lipase is synthesized by the cells that underlie the capillaries. The cells receive the fatty acids that are hydrolyzed from triacylglycerols

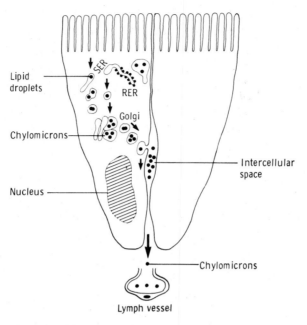

Figure 7.7. Schematic representation of the transport of lipids through the enterocytes and the formation of chylomicrons, where SER represents smooth endoplasmic reticulum and RER represents rough endoplasmic reticulum. The figure is reproduced from Brindley (1984) by the permission of Butterworth's Scientific Ltd.

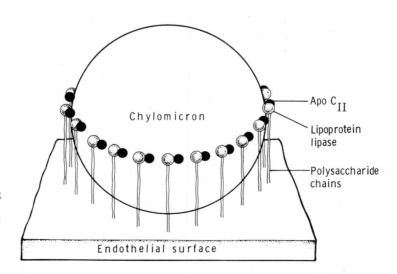

Figure 7.8. Schematic representation of the binding of a chylomicron to lipoprotein lipase. The figure is reproduced from Brindley (1984) by the permission of Butterworth's Scientific Ltd.

by the lipase (Chapter 13). The body is able to direct chylomicrons and VLDL to particular organs through its tissue-specific control of lipoprotein lipase activity. This subject will be considered in more detail later in this chapter.

The metabolic fate of fatty acids depends on the particular cells into which they are delivered. Fatty acids are readily used for energy production by muscle tissues. If the uptake exceeds the immediate requirement for β-oxidation, then the fatty acids are temporarily stored as triacylglycerols within the muscle cells. Storage is, of course, the major fate of fatty acids that enter adipocytes.

Partitioning of Fatty Acids Between the Portal Blood and the Lymphatic System

The description so far has dealt with the metabolism and transport of glycerolipids and cholesterol. Most of the long-chain fatty acids are transported from the intestine by incorporation into the esterified lipids of the chylomicrons or VLDL. These two lipoproteins also transport cholesterol, the fat-soluble vitamins, and other hydrophobic compounds and drugs. However, fatty acids with 10- to 12-carbon chain lengths are found both in the triacylglycerols of chylomicrons and in their unesterified forms bound to albumin in portal blood (Table 7.3). Short-chain-length fatty acids are transported almost exclusively by the latter route to the liver.

There are a number of reasons for the partitioning of fatty acids between portal blood and the fluid in the lymphatics on the basis of their chain lengths (Table 7.3). First, the short- and medium-chain acids which are often located at the 3-position are more readily hydrolyzed from triacylglycerols. This means that they are unlikely to be retained in the monoacylglycerols that are re-esterified in the enterocytes. The enterocytes also contain ester hydrolases that selectively degrade short- and medium-chain acylglycerols. Second, the fatty acids in the intestinal lumen partition between the micellar and aqueous phases. The shorter the chain length, the greater the tendency to partition into the aqueous phase and to become separated from the bulk of the lipid. The uptake of short-chain fatty acids into the enterocytes can take place against a concentration gradient. Finally, the enzymes responsible for re-esterification in the entero-

Table 7.3. Partition of Fatty Acids after Absorption Between the Portal Blood and the Lymphatic System

Chain length	Portal route (unesterified)	Lymphatics (esterified)
$C_2 - C_8$	Majority	Little
C_{10}	Majority	Significant
C_{12}	Significant	Significant
$>C_{12}$	Little	Majority

cytes discriminate against the short- and medium-chain-length fatty acids.

Use is made of the body's ability to absorb fatty acids by the portal route in conditions where the digestion, absorption, or transport of triacylglycerols in chylomicrons is impaired. The clinical management of these malabsorption syndromes (for example, pancreatic or biliary insufficiency, damage to the intestinal mucosa, abetalipoproteinemia) is facilitated by feeding medium-chain triacylglycerols, from which the fatty acids can be efficiently absorbed by the portal route. This energy-rich diet supplies the liver with a readily oxidizable substrate.

CONTROL OF TRIACYLGLYCEROL SYNTHESIS

The control of triacylglycerol synthesis varies from tissue to tissue. In the enterocytes of the small intestine, it is assumed that the load of dietary fatty acids is the major factor that controls the rate of triacylglycerol synthesis. The availability of 2-monoacylglycerol governs the balance between the monoacylglycerol pathway and the pathways of de novo synthesis. The major function of the enterocyte in this respect is to transport dietary fats and to synthesize chylomicrons.

In the liver the situation is more complex. This organ is required to incorporate into triacylglycerols fatty acids when they are derived primarily from de novo synthesis, from the diet, or from lipolysis in adipose tissue (Figure 7.9). In terms of metabolic and hormonal

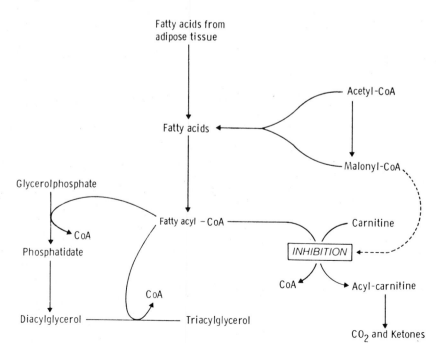

Figure 7.9. The relationship between fatty acid synthesis, oxidation, and esterification in the liver.

control, these represent quite different conditions. In addition to triacylglycerol synthesis, fatty acids can be used for β-oxidation and for the synthesis of phospholipids. These lipids are needed for membrane turnover and for the secretion of bile, VLDL, and HDL components.

The major function of triacylglycerol synthesis in heart and skeletal muscle is to enable fatty acids to be stored temporarily if their supply is greater than the rate of β-oxidation.

Long-term storage of triacylglycerols occurs in adipose tissue. In adipocytes the activities of the enzymes that synthesize triacylglycerols are controlled reciprocally with that of the hormone-sensitive lipase, which is responsible for fatty acid mobilization (Saggerson 1985). However, because of the high fatty acid availability during lipolysis, some of these can be recycled back to triacylglycerol.

Control of Phosphatidate Synthesis in the Liver

One of the obvious sites for the enzymic regulation of fatty acid esterification is at the level of the glycerolphosphate and dihydroxyacetone phosphate acyltransferases, since it is here that fatty acids become committed to glycerolipid synthesis. The maximum velocity of acyl-CoA synthetase is very much higher than those of the acyltransferases, but the flux into acyl-CoA esters is, of course, dependent upon fatty acid availability. At present there is no evidence for separate pools of acyl-CoA esters that selectively supply one pathway of fatty acid metabolism. Although fatty acids can be desaturated and elongated, the major competition to glycerolipid synthesis is provided by β-oxidation. Regulation of the relative activites of glycerolphosphate acyltransferase and carnitine palmitoyltransferase should therefore control this branch point (Figure 7.9). The competition for acyl-CoA esters might explain why there is such an active glycerolphosphate acyltransferase with low K_m values for glycerolphosphate and acyl-CoA esters in the mitochondria. In addition, there might be competition between the peroxisomal β-oxidation system and the specific dihydroxyacetone phosphate acyltransferase that is present in this organelle. However, as explained before, the metabolic fate of the majority of the phosphatidate and lysophosphatidate that can potentially be formed in mitochondria and peroxisomes respectively remains unclear.

We do know that carnitine palmitoyltransferase can be acutely regulated by inhibition with malonyl-CoA (Chapter 4). At present no such regulator of the acyltransferases has been found. Consequently, it is tempting to assume that the esterification system acquires those fatty acids not required by carnitine palmitoyltransferase. It might be wise, however, to reflect that the discovery of the malonyl-CoA inhibition of this enzyme is relatively recent. Previously, the prevailing view was that fatty acid oxidation responded to fatty acid flux without being regulated itself.

The mitochondrial glycerolphosphate acyltransferase is thought to be increased by insulin; it is decreased more in starvation and in diabetes than the acyltransferase of the endoplasmic reticulum. The latter enzyme should provide most of the phosphatidate that is used for the synthesis of triacylglycerols. Theoretically, the acyltransferase could operate at a high rate of fatty acid flux, and it presumably provides the liver with its large capacity for triacylglycerol synthesis. Its activity changes relatively little in different physiological conditions, although small decreases have been observed during starvation. The physiological significance of such changes is not clear, since the capacity of the liver to synthesize triacylglycerols is not decreased in this condition. This fact can be demonstrated by blocking β-oxidation, which diverts an increased flux of fatty acids into triacylglycerol synthesis.

Control of the Conversion of Phosphatidate to Triacylglycerol in Liver

The activity of phosphatidate phosphohydrolase in the liver changes much more dramatically than activity of other enzymes of triacylglycerol synthesis. The synthesis of phosphatidate phosphohydrolase is stimulated over a period of hours by glucocorticoids (for example, cortisol or corticosterone) and high concentrations of cAMP. It is also known that high concentrations of glucocorticoids in vivo increase hepatic triacylglycerol synthesis, produce a fatty liver, and stimulate VLDL secretion (Brindley and Sturton 1982).

The effects of glucocorticoids and cAMP on the synthesis of phosphatidate phosphohydrolase are antagonized by insulin. These hormonal interactions account for the high activities of this enzyme that are observed in the liver during conditions when metabolism is controlled to a greater extent by the stress hormones (adrenalin, glucagon, glucocorticoids) than by insulin (Table 7.4). In addition to these long-term effects (of glucocorticoids, cAMP, and insulin), vasopressin (and probably other hormones that mobilize intracellular Ca^{2+}, including angiotensin II and α_1-agonists) can rapidly stimulate phosphatidate phosphohydrolase activity. The action of vasopressin is also accompanied by an increase in triacylglycerol synthesis (Brindley 1985).

The increased phosphohydrolase activities that are illustrated in Table 7.4 provide the liver with an increased potential to synthesize triacylglycerol, which protects the liver against the large influxes of fatty acids that can be caused by high rates of fatty acid mobilization from adipose tissue. Phosphatidate phosphohydrolase exists both in the cytosol and on the endoplasmic reticulum. It is believed that the cytosolic form of the enzyme is physiologically inactive until it

Table 7.4. Changes in Phosphatidate Phosphohydrolase Activity in the Liver That Can Be Associated with an Increased Effect of Glucocorticoids and Other Stress Hormones Relative to Insulin

Treatment or condition	Duration	Increase in phosphohydrolase activity
Metabolic stress		
Starvation	6–40 h	1.3- to 2.3-fold
Surgical stress	6 h	3-fold
Subtotal hepatectomy	6 h	5.5-fold
Diabetes (mildly ketotic)	10 weeks	1.4-fold
Diabetes (ketotic)	48 h	2.8-fold but reversed by insulin
Hypoxia	24 h	2.5-fold
Toxic conditions produced by		
Hydrazine	4–24 h	2- to 4-fold
Morphine (mouse)	60–180 min	2-fold
Hormone injections		
Corticotropin	6 h	3.3- to 4.3-fold
Cortisol	5 days	2.4-fold
Genetic obesity (ob/ob mouse)	Long-term	2-fold
Ingestion of some nutrients		
Fructose, sorbitol, or glycerol	6 h	1.9- to 2.3-fold
Ethanol	7 h	6.9-fold

Source: Brindley and Sturton (1982);
Note: The results are from work with rats with the exception of morphine injections and the ob/ob mouse.

translocates onto the membranes on which phosphatidate is being synthesized. The signal for this translocation is the increased availability of fatty acids, which also stimulates the phosphohydrolase activity (Figure 7.10) (Brindley 1985).

The ability of fatty acids to promote the translocation of phosphatidate phosphohydrolase from the cytosol to the membrane-associated compartment appears to be under hormonal control. cAMP displaces the enzyme from the membranes, but this effect can be overcome by higher concentrations of fatty acids (Table 7.5). By contrast, insulin should decrease the intracellular concentration of cAMP and therefore the concentration of fatty acids required to promote the translocation. The mechanisms that cause these changes are not yet established, but they may involve the reversible phosphorylation of the phosphohydrolase. This type of control for phosphatidate phosphohydrolase is reminiscent of the regulation of CTP:phosphocholine cytidylyltransferase, which regulates the synthesis of phosphatidylcholine (Chapter 8). Phosphatidate phosphohydrolase and CTP:phosphocholine cytidylyltransferase can be classified as *ambiquitous*, which means that they exist in different locations of the cell and can regulate metabolism by moving from one location to another

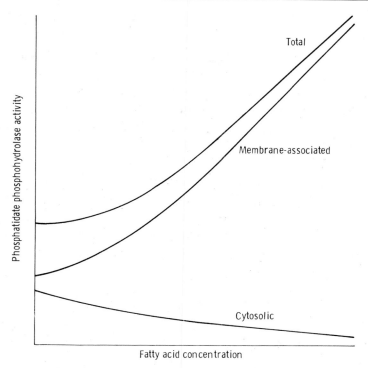

Figure 7.10. Activation and translocation of phosphatidate phosphohydrolase by fatty acids in isolated hepatocytes. The figure shows that a fatty acid (oleate) increases the total activity of phosphatidate phosphohydrolase and also increases its association with membranes rather than with the cytosol. It is believed that the membrane-associated enzyme is physiologically active and that the cytosolic enzyme forms a reservoir of potential activity.

Table 7.5. Effects of a cAMP Analogue and Oleate on the Location of Phosphatidate Phosphohydrolase in Isolated Hepatocytes

| Additions | Phosphohydrolase activity (%) | |
	Cytosolic	Membrane-associated
None	68	32
*CPT-cAMP (0.5mM)	86	14
Oleate (1mM)	48	52
CPT-cAMP (0.5mM) + Oleate (1mM)	42	60

* 8-(4-Chlorophenylthio)adenosine-3′,5′-cyclic monophosphate, a potent analogue of cAMP.

(Wilson 1980). The reason for coordinating the synthesis of phosphatidylcholine and triacylglycerol can readily be appreciated in terms of the synthesis of VLDL, which contain 20% and 60% (by weight), respectively, of these lipids (Table 7.1).

Diacylglycerol is a Precursor of Triacylglycerol, Phosphatidylcholine, and Phosphatidylethanolamine

The synthesis of phosphatidylcholine and phosphatidylethanolamine competes for the diacylglycerol that is produced by phosphatidate phosphohydrolase (Figure 7.3). The choline- and ethanolamine-phosphotransferases have higher affinities for diacylglycerol than does diacylglycerol acyltransferase. Therefore, at low fatty acid availability and a low rate of formation of diacylglycerol, the major flux from diacylglycerol is directed to phospholipid synthesis. This maintains membrane turnover and bile secretion.

It has been suggested that the activity of diacylglycerol acyltransferase could be decreased by a cAMP-dependent phosphorylation that would exaggerate this discrimination against triacylglycerol synthesis and favor phospholipid formation in conditions of stress (Brindley and Sturton 1982). If this were to occur it would be overcome by increased fatty acid availability, which would supply acyl-CoA for the acyltransferase. Thus, as the supply of fatty acid increases, the relative proportion of diacylglycerol converted to triacylglycerol would increase. The synthesis of phosphatidylcholine and phosphatidylethanolamine might eventually be limited by the availability of CDP-choline and CDP-ethanolamine; however, the abundance of fatty acids, would also stimulate the cytidylyltransferase, which would attempt to maintain CDP-choline synthesis (Chapter 8) provided that the phosphocholine was available.

In order to try to relate these changes in hepatic triacylglycerol synthesis to other metabolic pathways and to metabolism in other organs, two very different metabolic situations will be considered, namely, the conditions of high and low availabilities of insulin relative to glucagon, adrenalin, corticotropin, and glucocorticoids.

METABOLISM OF TRIACYLGLYCEROLS WHEN THE AVAILABILITY OF INSULIN IS HIGH

A high concentration of insulin is normally seen after the consumption of meals rich in glucose in the form of starch. Such a diet also produces good insulin control, since the tissues normally respond well to insulin action. If insulin predominates relative to glucagon, the catecholamines, and corticotropin in the control of metabolism, then glycolysis, and the synthesis of glycogen and fatty acid in the liver, are stimulated (Chapter 5). At the same time, β-oxidation and gluconeogenesis are suppressed (Figures 7.9 and 7.11). Glucocorticoids, which are also stress hormones, can have a permissive effect on some of these processes provided that insulin is available. Thus, they augment insulin action in stimulating the synthesis of glycogen and fatty acids.

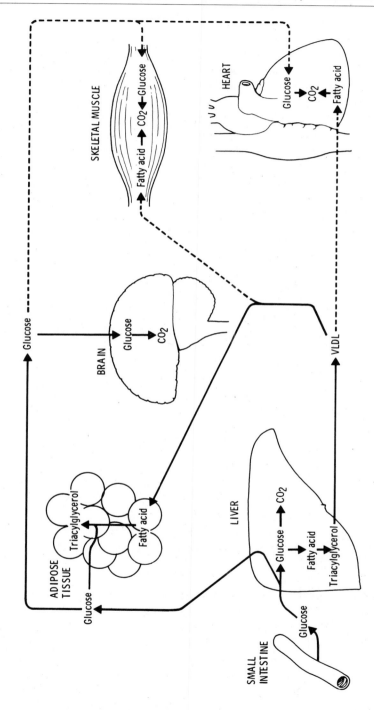

Figure 7.11. Some effects of insulin on metabolism. The solid lines represent major routes of metabolism.

The fatty acids that are produced are readily incorporated into triacylglycerols. Glycerolphosphate and dihydroxyacetone phosphate are provided by glycolysis, and fatty acids are not diverted into β-oxidation because of the inhibition of carnitine palmitoyltransferase by malonyl-CoA (Figure 7.9). Normally, the intracellular concentration of acyl-CoA esters remains relatively low, but because of effects of insulin and the relatively low concentration of cAMP, phosphatidate phosphohydrolase is readily translocated from the cytosol to the membrane-associated compartment, which facilitates glycerolipid synthesis. Furthermore, if glucocorticoids are relatively high, then the phosphohydrolase activity also increases and provides an increased potential for the synthesis of triacylglycerols.

Triacylglycerols do not normally accumulate in the liver but are efficiently packaged into VLDL and secreted into the blood (Figure 7.11). The main site of triacylglycerol hydrolysis and fatty acid uptake is in adipose tissue, since lipoprotein lipase activity in this organ is increased by insulin. This action of insulin can be augmented by the effects of glucocorticoids (Robinson et al. 1983; Cryer 1981). In lactation some of the VLDL (and chylomicrons) are diverted to the mammary gland, where prolactin increases lipoprotein lipase activity. The activity of lipoprotein lipase in skeletal and cardiac muscle is relatively low compared with adipose tissue, since muscle lipase activity is not maintained by insulin, but rather by glucocorticoids, with the possible involvement of glucagon, catecholamines, and thyroid hormones (Cryer 1981).

Insulin not only promotes the transfer of fatty acids (as triacylglycerols) from liver to adipose tissue, but it also stimulates glucose entry and the activities of the enzymes of triacylglycerol synthesis (Saggerson 1985). This leads to the efficient storage of triacylglycerol (Figure 7.9). The release of fatty acids from triacylglycerols in adipose tissue is suppressed because of the inactivation of hormone-sensitive lipase. This condition is maintained by the predominance of insulin over the stress hormones (Figure 7.12). The activation of the hormone-sensitive lipase by a cAMP-dependent phosphorylation is therefore prevented (Belfrage et al. 1984).

TRIACYLGLYCEROL METABOLISM IN CONDITIONS OF METABOLIC STRESS

Stress is characterized by the low activity of insulin relative to glucagon, the catecholamines, corticotropin, and the glucocorticoids in regulating metabolism. This type of hormonal balance occurs in starvation, diabetes, trauma, and toxic conditions. To a lesser extent, nutrients such as fructose, ethanol, and fat present the body with an energy load without releasing insulin. In fact, fructose and ethanol can provoke the release of stress hormones, and high-fat diets can cause tissues to become insulin-insensitive.

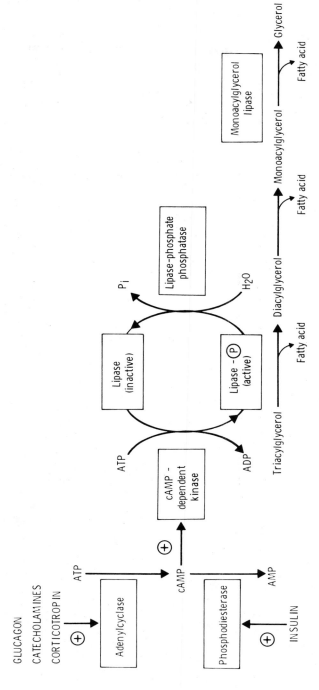

Figure 7.12. The control of the activity of hormone-sensitive lipase in adipose tissue. The plus sign indicates a stimulation.

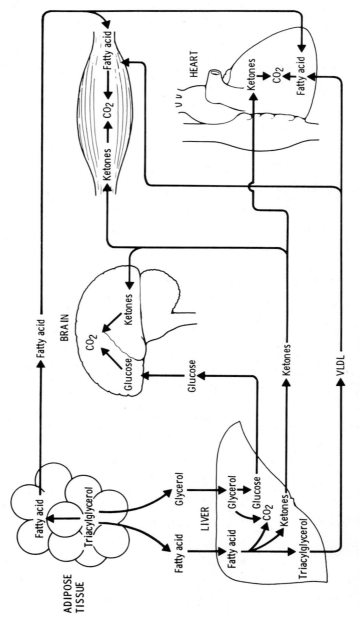

Figure 7.13. Some effects of glucocorticoids and high cAMP concentrations on metabolism.

The acute-acting stress hormones (glucagon, adrenalin) stimulate adenyl cyclase (adenylate cyclase) in a variety of tissues. In adipose tissue this stimulation leads to phosphorylation and activation (Belfrage et al. 1984) of the hormone-sensitive lipase (Figure 7.12). At the same time, the enzymes of triacylglycerol synthesis are inhibited (Saggerson 1985), and fatty acids and glycerol are consequently released into the circulation. Some of the fatty acids enter muscle tissue to become oxidized and provide energy (Figure 7.13). However, a large proportion of fatty acids is taken up by the liver. The high concentration of acyl-CoA esters that results, together with the increase in cAMP concentration, decreases the activity of acetyl-CoA carboxylase (Chapter 5), and the concentration of malonyl-CoA falls (Figure 7.9). Consequently, the inhibition of β-oxidation is relieved, and more of the fatty acids are metabolized to CO_2 and ketones. The oxidation process at low fatty acid supply can also be facilitated by the effect of cAMP in displacing phosphatidate phosphohydrolase from the membranes (Table 7.5), thereby inhibiting triacylglycerol synthesis.

However, in severe stress the supply of fatty acids and glycerol to the liver normally increases dramatically, as do the concentrations of acyl-CoA esters and glycerolphosphate in the liver. The high fatty acid availability overcomes the effects of cAMP in displacing phosphatidate phosphohydrolase (Table 7.5) and CTP:phosphocholine cytidylyltransferase (Chapter 8) from the membranes, and the synthesis of triacylglycerols and phosphatidylcholine is promoted. A fatty liver can often develop in stress conditions, when the capacity to secrete VLDL cannot cope with the increased rate of triacylglycerol synthesis. Fatty liver occurs especially in several toxic conditions where the synthesis of the apoproteins required in VLDL formation is impaired. Normally, the fatty condition is reversed when the fatty acid mobilization from adipose tissue subsides and the triacylglycerols can be secreted, or hydrolyzed so that the fatty acids can be oxidized by the liver.

The VLDL that are released during severe stress are preferentially metabolized by muscle tissue, since its lipoprotein lipase activity is increased by glucocorticoids and other stress hormones. The relative lack of insulin ensures that lipoprotein lipase activity is low in adipose tissue; the fatty acids are therefore prevented from recycling back into the storage depots (Figure 7.13).

One of the main functions of the liver in stress conditions and diabetes is to distribute energy to other organs in the form of glucose, ketones, and triacylglycerols (Table 7.6). It is significant that the control of phosphatidate phosphohydrolase in the liver by glucocorticoids, cAMP, the Ca^{2+}-mobilizing hormones, and insulin resembles

Table 7.6. Export of Energy from the Liver in Stress Conditions

Metabolite	Taken up by
Glucose	Brain and erythrocytes
Ketones	Skeletal and cardiac muscle Brain
Triacylglycerols (VLDL)	Skeletal and cardiac muscle

that of enzymes involved in amino acid breakdown and gluconeo-genesis (for example, tyrosine aminotransferase, argininosuccinate synthetase, argininosuccinate lyase, and phosphoenolpyruvate carboxykinase). Lipolysis from adipose tissue supplies the liver with fatty acids and glycerol for triacylglycerol synthesis, fatty acids for ketogenesis, and glycerol for gluconeogenesis (Figure 7.13). The hypertriglyceridemia and hyperglycemia that occur in stress and diabetes therefore seem to have a similar physiological basis. They result from increased hepatic secretion of triacylglycerol and glucose and their decreased removal from the circulation. The latter condition occurs because the uptake systems for glucose in muscle and adipose tissue and for triacylglycerols through lipoprotein lipase in adipose tissue are insulin-dependent.

FUTURE DIRECTIONS

Athough we have made progress in understanding the complex control of triacylglycerol metabolism, it is clear that there are many gaps in our knowledge. For instance, what are the major functions of the mitochondrial and peroxisomal systems that synthesize phosphatidate or lysophosphatidate? What are the relative contributions of glycerolphosphate and dihydroxyacetone phosphate to glycerolipid synthesis under different physiological conditions in different organs? Are the initial acyltransferases of glycerolipid synthesis controlled to a greater extent than is known at present? How are the enzymes of triacylglycerol synthesis regulated differentially in different tissues?

In order to answer some of these questions, a number of technical problems have to be overcome. The increased use of tissue- and cell-culture techniques has already meant that the specific effects of hormones and metabolites on triacylglycerol metabolism can be investigated. However, it is often difficult to reproduce or to appreciate the situation that exists in vivo. Care must be taken in interpreting the results and in deciding what is, or is not, an artifact. As always, the conclusions have to be compatible with what occurs in vivo.

One of the biggest problems in lipid metabolism has always been to isolate, characterize, purify, and to raise antibodies against the enzymes which are often tightly associated with membranes. Until this is done, the mechanisms by which the enzymes act and are controlled will not be fully understood. It is also difficult to work with substrates that are insoluble in water. In addition, a strict kinetic analysis of reaction rates is normally not possible.

If we can overcome some of these technical problems, then the rewards are potentially high. The regulation of triacylglycerol metabolism is an important part of intermediary metabolism. A greater appreciation of this regulation would help to alleviate or prevent a number of clinical conditions, including fatty liver, diabetes, obesity, hyperlipidemias, and atherosclerosis.

PROBLEMS

1. Describe the routes by which phosphatidate and lysophosphatidate can be synthesized in the liver. In which part of the hepatocytes do these reactions take place, and what is known about the significance of these different routes of metabolism?
2. Compare and contrast the digestion, absorption, and transport of the short-, medium-, and long-chain fatty acids that are present in dietary triacylglycerols of nonruminant animals.
3. What differences are there between the control of

the synthesis of triacylglycerols in liver and adipose tissue?
4. Describe how the synthesis of triacylglycerols in the liver and their transport in VLDL are controlled after the consumption of a meal rich in starch.
5. Describe the mechanism and control processes by which fatty acids are mobilized from adipose tissue in ketotic diabetes. What are the major metabolic fates of these acids?

BIBLIOGRAPHY

All entries are suggested for further reading.

Belfrage, P.; Fredrikson, G.; Stralfers, P.; and Tornqvist, H. 1984. In *Lipases*, eds. B. Borgström and H. Brockman, 365–416. Amsterdam: Elsevier.

Brindley, D. N. 1984. In *Fats in animal nutrition*, ed. J. Wiseman, 85–103. London: Butterworth's Scientific Ltd.

Brindley, D. N. 1985. Intracellular translocation of phosphatidate phosphohydrolase and its possible role in the control of glycerolipid synthesis. *Prog. Lipid Res.* 23: in press.

Brindley, D. N., and Sturton, R. G. 1982. "Phosphatidate metabolism and its relation to triacylglycerol biosynthesis." In *Phospholipids: New comprehensive biochemistry*, Vol. 4, eds. J. N.

Hawthorne and G. B. Ansell. Amsterdam: Elsevier Biomedical Press.

Cryer, A. 1981. Tissue lipoprotein lipase activity and its action in lipoprotein metabolism. *Int. J. Biochem.* 13:525–41.

Robinson, D. S.; Parkin, S. M.; Speake, B. K.; and Little, J. A. 1983. In *The adipocyte and obesity*, eds. A. Angel, C. H. Hollenberg, and D. A. K. Roncari, 127–36. New York: Raven Press.

Saggerson, E. D. 1985. In *New perspectives in adipose tissue*, eds. A. Cryer and R. L. R. Van, Chapter 5. London: Butterworth's Scientific Ltd.

Wilson, J. E. 1980. Brain hexokinase, the prototype ambiquitous enzyme. *Curr. Top. Cell. Regul.* 16:1–44.

CHAPTER 8

Phospholipid Metabolism in Eucaryotes

Dennis E. Vance

Our understanding of the biosynthesis of phospholipids in animals and other eucaryotic cells has lagged considerably behind that of fatty acid, cholesterol, and eicosanoid metabolism. The state of the art is similar to that of triacylglycerol, sphingolipid, and polyunsaturated fatty acid metabolism. Two common features among these areas are that most of the substrates are lipidic and that the enzymes are largely intrinsic membrane proteins. Despite innovative approaches and improved purification procedures, the solubilization and purification of membrane-bound enzymes is an extremely difficult and time-consuming task. Nonetheless, progress has been steady, and the mysteries of phospholipid metabolism are being solved. The object of this chapter is to provide a comprehensive overview of what is known about phospholipid metabolism and the direction of current research.

PHOSPHATIDYL-CHOLINE BIOSYNTHESIS
Historical Developments

The discovery of phosphatidylcholine is attributed to Gobley in 1847 (Pelech and Vance 1984). He demonstrated the presence of a phospholipid in egg yolk; this was later named *lecithin* after the Greek equivalent for egg yolk (*lekithos*). In the 1860s Diakonow and Strecker demonstrated that lecithin contained two fatty acids linked to glycerol and that choline was attached to the third hydroxyl by a phosphodiester linkage. Chemical synthesis in 1927 by Grün and Limpächer confirmed the proposed structure. With the structure in hand, the biochemistry could begin. The first significant step was the discovery by Charles Best in 1932 that animals had a dietary requirement for choline. Choline-deficient animals developed fatty livers and a number of other pathological conditions. However, it was not until 1955 that the essential nature of choline for cell survival and growth was demonstrated in culture by Eagle.

choline is essential nutrient

The major pathway for phosphatidylcholine biosynthesis occurs via CDP-choline.

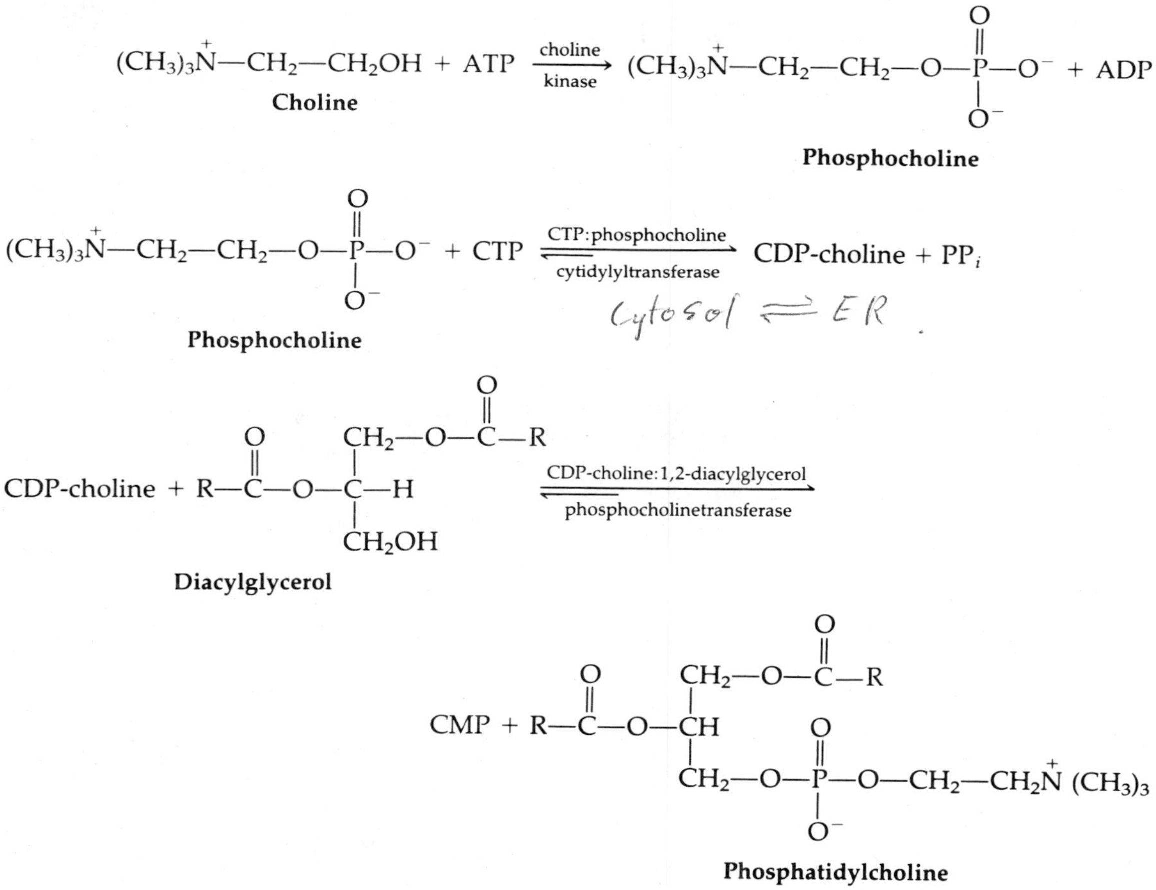

$$(CH_3)_3\overset{+}{N}-CH_2-CH_2OH + ATP \xrightarrow[\text{kinase}]{\text{choline}} (CH_3)_3\overset{+}{N}-CH_2-CH_2-O-\overset{\overset{O}{\|}}{\underset{\underset{O^-}{|}}{P}}-O^- + ADP$$

Choline   **Phosphocholine**

$$(CH_3)_3\overset{+}{N}-CH_2-CH_2-O-\overset{\overset{O}{\|}}{\underset{\underset{O^-}{|}}{P}}-O^- + CTP \underset{\text{cytidylyltransferase}}{\overset{\text{CTP:phosphocholine}}{\rightleftharpoons}} \text{CDP-choline} + PP_i$$

Phosphocholine

Cytosol $\rightleftharpoons$ ER .

$$\text{CDP-choline} + R-\overset{\overset{O}{\|}}{C}-O-\overset{\overset{CH_2-O-\overset{\overset{O}{\|}}{C}-R}{|}}{\underset{\underset{CH_2OH}{|}}{C}}-H \underset{\text{phosphocholinetransferase}}{\overset{\text{CDP-choline:1,2-diacylglycerol}}{\rightarrow}}$$

Diacylglycerol

$$CMP + R-\overset{\overset{O}{\|}}{C}-O-\overset{\overset{CH_2-O-\overset{\overset{O}{\|}}{C}-R}{|}}{\underset{\underset{CH_2-O-\overset{\overset{O}{\|}}{P}-O-CH_2-CH_2\overset{+}{N}(CH_3)_3}{|}}{CH}}$$

Phosphatidylcholine

After the development of radioactive isotopes, it was demonstrated that various animal tissues could make phosphatidylcholine from fatty acids, glycerol phosphate, and choline. In 1952, the well-known DNA enzymologist Arthur Kornberg and Wittenberg discovered choline kinase in yeast, liver, and other tissues. Subsequently, in the mid 1950s, Eugene Kennedy and co-workers demonstrated the requirement of CTP for phosphatidylcholine biosynthesis and described the reactions catalyzed by CTP:phosphocholine cytidyl-yltransferase and CDP-choline:1,2-diacylglycerol phosphocho-linetransferase. CTP is required not only for phosphatidylcholine synthesis but also for the de novo synthesis of all phospholipids (procaryotic and eucaryotic) at some step in the biosynthetic pathway.

An alternative pathway for phosphatidylcholine biosynthesis, of quantitative significance only in liver, is the conversion of phospha-tidylethanolamine (PE) to phosphatidylcholine (PC) via the methyla-tion pathway. The first suggestion of this pathway came in 1941 when Stetten fed [^{15}N]ethanolamine to rats and isolated [^{15}N]choline. Several decades elapsed before Bremer and Greenberg demonstrated the presence of a microsomal enzyme which converted phosphati-dylethanolamine to phosphatidylcholine via the transfer of methyl groups from S-adenosylmethionine (AdoMet), according to the fol-lowing reactions:

$$\text{PE} + \text{AdoMet} \longrightarrow N\text{-methyl-PE} + \text{AdoHcy}$$

$$N\text{-Methyl-PE} + \text{AdoMet} \longrightarrow N,N\text{-dimethyl-PE} + \text{AdoHcy}$$

$$N,N\text{-dimethyl-PE} + \text{AdoMet} \longrightarrow \text{PC} + \text{AdoHcy}$$

where AdoHcy is S-adenosylhomocysteine.

Once phosphatidylcholine is made, its fatty acid substituents can be modified by a deacylation-reacylation cycle, as originally described by Lands and co-workers in the early 1960s.

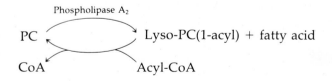

This cycle seems to be of particular importance for the introduction of polyunsaturated fatty acids into phosphatidylcholine. Alternatively, in 1958 Marinetti demonstrated that the product of deacylation, lysophosphatidylcholine, can be condensed with another lysophos-phatidylcholine to form phosphatidylcholine and glycerophos-phocholine.

$$2 \text{ Lysophosphatidylcholine} \longrightarrow \text{phosphatidylcholine} + \text{glycerophosphocholine}$$

Although such a reaction can clearly occur, its importance to cell physiology has not been demonstrated. A calcium-mediated base-exchange activity for the synthesis of phosphatidylcholine was demonstrated in the late 1950s by Hübscher, but this reaction has not been shown to be of quantitative significance in any tissue.

Choline Transport and Oxidation

A high-affinity (apparent K_m of about $10\mu M$) transport process for choline has been described in many systems. Despite the central importance of choline transport to the biosynthesis of phosphatidylcholine and the neurotransmitter acetylcholine, there are no detailed studies on the protein involved or its mechanism of action.

Once choline is inside the cell, its normal fate is phosphorylation by choline kinase. The major exception is liver, where choline is also oxidized to betaine. This reaction occurs in mitochondria and is catalyzed by choline dehydrogenase and betaine aldehyde dehydrogenase.

Choline $\qquad\qquad HOCH_2—CH_2—\overset{+}{N}(CH_3)_3$

$\qquad\qquad\qquad\qquad\qquad\downarrow$ Choline dehydrogenase

Betaine Aldehyde $\qquad HC—CH_2—\overset{+}{N}(CH_3)_3$ (with O double bond on HC)

$\qquad\qquad\qquad\qquad\qquad\downarrow$ Aldehyde dehydrogenase

Betaine $\qquad\qquad {}^-OC—CH_2—\overset{+}{N}(CH_3)_3$ (with O double bond on C)

(margin hand-drawn structures:)

$$O=N \begin{smallmatrix} CH_3 \\ \\ CH_3 \quad CH_3 \end{smallmatrix}$$

Trimethylamine oxide (TMA)

$\downarrow$

$$N \begin{smallmatrix} CH_3 \\ \\ CH_3 \quad CH_3 \end{smallmatrix}$$

(TMAO)

Betaine is an important donor of methyl groups for methionine biosynthesis and the one-carbon pool (Vance 1983). Choline dehydrogenase is tightly bound to the inner membrane of rat liver mitochondria. The enzyme was purified after extraction with 2.0% Triton X-100 (Pelech and Vance 1984) and has a molecular weight of 70,000. The only other tissue with significant choline dehydrogenase activity is the kidney. Betaine aldehyde dehydrogenase has not been purified.

Choline Kinase

Unlike many of the other phospholipid biosynthetic enzymes, choline kinase is cytosolic. The enzyme has been highly purified from liver and lung; its molecular weight is in the range 120,000–160,000. Several isoenzymes of choline kinase have been detected in rat liver, but the physiological function for multiple enzymic forms is not known. Choline kinase has a slightly alkaline pH optimum (pH 8.0–9.0), a high apparent K_m for ATP (2.0mM), and low apparent

K_m for choline $(0.03mM)$ (Pelech and Vance 1984). Most studies have indicated that choline kinase and ethanolamine kinase are separate enzymes. However, the enzyme was recently purified from rat kidney (Ishidate, Nakagomi, and Nakazawa 1984). The kidney enzyme has a molecular weight of 75,000–80,000 and is composed of two subunits. The pure kidney enzyme phosphorylated both choline and ethanolamine.

CTP:Phosphocholine Cytidylyltransferase

CTP:phosphocholine cytidylyltransferase has been recovered from cytosol and microsomes from a large number of different cells and tissues. The cytosolic form has been purified from rat liver in low yield (Pelech and Vance 1984). The enzyme has a molecular weight of 200,000 and requires lipid for activity. Lysophosphatidylethanolamine is the most potent activator. The enzyme is not activated by various detergents, lysophosphatidylcholine, fatty acids, or fatty acyl-CoA. The other phospholipids which activate the enzyme are phosphatidylglycerol, phosphatidylserine, and phosphatidylinositol, all of which are anionic at neutral pH. Phosphatidylethanolamine and phosphatidylcholine do not activate the enzyme. The enzymes from lung, brain, muscle, and all other tissues and cells examined have similar lipid requirements. The pH optimum for cytidylyltransferase is between 6.0 and 7.0. The apparent K_m for CTP depends on the amount of phospholipid associated with the enzyme. The highly purified enzyme in the presence of saturating phospholipid has an apparent K_m for CTP of 0.2 to $0.3mM$. Smaller amounts of lipid result in a higher apparent K_m for CTP (approximately $5mM$). The apparent K_m for phosphocholine $(0.2mM)$ is less sensitive $(0.2–0.7mM)$ to the amount of phospholipid. The enzyme has no activity for the biosynthesis of CDP-ethanolamine.

The microsomal form of the enzyme is immunologically similar to the cytosolic enzyme (Pelech and Vance 1984). Its kinetic constants are similar to those of the purified enzyme in the presence of phospholipid. Since cytidylyltransferase is inactive without phospholipid, it seems likely that the microsomal enzyme is the active species in the cell, and the cytosolic enzyme acts as a reservoir. A great deal of evidence which supports this hypothesis is summarized in a later section.

CDP-Choline: 1,2-Diacylglycerol Phosphocholine-transferase

The final reaction in phosphatidylcholine biosynthesis is catalyzed by CDP-choline:1,2-diacylglycerol phosphocholinetransferase, which is an intrinsic protein in the microsomal membrane. The enzyme activity is susceptible to inactivation by trypsin and thus has been localized to the cytosolic side of the microsomes. Under the same conditions, enzymes located within the lumen of microsomes (for example, glucose-6-phosphate phosphatase) are not degraded by trypsin. Phosphocholinetransferase is separate from the corresponding enzyme which makes CDP-ethanolamine. The enzyme shows a preference for diacylglycerol species with palmitate on C-1, and linoleate on C-2, of the glycerol moiety. Although the enzyme has been solubilized from the microsomal membrane, it remains to be purified to homogeneity.

Phosphatidyl-
ethanolamine-*N*-
Methyltransferase

The conversion of phosphatidylethanolamine to phosphatidylcholine takes place on the microsomal membrane and is catalyzed by an intrinsic enzyme exposed to the cytosolic surface. The reaction involves the sequential transfer of three methyl groups; whether one, two, or three enzymes are involved is controversial. The enzyme is distinctive in that the pH optima for all three reactions is 10.2 (Pelech and Vance 1984).

One problem that has troubled workers with this enzyme is a confusion between the steady concentration of the first product (*N*-methyl-phosphatidylethanolamine) and the amount of this product formed during an incubation (Audubert and Vance 1983). With phosphatidylethanolamine as the substrate, *N*-methyl-phosphatidylethanolamine is the only product at the start of the reaction. However, once sufficient *N*-methyl-phosphatidylethanolamine has accumulated (and this happens within a few seconds), the product is converted to *N*,*N*-dimethyl-phosphatidylethanolamine and phosphatidylcholine, as indicated.

$$PE \xrightarrow{\quad CH_3{}^{\circ}\quad} PE{-}CH_3{}^{\circ} \xrightarrow{\quad CH_3{}^{\bullet}\quad} PE\!\!\begin{array}{c} {}^{\nearrow}CH_3{}^{\circ} \\ {}^{\searrow}CH_3{}^{\bullet} \end{array} \xrightarrow{\quad CH_3{}^{\square}\quad} PE{-}CH_3\!\!\begin{array}{c} {}^{\nearrow}CH_3{}^{\circ} \\ {}^{\bullet} \\ {}^{\searrow}CH_3{}^{\square} \end{array}$$

Hence, in order to estimate the amount of *N*-methyl-phosphatidylethanolamine formed during an incubation, one must take into account the subsequent conversion of *N*-methyl-phosphatidylethanolamine to dimethyl-phosphatidylethanolamine and phosphatidylcholine. Since the assay is usually performed with [CH_3-^{3}H]S-adenosylmethionine, an equation must be used for calculating the amount of *N*-methyl-phosphatidylethanolamine formed in an incubation (dpm = disintegrations per minute):

$$\text{nmol } N\text{-methyl-PE} = \frac{\text{dpm in } N\text{-methyl-PE} + \tfrac{1}{2} \text{ dpm in } N,N\text{-dimethyl-PE} + \tfrac{1}{3} \text{ dpm in PC}}{\text{specific radioactivity of AdoMet}}$$

Many claims in the literature about the pH optimum, apparent K_m, and number of enzymes involved in conversion of phosphatidylethanolamine to phosphatidylcholine were based on data which were not corrected for the subsequent conversion of N-methylphosphatidylethanolamine to dimethylphosphatidylethanolamine and phosphatidylcholine (Audubert and Vance 1983).

Hirata and Axelrod (1980) have proposed that the methylation of phosphatidylethanolamine to phosphatidylcholine may be involved in certain physiological processes such as neutrophil chemotaxis and hormone activation of adenylate cyclase. These conclusions have been challenged by several research groups who have blocked the methylation reaction by drugs without adversely affecting the biological response. Whether or not phosphatidylethanolamine methylation has a major function other than the formation of phosphatidylcholine in the liver remains a topic of current interest.

REGULATION OF PHOSPHATIDYL-CHOLINE BIOSYNTHESIS
The Rate-Limiting Reaction

Considerable evidence has demonstrated that in most instances the rate of the cytidylyltransferase reaction limits the rate of phosphatidylcholine biosynthesis. The first evidence in favor of this conclusion was the measurement of pool sizes of the aqueous precursors in rat liver (choline = $0.23 mM$, phosphocholine = $1.3 mM$, CDP-choline = $0.03 mM$). These measurements assumed that 1 g wet tissue is 1 ml and there was no compartmentation of the pools. The latter assumption was probably not correct, since phosphatidylcholine biosynthesis occurs in both the cytosol and on the endoplasmic reticulum. Nevertheless, the relative concentrations are probably correct. Phosphocholine is clearly the most concentrated precursor, whereas CDP-choline is lowest in concentration. This buildup of phosphocholine indicates the bottleneck in the pathway is the reaction catalyzed by cytidylyltransferase.

Pulse-chase experiments demonstrate this bottleneck more vividly. After a $\frac{1}{2}$-h pulse of hepatocytes in culture with [CH_3-^3H]choline, most (more than 95%) of the radioactivity in the precursors of phosphatidylcholine was in phosphocholine (Figure 8.1), the remainder in choline and CDP-choline. A subsequent chase with unlabeled choline in the medium showed that the labeled phosphocholine was quantitatively converted to phosphatidylcholine (Figure 8.1). The radioactivity in CDP-choline remained low during the chase period since its concentration is very low and it is rapidly converted to phosphatidylcholine. The low radioactivity in choline suggests that the choline is immediately phosphorylated after it enters the cell.

Two additional points should be made. If a cell or tissue is in a steady state, pool sizes and reaction rates are not changing. Thus, although the rate of PC synthesis is determined by the cytidylyltrans-

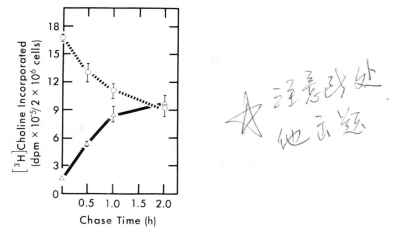

Figure 8.1. Incorporation of [CH₃-³H]choline into phosphocholine and phosphatidylcholine as a function of time. Hepatocytes from rat liver were incubated for 30 min with [CH₃-³H]choline. Subsequently, the cells were washed thoroughly and incubated (chased) for various times with unlabeled choline. The incorporation of radioactivity into phosphocholine (○••••••○) and phosphatidylcholine (△▬▬△) is shown. Adapted from Figure 1 from Pelech et al. 1983, *J. Biol. Chem.* 258:6783, with permission.

ferase reaction, the rates of the choline kinase, phosphocholinetransferase, and cytidylyltransferase reactions will all be the same. If this were not the case, we would see changes in the pool size of precursors. For example, if the choline kinase reaction were slower than the cytidylyltransferase, there would be a gradual depletion of phosphocholine and eventually the supply of phosphocholine might become limiting. Conversely, if the choline kinase reaction were faster, there would be an increase in the amount of phosphocholine. Thus, in most instances, cytidylyltransferase sets the pace, but the other reactions proceed at the same rate. The other point is that the choline kinase reaction commits choline to phosphatidylcholine biosynthesis, but usually is not the rate-limiting step.

Enzyme Translocation   As mentioned earlier, the cytidylyltransferase is recovered from tissues in both the cytosol and microsomes. There is usually some lipid present in the cytosol; hence, there is measurable activity of cytidylyltransferase without addition of phospholipid to the assay. Homogenization of rat liver in 0.145M NaCl results in the isolation of 75% of the activity in cytosol, with 25% in the microsomes. Therefore, most scientists have focused on the cytosolic activity. However, recent studies, particularly with HeLa cells, have shown a direct correlation between the cytidylyltransferase activity associated with

[handwritten margin notes: "cytidyltransferase | microsome ↓ ↑ rate of synthesis cytidyltransferase in cytosome ↓ ↓ rate of synthesis"]

microsomes and the rate of phosphatidylcholine synthesis in intact cells and tissues (Vance and Pelech 1984). Under normal culture conditions of HeLa cells, 65% of the enzyme activity is recovered in the cytosol, and 35% in the microsomes. When the HeLa cells were cultured in the presence of $1mM$ oleate, phosphatidylcholine synthesis was stimulated at least 10-fold. When these cells were fractionated, all the cytidylyltransferase was associated with microsomes, and none was in the cytosol. Thus, the rate of phosphatidylcholine synthesis in most tissues and cultured cells appears to be regulated by the distribution of cytidylyltransferase between cytosol and microsomes. The cytidylyltransferase in cytosol seems to be an inactive reservoir that can be translocated to the microsomes (endoplasmic reticulum in the cell), where the enzyme is stimulated by certain phospholipids. Obviously, a rapid increase in the rate of phosphatidylcholine synthesis can occur without an increase in protein synthesis.

Regulatory Mechanisms

The regulation of phosphatidylcholine biosynthesis is focused on cytidylyltransferase and the translocation of this enzyme between cytosol and microsomes. Two regulatory mechanisms have been identified which modulate the binding of cytidylyltransferase to microsomes (Pelech and Vance 1984).

First, covalent phosphorylation/dephosphorylation appears to be involved in modulation of cytidylyltransferase activity in liver. As is the case with many other anabolic processes (for example, acetyl-CoA carboxylase), cytidylyltransferase appears to be inactivated by a cAMP-dependent protein kinase and activated by a protein phosphate phosphatase. Whether or not the covalent modification is directly on cytidylyltransferase is unclear, since incorporation of [^{32}P]phosphate into the pure enzyme has not been demonstrated. Possibly an activator or inhibitor protein is the substrate for the kinase. Other protein kinases may also affect cytidylyltransferase; this possibility is currently under investigation.

The second well-established regulatory mechanism is stimulation of phosphatidylcholine synthesis by fatty acids or fatty acyl-CoAs (Pelech and Vance 1984). These compounds do not directly activate cytidylyltransferase but rather cause the enzyme to be translocated to microsomes, where the enzyme is activated by certain phospholipids. This activation by fatty acid ensures a coupling between the synthesis of diacylglycerol (Chapter 7) and a sufficient supply of CDP-choline. The fatty acid will be activated to the acyl-CoA and provide an increased supply of diacylglycerol for phosphatidylcholine synthesis. In liver, excess fatty acid is used for triacylglycerol synthesis after the requirement for phosphatidylcholine biosynthesis is satisfied.

In addition to modulation of cytidylyltransferase activity, the supply of substrate has been implicated as another means of altering phosphatidylcholine biosynthesis. In most cells the level of CTP is low (0.03 to 0.1mM) and below the apparent K_m for this nucleotide by cytidylyltransferase (0.2 to 0.3mM). Poliovirus infection stimulates phosphatidylcholine synthesis in HeLa cells and there is evidence that this occurs by an increase of CTP concentration in the cytoplasmic compartment (Pelech and Vance 1984). Similarly, in immature roosters, there is evidence that the supply of phosphocholine (normally 0.16mM in immature rooster liver) alters the rate of phosphatidylcholine synthesis in liver. The reason for the nearly 10-fold lower concentration of phosphocholine in rooster liver than in rat liver (1.3mM) is not known. Perhaps the rooster chow is low in choline.

Since the classical studies of Charles Best in 1932, it has been known that choline deficiency can result in accumulation of triacylglycerol in liver and, as a result, fatty liver. The mechanism for this effect has never been clearly elucidated. It appears that phosphatidylcholine biosynthesis is impaired, since the concentration of this lipid decreases. The fatty liver may partly result from a decreased secretion of lipoproteins, but this has not been clearly shown.

Lung Surfactant

Lungs secrete a lipid-and-protein-containing substance called *surfactant*, which prevents collapse of the alveoli when air is expelled. The major component of surfactant is dipalmitoylphosphatidylcholine. Respiratory distress syndrome, which occurs in some newborn children, appears to be due to a problem in surfactant production or secretion. Hence, this disease has greatly stimulated research on phosphatidylcholine synthesis in lung. Studies on the secretion of surfactant have been very difficult, since there are 40 different cell types in mammalian lung, and the type II cells which secrete surfactant comprise only 10% of the weight of the lung. Nevertheless, most workers agree that there is an increased synthesis of phosphatidylcholine after parturition which results from enhanced activity of the cytidylyltransferase, due to translocation of the enzyme from cytosol to microsomes. Whether dipalmitoyl phosphatidylcholine is made via the CDP-choline pathway or arises mainly via deacylation/reacylation of phosphatidylcholine remains undecided despite great efforts on this problem.

SPHINGOMYELIN BIOSYNTHESIS

Early work by Kennedy and co-workers suggested that sphingomyelin might arise via transfer of phosphocholine from CDP-choline to ceramide. Recent studies have clearly shown this to be unlikely and that instead the donor of phosphocholine is phosphatidylcholine.

$$\text{Phosphatidylcholine} + \text{ceramide} \rightleftharpoons \text{sphingomyelin} + \text{diacylglycerol}$$

Evidence for this reaction comes from pulse-chase experiments with [³H]choline which show tremendous flux of tritium through CDP-choline to phosphatidylcholine with no label transferred into sphingomyelin. After phosphatidylcholine becomes labeled, a precursor-product relationship can be demonstrated between phosphatidylcholine and sphingomyelin. In addition, an enzyme activity in plasma membranes has been characterized which transfers phosphocholine from phosphatidylcholine to ceramide. Finally, a temperature-sensitive mutant of Chinese hamster ovary cells defective in CDP-choline synthesis continues to make sphingomyelin but not phosphatidylcholine. Now that the biosynthetic pathway is firmly established, work on the characterization of sphingomyelin synthase and regulation of sphingomyelin biosynthesis can commence.

PHOSPHATIDYL-ETHANOLAMINE BIOSYNTHESIS
Historical Developments

Thudichum published a book on the chemical composition of the brain in 1884 in which he described "kephalin," a nitrogen-and-phosphorus-containing lipid different from lecithin. In 1913, Renall and Bauman independently isolated ethanolamine from kephalin. In 1930, Rudy and Page isolated what was probably the first pure preparation of phosphatidylethanolamine. The structure was confirmed by chemical synthesis in 1952 by Baer and colleagues.

The biosynthesis of phosphatidylethanolamine can occur via four pathways (Figure 8.2). The route via CDP-ethanolamine constitutes the de novo synthesis of phosphatidylethanolamine. The other pathways arise as a result of the modification of an existing phospholipid. The CDP-ethanolamine pathway was first described by Kennedy and Weiss in 1956, who identified two enzymes, CTP:phosphoethanolamine cytidylyltransferase and CDP-ethanolamine:1,2-diacylglycerol phosphoethanolaminetransferase. The decarboxylation of phosphatidylserine to yield phosphatidylethanolamine is the only route for phosphatidylethanolamine synthesis in E. coli (Chapter 3) and was shown in 1960 to occur in animal cells. Previously, it was thought that serine could be converted to ethanolamine. However, serine is not a direct precursor of ethanolamine; rather, the ethanolamine moiety appears to be generated only via phosphatidylserine decarboxylation. The Ca^{2+}-mediated base exchange enzyme and the reacylation enzyme were both discovered in the early 1960s at the same time as the corresponding enzymes for phosphatidylcholine synthesis were described.

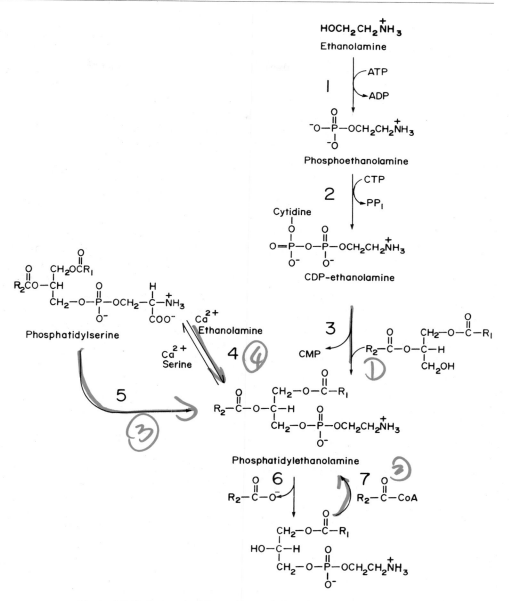

Figure 8.2. Pathways for biosynthesis of phosphatidylethanolamine. The numbers indicate the enzymes involved: 1, ethanolamine kinase; 2, CTP:phosphoethanolamine cytidylyltransferase; 3, CDP-ethanolamine:1,2-diacylglycerol phosphoethanolamine-transferase; 4, base-exchange enzyme; 5, phosphatidylserine decarboxylase; 6, phospholipase A_2; 7, acyl-CoA:lysophosphatidylethanolamine acyltransferase.

Ethanolamine Transport

Strangely, there seems to be little literature on the mechanism of ethanolamine transport into the cell. Whether or not active transport is required is not known. This is clearly an area which requires further study.

Ethanolamine Kinase

Ethanolamine kinase activity is recovered only from the cytosol of many different sources. The only purification reported is for the enzyme from soya bean, which has a monomer molecular weight of 18,000 and which also occurs as a dimer with a molecular weight of 37,000. Both species have ethanolamine kinase activity but do not phosphorylate choline. The apparent K_m for ethanolamine is $8\mu M$, and the enzyme has a pH optimum of 8.0. In liver, several enzymes with ethanolamine kinase activity have been found, one of which has a molecular weight of 36,000. It is generally believed that ethanolamine kinase is a different enzyme from choline kinase.

CTP: Phosphoethanolamine Cytidylyltransferase

CTP: phosphoethanolamine cytidylyltransferase was purified from rat liver cytosol by Sundler in 1975. Unlike the phosphocholine cytidylyltransferase, there is no evidence for the presence of this enzyme on microsomes, and the enzyme has no lipid requirement for activity. It has a molecular weight of 100,000 and is composed of two subunits. The enzyme has a pH optimum of 8.0, and the apparent K_m for CTP is $53\mu M$ and for phosphoethanolamine, $65\mu M$.

CDP-Ethanolamine: 1,2-Diacylglycerol Phosphoethanolamine-transferase

Phosphoethanolaminetransferase is an intrinsic protein in the microsomal membrane that appears to be different from the phosphocholinetransferase. Although it has been solubilized, a significant purification has not been achieved. The enzyme shows a distinct preference for diacylglycerol species that have 1-saturated, 2-docosahexaenoyl (22:6) fatty acid substituents.

Phosphatidylserine Decarboxylase

As mentioned in Chapter 3, phosphatidylserine decarboxylase has been purified from *E. coli*. It has been localized to the inner membrane of liver mitochondria, but has not been extensively purified.

Regulation of Phosphatidylethanolamine Biosynthesis

There was little knowledge about the regulation of phosphatidylethanolamine biosynthesis until Åkesson and Sundler tackled the problem in the 1970s. From studies initially on intact animals and subsequently on cultured hepatocytes, they provided evidence that the phosphoethanolamine cytidylyltransferase was usually rate limiting for phosphatidylethanolamine synthesis. The concentrations of the precursors to phosphatidylethanolamine in freeze-clamped liver were 1.09, 3.83, and 0.24 μmol/liver for ethanolamine, phosphoethanolamine, and CDP-ethanolamine, respectively. Hence, from the high ratio of phosphoethanolamine to CDP-ethanolamine, the conversion of phosphoethanolamine to CDP-ethanolamine appears to be the bottleneck in this pathway. In agreement with this proposal, incubation of freshly prepared rat hepatocytes with various concentrations of ethanolamine (0–0.25mM) resulted in a corresponding increase in the amount of phosphoethanolamine in the cell, with little change in CDP-ethanolamine concentration (Sundler and Åkesson 1975) (Figure 8.3). If the phosphoethanolaminetransferase were limiting the rate of phosphatidylethanolamine synthesis, an accumulation of CDP-ethanolamine would have been observed. Thus, at least in rat liver, the rate of phosphatidylethanolamine synthesis appears to be regulated by the phosphoethanolamine cytidylyltransferase, which is analogous to the regulation of phosphatidylcholine synthesis. Unlike the cytidylyltransferase involved in phosphatidylcholine synthesis, there is almost no literature on how the phosphoethanolamine cytidylyltransferase is regulated.

The relative importance of the various pathways for phosphatidylethanolamine synthesis is a complex problem which has not been adequately addressed. The base-exchange pathway has been estimated to account for 8–9% of phosphatidylethanolamine synthesis in

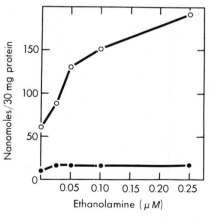

Figure 8.3. Effect of ethanolamine on the levels of phosphoethanolamine and CDP-ethanolamine. Hepatocytes were incubated with various concentrations of ethanolamine for 30 min. The cells were subsequently harvested and the amount of phosphoethanolamine (o——o) and CDP-ethanolamine (•——•) measured per 30 mg of cellular protein. Adapted from Figure 4 of Sundler and Åkesson (1975), with permission.

freshly isolated hepatocytes at physiological concentrations of ethanolamine ($20\mu M$). The contribution of phosphatidylserine decarboxylation was not evaluated in these hepatocytes. Early studies in rats suggested that phosphatidylserine decarboxylation might account for 5% of phosphatidylethanolamine made in liver. However, under certain cell culture conditions, most of the phosphatidylethanolamine in cells can arise from phosphatidylserine. With very few exceptions, the medium for culturing cells contains no ethanolamine. Small amounts of ethanolamine are provided in the serum which is usually added to growth medium. Since the direct decarboxylation of serine does not seem to occur,

$$\underset{\textbf{Serine}}{\text{HOCH}_2\text{—}\overset{\overset{\displaystyle\text{H}}{|}}{\underset{\underset{\displaystyle\text{COO}^-}{|}}{\text{C}}}\text{—}\overset{+}{\text{N}}\text{H}_3} \quad \xrightarrow{\quad \text{\textbar\textbar}\quad} \quad \underset{\textbf{Ethanolamine}}{\text{HOCH}_2\text{—}\text{CH}_2\text{—}\overset{+}{\text{N}}\text{H}_3 + \text{CO}_2}$$

and ethanolamine does not seem to arise by other pathways, the phosphatidylethanolamine in these cultured cells seems to derive largely from decarboxylation of phosphatidylserine. If the cells are supplied with ethanolamine, it is rapidly converted to phosphatidylethanolamine via the CDP-ethanolamine pathway. Hence, the cells have the potential to use ethanolamine for phosphatidylethanolamine synthesis. Interestingly, some mammary carcinoma and hydridoma cell lines require ethanolamine as a growth factor when cultured in the absence of serum. This is not a general requirement, since other hybridomas do not seem to require ethanolamine. Perhaps phosphatidylserine decarboxylation is impaired in the ethanolamine-requiring cells.

What is the origin of ethanolamine? There is not a clear answer to this question. Rats and presumably other mammals have a circulating serum level of ethanolamine of $20\mu M$ (Sundler and Åkesson 1975). It appears that ethanolamine must originate from the diet or arise from serine via phosphatidylserine decarboxylation and phosphatidylethanolamine degradation.

This is the only well-documented route for the biosynthesis of ethanolamine. Do plants contain a serine decarboxylase? The origin of ethanolamine was not addressed in a recent review article on plant phospholipids, yet plants have the capacity to make phosphatidylethanolamine via CDP-ethanolamine (Mudd 1980). The origin of ethanolamine is an example of a fundamental biochemical question which remains unanswered in the 1980s.

What regulates phosphatidylethanolamine biosynthesis is also an open question. What are the effects of hormones? Are cAMP or reversible phosphorylation reactions involved? Certain fatty acids seem to stimulate phosphatidylethanolamine biosynthesis (Sundler and Åkesson 1975), but the mechanism is not understood. Clearly, there is much to be done.

N-Acyl-Phosphatidyl-ethanolamine

Schmid and co-workers have found that ligation of one of the coronary arteries in a dog for 24 h resulted in the accumulation of a phosphatidylethanolamine derivative acylated on the amino group of ethanolamine (4–6% of total phospholipid in degenerated areas of the heart) with saturated and unsaturated fatty acids.

N-Acyl-phosphatidylethanolamine

Such derivatives of phosphatidylethanolamine have also been found in plants, granular cells of mammalian epidermis, and degenerating cultured kidney cells (BHK). It seems that this lipid may originate in large quantities as a result of cell death.

The biosynthesis of *N*-acyl-phosphatidylethanolamine can occur with microsomal or mitochondrial membrane preparations. The Ca^{2+}-requiring transacylase transfers a fatty acyl moiety primarily from the *sn*-1 position of phosphatidylcholine, phosphatidylethanolamine, or cardiolipin. The *N*-acyl-ethanolamine moiety can be removed by a phospholipase D which has no activity for phosphatidylcholine or phosphatidylethanolamine. The *N*-acyl-ethanolamine also accumulates in infarcted areas of canine hearts. In such areas, the mitochondria release abnormal amounts of Ca^{2+}, but *N*-acyl-ethanolamine prevents this release. Hence, accumulation of these lipids in damaged heart tissue may be a defense mechanism against cell damage as a result of ischemia.

PHOSPHATIDYL-SERINE
Historical Developments

Phosphatidylserine accounts for 5–15% of the phospholipids in eucaryotic cells. The lower concentration of phosphatidylserine compared with phosphatidylcholine and phosphatidylethanolamine is probably the reason why it was not discovered as a separate component of "kephalin" until 1941 by Folch. A structure was proposed by Folch in 1948 and confirmed by chemical synthesis in 1955 by Baer and Maurukas (Ansell and Spanner 1982).

Biosynthesis

Phosphatidylserine is made de novo in procaryotes via the CDP-diacylglycerol pathway (Chapter 3). The current evidence suggests that this biosynthetic route does not occur in animals or plants (Mudd 1980; Ansell and Spanner 1982; Baranska 1982), but clearly does in yeast (Baranska 1982).

The pathway for phosphatidylserine biosynthesis in animals is a base-exchange reaction

$$\text{Phosphatidylethanolamine} + \text{L-serine} \xrightarrow{\text{Ca}^{2+}} \text{phosphatidylserine} + \text{ethanolamine}$$

which was first described by Hübscher in 1959. He subsequently demonstrated that the enzyme was microsomal, required Ca^{2+}, had a K_m for serine of 0.5mM, and had pH optimum of 8.3 (Ansell and Spanner 1982). Since the reaction is an exchange of the head group, phosphatidylserine does not appear to be made directly from phosphatidic acid in animal tissues.

Brain tissue has a higher percentage of phospholipids as phosphatidylserine (15%) than other tissues (Baranska 1982). Hence, brain is probably the richest source of the phosphatidylethanolamine-serine exchange enzyme. This enzyme was solubilized and partially purified by Taki and Kanfer in 1978 (Ansell and Spanner 1982; Baranska 1982), who also established that the enzyme was separate from choline or ethanolamine phospholipid exchange activity. Moreover, the preparation had no phospholipase D activity. The enzyme prefers phosphatidylethanolamine as a substrate; phosphatidylcholine, phosphatidylinositol, and phosphatidylserine are essentially inactive as substrates. The phosphatidylethanolamine-serine exchange activity appears to be localized on the cytosolic face of liver microsomes (Bell, Ballas, and Coleman 1981). Similarly, most glycerolipid biosynthetic activities appear to occur on the cytosolic face of the endoplasmic reticulum (Bell, Ballas, and Coleman 1981). This raises the question of how the glycerolipids move to the inner leaflet of the endoplasmic reticulum and to other membranes in the cell (Chapter 14).

Regulation of Phosphatidylserine Biosynthesis

Regulation of phosphatidylserine biosynthesis is another important area of lipid metabolism in animals which is virtually untouched. Is the rate of phosphatidylserine synthesis influenced by the supply of serine, Ca^{2+}, or phosphatidylethanolamine? Phosphatidylethanolamine is the major lipid for exchange with serine, and phosphatidylserine can be decarboxylated to phosphatidylethanolamine in the mitochondria. Is there a futile cycle where phosphatidylethanolamine is converted to phosphatidylserine, which can be decarboxylated to phosphatidylethanolamine? If so, what function does the cycle serve? Perhaps, as indicated above, this is the only way ethanolamine can be made. How do hormones, energy supply, and other factors influence the rate of phosphatidylserine synthesis? There are many fundamental questions and few answers.

CDP-Diacylglycerol: Serine-*O*-Phosphatidyltransferase (Phosphatidylserine Synthase) in Yeast

Phosphatidylserine synthase catalyzes the same reaction as found in procaryotes (Chapter 3) and has been described in yeast by several different laboratories (Carson, Atkinson, and Waechter 1982).

$$\text{CDP-diacylglycerol} + \text{L-serine} \xrightarrow{\text{Mg}^{2+}} \text{phosphatidylserine} + \text{CMP}$$

The enzyme is microsomal and is specific for serine. Yeast mutants defective in phosphatidylserine synthase have been isolated that require ethanolamine, choline, or very high levels of serine for growth. Yeast will normally grow on serine, so phosphatidylserine is converted to phosphatidylethanolamine by decarboxylation and this phosphatidylethanolamine can be methylated to phosphatidylcholine. The gene (*CHO1*) coding for phosphatidylserine synthase from yeast has recently been cloned (Letts et al. 1983). When yeast mutants defective in phosphatidylserine synthase were transformed with plasmids that contain the *CHO1* gene, a sixfold higher amount of phosphatidylserine synthase was recovered compared with wild-type yeast (Letts et al. 1983). This also doubled the amount of phosphatidylserine in the cell with a corresponding decrease in phosphatidylinositol (Letts et al. 1983). The data suggest that the amount of phosphatidylserine synthase in the cells may be important in regulating the distribution of CDP-diacylglycerol between the biosynthesis of phosphatidylinositol and phosphatidylserine. Increased amounts of phosphatidylserine synthase in the plasmid-transformed yeast should facilitate the purification of this enzyme.

The phosphatidylethanolamine-serine exchange enzyme has not been reported in yeast. Perhaps cells have either one of the two pathways for phosphatidylserine synthesis, but not both.

INOSITOL PHOSPHOLIPIDS
Historical Developments

The first report of an inositol-containing lipid was in 1930 from *Mycobacteria* (Hawthorne 1982). That discovery is humorous because inositol lipids are rarely found in bacteria. Brain is the richest source of these lipids, as first discovered by Folch and Wooley in 1942 (Hawthorne 1982). In 1949, Folch described phosphatidylinositol phosphate (PI-P), which was later found to include phosphatidylinositol (PI) and phosphatidylinositol bisphosphate (PI-P$_2$). The chemical structures of PI, PI-P, and PI-P$_2$ (Figure 8.4) were determined by Ballou and co-workers from 1959 to 1961 (Hawthorne and Kemp 1964).

The scheme for the biosynthesis of the inositol phospholipids is as follows:

$$\text{Phosphatidic acid} + \text{CTP} \rightleftharpoons \text{CDP-diacylglycerol} + \text{PP}_i$$

$$\text{CDP-diacylglycerol} + \text{inositol} \rightleftharpoons \text{phosphatidylinositol} + \text{CMP}$$

$$\text{Phosphatidylinositol} + \text{ATP} \longrightarrow \text{phosphatidylinositol-4-phosphate} + \text{ADP}$$

$$\text{Phosphatidylinositol-4-phosphate} + \text{ATP} \longrightarrow \text{phosphatidylinositol-4,5-bisphosphate} + \text{ADP}$$

Agranoff et al. published the first experiments in 1958 on the incorporation of ^{3}H-inositol into phosphatidylinositol (Hawthorne and Kemp 1964). The scheme postulated phosphatidic acid, CDP-choline, and CDP-diacylglycerol as precursors. Subsequently, Paulus and Kennedy showed that CTP, rather than CDP-choline, was the preferred nucleotide donor. In several tissues, separate kinases have been demonstrated that use ATP for phosphorylation of phosphatidylinositol to phosphatidylinositol-4-phosphate and phosphatidylinositol-4-phosphate to phosphatidylinositol-4,5-bisphosphate.

Phosphatidylinositol is present in rat liver at a concentration of 1.7 μmol per gram of liver (Creba et al. 1983). Phosphatidylinositol phosphate and phosphatidylinositol bisphosphate are present at much lower concentrations (1–3% of phosphatidylinositol) (Creba et al. 1983).

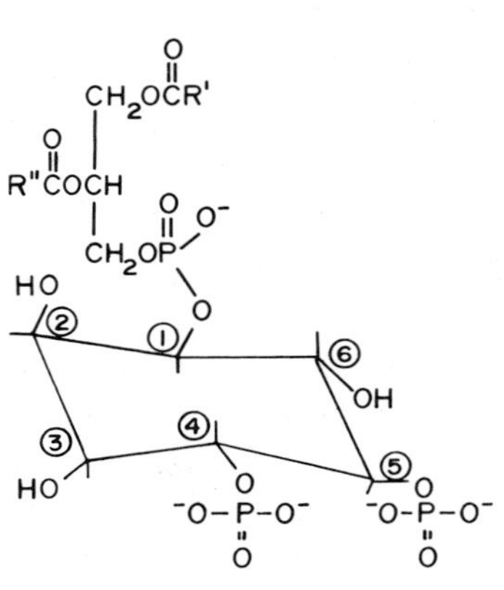

(handwritten margin note: two kinase to phosphorylate inositol derivative.)

Figure 8.4. Structure of phosphatidylinositol-4,5-bisphosphate (PI-P$_2$).

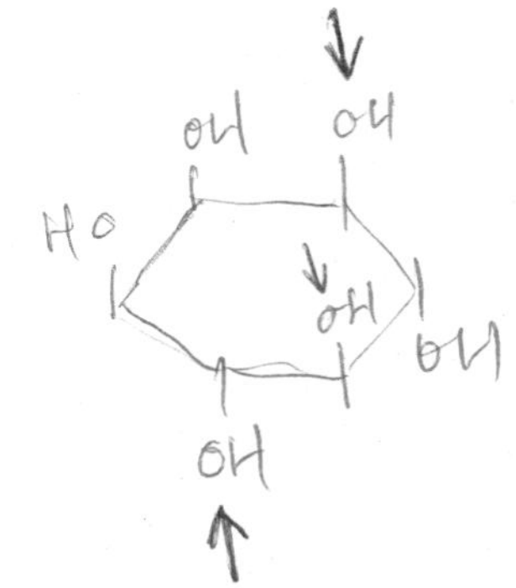

Enzymes *Myo*-inositol is taken up by isolated liver cells by a carrier-mediated system which is not active transport. Once inside the cell, the inositol is converted to phosphatidylinositol, as shown in the preceding section. The enzymes involved in the synthesis of phosphatidylinositol are on the cytosolic surface of the endoplasmic reticulum (Bell, Ballas, and Coleman 1981). Phosphatidylinositol kinase has been reported on plasma membranes from brain and liver and in lysosomes from liver. Phosphatidylinositol-4-phosphate kinase has been reported on plasma membrane, in Golgi, and in cytosol in the brain (Hawthorne 1982). Of the enzymes in inositol phospholipid biosynthesis, only CDP-diacylglycerol:inositol transferase from microsomes has been purified (Hawthorne 1982). The enzyme requires Mg^{2+} or Mn^{2+} and phospholipid.

Phosphatidylinositol is enriched with arachidonic acid when compared with phosphatidic acid. This enrichment appears to occur by deacylation-reacylation reactions similar to the ones described earlier for phosphatidylcholine.

Receptors and Turnover of Inositol Phospholipids Over 30 years ago, Hokin and Hokin observed what was later called the *phosphatidylinositol cycle* (Hawthorne 1982). Incubation of pigeon pancreas with the neurotransmitter acetylcholine causes the release of the digestive enzyme amylase. If ^{32}P is included in the incubation, there is a rapid labeling of phosphatidylinositol and phosphatidic acid. It is now known that binding of acetylcholine to its receptor results in hydrolysis of inositol lipids to diacylglycerol and inositol phosphates. The ^{32}P incorporation is the result of resynthesis and rephosphorylation of the lipids. This series of reactions is known as the phosphatidylinositol cycle (Figure 8.5). Many other agonists (for example, vasopressin, thrombin) cause similar rapid labeling of phosphatidylinositol in many different types of cells. Much work has been expended since 1953 to understand the relationship between receptor stimulation and inositol lipid metabolism. Despite intensive efforts, our understanding of these events had not progressed substantially until recently. It is now recognized that the initial event in inositol lipid metabolism occurs within 20–30 s of the binding of the agonist to the receptor and involves primarily the catabolism of phosphatidylinositol bisphosphate to diacylglycerol and inositol triphosphate (Figure 8.5). For example, 20% of phosphatidylinositol bisphosphate is degraded within 30 s of exposure of hepatocytes to vasopressin or exposure of platelets to thrombin (Fisher, van Rooijen, and Agranoff 1984). As shown in Figure 8.5, the diacylglycerol is converted to CDP-diacylglycerol, and the inositol triphosphate is degraded to inositol. The inositol reacts with CDP-diacylglycerol to yield phosphatidylinositol, which can be converted to phosphatidylinositol bisphosphate, thus completing the cycle.

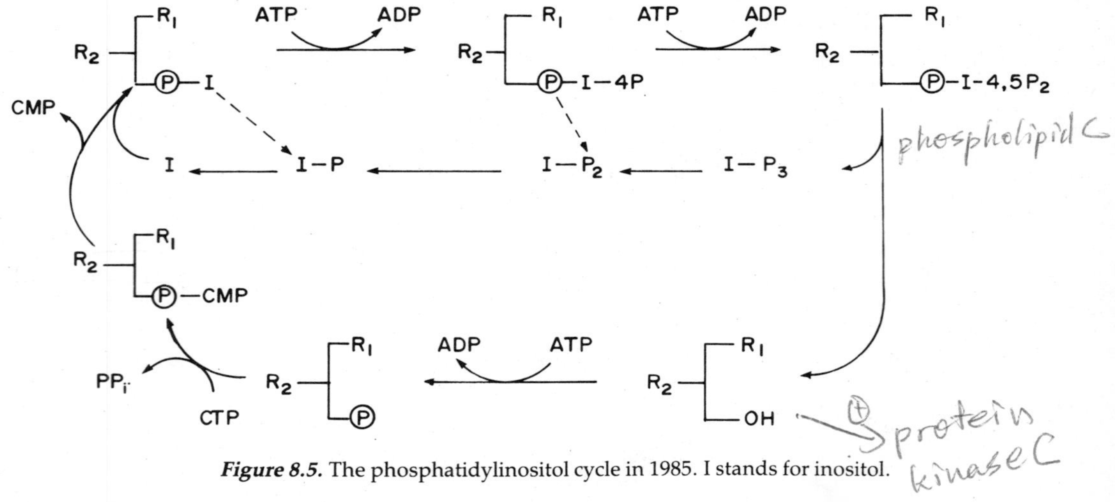

Figure 8.5. The phosphatidylinositol cycle in 1985. I stands for inositol.

There appear to be two reasons why scientists, until recently, overlooked the involvement of phosphatidylinositol bisphosphate and phosphatidylinositol phosphate in the receptor activation (Fisher, van Rooijen, and Agranoff 1984). First, the reactions occur extremely rapidly. Second, the polyphosphoinositides are not extracted by the normal lipid extraction procedures that involve $CHCl_3$ and CH_3OH. If concentrated HCl (1 mL) was included with 100 mL of $CHCl_3$ and 100 mL of CH_3OH, phosphatidylinositol bisphosphate and phosphatidylinositol phosphate were extracted (Creba et al. 1983).

The degradation of phosphatidylinositol bisphosphate (and stimulation of the phosphatidylinositol cycle) is caused by hormones that increase the concentration of cytosolic Ca^{2+} (Creba et al. 1983). The relationship between these two phenomena has been investigated intensely. By January 1985 a consensus appeared to be emerging—namely, that the agonist (for example, vasopressin) binds to a receptor which somehow activates a phospholipase C that degrades phosphatidylinositol bisphosphate to diacylglycerol and inositol triphosphate. The release of inositol triphosphate is postulated to be involved in the binding and mobilization of intracellular Ca^{2+}. The release of inositol triphosphate and the increase in cytosolic Ca^{2+} precedes the activation of phosphorylase, a well-described Ca^{2+}-mediated effect of vasopressin (Nishizuka 1984). Thus, inositol triphosphate is thought to be the intracellular signal which is responsible for the vasopressin-induced increase in cytosolic Ca^{2+} from intracellular Ca^{2+} stores.

Two closely related phenomena are under investigation. First, the other product of phosphatidylinositol bisphosphate degradation is diacylglycerol. This lipid activates the recently discovered protein kinase C (Nishizuka 1984) (which also requires Ca^{2+} and phosphatidylserine for activity). The hypothesis is that diacylglycerol generated from phosphatidylinositol bisphosphate may be the natural signal for activation of protein kinase C. However, no physiological responses have been clearly linked to activation of protein kinase C. Moreover, it is not known whether sufficient diacylglycerol is produced for activation of the kinase. This is clearly a rapidly developing area of research which will require students to search the current literature for the more recent developments.

The second recent observation that relates to the phosphatidylinositol cycle is the discovery that transforming proteins produced by two viruses will phosphorylate phosphatidylinositol. The Rous sarcoma virus codes for a protein, $pp60^{V\text{-}SRC}$, which phosphorylates tyrosine residues on certain proteins. This enzyme also will phosphorylate phosphatidylinositol, diacylglycerol, and glycerol. Similarly, the transforming protein coded by avian sarcoma virus, $UR2,p68^{V\text{-}ros}$, phosphorylates specific tyrosine residues of proteins and phosphatidylinositol. It is plausible that phosphorylation of phosphatidylinositol by these kinases would stimulate the phosphatidylinositol cycle. The release of inositol triphosphate and/or diacylglycerol could be key steps in the unrestricted cell growth that results from the transformation of cells by either of these viruses. Once again, interested students must go to the current literature to discover the status of these proposals.

POLYGLYCERO-PHOSPHOLIPIDS
Historical Developments

Diphosphatidylglycerol, commonly known as *cardiolipin*, was discovered in 1942 in beef heart by Pangborn (Hostetler 1982). The correct structure (Figure 8.6) was proposed in 1956–57 and confirmed by chemical synthesis in 1965–66 by de Haas and van Deenen (Hostetler 1982). Phosphatidylglycerol (Figure 8.6) was first isolated in 1958 from algae by Benson and Mauro. The structure was confirmed as *sn*-1,2-diacylglycerol-3-phospho-*sn*-1'-glycerol by Haverkate and van Deenen in 1964–65. The third lipid in this class, bis(monoacylglycerol)phosphate was recovered from pig lung by Body and Gray in 1967. The two fatty acyl esters were later assigned to the two primary hydroxyl groups (Figure 8.6). The stereochemistry was shown to differ from the other two lipids since bis(monoacylglycerol)phosphate contains *sn*-(monoacyl)glycerol-1-phospho-*sn*-1'-(monoacyl)glycerol (Hostetler 1982), rather than a *sn*-glycerol-3-phospho linkage.

These three lipids are widely distributed in animals, plants, and microorganisms. In animals diphosphatidylglycerol is found in highest concentration in cardiac muscle (9–15% of phospholipids)

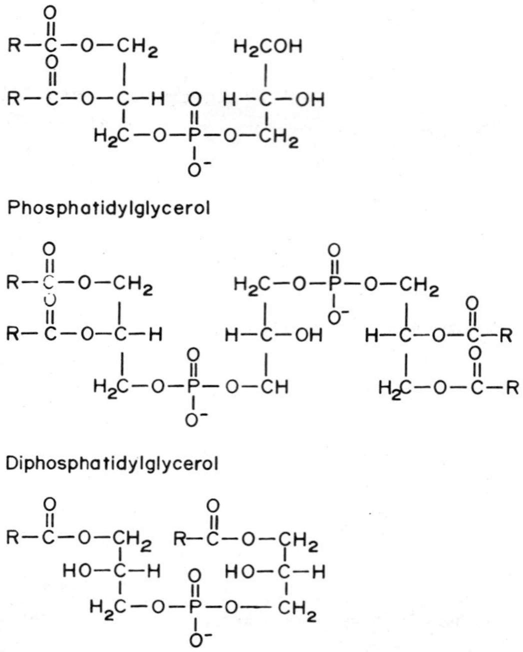

Phosphatidylglycerol

Diphosphatidylglycerol

Figure 8.6. Structures of polyglycerophospholipids.

Bis(monoacylglycerol)phosphate

(Hostetler 1982). Phosphatidylglycerol is generally present at a concentration of less than 1% of total phospholipids, except in lung, where it comprises 2–5% of the phospholipid. In pulmonary surfactant and alveolar type-II cells, phosphatidylglycerol is 7–11% of the total lipid phosphorus (Hostetler 1982). Bis(monoacylglycerol)phosphate is less than 1% of the phospholipids in normal animal tissue, except for alveolar (lung) macrophages, where it is 14–18% of the phospholipid. Diphosphatidylglycerol and phosphatidylglycerol are quantitatively more important lipids in plants and bacteria than in animals; in two blue-green algae, phosphatidylglycerol is the only phospholipid (Hostetler 1982). Similarly, in *Acholeplasma laidlawii*, phosphatidylglycerol is the only phosphoglyceride present (Hostetler 1982).

The biosynthesis of phosphatidylglycerol was demonstrated by Kennedy and co-workers in 1963 in chicken liver as follows:

CDP-diacylglycerol + *sn*-glycerol-3-phosphate ⟶ phosphatidylglycerol phosphate + CMP

Phosphatidylglycerol phosphate ⟶ phosphatidylglycerol + P_i

This pathway was also described in *E. coli* in 1964 by Kanfer and Kennedy.

Establishment of the pathway for diphosphatidylglycerol biosynthesis required several years. It is now well accepted that in mitochondria there is a transfer of phosphatidic acid from CDP-diacylglycerol to phosphatidylglycerol, according to the following reaction:

Phosphatidylglycerol + CDP-diacylglycerol $\longrightarrow$ diphosphatidylglycerol + CMP

The reaction for diphosphatidylglycerol synthesis in *E. coli* (Chapter 3) differs and involves the condensation of two molecules of phosphatidylglycerol.

2 phosphatidylglycerol $\longrightarrow$ diphosphatidylglycerol + glycerol

The latter pathway for diphosphatidylglycerol synthesis does not seem to occur in mammalian mitochondria (Hostetler 1982).

Bis(monoacylglycerol)phosphate can be formed from phosphatidylglycerol or diphosphatidylglycerol, but the enzyme system has not been well characterized (Hostetler 1982). Since the phosphate group is on the *sn*-3 position of glycerol in phosphatidylglycerol and on the *sn*-1 position in bis(monoacylglycerol)phosphate, some rearrangement must occur.

Enzymes and Subcellular Location   Phosphatidylglycerol and its biosynthetic enzymes are found in the mitochondria and microsomes of animal cells. Diphosphatidylglycerol is biosynthesized exclusively in the mitochondria and is found only in this organelle. Bis(monoacylglycerol)phosphate comprises 7% of the phospholipid from rat liver lysosomes, where enzymes for its biosynthesis are located.

None of the mammalian enzymes involved in polyglycerophospholipid synthesis has been purified. However, phosphatidylglycerol phosphate synthase has been purified from *E. coli* (Chapter 3). All three enzymes involved in diphosphatidylglycerol biosynthesis in mitochondria have been solubilized from the membranes and partially purified by McMurray and co-workers (Hostetler 1982). Unfortunately, the small amounts of enzymes present and their membranous environment have made progress on these enzymes most difficult.

Phosphatidylglycerol and diphosphatidylglycerol appear to be degraded by phospholipases, as described in Chapter 10. Bis(monoacylglycerol)phosphate, however, is relatively resistant to degradation by phospholipases. This may be due to its unusual stereochemistry (Hostetler 1982). Perhaps its role in the lysosomes is for stabilization of the membrane against degradation by endogenous lysosomal phospholipases.

Bis(monoacylglycerol)-
phosphate
Accumulation in
Niemann-Pick Disease

Whereas in normal humans bis(monoacylglycerol)phosphate constitutes only 0.8% of the phospholipids in liver, in patients with Niemann-Pick disease bis(monoacylglycerol)phosphate accumulates to a concentration of between 8 and 14% of the phospholipids (Hostetler 1982). This disease is primarily characterized by an accumulation of sphingomyelin and a deficiency of sphingomyelinase (Chapter 12). The metabolic basis for the accumulation of bis(monoacylglycerol)phosphate in this disease and in several other disorders has not been explained. Clearly, the biosynthesis of bis(monoacylglycerol)phosphate, its catabolism, and its function are other important areas of lipid biochemistry which deserve further study.

PHOSPHONOLIPIDS
Historical Perspective

The natural occurrence of an analogue of phosphoethanolamine (2-aminoethylphosphonic acid) that contains a carbon-phosphorus bond was reported in 1959 by Horiguchi and Kandatsu. Rouser and colleagues in 1963 reported the first phosphonolipid, N-acyl-sphingosyl-1-O-aminoethylphosphonate (Figure 8.7), in a sea anemone.

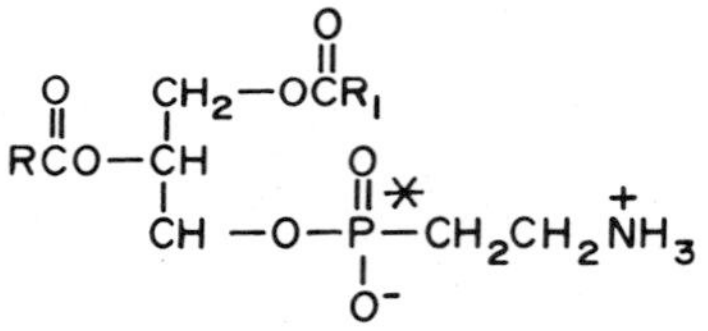

Diacylglyceroaminoethylphosphonate

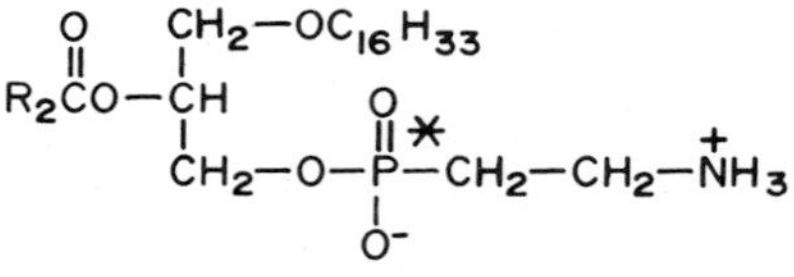

1-O-Hexadecyl-2-acyl-glyceroaminoethylphosphonate

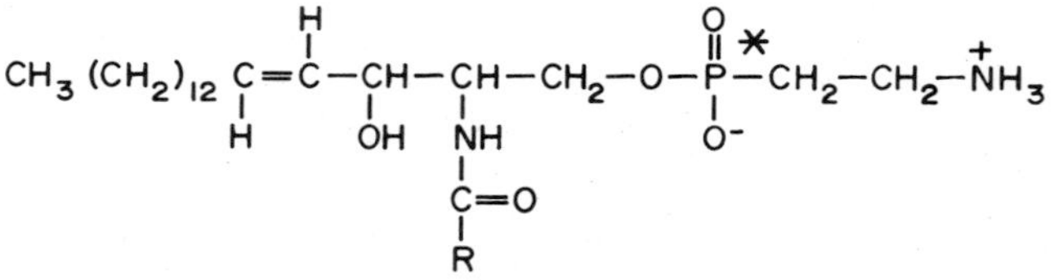

N-Acyl-sphingosyl-1-O-aminoethylphosphonate

Figure 8.7. Structures of several phosphonolipids. The asterisk indicates the carbon-phosphorus bonds.

The phosphonolipids are present in only trace amounts in mammalian tissues, but occur extensively in molluscs, coelenterates, and protozoa. Glycerophosphonolipids (Figure 8.7) were first isolated from the protozoon *Tetrahymena pyriformis* by Liang and Rosenberg in 1966 (Hori and Nozawa 1982). Glycerophosphonolipids are present in high concentrations in *Tetrahymena* (about 23% of phospholipid) and are the major phospholipid in the ciliary membrane (Hori and Nozawa 1982). The glycerophosphonolipids are absent from molluscs and coelenterates, which instead contain large amounts of *N*-acyl-sphingosyl-aminoethylphosphonate (5–20% of the phospholipids) (Hori and Nozawa 1982).

Biosynthesis   The unique feature of the phosphonolipids is the carbon-phosphorus single bond. Aminoethylphosphonic acid appears to arise from phosphoenolpyruvate according to the reaction

$$
\begin{array}{ccc}
\underset{\textbf{Phosphoenolpyruvate}}{
\begin{array}{c}
\mathrm{COO^-}\ \ \mathrm{O} \\
|\ \ \ \ \ \ \| \\
\mathrm{C{-}O{-}P{-}O^{=}} \\
\|\ \ \ \ \ \ | \\
\mathrm{CH_2}\ \ \ \mathrm{O^-}
\end{array}}
& \xrightarrow{\ CO_2\ } &
\underset{\textbf{2-Phosphonoacetaldehyde}}{
\begin{array}{c}
\mathrm{O} \\
\| \\
\mathrm{HC} \\
| \\
\mathrm{CH_2} \\
| \\
\mathrm{^-O-P=O} \\
| \\
\mathrm{O^-}
\end{array}}
& \xrightarrow{\ -\overset{+}{N}H_3\ } &
\underset{\textbf{2-Aminoethylphosphonic acid}}{
\begin{array}{c}
\overset{+}{\mathrm{N}}\mathrm{H_3} \\
| \\
\mathrm{CH_2} \\
| \\
\mathrm{CH_2} \\
| \\
\mathrm{^-O-P=O} \\
| \\
\mathrm{O^-}
\end{array}}
\end{array}
$$

Once aminoethylphosphonate is formed, it is apparently incorporated into phospholipidlike compounds via the CMP derivative.

$$
\mathrm{CTP} + {}^-\mathrm{O}{-}\underset{\underset{\mathrm{O_-}}{|}}{\overset{\overset{\mathrm{O}}{\|}}{\mathrm{P}}}{-}\mathrm{CH_2CH_2}{-}\overset{+}{\mathrm{N}}\mathrm{H_3} \longrightarrow \mathrm{CMP}{-}\mathrm{O}{-}\underset{\underset{\mathrm{O_-}}{|}}{\overset{\overset{\mathrm{O}}{\|}}{\mathrm{P}}}{-}\mathrm{CH_2CH_2}{-}\overset{+}{\mathrm{N}}\mathrm{H_3} + \mathrm{PP}_i
$$

    2-Aminoethylphosphonic acid   **CMP-Aminoethylphosphonic acid**

$$
\mathrm{CMP}{-}\mathrm{O}{-}\underset{\underset{\mathrm{O^-}}{|}}{\overset{\overset{\mathrm{O}}{\|}}{\mathrm{P}}}{-}\mathrm{CH_2CH_2}\overset{+}{\mathrm{N}}\mathrm{H_3} + \text{diacylglycerol} \longrightarrow
$$

                                   diacylglyceroaminoethylphosphonate + CMP

CMP-Aminoethylphosphonic acid

These reactions are analogous to those for the biosynthesis of phosphatidylethanolamine via CDP-ethanolamine (Figure 8.2). It has not been determined whether the same enzymes are involved for the biosynthesis of both lipid classes (Hori and Nozawa 1982). Detailed studies on the biosynthesis of N-acyl-sphingosylaminoethylphosphonate have not been reported (Hori and Nozawa 1982).

The phosphonolipids seem to be degraded by phospholipase C activities. The subsequent degradation of the phosphono (carbon-phosphorus) bond presents a unique problem to the biochemist. Interestingly, an enzyme which cleaves this bond has not been reported in any of the sources that contain these compounds in significant quantities (Hori and Nozawa 1982). However, such an enzyme, *phosphonatase*, which degrades aminoethylphosphonate, has been isolated from *Bacillus cereus* even though bacteria do not contain phosphonolipids. The aminoethylphosphonate is first transaminated to form the aldehyde analogue by a transaminase; then phosphonatase cleaves the carbon-phosphorus bond to yield acetaldehyde and inorganic phosphate (Hori and Nozawa 1982).

$$\overset{+}{H_3N}-CH_2CH_2-\overset{\overset{\displaystyle O}{\|}}{\underset{\underset{\displaystyle O^-}{|}}{P}}-O^- \xrightarrow[\text{pyruvate} \quad \text{alanine}]{} HC-CH_2-\overset{\overset{\displaystyle O}{\|}}{\underset{\underset{\displaystyle O^-}{|}}{P}}-O^- \longrightarrow \overset{\overset{\displaystyle O}{\|}}{HC}-CH_3 + P_i$$

2-Aminoethylphosphonate   **2-Phosphonoacetaldehyde**   **Acetaldehyde**

The specific function of the phosphonolipids as distinct from other phospholipids has not been clearly defined. Studies on the subcellular distribution of the glycerophosphonolipids in *Tetrahymena* showed an enrichment of these lipids in the cilia and pellicles of these organisms relative to internal membranes such as mitochondria and microsomes (Hori and Nozawa 1982). This may be a protective adaptation due to the relative resistance of the phosphonolipids to chemical and enzymatic degradation. *Tetrahymena* do not have the protective coating that fungal and many bacterial cells do.

Growth of *Tetrahymena* at low temperatures causes an enrichment of polyunsaturated fatty acids (particularly 18:2) in the phosphonolipids. However, the phosphono bond shows no significant difference from phosphate bonds in the physical behavior of membrane lipids as a function of temperature. Hence, the temperature response is similar to that of phospholipids seen in bacteria (Chapter 3) and eucaryotes (Chapter 2) which do not contain phosphonolipids. As a result, no selective advantage is apparent for the phosphonolipids in temperature acclimatization of membranes.

The selection of mutants of *Tetrahymena* without phosphonolipids would demonstrate that the lipids are not essential for life of this organism. Such mutants would allow comparative studies with wild-type strains which might provide definitive information on the specialized function, if any, of the phosphonolipids.

PROBLEMS

1. You are interested in determining the apparent K_m for S-adenosylmethionine in the conversion of phosphatidylethanolamine to N-methyl-phosphatidylethanolamine. You incubate liver microsomes with various concentrations of $[CH_3\text{-}^3H]$S-adenosylmethionine. At the end of the reaction, you extract the lipid and separate the various phospholipids by thin-layer chromatography. You identify N-methyl-phosphatidylethanolamine, elute the lipid into a scintillation vial, and determine the radioactivity. By plotting the radioactivity in N-methylphosphatidylethanolamine against concentrations of S-adenosylmethionine, would you be able to calculate an apparent K_m for S-adenosylmethionine?

2. Forskolin is a recently discovered terpene which stimulates formation of cAMP. Suggest an experiment which would indicate whether forskolin affects phosphatidylcholine biosynthesis in cultured hepatocytes.

3. You are interested in studying the regulation of phosphatidylethanolamine biosynthesis in cultured HeLa cells. In your initial studies, you were very surprised to find that after a 30-min pulse with $[^3H]$ethanolamine ($1 \mu M$), 90% of the radioactivity was found in phosphatidylethanolamine (a similar experiment with $[^3H]$choline yielded only 5% of the radioactivity in phos-

phatidylcholine after a 30-min pulse). Provide an explanation for the rapid incorporation of radioactivity into phosphatidylethanolamine in these cells.

4. In studies on the regulation of phosphatidylserine biosynthesis in hepatocytes you were surprised to find radioactivity from $[^3H]$serine in phosphatidylcholine. You had not observed this previously when you did similar studies with HeLa cells. Provide an explanation for the apparent discrepancy.

5. In studies on the biosynthesis of phosphatidylethanolamine in *Tetrahymena*, incubation of these protozoa with $[^3H]$ethanolamine yielded large amounts of label in phosphatidylethanolamine. Since these organisms might somehow convert the label into acetate, you wished to prove that all the label was indeed in the ethanolamine moiety of phosphatidylethanolamine. Hence, you digested labeled phosphatidylethanolamine (purified by thin-layer chromatography) from these cells with phospholipase D. As you expected, all the radioactivity was released into an aqueous extract. However, reanalysis of the phospholipase D digest showed that 25% of the phosphatidylethanolamine (which was unlabeled) was still present and had not been digested by the enzyme. Can you explain this apparent anomaly?

BIBLIOGRAPHY

Entries preceded by an asterisk are suggested for further reading.

*Ansell, G. B., and Spanner, S. 1982. "Phosphatidylserine, phosphatidylethanolamine, and phosphatidylcholine." In *Phospholipids*, ed. J. N. Hawthorne and G. B. Ansell, ch. 1. Amsterdam: Elsevier.

Audubert, F., and Vance, D. E. 1983. Pitfalls and problems in studies on the methylation of phosphatidylethanolamine. *J. Biol. Chem.* 258:10695–701.

*Baranska, J. 1982. Biosynthesis and transport of phosphatidylserine in the cell. *Adv. Lipid Res.* 19:163–84.

*Bell, R. M.; Ballas, L. M.; and Coleman, R. A. 1981. Lipid topogenesis. *J. Lipid Res.* 22:391–403.

Carson, M. A.; Atkinson, K. D.; and Waechter, C. J.

1982. Properties of particulate and solubilized phosphatidylserine synthase activity from *Saccharomyces cerevisiae*: Inhibitory effect of choline in the growth medium. *J. Biol. Chem.* 257:8115–21.

Creba, J. A.; Downes, C. P.; Hawkins, P. T.; Brewster, G.; Michell, R. H.; and Kirk, C. J. 1983. Rapid breakdown of phosphatidylinositol-4-phosphate and phosphatidylinositol 4,5-bisphosphate in rat hepatocytes stimulated by vasopressin and other calcium-mobilizing hormones. *Biochem. J.* 212:733–47.

*Fisher, S. K.; van Rooijen, L. A. A.; and Agranoff, B. W. 1984. Renewed interest in the polyphosphoinositides. *Trends Biochem. Sci.* 9:53–56.

*Hawthorne, J. N. 1982. "Inositol phospholipids." In *Phospholipids*, ed. J. N. Hawthorne, and G. B. Ansell, ch. 7. Amsterdam: Elsevier.

Hawthorne, J. N., and Kemp, P. 1964. The brain phosphoinositides. *Adv. Lipid Res.* 2:127–66.

Hirata, F., and Axelrod, J. 1980. Phospholipid methylation and biological signal transmission. *Science* 209:1082–90.

*Hori, T., and Nozawa, Y. 1982. "Phosphonolipids." In *Phospholipids*, ed. J. N. Hawthorne and G. B. Ansell, ch. 3. Amsterdam: Elsevier.

*Hostetler, K. Y. 1982. "Polyglycerophospholipids: Phosphatidylglycerol, diphosphatidylglycerol, and bis(monoacylglycero)phosphate." In *Phospholipids*, ed. J. N. Hawthorne and G. B. Ansell, ch. 6. Amsterdam: Elsevier.

Ishidate, K.; Nakagomi, K.; and Nakazawa, Y. 1984. Complete purification of choline kinase from rat kidney and preparation of rabbit antibody against rat kidney choline kinase. *J. Biol. Chem.* 259:14,706–710.

Letts, V. A.; Klig, L. S.; Bae-Lee, M.; Carmen, G. M.; and Henry, S. A. 1983. Isolation of the yeast structural gene for the membrane-associated enzyme phosphatidylserine synthase. *Proc. Natl. Acad. Sci. USA* 80:7279–83.

*Mudd, J. B. 1980. "Phospholipid biosynthesis." In *The biochemistry of plants* ed. P. K. Stumpf and E. E. Conn, ch. 9. New York: Academic Press.

*Nishizuka, Y. 1984. The role of protein kinase C in cell surface signal transduction and tumour promotion. *Nature* 308:693–98.

*Pelech, S. L., and Vance, D. E. 1984. Regulation of phosphatidylcholine biosynthesis. *Biochim. Biophys. Acta* 779:217–51.

Sundler, R., and Åkesson, B. 1975. Regulation of phospholipid biosynthesis in isolated rat hepatocytes: effect of different substrates. *J. Biol. Chem.* 250:3359–67.

Vance, D. E. 1983. "Metabolism of glycerolipids, sphingolipids, and prostaglandins." In *Biochemistry*, ed. G. Zubay, ch. 14. Reading, Mass.: Addison-Wesley.

*Vance, D. E., and Pelech, S. L. 1984. Enzyme translocation in the regulation of phosphatidylcholine biosynthesis. *Trends Biochem. Sci.* 9:17–20.

CHAPTER 9

Metabolism, Regulation, and Function of Ether-Linked Glycerolipids

Fred Snyder

AN OVERVIEW	Two types of ether-linked aliphatic moieties are found in glycero-	
General Chemical	lipids. Those with alkyl groups are commonly called *glyceryl ethers*,	
Structures	and those with alk-1-enyl groups are referred to as *plasmalogens*	

AN OVERVIEW
General Chemical
Structures

Two types of ether-linked aliphatic moieties are found in glycerolipids. Those with alkyl groups are commonly called *glyceryl ethers*, and those with alk-1-enyl groups are referred to as *plasmalogens* or *vinyl ethers*. The double bond in the alk-1-enyl linkage of native plasmalogens has a cis configuration. Usually the ether-linked aliphatic moieties are rather simple in most cells, with 16:0, 18:0, and 18:1 carbon chains commonly encountered at the *sn*-1 position.

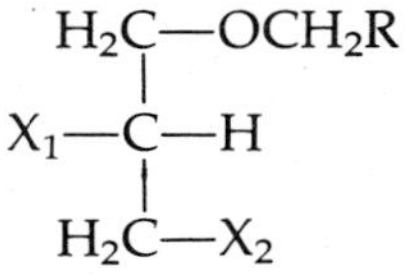

$$H_2C—OCH_2R$$
$$X_1—C—H$$
$$H_2C—X_2$$

Alkyl-type lipid

$$H_2C—OCH=CHR$$
$$X_1—C—H$$
$$H_2C—X_2$$

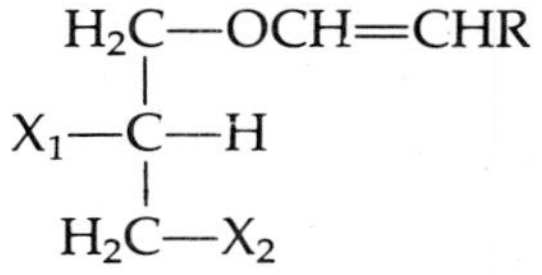

Alk-1-enyl–type lipid

X_1 = acyl or hydroxyl

X_2 = acyl, phosphate, or phosphorus nitrogenous base

Both types of ether-linked aliphatic chains occur in neutral lipids and phospholipids. Plasmanyl- (alkyl) and plasmenyl (alk-1-enyl) are the terms recommended by IUB-IUPAC for the 1-alkyl-2-acyl and 1-alk-1-enyl-2-acyl equivalents of phosphatidyl, which is used to designate the diacylglycerol phosphate grouping in glycerophospholipids. Thus, plasmanic acid and plasmenic acid are the alkyl and alk-1-enyl analogues of phosphatidic acid, respectively, whereas plasmenylethanolamine is 1-alk-1-enyl-2-acyl-*sn*-glycero-3-phosphoethanolamine, (an ethanolamine plasmalogen). In mammals the alkyl and alk-1-enyl chains are located at the *sn*-1 position of glycerolipids (referred to as the L stereoisomer based on Baer-Fischer nomenclature); an exception for the opposite stereochemical configuration of native ether lipids is the unusual *sn*-2,3-dialkyl–type lipids detected in halophilic bacteria. Dialkyl phospholipids, presumably with alkyl groups at the *sn*-1 and *sn*-2 positions, have also been found in bovine heart and as minor components of some other tissues.

Historical Background

The presence of alkyl ether lipids in liver oils of various saltwater fish was originally described in 1920 by the Japanese scientists Tsujimoto and Toyama. (The terms *chimyl* [16:0 alkyl], *batyl* [18:0 alkyl], and *selachyl* [18:1 alkenyl] alcohols are based on the fish species in which each type is most prominent in the alkylglycerols.) However, the complete proof of the chemical nature of the alkyl linkage at the *sn*-1 position in these glycerolipids did not become apparent until 13 years later from a study by Davies, Heilbron, and Jones from England.

In 1924, plasmalogens were detected accidentally by the Germans Feulgen and Voit. After first preserving a variety of fresh tissue slices, including aortas and kidneys of rats, horse muscle, and protozoa in a $HgCl_2$ solution, they erroneously treated the specimens the following day with a fuchsin–sulfurous acid solution without the normal fixation and related histological processing with solvents. Under these conditions the plasma of cells, but not the nuclei, was stained a red-violet color. This phenomenon led Feulgen to conclude that an aldehyde was present in the cytoplasm, and he called this substance "plasmal." If the histological preparations were treated with a solvent before exposure to the dye, no colored stain appeared in the cytoplasm. The unknown precursor of the cytosolic aldehyde that reacted with the dye was called plasmalogen, a name that continues to be retained as the generic term for alk-1-enyl–containing glycerolipid classes. Despite the efforts of many different researchers to identify the chemical structure of the plasmalogens described by Feulgen and Voit, it was not until the late 1950s that the precise structural features of the alk-1-enyl linkage in ethanolamine plasmalogens was proven through the combined efforts of Rapport, Marinetti, Gray, Debuch, and their co-workers.

During the late 1960s and early 1970s the development of new analytical and chromatographic methods for the characterization of both the alkyl- and alk-1-enyl-glycerols made it much easier for scientists to investigate the ether-linked lipids. These technological advances provided new approaches and more sensitive and specific methods for studies of ether lipids that led to our current knowledge about their widespread occurrence, metabolism, and function.

Between 1969 and 1972, the first cell-free enzyme systems that synthesized the alkyl and alk-1-enyl linkages were discovered along with other enzymatic steps associated with these pathways. As with the initial discovery of ether-linked lipids, elucidation of the biosynthesis of the alkyl ethers preceded that of the alk-1-enyl ethers. In fact, the close metabolic interrelationship between the alkyl and alk-1-enyl types of lipids did not become apparent until 1972, when it was documented in a cell-free system that the alkyl moiety of phospholipids was the direct precursor of the alk-1-enyl grouping.

Perhaps the most exciting development in the ether-lipid field has been the recent discovery of certain acetylated forms of alkylglycerolipids (originally described as *platelet activating factor*, or PAF, I, Figure 9.1) that possess potent biological activities. These bioactive phospholipids produced by various blood cells and other tissues exert many different types of biological responses. Some have been implicated in the pathogenesis of such diverse disease processes as hypertension, allergies, inflammation, and anaphylaxis, to name only a few examples.

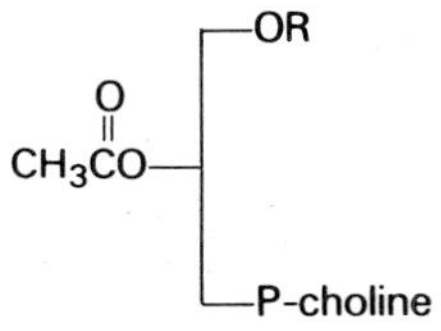

I Alkylacetyl-GPC; PAF

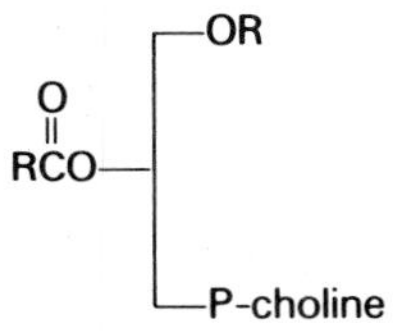

II Alkylacyl-GPC

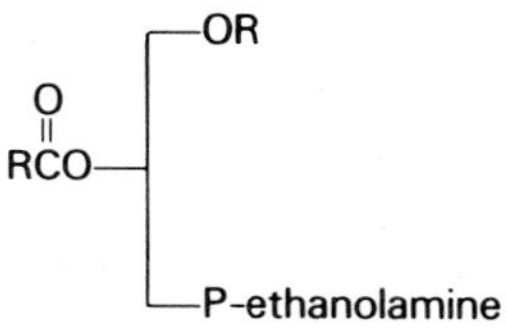

III Alkylacyl-GPE

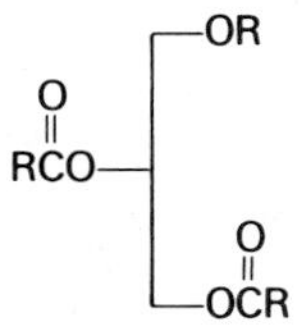

IV Alkyldiacylglycerol

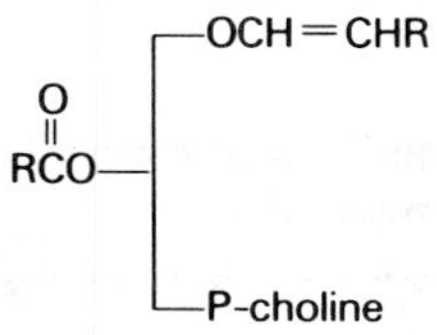

V Alk-1-enylacyl-GPC (plasmalogen)

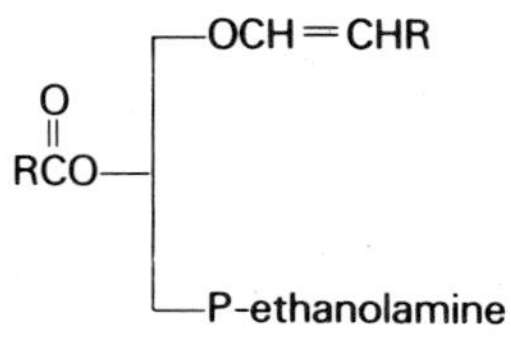

VI Alk-1-enylacyl-GPE (plasmalogen)

Figure 9.1. Ether-linked glycerolipids that occur in nature; GPC and GPE refer to *sn*-glycero-3-phosphocholine and *sn*-glycero-3-phosphoethanolamine, respectively. In structure VII, R = phytanyl.

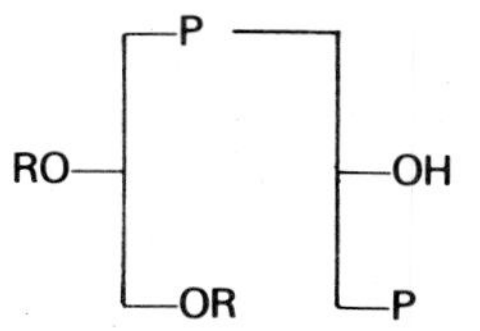

VII 2,3-Diphytanyl-*sn*-glycero-1-P-3′-*sn*-glycero-1-P

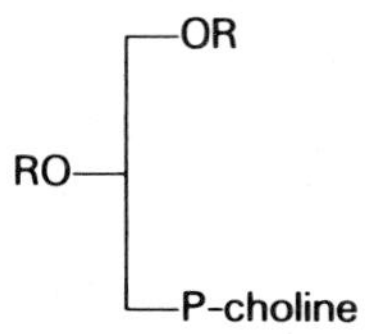

VIII Dialkyl-GPC

As with other progress in the ether lipid field, information about the possible functions of plasmalogens has lagged far behind that of the alkyl lipids. In fact, no specific cellular role has yet been identified for plasmalogens. However, the relatively high proportion of plasmalogens in the ethanolamine phospholipid fraction of nervous tissue and certain other cells suggests that these unique lipids must have an essential function in at least some biological systems. Their cellular importance has been emphasized most recently in studies of a genetic disorder in infants known as the Zellweger syndrome. This disease is a lethal, recessive autosomal inborn error of metabolism where cells of the brain and liver of infants are depleted of plasmalogens. Apparently, the deficiency of a specific enzyme activity (an acyltransferase) that synthesizes the immediate precursor (acyldihydroxyacetone-P) of the alkyl ether bond is responsible for the plasmalogen deficit. The fact that a deficiency in plasmalogen levels leads to the early death of infants afflicted with Zellweger syndrome suggests that plasmalogens are essential components of certain tissues during early embryonic and postnatal development.

Two complete books (Snyder 1972; Mangold and Paltauf 1981) and recent review articles (Horrocks and Sharma 1982; Snyder, Lee, and Wykle 1985) cover the detailed biochemical aspects of ether lipids in general. Recent research on the fast-moving developments in the PAF field and related biologically active glycerolipids has also been described in depth (Lee and Snyder 1985; Snyder 1985).

Occurrence in Nature   Ether-linked lipids occur throughout the animal kingdom and are even found as minor components in higher plants. Some mammalian tissues, and avian, marine, molluscan, protozoan, bacterial, and other species contain significant proportions of ether-linked lipids. The highest levels of ether lipids in mammals occur in nervous tissue, heart muscle, testes, kidney, preputial glands, tumor cells, erythrocytes, bone marrow, spleen, skeletal tissue, neutrophils, macrophages, and platelets. The dietary consumption of ether lipids by humans has essentially been ignored by nutritionists, but certain meats and seafoods contain relatively high amounts of these lipids.

Some membranes, such as the myelin sheath, are highly enriched in ethanolamine plasmalogens, with a mole ratio of alk-1-enyl moieties to ethanolamine lipid phosphorus of approximately 1:3. In general, the alkyl groupings are primarily associated with the choline glycerophospholipids, and the alk-1-enyl moieties are almost exclusively found in the ethanolamine-containing phospholipids, except in heart tissue, where choline plasmalogens are prominent.

Figure 9.1 illustrates the major types of ether-linked lipids found in nature. 1-Alkyl-2,3-diacyl-sn-glycerols (IV, Figure 9.1) are characteristic components of tumor lipids, and 1-alk-1-enyl-2,3-diacyl-*sn*-glycerols (neutral plasmalogens) have also been detected in tumors and adipose tissue of mammals and in fish liver oil.

1-Alkyl-2-acyl-*sn*-glycero-3-phosphocholine (II, Figure 9.1) is a significant component of platelets, neutrophils, and macrophages and is a precursor of platelet activating factor (PAF, 1-alkyl-2-acetyl-*sn*-glycero-3-phosphocholine; I, Figure 9.1). Thus, this lipid class is probably a significant constituent of all cells known to produce PAF (for example, eosinophils, basophils, monocytes, kidney, endothelial cells, and mast cells). PAF is also found in saliva, urine, and amniotic fluid, which implicates additional cells that release PAF to these fluids.

Dialkylglycerophosphocholines (VIII, Figure 9.1) are known to occur as minor constituents of bovine heart. Moreover, heart tissue is unique with respect to its plasmalogen content, since this is the only mammalian tissue known to contain significant amounts of choline plasmalogens (V, Figure 9.1), rather than the usually encountered ethanolamine plasmalogens (VI, Figure 9.1).

As mentioned earlier, halophilic bacteria contain a dialkyl type of glycerolipid (VII, the diphytanyl ether analogue of phosphatidyl-glycerophosphate, Figure 9.1) that has an unusual stereochemical configuration, with the diether moieties located at the *sn*-2 and *sn*-3 positions. The biosynthetic pathway for the formation of the ether bond in halophiles is still unknown. Moreover, despite the significant amounts of ethanolamine plasmalogens (VI, Figure 9.1) in anaerobic bacteria such as *Clostridium butyricum*, no information is yet available about how the alk-1-enyl ether bond is biosynthesized in this organism. This is probably one of the few examples where it has not been possible to use bacteria rich in a specific lipid as a model for investigating the enzymes responsible for that lipid's production.

Analytical Tools for Ether Lipids

Adsorption chromatography, whether thin-layer or column, can resolve the neutral lipid fraction into the ether-linked lipid classes and the corresponding acyl analogues: the order of migration or elution is alk-1-enyl > alkyl > acyl. In contrast, phospholipid subclasses of ether and ester lipids are not easily resolved by adsorption chromatography, although individual molecular species can be isolated by high-performance liquid chromatography. Usually, in order to separate and quantitate the diacyl, alkylacyl, and alk-1-enylacyl fractions of a particular phospholipid class, it is first necessary to remove the phosphobase moiety using phospholipase C and then to make either the acetate or benzoate derivatives of the diacyl-, alkylacyl-, or alk-1-enylacylglycerols. These *diradylglycerol* derivatives are easily resolved by adsorption chromatography; the benzoates offer an advantage over acetates because they can be quantitated on the basis of their UV-absorbing property.

A number of chemical reactions are useful in identification and analysis of the ether-linked lipids (Table 9.1). Since the ether linkages are unaffected by chemical reduction with Vitride or LiAlH$_4$, these reducing agents are excellent for removing esterified groupings (for example, acyl or phosphobase) from lipids without loss of the ether linkage attached to the glycerol. The alkylglycerols and alk-1-enylglycerols produced by chemical reduction are important in the characterization of the ether chains and are a starting point for subsequent preparation of derivatives for chromatographic and mass spectral analysis. Either phospholipase A$_2$ or mild alkaline hydrolysis is usually used to hydrolyze the sn-2 acyl moiety of ether-linked phospholipids. Monomethylamine is also a useful reagent for removing the acyl moieties from the sn-2 position of ether-linked phospholipids; the products are the same as those obtained with mild alkaline hydrolysis (Table 9.1), except that N-methyl fatty acid amides are formed instead of fatty acids.

Complete hydrolysis of all esterified fatty acids in ether-linked neutral lipids (for example, alkyldiacylglycerols) is accomplished by either mild alkaline or monomethylamine hydrolysis. If selective hydrolysis of only the sn-3 acyl groups is desired, either pancreatic or Rhizopus lipases can be employed. Pancreatic or Rhizopus lipases (Table 9.1) are useful enzymes for removal of diacyl phospholipids that are contaminants of ether-linked lipid classes. The lipases from both sources exhibit the same positional specificity for the sn-1 and sn-3 acyl groups in glycerolipids.

Ether-linked phospholipids also serve as substrates in phospholipase C– or D–catalyzed reactions. The glycerolipid products formed by phospholipase C treatment of ether phospholipids can then be derivatized for subsequent analysis. Benzoate and acetate derivatives are especially important in the analysis of alkylglycerols, alk-1-enylglycerols, alkylacylglycerols, and alk-1-enylacylglycerols by high-performance liquid and gas-liquid chromatography. Acetolysis (with acetic acid and acetic anhydride) of any ether lipid replaces all ester groupings (acyl and phosphate) with acetate. Isopropylidene derivatives or trimethylsilyl ethers of alkylglycerols are also important derivatives for gas-liquid chromatographic analysis, especially when it is combined with mass spectrometry. Alk-1-enylglycerols can be analyzed in the same way after they have been converted to alkylglycerols by catalytic hydrogenation.

Acid hydrolysis of the alk-1-enyl linkage produces the corresponding fatty aldehyde. The free aldehydes can be measured directly by gas-liquid chromatography if they are analyzed immediately after

(handwritten annotations, top of page)

$$H_2C-O-(CH_2)_n CH_3$$
$$HO-CH$$
$$CH_2-O-\overset{O}{\underset{O}{P}}-O-choline$$

n = 18. has antitumor activity

Table 9.1. The Analysis of Ether-linked Lipids Commonly Found in Mammalian Cells: Chemical Reactions and Lipid Products

Starting lipid sample	Primary lipid products formed by chemical or enzymatic reactions				
	Vitride [NaAlH₂(OCH₂CH₂OCH₃)₂] or LiAlH₄ reduction	Mild alkaline hydrolysis	Phospholipase A₂	Phospholipase C	Pancreatic or Rhizopus lipase
H₂COR / O=RCOCH / H₂C-phosphobase — 1-Alkyl-2-acyl-sn-glycero-3-phosphobase	H₂COR / HOCH + RCH₂OH / H₂COH — 1-Alkyl-sn-glycerol + fatty alcohol	H₂COR / HOCH + RCOOH / H₂C-phosphobase — 1-Alkyl-2-lyso-sn-glycerol-3-phosphobase + fatty acid	H₂COR / HOCH + RCOOH / H₂C-phosphobase — 1-Alkyl-2-lyso-sn-glycerol-3-phosphobase + fatty acid	H₂COR / O=RCOCH / H₂COH — 1-Alkyl-2-acyl-sn-glycerol	No reaction
H₂COCH=CHR / O=RCOCH / H₂C-phosphobase — 1-Alk-1-enyl-2-acyl-sn-glycero-3-phosphobase	H₂COCH=CHR / HOCH + RCH₂OH / H₂COH — 1-Alk-1-enyl-sn-glycerol + fatty alcohol	H₂COCH=CHR / HOCH + RCOOH / H₂C-phosphobase — 1-Alk-1-enyl-2-lyso-sn-glycero-3-phosphobase + fatty acid	H₂COCH=CHR / HOCH + RCOOH / H₂C-phosphobase — 1-Alk-1-enyl-2-lyso-sn-glycero-3-phosphobase + fatty acid	H₂COCH=CHR / O=RCOCH / H₂COH — 1-Alk-1-enyl-2-acyl-sn-glycerol	No reaction
H₂COR / O=RCOCH / H₂COCR=O — 1-Alkyl-2,3-diacyl-sn-glycerol	H₂COR / HOCH + RCH₂OH / H₂COH — 1-Alkyl-sn-glycerol + fatty alcohol	H₂COR / HOCH + RCOOH / H₂COH — 1-Alkyl-sn-glycerol + fatty acid	No reaction	No reaction	H₂COR / O=RCOCH + RCOOH / H₂COH — 1-Alkyl-2-acyl-sn-glycerol + fatty acid

Note: Acid treatment of plasmalogens converts the alk-1-enyl moiety to the corresponding aldehyde; if HCl/methanol is used, dimethylacetals of fatty aldehydes are formed.

(handwritten annotations, bottom of page)

$$R-\overset{O}{\overset{\|}{C}}-O-\overset{H_2C-O-R}{\underset{H_2C-O-\overset{\|}{C}-R}{C}} \xrightarrow{\text{Acid}} R-\overset{O}{\overset{\|}{C}}-O-\overset{H_2C=O}{\underset{H_2C-O-\overset{\|}{C}-CH_3}{C}} + R-C \overset{O}{\underset{H}{}}$$

being generated. If methanol is present during the acid hydrolysis, a more stable derivative of the aldehyde, the dimethylacetal, is produced.

BIOSYNTHESIS OF ETHER LIPID PRECURSORS

The biosynthetic pathways of the alkyl- and acylglycerolipids from dihydroxyacetone-P is shown in Figure 9.2. Glycerolipids with ether bonds originate from long-chain fatty alcohols and acyldihydroxyace-tone-P via a unique reaction catalyzed by alkyldihydroxyacetone-P synthase (enzyme II, Figure 9.2). The acyldihydroxyacetone-P is synthesized by acyl-CoA:dihydroxyacetone-P acyltransferase (enzyme I, Figure 9.2). This product is the immediate precursor of alkylglycerolipids, but can alternately be converted to phosphatidic acid via the stepwise reactions catalyzed by an NADPH-dependent oxidoreductase (enzyme V, Figure 9.2) and an acyltransferase (enzyme VI, Figure 9.2). The alkyl analogue of phosphatidic acid is the precursor of both neutral lipids and phospholipids that possess ether bonds (Figure 9.3).

Acyl-CoA Reductase

Fatty alcohol precursors in ether lipid biosynthesis are derived from acyl-CoAs according to the following reaction catalyzed by a membrane-associated acyl-CoA reductase:

$$\text{RCOS—CoA} \xrightarrow[\text{2 NADPH + 2H}^+]{} \text{[RCHO]} \xrightarrow[\text{2 NADP}]{} \text{ROH + CoA}$$

acyl - CoA reductase (membrane-associated)

A cytosolic form of the reductase from bovine heart has been described. Other soluble aldehyde reductases are known to exist, but their lack of substrate specificity suggests they do not play a significant role in acyl-CoA reduction.

The acyl-CoA reductases associated with membrane systems use only acyl-CoA substrates, and in mammalian cells, they exhibit a specific requirement for NADPH. Although only traces of fatty aldehydes can be detected in these reactions, the use of trapping agents such as semicarbazide has documented that aldehydes are indeed formed as intermediates. Acyl-CoA reductase prefers saturated substrates over acyl-CoAs that are unsaturated; in fact, the enzyme in brain microsomes is not able to convert polyunsaturated moieties to fatty alcohols. Acyl-CoA reductase has been investigated in various tumors, mouse preputial glands, brains and hearts of rats, *Euglena gracilis*, and certain bacteria. Some evidence indicates that, at least in brain, acyl-CoA reductase might be localized in microperoxisomes instead of the microsomal fraction. Topographical studies

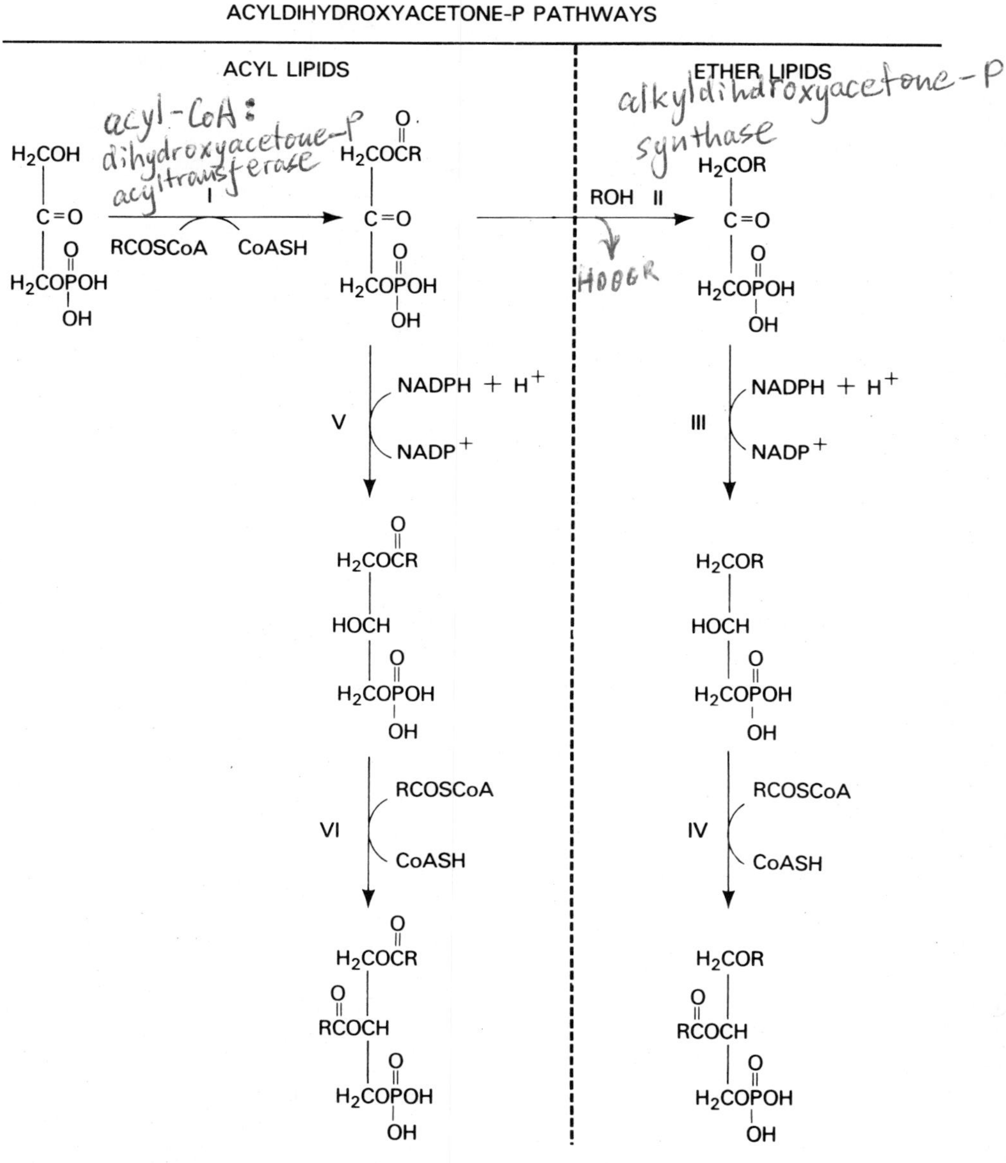

ACYLDIHYDROXYACETONE-P PATHWAYS

Figure 9.2. The biosynthesis of acyl and alkyl glycerolipids. The Roman numerals designate the enzymes responsible for catalyzing each reaction step: I, acyl-CoA:dihydroxyacetone-P acyltransferase; II, alkyldihidroxyacetone-P synthase; III, NADPH:alkyldihydroxyacetone-P oxidoreductase; IV, acyl-CoA:1-alkyl-2-lyso-*sn*-glycero-3-P acyltransferase; V, NADPH:acyldihydroxyacetone-P oxidoreductase; and VI, acyl-CoA:1-acyl-2-lyso-*sn*-glycero-3-P acyltransferase.

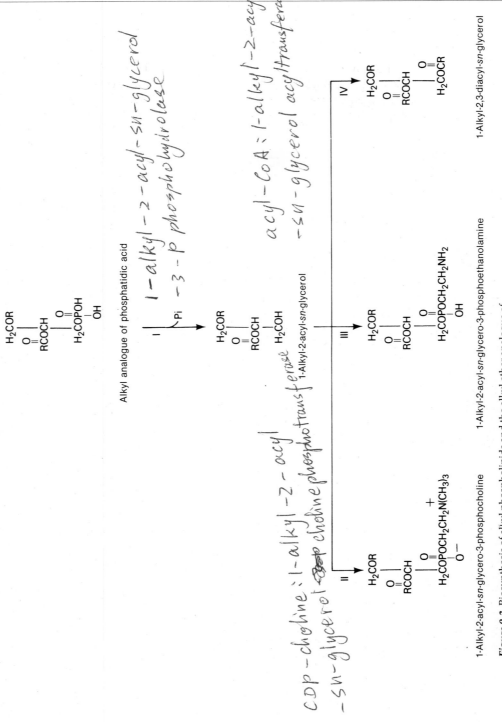

Figure 9.3. Biosynthesis of alkyl phospholipids and the alkyl ether analogue of triacylglycerols. The Roman numerals designate the enzymes responsible for catalyzing each reaction step: I, 1-alkyl-2-acyl-*sn*-glycerol-3-P phosphohydrolase; II, CDP-choline:1-alkyl-2-acyl-*sn*-glycerol cholinephosphotransferase; III, CDP-ethanolamine:1-alkyl-2-acyl-*sn*-glycerol ethanolaminephosphotransferase; and IV, acyl-CoA:1-alkyl-2-acyl-*sn*-glycerol acyltransferase.

of both microsomal vesicles and microperoxisomal particles have revealed that the active center of acyl-CoA reductase is located at the cytosolic surface of these membranes.

Formation of fatty alcohols has also been observed in rabbit harderian glands and *E. gracilis*, where neither acyl-CoA nor the free acid is a substrate for the reductase. In these systems, the NADPH-dependent reductase appears to be closely coupled with fatty acid synthase; it is thought that the fatty acid bound to acyl carrier protein is the substrate for this acyl reductase, rather than an acyl-CoA.

Dihydroxyacetone-P Acyltransferase

Presumably, dihydroxyacetone-P acyltransferase (enzyme I, Figure 9.2) is present in all cells that synthesize alkylglycerolipids, since the acylation of dihydroxyacetone-P is an obligatory step in the biosynthesis of the ether bond in glycerolipids. On the other hand, the quantitative importance of the dihydroxyacetone-P versus *sn*-glycerol-3-P routes in the biosynthesis of glycerolipids is not yet firmly established in most cells (Chapter 7).

Some current evidence suggests that the dihydroxyacetone-P acyltransferase and related enzymes of the ether lipid pathway are localized in microperoxisomes. Therefore, the original discovery and subsequent finding of these enzymes in microsomes and/or mitochondria could be explained by the fact that microperoxisomes sediment with microsomes, and large peroxisomes sediment with mitochondria under the usual preparation conditions of these subcellular fractions. Nevertheless, Ehrlich ascites cells are essentially devoid of peroxisomes, yet are a rich source of ether-linked lipids and the enzyme alkyldihydroxyacetone-P synthase (enzyme II, Figure 9.2), which synthesizes the alkyl ether bond. It is clear that much more work is required in this area before the quantitative significance and role of the different organelle systems in the biosynthesis of ether-linked glycerolipids among different cell types can be established.

Investigations of the topographical location of dihydroxyacetone-P acyltransferase in membrane preparations from rabbit harderian glands and rat brains indicate that unlike most other enzymes in glycerolipid metabolism, dihydroxyacetone-P acyltransferase appears to be located on the internal side of the membrane vesicles. Moreover, two types of dihydroxyacetone-P acyltransferases apparently exist in mammals, a peroxisomal and a microsomal type. The peroxisomal activity is specific for dihydroxyacetone-P and does not utilize *sn*-glycerol-3-P as a substrate, whereas the microsomal dihydroxyacetone-P acyltransferase from fat cells, liver, and several other rat tissues can acylate either dihydroxyacetone-P or *sn*-glycerol-3-P. Furthermore, the properties of the enzyme in peroxisomes differ from the microsomal form in that the latter is inhibited by N-ethylmaleimide. Substrate competition experiments also support these differences.

BIOSYNTHESIS OF ALKYL LIPIDS
Biosynthesis of the Ether Bond

Formation of the alkyl ether bond in glycerolipids is catalyzed by alkyldihydroxyacetone-P synthase. The reaction (enzyme II, Figures 9.2 and 9.4) is unique among biochemical reactions, since it is the only one known where a fatty alcohol is substituted for an acyl moiety, the latter being esterified to dihydroxyacetone-P. Alkyldihydroxy-acetone-P synthase has been primarily investigated in microsomal preparations; however, as with the dihydroxyacetone-P acyltransferase, recent evidence indicates that the synthase activity is also located in the peroxisomes, at least in some cells.

The following are important characteristics of the forward reaction that produces alkyldihydroxyacetone-P:

1. The pro-R hydrogen at C-1 of the dihydroxyacetone-P moiety of acyldihydroxyacetone-P exchanges with water.
2. The configuration of the C-1 carbon of the dihydroxyacetone-P moiety of acyldihydroxyacetone-P is preserved.
3. The acyl group of acyldihydroxyacetone-P is cleaved before addition of the fatty alcohol.
4. Either fatty acids or fatty alcohols can bind to the activated enzyme–dihydroxyacetone-P intermediate to produce acyldihydroxyacetone-P or alkyldihydroxyacetone-P, respectively.
5. A Schiff's base intermediate is not formed.
6. The oxygen in the ether linkage of alkyldihydroxyacetone-P is donated by the fatty alcohol.
7. Both oxygens in the acyl linkage of acyldihydroxyacetone-P are found in the fatty acid product.
8. Acyldihydroxyacetone-P acylhydrolase activity is not associated with purified alkyldihydroxyacetone-P synthase.
9. Alkyldihydroxyacetone-P synthase activity is sensitive to modifiers of sulfhydryl and amino functional groups.
10. A nucleophilic cofactor (possibly an amino acid functional group at the active site) covalently binds the dihydroxyacetone-P portion of the substrate acyldihydroxyacetone-P.
11. A ketone function is an essential feature of the substrate, acyldihydroxyacetone-P.

These observations have led to the proposed ping-pong reaction mechanism for alkyldihydroxyacetone-P synthase illustrated in Figure 9.4.

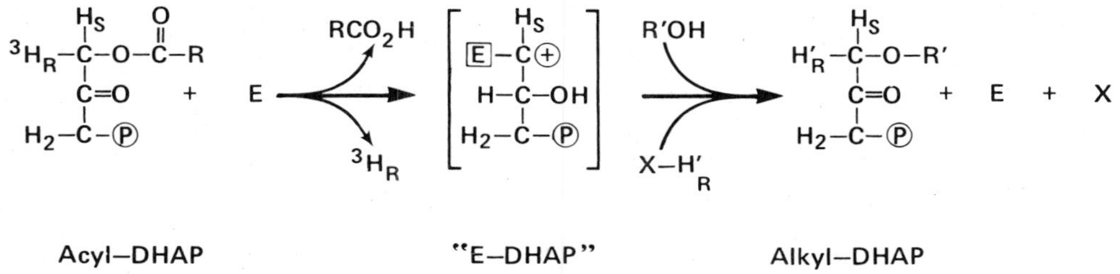

Acyl—DHAP   "E—DHAP"   Alkyl—DHAP

Figure 9.4. The molecular reaction catalyzed by alkyldihydroxyacetone-P synthase in the formation of the alkyl ether bond has a ping-pong mechanism. DHAP is dihydroxyacetone-P. Upon binding of acyl-DHAP to alkyl-DHAP synthase $\boxed{E}$, the pro-*R* hydrogen at C-1 is exchanged by an enolization of the ketone, followed by release of the acyl moiety to form an activated E-DHAP. C-1 is thought to carry a positive charge that may be stabilized by an essential sulfhydryl group of the enzyme; the incoming alkoxide ion reacts at C-1 to form alkyl-DHAP. X is a nucleophilic cofactor, possibly an amino acid functional group at the active site that covalently binds the DHAP portion of the substrate.

Alkyldihydroxyacetone-P synthase has been solubilized from Ehrlich ascites cell microsomes and subsequently purified approximately 1000-fold. Kinetic experiments with the partially purified enzyme have demonstrated that the reaction mechanism of alkyldihydroxyacetone-P synthase is of a ping-pong type, with an activated enzyme–dihydroxyacetone-P intermediate playing a central role. The existence of this intermediate explains the reversibility of the reaction, since the enzyme–dihydroxyacetone-P complex can react with either fatty alcohols (forward reaction) or fatty acids (back reaction). This mechanism is also consistent with the overall characteristics of alkyldihydroxyacetone-P synthase listed earlier in this section.

Alkyldihydroxyacetone-P synthase exhibits a very broad specificity for fatty alcohols of different alkyl chain lengths. On the other hand, the specificity of this enzyme for acyldihydroxyacetone-P with different acyl chains is less well understood, primarily because of the unavailability of acyldihydroxyacetone-P with different acyl chains.

Topographical studies of alkyldihydroxyacetone-P synthase in membranes from rabbit harderian glands revealed that this enzyme activity is located on the lumenal side of microsomal vesicles. However, since this gland functions primarily as a secretory organ, it would not be appropriate to formulate a general conclusion about the topographical location of alkyldihydroxyacetone-P synthase in other membrane systems.

Biosynthesis of the Alkyl Analogue of Phosphatidic Acid

Once alkyldihydroxyacetone-P is synthesized, it can then be readily converted to the alkyl analogue of phosphatidic acid (Figure 9.2) in a two-step reaction sequence involving NADPH:alkyldihydroxyacetone-P oxidoreductase and acyl-CoA:1-alkyl-2-lyso-*sn*-glycerol-3-P acyltransferase (enzymes III and IV, Figure 9.2). The NADPH-dependent oxidoreductase appears to reduce both the alkyl and acyl analogues of dihydroxyacetone-P. Dietary ether lipids can also enter this pathway, since alkylglycerols are known to be phosphorylated by an ATP:alkylglycerol phosphotransferase to form 1-alkyl-2-lyso-sn-glycerol-3-P.

Biosynthesis of Neutral Lipids and Phospholipids with O-Alkyl Bonds

The alkyl analogue of phosphatidic acid plays a central role in the formation of lipids that contain ether bonds. Reaction steps beginning with 1-alkyl-2-acyl-sn-glycerol-3-P in the routes leading to the more complex ether lipids (Figure 9.3) are thought to be catalyzed by the same enzymes involved in the pathways established by Kennedy and co-workers in the late 1950s for the diacylglycerolipids (Chapters 7 and 8).

1-Alkyl-2-acyl-sn-glycerols, derived from the alkyl analogue of phosphatidic acid by the action of a phosphohydrolase (enzyme I, Figure 9.3), serve as substrates for cholinephosphotransferase (enzyme II, Figure 9.3), ethanolaminephosphotransferase (enzyme III, Figure 9.3), or acyl-CoA acyltransferase (enzyme IV, Figure 9.3) in the formation of the alkyl analogues of phosphatidylcholine, phosphatidylethanolamine, or triacylglycerols respectively (Figure 9.3). Thus, the alkylacylglycerols participate at a very crucial branch point in the ether lipid pathway, much like their counterparts, the diacylglycerols.

Alkyldiacylglycerols produced by acyl-CoA:alkylacylglycerol acyltransferase (enzyme IV, Figure 9.3) are characteristic markers in most tumors from animals and humans. The two major classes of ether-linked phospholipids formed via the choline- and ethanolamine-phosphotransferases serve as precursors for two other important classes of ether lipids. One of the products, 1-alkyl-2-acyl-sn-glycero-3-phosphocholine, is the storage form of the ether lipid precursor of platelet activating factor, a potent biologically active phospholipid discussed later. The other product, an alkyl analogue of phosphatidylethanolamine, serves as the direct precursor of ethanolamine plasmalogens in an unusual desaturation reaction in which this intact alkylacyl phospholipid is the substrate for a Δ^1-alkyl desaturase, a microsomal mixed-function oxidase that is described in the next section.

BIOSYNTHESIS OF PLASMALOGENS

The enzyme system responsible for the biosynthesis of ethanolamine plasmalogens from alkyl lipids was characterized over a decade ago; the reverse of this reaction (that is, conversion of an alk-1-enyl moiety to the alkyl) has never been demonstrated. The Δ^1-alkyl desaturase, which produces the alk-1-enyl grouping, is a somewhat unusual enzyme, since it can specifically and stereospecifically abstract hydrogen atoms from C-1 and -2 of the O-alkyl chain of an intact phospholipid molecule, 1-alkyl-2-acyl-sn-glycero-3-phospho-ethanolamine. Only the intact ethanolamine phospholipid is known to serve as a substrate for the Δ^1-alkyl desaturase.

Δ^1-Alkyl desaturase, like the acyl-CoA desaturases (Chapter 6), shows the typical requirements of a microsomal mixed-function oxidase: molecular oxygen, a reduced pyridine nucleotide, cytochrome b_5, and a terminal desaturase protein that is sensitive to cyanide. The reaction mechanism responsible for the biosynthesis of the ethanolamine plasmalogens is unknown. However, it is clear from an investigation with a tritiated fatty alcohol, that only the 1S and 2S (erythro) labeled hydrogens are lost during the formation of the cis alk-1-enyl moiety of ethanolamine plasmalogens. The Δ^1-alkyl desaturase does not utilize 1-alkyl-2-acyl-sn-glycero-3-phosphocholine as a substrate. Thus, the biosynthesis of significant quantities of choline plasmalogens that occurs in heart remains an enigma, although it has been speculated that base exchange, the phosphatidylethanolamine methylation pathway, or a coupled phospholipase C and CDP-choline cholinephosphotransferase could account for the rare occurrences of significant quantities of choline plasmalogens in certain tissues, such as heart.

CATABOLIC PATHWAYS
Ether Lipid Precursors

Fatty alcohols are oxidized to fatty acids via NAD^+:fatty alcohol oxidoreductase, a microsomal enzyme that is found in most cells and probably accounts for the extremely low levels of fatty alcohols generally found in mammalian tissues. The detection of fatty aldehyde intermediates in this reaction (by trapping them as semicarbazide derivatives) suggests that the fatty alcohol oxidoreductase catalyzes a two-step reaction. The difficulty in purifying the membrane-associated enzymes of fatty alcohol metabolism has prevented much progress in this area, except for the demonstration of the reaction sequence and the characterization of some of the enzymatic properties in the complex environment of their membrane locations.

Dihydroxyacetone-P can be channeled away from the ether lipid pathway through the formation of sn-glycerol-3-P via NADH:-glycerol-3-P dehydrogenase. An alternate bypass of alkyldihydroxy-acetone-P formation occurs if acyldihydroxyacetone-P is reduced by the NADPH-dependent oxidoreductase (enzyme V, Figure 9.2), since the product, 1-acyl-2-lyso-sn-glycerol-3-P, is directed into diacylglycerolipids. The metabolic removal and/or formation of fatty alcohols, dihydroxyacetone-P, or acyldihydroxyacetone-P from the ether lipid precursor pool represent important control points in the ether lipid pathway. Regulatory aspects of these reactions are discussed later in this chapter.

Alkyl Cleavage Enzyme

Oxidative cleavage of the O-alkyl linkage in glycerolipids is catalyzed by a microsomal tetrahydropteridine ($Pte \cdot H_4$)-dependent alkyl monooxygenase. Alkyl cleavage activities are highest in livers and lowest in tumors and other tissues that contain significant quantities of alkyl lipids; in fact, the levels of ether lipids appear to be inversely related to the alkyl cleavage enzyme activity. The reaction responsible for the cleavage of the O-alkyl bond is thought to occur in the following manner, with the formation of a transient hemiacetal intermediate:

$$\begin{array}{c} H_2COCH_2CH_2R \\ | \\ HOCH \\ | \\ H_2COH \end{array} + Pte \cdot H_4 \xrightarrow{O_2} \left[\begin{array}{c} OH \\ | \\ H_2C{-}OC{-}CH_2R \\ H \\ | \\ HOCH \\ | \\ H_2COH \end{array} \right] \longrightarrow RCH_2CHO + glycerol + Pte \cdot H_2 + H_2O$$

1-Alkyl-*sn*-glycerol   **Hemiacetal intermediate**

Fatty aldehydes produced in the cleavage reaction can be either oxidized to the corresponding acid or reduced to the alcohol by the appropriate enzymes; the $Pte \cdot H_4$ is regenerated from the $Pte \cdot H_2$ by an NADPH-linked pteridine reductase, a cytosolic enzyme. Oxidative attack on the ether-linked grouping in lipids is similar to that described for the hydroxylation of phenylalanine.

Structural requirements of glycerolipid substrates utilized by the alkyl cleavage enzyme (Table 9.2) are (a) an O-alkyl moiety at the *sn*-1 position, (b) a free hydroxyl group at the *sn*-2 position, and (c) a free hydroxyl or phosphobase group at the *sn*-3 position. If the hydroxyl group at the *sn*-2 position is replaced by a ketone or acyl grouping, or when a free phosphate is at the *sn*-3 position, the O-alkyl moiety at the *sn*-1 position is not cleaved by the $Pte \cdot H_4$-dependent monooxygenase. The 1-alkyl-2-lysophospholipids are substrates for the cleavage enzyme, but they are attacked at much slower rates than the alkylglycerols; alkylglycols (CH_2OHCH_2OR) are also cleaved in the same manner as the alkylglycerols, but also more slowly.

Plasmalogenases

Microsomal enzymatic activities have been described that hydrolyze the alk-1-enyl grouping of plasmalogens; the products are a fatty aldehyde and either 1-lyso-2-acyl-*sn*-glycero-3-phosphoethanolamine

Table 9.2. Ether Lipids as Substrates for Various Lipases

Enzyme	Products formed from:		
	1-Alkyl-(or 1-alk-1-enyl)-2,3-diacyl-sn-glycerols	1-Alkyl-(or 1-alk-1-enyl)-2-acyl-sn-glycero-3-phosphobase	1-Alkyl-(or 1-alk-1-enyl)-2-lyso-sn-glycero-3-phosphobase
	Products formed		
Pancreatic lipase or Rhizopus lipase*	1-Alkyl-(or 1-alk-1-enyl)-2-acyl-sn-glycerols + RCOOH	No reaction	No reaction
Phospholipase A$_2$	No reaction	1-Alkyl-(or 1-alk-1-enyl)-2-lyso-sn-glycero-3-phosphobase + RCOOH	No reaction
Phospholipase C	No reaction	1-Alkyl-(or 1-alk-1-enyl)-2-acyl-sn-glycerol + phosphobase	1-Alkyl-(or 1-alk-1-enyl)-sn-glycerol + phosphobase
Phospholipase D	No reaction	1-Alkyl-(or 1-alk-1-enyl)-2-acyl-sn-glycero-3-P + base	1-Alkyl-(or 1-alk-1-enyl)-sn-glycero-3-P + base
Lysophospholipase D	No reaction	No reaction	1-Alkyl-(or 1-alk-1-enyl)-sn-glycero-3-P + base
Pte·H$_4$-dependent monooxygenase (specific for alkyl group)	No reaction	No reaction	sn-Glycero-3-phosphobase + RCHO
Plasmalogenase (specific for alk-1-enyl group)	2,3-Diacyl-sn-glycerols + RCHO	1-Lyso-2-acyl-sn-glycero-3-phosphobase + RCHO	sn-Glycero-3-phosphobase + RCHO

* *Rhizopus arrhizus or delamar.*

(or choline) or *sn*-glycero-3-phosphoethanolamine (or choline), depending on the chemical structure of the original substrate presented to the plasmalogenase (Table 9.2). A number of plasmalogenase activities have been described in microsomal preparations of liver and brain membranes from rats, cattle, and dogs. The possibility that a plasmalogenase (reaction I below) works in concert with lysophospholipase A_2 (reaction II) to release the rich source of arachidonic acid generally associated with plasmalogens is an intriguing thought, since control of the release of arachidonic acid from such phospholipids could be regulated through the substrate specificities of two separate enzymes. A possible sequence of these hypothetical reactions is illustrated here.

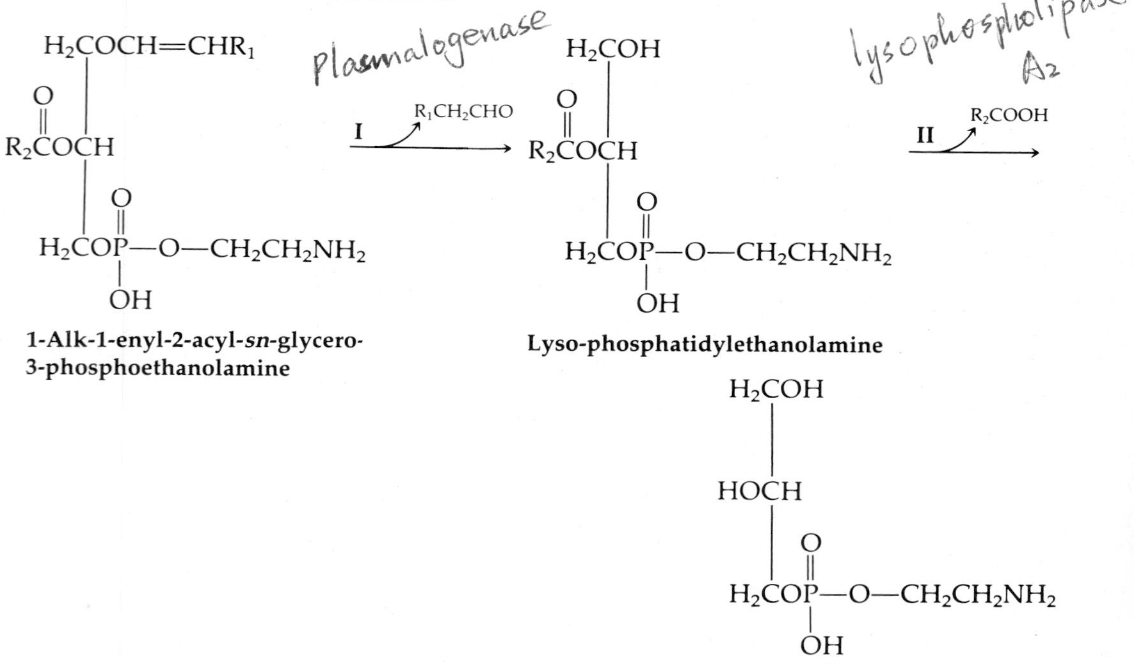

1-Alk-1-enyl-2-acyl-*sn*-glycero-
3-phosphoethanolamine

Lyso-phosphatidylethanolamine

Glycerophosphoethanolamine

Plasmalogenases have not been purified. Therefore, the role of plasmalogenases in lipid metabolism still remains obscure, but it is likely that these enzymes are effective regulators of the cellular levels of plasmalogens and could even be involved in arachidonic acid metabolism, as already mentioned.

Lipases   In general, the ester groupings associated with either the alkyl or alk-1-enyl glycerolipids are hydrolyzed by lipolytic enzymes with the same degree of specificity as their acyl counterparts. However, the presence of an ether linkage at the sn-1 position of the glycerol moiety impairs the overall reaction rate to the extent that certain lipases have been successfully used to remove diacyl contaminants in the purification of some ether-linked phospholipids. Table 9.2 lists various lipases that have been investigated with ether lipids as substrates and identifies the products of such reactions. The only lipase (other than those that cleave the ether linkages) known to exhibit an absolute specificity for ether-linked lipids is lysophospholipase D, an enzyme discovered by Wykle that exclusively recognizes 1-alkyl-2-lyso-sn-glycero-3-phosphobases or 1-alk-1-enyl-2-lyso-sn-glycero-3-phosphobases as substrates.

The lypolytic enzymes of ether lipid metabolism have received far less attention than those associated with the biosynthetic pathways. Certainly, more knowledge about this area of research is required before regulation of the metabolic steps that degrade the ether-linked lipids can be understood.

METABOLISM OF PLATELET ACTIVATING FACTOR (PAF)
Biosynthesis of PAF

In 1979, a potent biologically active phospholipid was discovered that could aggregate platelets at a concentration of $10^{-11}M$ and induce an antihypertensive response in rats when as little as 60 ng were administered intravenously to hypertensive rats. The chemical formula of the semisynthetic lipid tested in these experiments was 1-alkyl-2-acetyl-sn-glycero-3-phosphocholine, as follows.

$$H_2COR$$

$$CH_3\overset{O}{\overset{\|}{C}}OCH$$

$$H_2CO\overset{O}{\overset{\|}{P}}-O-CH_2CH_2\overset{+}{N}(CH_3)_3$$
$$O^-$$

1-Alkyl-2-acetyl-sn-glycero-3-phosphocholine

This compound was ultimately shown to be identical with PAF isolated from IgE-stimulated rabbit basophils. Studies of enzyme activities involved in PAF metabolism have demonstrated that a variety of blood cells, many healthy tissues, and tumor cells can synthesize and degrade PAF. The ubiquitous distribution of these enzymes and the diverse biological properties of PAF indicate that these bioactive phospholipids in mammals serve as multifunctional cellular mediators.

PAF can be synthesized by two different routes, acetyl-CoA:1-alkyl-2-lyso-*sn*-glycero-3-phosphocholine acetyltransferase or CDP-choline:1-alkyl-2-acetyl-*sn*-glycerol cholinephosphotransferase.

$$\begin{array}{ccc}
\text{H}_2\text{COR} & & \text{H}_2\text{COR} \\
| & & \quad\text{O} \quad | \\
\text{HOCH} \quad + \text{CH}_3\text{COSCoA} \xrightarrow{\text{acetyltransferase}} & & \overset{||}{\text{CH}_3\text{COCH}} \quad + \text{CoASH} \\
| & & | \\
\text{H}_2\text{COP}\!-\!\text{O}\!-\!\text{CH}_2\text{CH}_2\overset{+}{\text{N}}(\text{CH}_3)_3 & & \text{H}_2\text{COP}\!-\!\text{O}\!-\!\text{CH}_2\text{CH}_2\overset{+}{\text{N}}(\text{CH}_3)_3
\end{array}$$

1-Alkyl-2-lyso-*sn*-glycero-3-phosphocholine   PAF

$$\begin{array}{ccc}
\text{H}_2\text{COR} & & \text{H}_2\text{COR} \\
\quad\text{O}\quad| & & \quad\text{O}\quad| \\
\overset{||}{\text{CH}_3\text{COCH}} \quad + \text{CDP-choline} \xrightarrow{\text{cholinephosphotransferase}} & & \overset{||}{\text{CH}_3\text{COCH}} \quad + \text{CMP} \\
| & & | \\
\text{H}_2\text{COH} & & \text{H}_2\text{COP}\!-\!\text{O}\!-\!\text{CH}_2\text{CH}_2\overset{+}{\text{N}}(\text{CH}_3)_3
\end{array}$$

1-Alkyl-2-acetyl-*sn*-glycerol   PAF

Both enzyme activities occur at relatively high levels in most cells; however, the acetyltransferase activity appears to be stimulated by agents that induce the cellular production of PAF (for example, Ca^{2+} ionophore A23187 or zymosan), whereas the cholinephosphotransferase does not seem to be affected by such agents. On the other hand, in rat kidney medulla the activity of the cholinephosphotransferase that synthesizes PAF is much higher than the acetyltransferase; therefore, the cholinephosphotransferase in the renal medulla could be significant in the generation of the hypotensive activity associated with PAF. It is interesting that 1-acyl-2-lyso-*sn*-glycero-3-phosphocholine also can be acetylated by the acetyltransferase. In fact, when labeled acetate is used as a precursor in stimulated cells, the acyl analogue of PAF can also be formed. Although the quantitative importance or the physiological roles of the two pathways for PAF biosynthesis are unknown at the present time, it is clear that their significance is probably dependent upon the specific cell types and the physiological or pharmacological state of the system under investigation.

The cholinephosphotransferase involved in the biosynthesis of PAF exhibits a high degree of specificity for 1-alkyl-2-acetyl-*sn*-glycerols or 1-alkyl-2-propionyl-*sn*-glycerols. An interesting characteristic of this enzyme that distinguishes it from the classical cholinephosphotransferase that utilizes long-chain diradylglycerols as substrates is that unlike the latter, the CDP-choline:1-alkyl-2-acetyl-*sn*-glycerol cholinephosphotransferase is not inhibited by dithiothreitol.

Recent evidence indicates that the 1-alkyl-2-acetyl-*sn*-glycerol used by the dithiothreitol-insensitive cholinephosphotransferase is formed by an acetyltransferase (reaction I below) that transfers the acetyl moiety of acetyl-CoA to 1-alkyl-2-lyso-*sn*-glycerol-3-phosphate. This phosphorylated intermediate appears to be the substrate for a phosphohydrolase (reaction II). Thus, the following reaction sequence is required for the formation of the important alkylacetylglycerol precursor of PAF synthesized by the dithiothreitol-insensitive cholinephosphotransferase pathway.

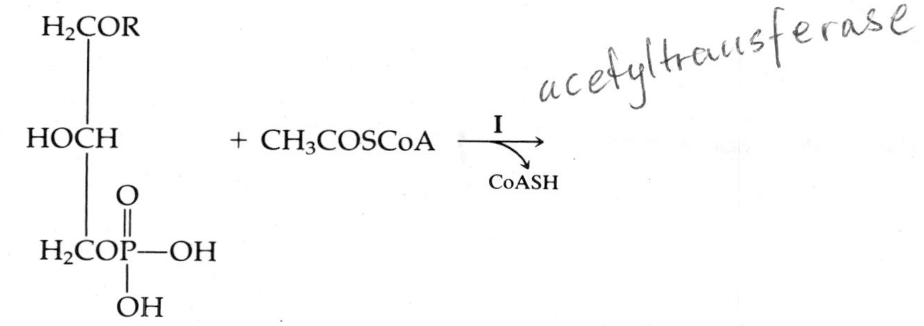

$$
\begin{array}{l}
\text{H}_2\text{COR} \\
|\\
\text{HOCH} \quad\quad + \text{ CH}_3\text{COSCoA} \xrightarrow[\text{CoASH}]{\;I\;} \text{\em acetyltransferase} \\
\quad\quad\;\;\text{O} \\
\quad\quad\;\;\|\\
\text{H}_2\text{COP—OH} \\
\quad\quad\;\;\text{OH}
\end{array}
$$

1-Alkyl-2-lyso-*sn*-glycerol-3-phosphate

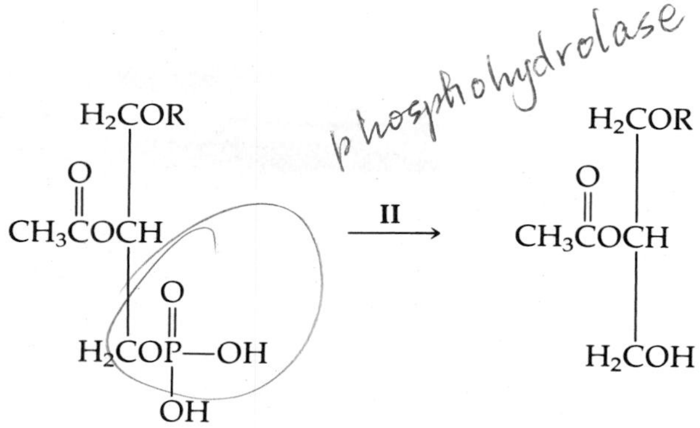

$$
\begin{array}{ccc}
\text{H}_2\text{COR} & & \text{H}_2\text{COR} \\
\;\text{O} & & \;\text{O} \\
\;\| & \xrightarrow{\;II\;} & \;\| \\
\text{CH}_3\text{COCH} & \text{\em phosphohydrolase} & \text{CH}_3\text{COCH} \\
\;| & & \;| \\
\;\text{O} & & \\
\;\| & & \\
\text{H}_2\text{COP—OH} & & \text{H}_2\text{COH} \\
\quad\text{OH} & &
\end{array}
$$

1-Alkyl-2-acetyl-*sn*-glycerol-3-phosphate   **1-Alkyl-2-acetyl-*sn*-glycerol**

Inactivation of PAF   Inactivation of PAF is accomplished via 1-alkyl-2-acetyl-*sn*-glycero-3-phosphocholine acetylhydrolase, the enzyme that catalyzes the hydrolysis of the acetate moiety to produce the inactive 2-lyso form of PAF (Figure 9.5). This type of hydrolytic activity is catalyzed by an

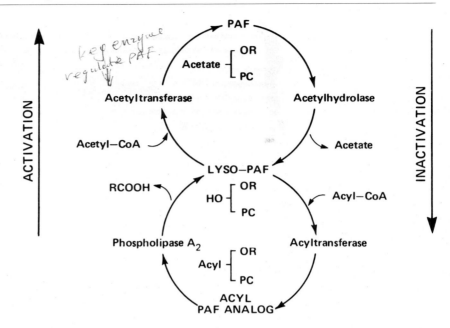

key enzyme regulate PAF. (handwritten)

Figure 9.5. The activation-inactivation cycle for PAF. PC is phosphocholine.

(handwritten structures at left)

① acetylhydrolase
② phospholipase A₂

acetylhydrolase that occurs in both an intracellular (cytosolic fraction) and extracellular form (serum). The serum enzyme is resistant to proteases and has a higher molecular weight than the cytosolic enzyme. Since all other characteristics of the two enzyme activities appear to be identical, it is likely that the intracellular form is processed in the cell for export (for example, by glycosylation or protein modification) to the blood compartments.

Acetylhydrolase's (serum or intracellular forms) properties clearly indicate that it differs from the usual type of activity described for phospholipase A₂ in tissues, although phospholipase A₂ can also hydrolyze the acetate moiety of PAF. Other phospholipids (such as acylacetyl types) with an acetate at the sn-2 position also appear to be substrates for the acetylhydrolase, whereas those phospholipids with long-chain acyl moieties at the sn-2 position are not. Acetylhydrolase activities are widely distributed in a variety of cells and tissues throughout the animal kingdom. Interestingly, genetically hypertensive rats have higher activities of acetylhydrolase than do normotensive animals; similar results have been observed in white, but not black, hypertensive patients. Therefore, it is possible that this enzyme could be important in the pathogenesis of hypertension.

The PAF Cycle   1-Alkyl-2-lyso-sn-glycero-3-phosphocholine (lyso-PAF, Figure 9.5), the product of the acetylhydrolase reaction, is rapidly reacylated by an acyltransferase to produce 1-alkyl-2-acyl-sn-glycero-3-phosphocholine, the stored precursor form of PAF. The acylation step shows a preference for arachidonic acid in platelets, neutrophils, and macrophages. Thus, when cell stimuli induce the activation portion of the PAF cycle (Figure 9.5), both PAF and arachidonic acid metabolites are generated as cell mediators. Unfortunately, the relationships and significance of the simultaneous release of both types of potent chemical mediators by a single stimulus are unknown at the present time.

PAF is formed from 1-alkyl-2-acyl-sn-glycero-3-phosphocholine through the combined actions of phospholipase A_2 and acetyl-CoA:1-alkyl-2-lyso-sn-glycero-3-phosphocholine acetyltransferase (Figure 9.5). Acetyltransferase activity is stimulated in activated cellular states, (for example, during the phagocytosis of zymosan by macrophages or treatment with Ca^{2+} ionophore A-23187), whereas the activity is decreased in cells that exhibit an impaired release of PAF (for example, thioglycollate-treated macrophages). The acetyltransferase of this cycle differs in a number of properties, including its substrate specificity, from the acyltransferase that transfers long-chain fatty acids from acyl-CoAs to 1-alkyl-2-lyso-sn-glycero-3-phosphocholine. Both alkyl- and acyl-lysophospholipids serve as acceptor molecules for the acetate transfer catalyzed by the acetyltransferase.

Other enzymes that can play an important role in the removal of lyso-PAF as an intermediate in PAF metabolism (formed by acetylhydrolase) are the $Pte \cdot H_4$-dependent alkylglycerolipid monooxygenase and the lysophospholipase D shown in Table 9.2. Both enzymes utilize only the lyso ether-linked phospholipids as substrates; therefore, their actions must be coordinated with either the acetylhydrolase or phospholipase A_2 of the activation-inactivation cycle (Figure 9.5).

REGULATION OF ETHER LIPID METABOLISM   Results based on experiments with several different cell types and tissues have indicated that ether-linked lipids have a relatively high turnover rate, especially those molecular species that possess polyunsaturated acyl moieties at the sn-2 position. Factors involved in the regulation of the ether pathways have been investigated, but such studies have mainly been of the descriptive type and have not dealt with molecular mechanisms of enzyme regulation. Nevertheless, a number of general conclusions have been reached about conditions that influence the levels and turnover of ether lipids; also, specific enzymatic steps in the metabolic pathways of ether lipids have been identified as important control points, and these steps will undoubtedly be the focus of future regulatory studies.

Key control points that must be considered in the metabolism of ether-linked lipids are (a) those enzymes that catalyze the formation and catabolism of the ether lipid precursors (fatty alcohols and dihydroxyacetone-P), (b) the enzyme responsible for the synthesis of alkyldihydroxyacetone-P (alkyldihydroxyacetone-P synthase), and (c) those enzymes responsible for the degradation of the ether lipids. Glycolysis also plays an important role in controlling the levels of ether lipids; presumably, the high glycolytic rate of tumors and the generation of dihydroxyacetone-P explain the relatively high levels of ether lipids in tumor cells. Factors that increase the level of dihydroxyacetone-P for utilization in the acyldihydroxyacetone-P pathway could lead to an increased proportion of ether-linked lipids; such a correlation has been observed in a series of transplantable hepatomas that possess high rates of glycolysis, low glycerol-P dehydrogenase activities, and high levels of ether-linked lipids. However, since exceptions have been noted, other factors are also obviously equally important in regulating the amounts of ether lipids. A general conclusion, from numerous reports in the literature, is that higher levels of ether lipids are reached through the coordinated actions of increased activities of the biosynthetic enzymes and the decreased activities of the degradative enzymes in the metabolic pathway for ether-linked lipids.

As with other ether lipids, little information is yet available about the regulatory aspects of PAF metabolism. It is clear that stimuli that increase the production and secretion of PAF also enhance the activity of acetyl-CoA:1-alkyl-2-lyso-sn-glycero-3-phosphocholine acetyltransferase at least 3- to 10-fold compared with unstimulated cells. Surprisingly, acetylhydrolase activities are also increased by agents known to stimulate PAF production. However, the temporal relationship for acetyltransferase and acetylhydrolase stimulation has not yet been investigated. Nevertheless, it appears that the activation-inactivation cycle for PAF plays an important role in the regulation of PAF levels in most cells. There is evidence that the acetyltransferase activity is under the control of a protein kinase and a protein phosphohydrolase; phosphorylation activates, and dephosphorylation inactivates, acetyl-CoA:1-alkyl-2-lyso-sn-glycero-3-phosphocholine acetyltransferase (Figure 9.6). Thus, the various agents that increase PAF production could exert their effects by stimulating this protein kinase activity.

In contrast, the alternate pathway for PAF synthesis, which is catalyzed by the dithiothreitol-insensitive cholinephosphotransferase, does not appear to be affected by known stimuli. However, since this phenomenon has been observed only with human neutrophils, it is not possible to generalize about the regulatory aspects or significance of the alternate cholinephosphotransferase route for PAF

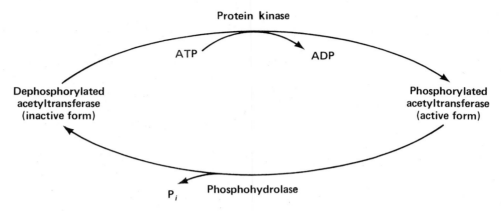

Figure 9.6. Regulation of PAF biosynthesis by a protein kinase and a protein phosphohydrolase. Acetyl-CoA:1-alkyl-2-lyso-*sn*-glycero-3-phosphocholine acetyltransferase is activated by phosphorylation and inactivated by dephosphorylation.

synthesis. This pathway appears to be the major synthetic route for PAF biosynthesis in the medulla of rat kidneys.

Understanding the factors responsible for the regulation of PAF metabolism is difficult because PAF and some arachidonic acid metabolites share a common precursor, 1-alkyl-2-acyl-*sn*-glycero-3-phosphocholine. Also, the influence of PAF on arachidonic acid metabolism complicates any interpretation of the regulatory controls involved in this important area of ether lipid metabolism.

FUNCTIONS OF ETHER LIPIDS AND THEIR BIOLOGICAL ACTIVITIES

Cellular functions of ether-linked glycerolipids are still poorly understood, but they seem to play a role both as membrane components and as cellular mediators. Both the alkyl and alk-1-enyl phospholipids that contain long-chain acyl groups at the *sn*-2 position are thought to be essential structural components of some membrane systems, (for example, the myelin sheath), and some ether lipids appear to have the capacity to serve as protective storage reservoirs for polyunsaturated fatty acids. The apparent protective nature of ether-linked groups is due to their ability to retard the rate of hydrolysis of acyl moieties at the *sn*-2 position in the same molecule by phospholipase A_2. The preferential sequestering of polyunsaturated fatty acids in ether-linked phospholipids has been observed in essential fatty acid deficiency; under this condition the ether lipids of testes from deficient rats retained arachidonic acid, whereas the diacyl phospholipids were rapidly depleted of their arachidonoyl moieties.

The multifaceted responses generated by PAF in vivo and in target cells has emphasized the important role of such lipids as diverse regulators of metabolic and cellular processes. Biological activities of PAF have been described for target cells (platelets and neutrophils) and for more complex systems, such as isolated tissues or in vivo. Specific receptors for PAF have been detected in target cells, and a stereospecific requirement of the receptors for native PAF (1-alkyl-2-acetyl-*sn*-glycero-3-phosphocholine) has been established. In neutrophils, PAF causes aggregation and degranulation responses, induces chemotaxis and chemokinesis, increases cell adherence, enhances respiratory bursts and superoxide production, and stimulates the production of arachidonic acid metabolites formed via the lipoxygenase pathway. PAF also causes aggregation and degranulation of platelets and stimulates the uptake of calcium, protein phosphorylation, the phosphatidylinositol cycle, and arachidonic acid metabolism (potentiates the formation of hydroxy fatty acids and thromboxanes). In the more complex systems investigated, PAF increases vascular permeability, induces the constriction of the ileum and lung strips, stimulates hepatic glycogenolysis, causes bronchoconstriction, increases pulmonary resistance, causes pulmonary hypertension and edema, decreases dynamic lung compliance, produces neutropenia and thrombocytopenia, causes intestinal necrosis, and produces systemic hypotension. Many of the biological activities described for PAF are associated with hypersensitivity reactions, such as allergies and inflammation. The myriad events that can occur when PAF reaches sufficiently high levels in blood cumulate in anaphylactic shock.

FUTURE DIRECTIONS

Future studies in the ether lipid field are needed to address the many unanswered questions about their regulatory controls and functions. Significant progress has been made in our understanding of the mechanism of formation of the alkyl ether bond and of the biological activities, metabolism, and some regulatory aspects of the alkyl-acetylated phospholipids, but nothing is yet known about the mechanism of action of these unique bioactive phospholipids. A group of unnatural ether-linked phospholipids that are analogues of PAF possess highly selective antineoplastic properties; therefore, future research on their mode of action could also provide exciting dividends for researchers who pursue this area.

Despite the large quantities of ethanolamine plasmalogens found in nervous tissue and other cells, we have no idea how they function. Prospective investigators concerned with the biological role of plasmalogens will need to consider the molecular mechanism of the desaturation step responsible for their biosynthesis and the way in which plasmalogen levels are regulated. Another segment of research on ether lipids that has virtually been untouched is their dietary significance.

In order to understand more fully the regulation of the synthesis and degradation of the ether-linked lipids, it will be necessary to purify key enzymes of the biosynthetic and catabolic pathways so that their reaction mechanisms can be delineated. Antibodies against these enzymes could be prepared for use as probes to evaluate specific metabolic steps in cellular systems. Information is also lacking about the physical properties of specific molecular species of ether lipids, so their role as membrane constituents is not yet fully appreciated.

Knowledge about the cellular functions of ether lipids will be enhanced as scientists from different disciplines enter research that encompasses the various types of ether-linked lipids. The effect of such interdisciplinary efforts has been well demonstrated in the PAF field. Before 1979, few scientists were working in either of the areas of ether lipid biochemistry or PAF biology. However, the two fields were merged after the discovery that PAF was an alkyl phospholipid, and the number of investigators increased exponentially because of the realization that these biologically active phospholipids are such potent and diverse cellular mediators. As a result, progress in the PAF field during the past 4 years has mushroomed. The continued escalation of research on the subject of PAF will contribute enormously toward our understanding of all aspects of the biochemistry of the ether lipids, including their functional role in cells.

PROBLEMS

1. A sample of commercial platelet activating factor (1-alkyl-2-acetyl-sn-glycero-3-phosphocholine) was contaminated with the unnatural biologically inactive D stereoisomer (that is, phosphocholine at the sn-1 position). What would you do to obtain a biologically active preparation of PAF, free of the contaminant in the commercial sample?

2. A glycerophospholipid was subjected to the following reactions; (+) indicates that a reaction occurred, whereas (−) indicates that there was no reaction. (a) HCl at room temperature (+), (b) Grignard reagent (+), (c) phospholipase A_2 (+), (d) pancreatic lipase (−), (e) LiAlH$_4$ or Vitride reducing reagent (+), (f) NaBH$_4$ (−), (g) acid/methanol with heat (+), (h) phospholipase D (+), and (i) phospholipase C (+). After hydrogenation of the lipid products formed by LiAlH$_4$ reduction, an isopropylidene deriva-

tive could be prepared. Based on these observations, give the chemical name (based on IUPAC-IUB Lipid Nomenclature) and the general structural formula of a phospholipid class that could qualify as the unknown structure.

3. [1-^{14}C, 1-^{3}H]Hexadecanol with a ^{3}H/^{14}C ratio of 1.0 was incorporated into various cellular lipids. Assuming that no enzymatic isotopic discrimination is involved and that no reutilization of any metabolite occurs, indicate what the ^{3}H/^{14}C ratio would be in the following lipid moieties: (a) acyl moieties, (b) alkyl moieties, and (c) alk-1-enyl moieties.

4. Specify the location (carbon-chain number) and the radiolabel for hexadecanol that could be incorporated into alkyl and alk-1-enyl moieties and/or wax esters, but that would not radiolabel any acyl moieties.

BIBLIOGRAPHY

All entries are suggested for further reading.

Horrocks, L. A., and Sharma, M. 1982. "Plasmalogens and O-alkyl glycerophospholipids." In *Phospholipids*, ed. J. N. Hawthorne and G. B. Ansell, 51–93. Amsterdam: Elsevier.

Lee, T.-c., and Snyder, F. 1985. "Function, metabolism, and regulation of platelet activating factor and related ether lipids." In *Phospholipids and cellular regulation*, ed. J. F. Kuo. Boca Raton, Fla.: CRC Press. In press.

Mangold, H. K., and Paltauf, F., eds. 1981. *Ether lipids: Biochemical and biomedical aspects*. New York: Academic Press.

Snyder, F., 1972. *Ether lipids: Chemistry and biology*. New York: Academic Press.

Snyder, F. 1985. Chemical and biochemical aspects of "platelet activating factor," a novel class of acetylated ether-linked choline phospholipids. *Med. Res. Rev.* 5:107–140.

Snyder, F.; Lee, T.-c.; and Wykle, R. L. 1985. "Ether-linked glycerolipids and their bioactive species: Enzymes and metabolic regulation." In *The enzymes of biological membranes*. Vol. 2, ed. A. Martonosi. New York: Plenum. 1–58.

CHAPTER 10

Phospholipases

Moseley Waite

Moseley Waite

OVERVIEW
Definition of
Phospholipases

The *phospholipases* are a group of enzymes widely distributed through-out nature whose generic name indicates their common property of catalyzing the hydrolysis of phospholipids. Since the substrates hydrolyzed normally exist in an aggregated state, these enzymes also have the common property of acting at a water-lipid interface. In this regard, they are distinct from the general class of esterases that exhibit normal saturation kinetics, as shown in Figure 10.1. In general, phospholipases have a weak capacity to hydrolyze soluble monomeric phospholipids below their *critical micellar concentration* (cmc) and become fully active only when aggregated structures of lipids are formed above their cmc. The activity of phospholipases on aggregated phospholipids can be more than a 1000-fold higher than that observed with soluble substrates. In this chapter, the term *aggregated phospholipid* refers to phospholipid above its cmc without reference to a specifically defined form such as micelle, bilayer, or hexagonal array.

Beyond these general similarities in properties of phospholipases there is a great diversity in their characteristics. These differences are found in their site of attack on the phospholipid molecule, distribution, function, mode of action, and hydrophobicity. The classification of the phospholipases based on their site of attack is given in Figure 10.2.

The phospholipases A are acyl hydrolases classified according to their hydrolysis of the 1-acyl ester (phospholipase A_1) or the 2-acyl ester (phospholipase A_2). A few phospholipases hydrolyze both

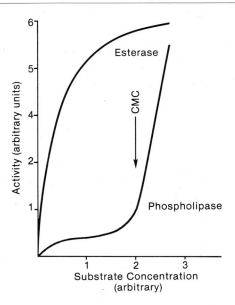

Figure 10.1. Dependence of phospholipase and nonspecific esterase activity on substrate concentration. Esterase exhibits Michaelis-Menten kinetics on soluble substrates, whereas phospholipase becomes fully active above the cmc of the substrate.

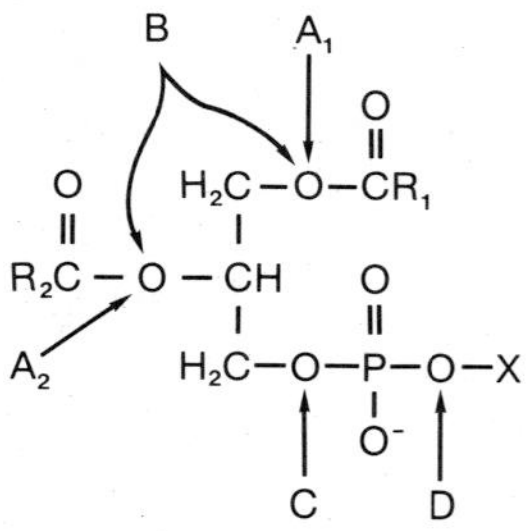

Figure 10.2. Site of hydrolysis by phospholipases.

acyl groups and are termed the phospholipases B. In addition, lyso-phospholipases remove one or the other acyl groups from mono-acyl(lyso)phospholipids. Phospholipases B have high lysophospho-lipase activity, as might be expected. Cleavage of the glycerol-phosphate bond is catalyzed by phospholipase C, while the removal of the base group is catalyzed by phospholipase D. The phospho-lipases C and D are phosphodiesterases.

As a broad generalization, two types of functions can be ascribed to phospholipases. First, many phospholipases are digestive enzymes found in high concentration in venoms, bacterial secretions, and digestive juices of higher organisms. These enzymes are soluble in an aqueous environment even though they avidly adsorb to lipid inter-faces. Second, many phospholipases are regulatory enzymes. This role is quite broad and still poorly defined for many phospholipases. Many of the regulatory phospholipases are membrane associated.

Perhaps one of the clearest examples of a regulatory function is that found in the arachidonate cascade in which a phospholipase(s) regulates the formation of the bioactive eicosanoids at the level of deacylation from the phospholipid (see Chapter 11 for a complete description of the eicosanoids). Likewise, the phospholipase C that degrades phosphatidylinositols in the phosphatidylinositol cycle is thought to be regulatory in Ca^{2+} metabolism and in activation of protein kinase C (Chapter 8). Both classes of phospholipases appear to have hydrophobic regions that avidly bind to phospholipid interfaces.

Assay of Phospholipases

The subject of phospholipase assays has been concisely reviewed by van den Bosch and Aarsman (1979). For thorough coverage of this aspect of phospholipases, their paper is recommended. There are many approaches to the assay of phospholipases. Factors such as sensitivity, site of attack, and equipment available will dictate to an extent the assay of choice.

The simplest procedure is to measure the proton released during hydrolysis with a titration apparatus; in this way, a rapid continuous recording of activity can be made. The usual substrate used in this assay is egg yolk lipid emulsion prepared in the absence of a buffer. This undoubtedly is the most economical approach; however, this technique lacks the sensitivity and specificity of other types of assays. The lower limit of activity is about 0.1 μmol of substrate hydrolyzed per minute. Therefore, this assay has been limited primarily to the better-characterized digestive phospholipases that have high turnover numbers, such as snake venom and pancreatic phospholipases.

A second procedure that is equally rapid yet has greater sensitivity and specificity employs phospholipid analogues that have fluorescent moieties or moieties that can be reacted with chromophores such as Ellmann's reagent. The assays are usually performed on monomeric substrates. An example of the former substrate is 1,2-bis[4-(1-pyreno)butynoyl]-*sn*-glycero-3-phosphorylcholine); hydrolysis of this compound yields the soluble monomeric product 4-(1-pyreno)butyrate, which emits energy at 382 nm when excited at 332 nm, while the micellar substrate emits energy at 480 nm (Hendrickson and Rank 1981).

Examples of compounds whose products are reactive with

4-(1-pyreno) butyrate

Ellmann's reagent are phosphatidylcholines with thioacylesters that upon hydrolysis release a reactive sulfhydryl group (van den Bosch and Aarsman 1979).

$$^-OOC \underset{S \text{---} S}{\overset{NO_2 \qquad NO_2}{\bigcirc \quad \bigcirc}} COO^-$$

5, 5' - Dithiobis (2-Nitrobenzoate)
(Ellmann's Reagent)

These assays can be at least 100 times more sensitive than titration procedures and can be quite useful in the assay of intracellular phospholipases that have low turnover numbers. The lack of availability of these substrates, plus complications in the assay when aggregated phospholipids are used, can limit their usefulness in some cases. For example, the low activity of phospholipases on monomeric phospholipids offsets the enhanced sensitivity of the assay.

A third approach is the use of phospholipid molecules with isotopes incorporated preferentially into specific positions in the molecule (van den Bosch and Aarsman 1979). The various products of hydrolysis are separated from the substrate by extraction and chromatographic procedures. These substrates can be prepared by either chemical or enzymatic methods. By the appropriate choice of labeling, the specificity of the enzymes can readily be established, and as little as a few picomoles of substrate produced can be detected. The use of isotopes has been most useful in the measurement of phospholipase activity in the membranes of whole cells or isolated subcellular fractions. In this case, the label is incorporated into membranous phospholipid in vivo followed by stimulation of the cell or organelle by an agent that activates the phospholipase. This approach has proven to have the advantages of specificity plus sensitivity. While radioactive labeling of substrates has been utilized widely in the study of phospholipases with low turnover numbers, the method is limited by the laborious nature of the assay and the expense of the isotopic substrates.

Interaction of Phospholipases with Interfaces

The distinction of phospholipases from the general esterases implies that special consideration must be given to the interfacial interaction of the enzymes with aggregated lipids. The increase in activity found in the hydrolytic rate when phospholipids are present above their cmc (Figure 10.1) indicates that all phospholipases have a certain hydrophobic character regardless of their origin or type. Therefore, two aspects of enzyme activity with phospholipid aggregates must be considered: (1) the hydrophobic interaction and (2) the formation of the catalytic Michaelis complex. It is generally thought that four factors are responsible for the favored hydrolysis at interfaces:

1. increased effective substrate concentration
2. orientation of the phospholipid molecule at the interface
3. enhanced diffusion of the products from the enzyme
4. conformational change of the enzyme upon binding to the interface

The first factor can be understood most readily by the example of dihexanoyl phosphatidylcholine, which has a cmc of about 10 mM. Above this interfacial concentration, micelles form that have an effective concentration in the surface that is several molar, or three orders of magnitude higher than the free monomers in solution. Since the enzyme binds to the interface, the high local concentration of substrate saturates the enzyme. Once the enzyme is bound to the micelles (or in the case of long-chain phospholipids, bilayers, or hexagonal arrays), the increased concentration of substrate available markedly increases activity, assuming that the enzyme does not dissociate from the aggregate after the hydrolysis of a phospholipid molecule.

The second factor was established using ^{1}H-NMR measurements that demonstrated that the orientation of the acyl esters is restricted in a bilayer system. Presumably, this restriction of the phospholipid facilitates interaction of the substrate molecule with the active site in the enzyme. Figure 10.3 shows that the 1-acyl ester is more deeply buried in the hydrophobic region than is the 2-acyl ester. It is not yet clear just why such an orientation of the acyl esters favors enzymatic hydrolysis or why this would be the case for all types of acyl hydrolases. Likewise, polar head group interaction influences the

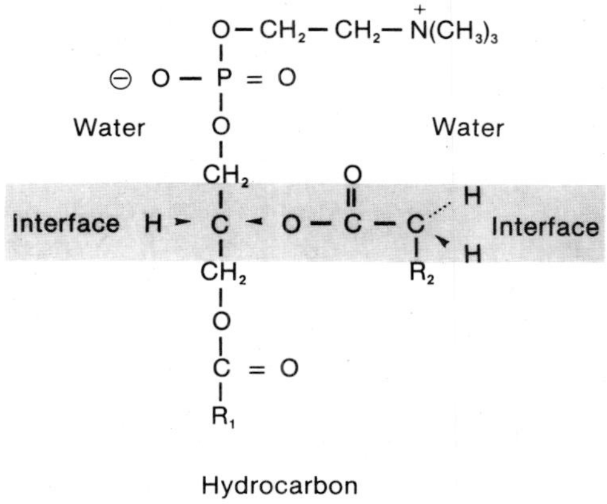

Figure 10.3. Orientation of ester bonds of phosphatidylcholine acyl chains. Adapted from Dennis (1983).

activity of the phosphodiesterase-type phospholipases C and D. However, little information on the relationship of these phospholipases to the orientation of the moieties under attack is available.

The third factor postulated to favor activity, enhanced diffusion of the products from the enzyme, is dependent on both the nature of the substrate and the site of attack. When long-chain phospholipids, as found in membranes, are substrates for hydrolysis, the products have a low cmc and would remain associated with the hydrophobic region of the enzyme if only an aqueous environment were present. This is true for all types of phospholipases, although in the case of phosphodiesterases one of the products becomes water soluble and less hydrophobic binding occurs between products and enzyme than occurs with the other acyl hydrolases. The presence of the substrate aggregate favors the solubilization of the nonpolar product in a hydrophobic environment, thereby favoring diffusion of product from the enzyme. If, however, short-chain phospholipids were substrates, the products would be relatively water soluble, so the lipid diffusion factor would be expected to be less predominant.

The fourth point, conformational changes in phospholipases, has been postulated to account in part for the activity on aggregated lipid (Verheij, Slotboom, and deHaas 1981). The structural and spectral properties of some digestive acyl hydrolases have been found to change when Ca^{2+} or substrate binds. Also, kinetic studies with the pancreatic enzyme suggest that an activation process, probably linked to a conformational change, is necessary for maximal activity. This process could be related to a dimerization of the enzyme that has been shown for the activation of a snake venom phospholipase (Dennis 1983).

The nature of the aggregated lipid markedly influences the activity of phospholipases on the basis of the following parameters:

1. charge at the lipid-water interface
2. packing of the molecules within the aggregate
3. polymorphism of aggregate
4. fluidity of phospholipid

The ionic charge of the lipid has long been known to influence the activity of phospholipases. The interfacial charge is a reflection of the surface pH, which can be different from the bulk pH and, therefore, can influence the pH optimum of the phospholipase. There are various techniques to measure surface charge, for example, microelectrophoresis (Dawson 1966). Ionic amphipaths as well as the ionic content of the aqueous environment influence this surface charge. In addition to the charge imparted by the polar head group of the individual phospholipid molecule, the chemical structure of the polar head group is also important in the formation of the Michaelis complex, as shown by the ability of phospholipases to distinguish among phospholipids in aggregates of the same charge.

The effect of molecular packing of lipid molecules within the aggregate on phospholipase activity has long been recognized as a factor which regulates activity. While this effect is most pronounced for the acyl hydrolases, it also occurs with the phosphodiesterases. The influence of packing is most readily determined using mono-molecular films of phospholipid with defined surface pressures (Verger and Pattus 1982). In this system, the phospholipids are present as a monomolecular film at the air-water interface. The pressure of the film and, therefore, the area occupied by the indi-vidual phospholipid molecules can be regulated by a mechanical bar placed at the end of the film (Verheij, Slotboom, and deHaas 1981). In an interesting study, the ability of various phospholipases to penetrate films of various pressures was used to estimate the surface pressure and concentration of phospholipids in erythrocytes (Demel et al. 1981). The monolayer system, therefore, appears to be a good model for the study of natural bilayers.

The polymorphic states that have been considered primarily in-clude micelles, various bilayer liposome structures, and hexagonal arrays. The first two of these are of considerable physiological interest, whereas the activity of phospholipases on hexagonal arrays has not yet been ascribed to a particular function in natural systems. The micellar mixture of phospholipids and bile salts found in the intestine is a prime example of the degradation of micellar phospho-lipid. Likewise, the attack of membranous bilayer systems by both digestive and regulatory phospholipases occurs widely. The attack of hexagonal arrays, while little studied, could possibly have some interesting functions including an involvement in membrane fusion. For example, a phospholipase that preferentially attacks a transitory hexagonal array at a point of fusion could be responsible for removal of the fusative lipid and the reestablishment of a bilayer membrane.

As is true for many membrane-associated enzymes, the state of fluidity also regulates phospholipase activity. It is well established that many phospholipases are most active at the phase transition of a pure phospholipid or in mixed phospholipid systems that exhibit phase transition. It appears that phospholipases recognize and pene-trate fissures between gel and liquid crystalline phases that favor activity. Interestingly, phospholipases in some model systems appear to degrade phospholipid in the gel phase more rapidly than in a liquid-crystal structure (Verheij, Slotboom, and deHaas 1981; Dennis 1983). For example, the phospholipase A_2 from *Naja naja naja* venom hydrolyzed bilayers of disaturated phosphatidylcholines more rap-idly below, than above, the phase transition (Verheij, Slotboom, and deHaas 1981).

Kinetic analysis of phospholipases using aggregated phospholipid is very complex and must be approached using models other than those used with water-soluble substrates. The commonly used Michaelis-Menten–type kinetics, although often reported, have little value except for comparative purposes. As noted previously, varying substrate concentration above the cmc determines only the number of lipid aggregates present, not the concentration of phospholipid within the aggregate. Two basic approaches have evolved to circumvent this problem: the monolayer system, in which the concentration of phospholipid is regulated by the pressure applied (Verger and Pattus 1982) and a mixed-micelle system, in which the phospholipids are diluted within the micelle by the addition of increasing amounts of detergent (Dennis 1983). In the latter system, the concentration of the phospholipid in an individual micelle can be regulated by the ratio of phospholipid to detergent, while absolute phospholipid concentration can be varied with a constant phospholipid to detergent ratio. The use of a constant ratio of phospholipid to detergent varies only the number of micellar particles, not the concentration of phospholipid within the micelle. The experiments with mixed micelles allow the determination of K_S^A, the K_m of the initial binding step in which the enzyme binds to the micelle, and K_m^B, the surface K_m that measures the affinity of the active site of the enzyme for the substrate molecule. The K_S^A is expressed in moles per liter, whereas the K_m^B is expressed as moles per square centimeter, a reflection of activity in two dimensions rather than the more usual three dimensions.

THE PHOSPHOLIPASES

Many phospholipases have been at least partially purified and characterized. Since it is impossible to cover all enzymes, only a few examples will be given. These represent a wide range of functions and organisms.

Phospholipase A_1

Phospholipases A_1 are widely distributed throughout nature. In addition, some lipases that preferentially degrade triacylglycerols (Chapter 7) are also capable of hydrolyzing the 1-acyl ester bond of phospholipids. In general, phospholipases A_1 have a rather broad specificity and act well on lysophospholipids. For this reason, the functions and naming of these enzymes remain somewhat obscure. Phospholipases A_1 have been recognized only since the 1950s. The first report of the purification of a phospholipase A_1 was by Scandella and Kornberg (1971), one of the two phospholipases A_1 from E. coli to be described further.

Two phospholipases A_1 have been purified from E. coli, one that is resistant, and one that is sensitive, to detergents (Dennis 1983). The former is localized in the outer membrane, whereas the latter is found on the cytoplasmic membrane and in soluble fractions. Both have been purified and extensively studied. The detergent-resistant enzyme has a broad substrate specificity, hydrolyzing all phospholipids tested, while the detergent-sensitive enzyme preferentially degrades phosphatidylglycerol. The detergent-sensitive enzyme also acts as a transacylase. Transacylation reactions presumably occur when an acyl-enzyme intermediate is formed in a two-step reaction. In the second step, the acyl group is transferred nonspecifically to the hydroxyl acceptor of a soluble alcohol or of a lipid such as a monoacyl lipid. The *E. coli* transacylase-acting phospholipase forms methyl esters of the fatty acid in the presence of methanol. *E. coli* mutants deficient in either one or both phospholipases have been developed. These mutants have normal growth characteristics and phospholipid turnover that leaves open the question of the function of these enzymes.

Rat liver lysosomes contain a soluble phospholipase A_1 that is a glycoprotein and has multiple forms, perhaps due to variation in the carbohydrate content (Dennis 1983; van den Bosch 1980). It has optimal activity at pH 4.0 and does not require Ca^{2+} for activity. The apparent substrate specificity is highly dependent on the physical structure of the substrate. As shown in Figure 10.4, phosphatidylethanolamine is preferentially degraded when no Triton X-100 (or WR1339) is present. Triton stimulates the activity toward all phospholipids tested except phosphatidylethanolamine. In that case,

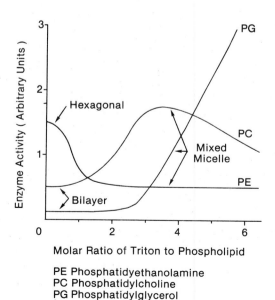

Figure 10.4. Effect of Triton on lysosomal phospholipase substrate specificity. Triton stimulates the hydrolysis of phosphatidylcholine and phosphatidylglycerol liposomes but inhibits hydrolysis of phosphatidylethanolamine hexagonal arrays. Adapted from Robinson and Waite (1983).

PE Phosphatidyethanolamine
PC Phosphatidylcholine
PG Phosphatidylglycerol

activity is inhibited, which suggests that in the absence of Triton the hexagonal arrays formed by phosphatidylethanolamine are more readily attacked than the bilayer liposomes of the other phospholipids. When Triton is added and mixed micelles are formed, the enzyme is optimally active on phosphatidylglycerol. In addition to forming mixed micelles, Triton dilutes the high negative charge of phosphatidylglycerol, favoring activity.

Lysosomal phospholipase A_1 activity is inhibited by Ca^{2+} when pure phosphatidylethanolamine is used (Figure 10.5). This inhibition probably relates to the effect of Ca^{2+} on the surface charge. The enzyme is optimally active on phosphatidylethanolamine that has a slight negative charge. Changing the charge in either direction by the

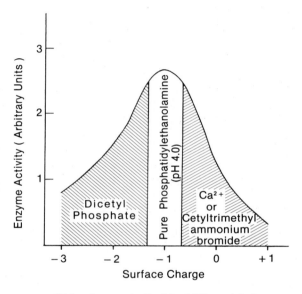

Striped areas indicate addition of Ca^{2+} or amphipath to phosphatidylethanolamine

Figure 10.5. Surface-charge dependency of lysosomal phospholipase A_1.

$$[CH_3(CH_2)_{14}CH_2O—]_2\overset{\displaystyle O^-}{\underset{\displaystyle \|}{P}}{=}O$$

Dicetylphosphate

$$CH_3(CH_2)_{14}CH_2—N^+(CH_3)_3Br^-$$

Cetyltrimethylammonium bromide

The addition of charged amphipaths alters lipid surface charge as determined by microelectrophoresis, which results in lower rates of hydrolysis. Adapted from Robinson and Waite (1983).

addition of charged amphipaths or Ca^{2+} decreases the enzymatic hydrolysis of phosphatidylethanolamine.

The lysosomal phospholipase activity on neutral phosphatidylcholine, on the other hand, is increased by the addition of a small percentage of dicetylphosphate, which gives the liposome a slight negative charge. Ca^{2+} stimulates hydrolysis of the acidic phospholipids by partially neutralizing the negative charge. Although the function of this enzyme is not established, it probably degrades phagocytosed materials, as do other lysosomal enzymes.

There are two lipases that primarily degrade triacylglycerols in lipoproteins but also have phospholipase A_1 activity. Their function in lipoprotein metabolism is described in Chapters 7 and 13 and, therefore, will be mentioned only briefly here. The extrahepatic lipoprotein lipase is optimally active on triacylglycerol when activated by apoprotein C–2, although monoacylglycerols, diacylglycerols, and phospholipids are degraded as well. This enzyme has optimal activity at slightly alkaline pH values (8.0–9.0) and is stimulated by Ca^{2+}. However, in the absence of apoprotein C–2, this enzyme maintains full activity on monoacylglycerol and phospholipids but loses almost all activity on triacylglycerol. It would appear, therefore, that the function of the activator is to promote interaction of the enzyme with the highly hydrophobic triacylglycerol. Physiologically, this enzyme is responsible for the degradation of the neutral lipids in triacylglycerol-rich chylomicrons and VLDL, although it could also be involved in the degradation of phospholipid in these and other lipoproteins.

The hepatic lipase is less well understood even though it has been purified. It does not appear to be activated to the same degree by apoproteins as does lipoprotein lipase and works well on triacylglycerol emulsions and on pure phospholipid and monoacylglycerol. Optimal activity of the hepatic lipase is found at pH 8.0–9.0, and the enzyme is stimulated about twofold by Ca^{2+}. Under the appropriate conditions, the preferred reaction is a transacylation in which two molecules of monoacylglycerol are converted to a molecule each of diacylglycerol and free glycerol. This strongly suggests that an acyl-enzyme intermediate is formed.

Phospholipids can also serve as acyl donors, as in the following reaction.

$$\text{Phospholipid} + \text{monoacylglycerol} \longrightarrow \text{lysophospholipid} + \text{diacylglycerol}$$

Although this reaction would be expected to be reversible, steric problems apparently prevent the diacylphospholipid from being formed. As a phospholipase, hepatic lipase has a marked preference for phosphatidylethanolamine, although it is not clear if this specificity is due to the physical state of the substrate or to the formation of the Michaelis complex. One of the interesting features of this enzyme is that when it is modified by proteolysis, its ability to hydrolyze triacylglycerol, but not monoacylglycerol or phospholipid, is reduced. The possible physiological role proteolysis might play is unclear, although perhaps it could regulate the type of particle attacked by the hepatic lipase. One possible function of the enzyme is to degrade part of the phospholipid in high density lipoprotein (HDL$_2$) yielding HDL$_3$, which may provide a key shuttle of cholesterol to the liver (Figure 10.6) (van'T Hooft, van Gent, and van Tol 1981). In this postulate, the phospholipid in HDL$_2$ (which is cholesterol-ester rich) is degraded by the hepatic lipase, which reduces the amount of surface lipid. This promotes fusion of the lipoprotein to the membrane and transfer of cholesterol ester to the parenchymal cell with the subsequent formation of HDL$_3$. While this proposal is not yet established, it does indicate a role of the hepatic lipase as a phospholipase A$_1$.

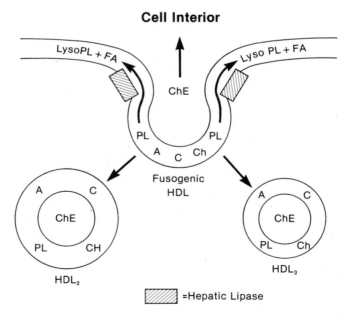

Figure 10.6. Postulated role of hepatic lipase in cholesterol ester uptake by hepatocytes. Hydrolysis of high density lipoprotein$_2$ (HDL$_2$) phospholipids produces fusogenic HDL, which deposits cholesterol ester in the hepatocyte. The resulting particle, HDL$_3$, is smaller in size than its precursor as the result of the process. Abbreviations: A and C, apoproteins; PL, phospholipid; Ch(E), cholesterol (ester); FA, free fatty acid. Adapted from van'T Hooft, van Gent, and van Tol (1981).

Phospholipase A₂   Phospholipases A_2 have been thoroughly studied and are well understood at the structural level. While phospholipases A_2 are ubiquitous in nature, most emphasis has been placed on the venom and pancreatic enzymes, since these are abundant, easily purified, and stable to a variety of manipulations. The phospholipases A_2 were the first of the phospholipases to be recognized. Over a century ago, Bokay recognized that phosphatidylcholine was degraded by some component in pancreatic juice (Bokay 1877–78). This component is now known to be the pancreatic phospholipase A_2. At the turn of the century, it was recognized that cobra venom had hemolytic activity directed toward the membranes of erythrocytes (Kyes 1902). A decade later, the lytic compound produced by the venom phospholipase was identified and termed *lysocithin* (later, *lysolecithin*). These studies spurred further investigation of this intriguing class of enzymes and their mechanism of attack on water-insoluble substrates. The study of the structure of these enzymes is an excellent example of the use of comparative protein analysis in evolutionary biochemistry. Sufficient quantities of both the bovine pancreatic and *Crotalus atrox* venom enzymes have been obtained for X-ray crystallographic analysis. Also, the peptide sequences of nearly 40 of this class of phospholipases are known, which demonstrates their structural, functional, and evolutionary closeness.

The pancreatic phospholipases are synthesized as zymogens that are activated by the cleavage of a heptapeptide by trypsin. Both the zymogen and the processed enzyme are active on monomers of phospholipid. Figure 10.7 shows the amino acid sequence of the bovine pancreatic pro-phospholipase A_2. The seven disulfide bonds provide the stability observed. Cleavage of the Arg-Ala bond at the position indicated exposes the necessary hydrophobic binding sites and allows interaction of the enzyme with lipid aggregates or monolayers. The processed enzyme has a molecular weight of about 14,000.

Verger, Mieras, and deHaas (1973) postulated a model for the binding and activation of the pancreatic enzyme to monolayers of phosphatidylcholine, as shown in Figure 10.8. The initial binding of the enzyme (E) in the subphase, designated k_P, is a slow phase that precedes the formation of the active enzyme (E*) adsorbed to the monolayer of phospholipid. The E* forms the catalytically active complex (E*S) with the interfacial substrate (S). In this monolayer system employing short-chain phosphatidylcholine, the short-chain fatty acid (P), diffuses into the subphase, since it is water soluble. More hydrophobic products would diffuse away from the enzyme but remain in the liposome or aggregate. Interestingly, the venom phospholipases bind more rapidly to the monolayer than the pancreatic phospholipases, as shown by their relatively low value for k_P.

Figure 10.7. Amino acid sequence of bovine pro-phospholipase A$_2$. Proteolytic cleavage removes the heptapeptide exposing the *N*-terminal alanine (1). From Verheij, Slotboom, and deHaas (1981).

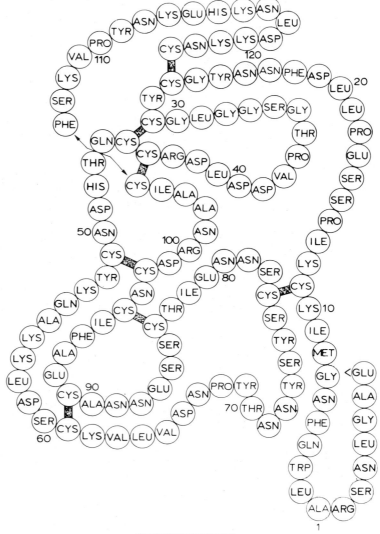

Figure 10.8. Model of pancreatic phospholipase A$_2$ binding to and hydrolysis of a monolayer of phospholipid. Abbreviations: E$^{(*)}$, enzyme (activated); S, substrate; P, product. E is in the aqueous subphase and E* is absorbed to the monomolecular film. From Verger, Mieras, and deHaas (1973).

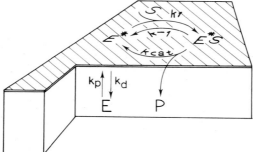

The hydrolysis of monomolecular films by phospholipases increases with increasing pressure to a critical point; beyond this point the enzyme is no longer capable of penetrating the film, and activity ceases (Figure 10.9). The increase in activity is expected, since the surface concentration of substrate increases. On the other hand, the amount of enzyme bound decreases with increasing pressure, since less space is available for its penetration.

There is clear evidence that the catalytic site is distinct from the lipid binding site in both the pancreatic and venom phospholipases, as has been demonstrated by chemical modification of the processed phospholipases and nuclear magnetic relaxation studies. Compounds such as bromophenacyl bromide that react with His-48 in the active site do not prevent binding of the substrate to the processed enzyme. Likewise, the presence of distinct catalytic and binding sites would account for both the zymogen and the processed enzyme acting on monomeric substrates, while only the processed enzyme binds and acts at interfaces. The conversion of zymogen to processed enzyme is thought to result in conformational changes that not only expose more hydrophobic residues for binding to lipid but also align the active site with the interface.

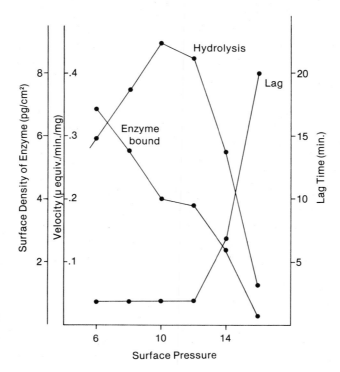

Figure 10.9. Dependence of enzyme binding, rate of hydrolysis, and lag in activity upon monolayer surface pressure. Adapted from Verger and Pattus (1982).

A different model is proposed by Dennis and co-workers (1983) based on the finding that the *Naja naja naja* enzyme requires an activator molecule that contains phosphorylcholine. For example, sphingomyelin (which is not a substrate for the phospholipase A_2) can stimulate the hydrolysis of phosphatidylethanolamine, ordinarily not a good substrate. Figure 10.10 illustrates the proposed mechanism of activation by mixed micelles. Two monomers of enzyme are shown to be required, one that binds the activator molecule and one that binds the substrate. Interaction of the two monomers activates the catalytic monomer which promotes hydrolysis. It is possible that the dimeric form, known to exist for the *Crotalus* enzymes, could interact with the activator and substrate sequentially with the same effect. Alternatively, Dennis has also suggested that a monomer could have two sites, so that dimer formation would not be required.

The venom and pancreatic phospholipases have an absolute requirement for Ca^{2+}, which is bound adjacent to His-48, as shown by Ca^{2+} blockage of the binding of bromophenacyl bromide to this residue. The binding of Ca^{2+} involves an octahedral arrangement with seven oxygen bridges and lowers the pK of the essential His-48 from 7 to 5.7.

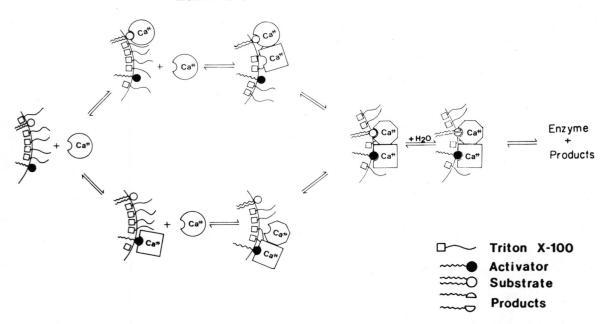

Figure 10.10. Proposed two-site dimer model for *N. naja naja* venom phospholipase. The activator monomer (square) or the inactive catalytic monomer (circle) can bind its phospholipid with subsequent binding of and interaction with the other monomer activating the catalytic monomer (octagon). As presented, both the activator and catalytic proteins have bound Ca^{2+}. From Dennis (1983).

A proton relay system that employs a molecule of water as the nucleophile attacking the ester bond has been proposed as the mechanism of catalysis. A serine is not involved, as is usually the case with esterases. Figure 10.11 shows a mechanism proposed by Verheij, Slotboom, and deHaas (1981). In this case, the Asp-99–His-48 pair removes a proton from bound water, producing the nucleophilic hydroxyl group. Ca^{2+} interacts with both the phosphate and the carbonyl groups of the ester undergoing hydrolysis as well as the carboxyl of Asp-49 and binds the free fatty acid formed until the fatty acid diffuses from the active site. Studies with $H_2^{18}O$ showed that the enzyme acts through an O-acyl cleavage mechanism. While some details are yet to be resolved, the proposal satisfies most experimental results.

The degree of homology of the structures of 32 venom and pancreatic phospholipases has been used for a classification system of the phospholipases. It has been proposed that the pancreatic phospholipases are closely related to the venom enzymes of the Elapids (*Naja* and *Bungarus* among others), while the phospholipases of the Viperids (*Crotalus*) appear to be unrelated. Dufton and Hider (1983) give a thorough discussion of the possible evolutionary reasons for

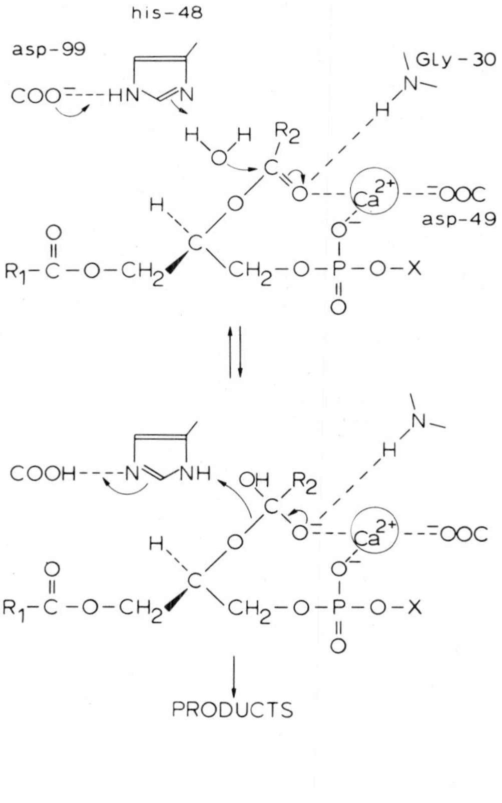

Figure 10.11. Proposed proton-relay mechanism of hydrolysis by venom and pancreatic phospholipases. From Verheij, Slotboom, and deHaas (1981).

the surprising dissimilarities among the venom phospholipases. Apparently, minor modifications in sequence have pronounced effects on their function. For example, the neurotoxic phospholipase from *Notechis scutatus*, Notexin, uniquely has methionine at position 71. Also, it has a higher number of positively charged and hydrophobic residues than the closely related enzyme *Notechis scutatus* II 1 (Dufton and Hider 1983).

The structural differences between the pancreatic and *Crotalus* phospholipases A_2 produce pronounced differences in the nature of the interaction between enzyme molecules at interfaces. The pancreatic enzymes are found as monomers and are thought to act as monomers. The phospholipases A_2 from *C. adamanteus*, however, form dimers at low concentrations of enzyme (about $10^{-9}M$) and have been shown to be active in the dimeric form by active ultracentrifugation, a technique that measures enzyme activity during centrifugation and therefore determines the molecular weight of the active enzyme species. Similar dimerization has been proposed for other venom phospholipases. No lag phase is observed when venom phospholipases degrade monolayers, possibly because the active enzyme dimers have already been formed. The monomeric pancreatic phospholipases, on the other hand, require an activation process and exhibit a lag phase prior to activity.

A phospholipase A_2 similar in molecular size to the venom and pancreatic enzymes has been purified from rat liver mitochondria (deWinter, Vianen, and van den Bosch 1982). It has an approximate molecular size of about 14,000, has an absolute requirement for Ca^{2+}, is inhibited by bromphenacyl bromide, and has an alkaline pH optimum. This phospholipase differs considerably from the venom and pancreatic enzymes in its hydrophobic and stability characteristics. Unlike the pancreatic and venom enzymes, the mitochondrial phospholipase is an integral membrane protein that requires acetone extraction for solubilization and glycerol for stabilization. It has a dual localization in the inner and outer mitochondrial membranes and is equally active on membranous and exogenously added phospholipids, in both cases showing a preference for phosphatidylethanolamine. The function of this phospholipase is unclear, but its activity in some way appears to be linked to the energy state of the mitochondria. Tightly coupled mitochondria do not exhibit phospholipase A_2 activity until the energy charge is lost. Hydrolysis of membranous lipid is rapid once the ATP and the respiratory control ratio drop to minimal levels, resulting in the degradation of 20–30% of the endogenous phosphatidylethanolamine (Parce, Cunningham, and Waite 1978).

Considerable interest has been shown over the past several years in the role phospholipases A_2 play in the arachidonate cascade

(Chapter 11). It is well established that arachidonate is released from phospholipid before it becomes a substrate for the cyclooxygenase or lipoxygenase systems. While the mechanism of release was controversial for some time, it has become clear that phospholipases A_2 contribute significantly to the process. It is not possible to generalize about the nature of the phospholipases A_2 involved, since they may differ, depending upon the cellular origin. For example, some cells deacylate phosphatidylinositol or phosphatidylcholine, whereas other cells have less specificity and deacylate all phospholipids. A major problem in this area of research is determining which phospholipase A_2 in the cell responds to a stimulant and specifically releases arachidonate for hormone synthesis. Jesse and Franson (1979) purified from platelets a phospholipase A_2 that was inhibited by indomethacin, a characteristic of a phospholipase that might be expected to be involved in the arachidonate cascade.

Phospholipase B and Lysophospholipases

The distinction between phospholipase B and lysophospholipases is not clear, since the phospholipase B from *Penicillium notatum* that has been purified and thoroughly characterized has high lysophospholipase activity. Such activity would be expected, since phospholipase B deacylates at both the *sn*-1 and *sn*-2 positions, with a lysophospholipid as an intermediate that is subsequently deacylated. Lysophospholipases, on the other hand, need not attack diacylphospholipids, although most have some activity on diacylphospholipids. These enzymes are widely distributed and found in microorganisms, bee venoms, and mammalian tissues. The phospholipase B purified from *P. notatum* and the lysophospholipases from beef liver will be discussed here.

One of the crucial factors in establishing the dual activity of an enzyme is the purification of the enzyme to homogeneity. This is particularly true of phospholipase B, since it is known that the combination of phospholipase A_1 and phospholipase A_2 will catalyze the complete deacylation of diacylphospholipid. Purification of phospholipase B was achieved by the group of Saito in 1973 (Kawasaki and Saito 1973). It is one of the larger phospholipases described thus far, with a molecular weight of about 90,000. It is optimally active on a wide range of substrates at pH 4.0 and does not have a metal-ion requirement. The proposed sequence of events for phospholipase B activity is as follows:

$$\text{Diacylphospholipid} \xrightarrow{\text{phospholipase } A_2} \text{fatty acid} + \text{1-acyl-lysophospholipid}$$

$$\text{1-Acyl-lysophospholipid} \xrightarrow{\text{lysophospholipase}} \text{glycerophosphoryl base} + \text{fatty acid}$$

The sum of the two activities of the enzyme, phospholipase A_2 and lysophospholipase, gives the total activity of phospholipase B. Detailed studies with substrate analogues have allowed a proposed mechanism for this enzyme:

1. The enzyme has two binding sites: site I, which binds diacylphospholipids, and site II, which binds monoacyl(lyso)phospholipids.
2. The acyl group at position 2 is transferred covalently to a moiety X in the active site of the enzyme when diacylphospholipid is bound to site I.
3. The lysophospholipid thus formed is transferred to site II.
4. The acyl group bound to moiety X is transferred to H_2O to liberate the fatty acid.
5. The acyl group at position 1 is transferred to moiety X and subsequently transferred to H_2O to effect the second deacylation.

While this proposed mechanism has not yet been fully verified, it is supported by a number of observations. Probes with chemical modifiers, however, have failed to distinguish between the two proposed active sites. Since acyl-enzyme intermediates are postulated, it would be expected that the enzyme could catalyze a transacylation in which two molecules of lysophospholipid are converted to diacylphospholipid plus glycerophosphoryl base. Although not yet demonstrated with the *P. notatum* phospholipase B, transacylation has been demonstrated with a number of lysophospholipases at concentrations of lysophospholipid above its cmc.

In the absence of detergents, phospholipase B is roughly 100 times more active on lysophospholipid than on diacyl lipid. However, when Triton X-100 is present, the hydrolysis of diacyl lipid is increased, while that of the lysolipid is inhibited, resulting in roughly equal activity on the two substrates. Under optimal conditions, lysophospholipase activity (no detergent) is 16-fold higher than phospholipase activity (plus detergent). The inhibition of lysophospholipases by detergents is a general effect and has been used to block this activity when measurement of phospholipase A activity is sought. This inhibition by detergents can be misleading, however, since detergents can stimulate certain phospholipases.

Two distinct lysophospholipases have been purified from beef liver. Lysophospholipase I is localized in the cytosol and the matrix space of the mitochondria, while lysophospholipase II is microsomal. These enzymes split both oxyester and thioester bonds and therefore are suitable for assay colorimetrically using Ellmann's reagent. As found with the phospholipase B from *P. notatum*, both have esterase activity on soluble acyl esters (for example, *p*-nitrophenyl acetate) and exhibit low activity on diacylphospholipids. Lysophospholipase II has comparable activity on micellar and monomeric substrates, indicating that aggregate binding does not modulate activity of this enzyme. In addition, micelles of short-chain phosphatidylcholine are more readily attacked by lysophospholipase II than are the bilayer liposomes of long-chain phosphatidylcholines, which further demonstrates the importance of the physical structure of the substrate. When acting on diacylphospholipid, lysophospholipase II removes the fatty acid from the sn-1 position, which distinguishes it from the *P. notatum* phospholipase B. Like the venom phospholipase A_2, it appears that lysophospholipase II proceeds by O-acyl cleavage of an acyl-enzyme intermediate.

At this time, the functions of phospholipases B and lysophospholipases remain speculative. It is reasonable to assume, however, that these enzymes protect cells from the buildup of lytic lysophospholipids. They may also function physiologically as transacylases, forming diacylphospholipids from 2 mol of monoacylphospholipid. Such transacylation activity between two lysophosphatidylcholines has been found in lung and could account for the synthesis of some portion of the dipalmitoylphosphatidylcholine that is abundantly present (Chapter 8). The concentrations of lysophosphatidylcholine required for transacylation are high, and whether sufficient concentrations exist for physiological transacylation remains unclear.

Phospholipase C Phospholipases C have been associated with bacteria since the classic demonstration by Macfarlane and Knight (1941) that the clostridial α-toxin was phospholipase C. The most common source of the enzyme is from culture filtrates of *Clostridium perfringens*. This enzyme exhibits microheterogeneity based on electrofocusing, and each form has a constant ratio of α-toxin and phospholipase C activity. Although the enzyme has a broad specificity, phosphatidylcholine is the preferred substrate. In studies to define the effect of substrate charge on enzyme activity better, Bangham and Dawson (1958) found that a slight positive charge favored activity. This finding would account, at least in part, for the fact that neutral phosphatidylcholine is a better substrate than the acidic phospholipids.

The most extensively studied phospholipases C are those from *Bacillus cereus*. Three enzymes have now been identified and purified from the culture media of this organism: one specific for phosphatidylinositol, one having broad specificity similar to the enzyme from *C. perfringens*, and a sphingomyelinase. The enzyme with broad specificity, which is active on phosphatidylcholine, has 2 mol of Zn^{2+} tightly bound to histidine of the enzyme. Removal of the Zn^{2+} causes the reversible loss of activity. If Zn^{2+} is replaced by Co^{2+}, the specificity of the enzyme changes somewhat; sphingomyelin, not normally degraded, becomes a substrate. This enzyme is extremely resistant to degradation even though it does not have disulfhydryl groups. Zn^{2+} apparently maintains the structure of the enzyme rather than being involved in catalysis directly. The phospholipase C with broad specificity does not appear to be too closely related to the phosphatidylinositol-specific phospholipase C from the same organism, based on its lack of Zn^{2+} and its molecular weight, 29,000 (versus 23,000). Although these two enzymes are devoid of toxic activity, the phosphatidylinositol-specific enzyme does cause the release of alkaline phosphatase from cellular membranes, resulting in *phosphatasemia*, the release of alkaline phosphatase into circulation.

Phospholipase C is also found in a wide range of mammalian tissues. One of the earliest reports of a mammalian phospholipase C came from Sloane-Stanley (1953), who demonstrated the release of inositol from phosphatidylinositol, catalyzed by a brain preparation. Phospholipases C have now been partially purified from muscle, brain, platelets, and ram seminal vesicles. Although more than one type has been identified, all except the lysosomal phospholipase C are specific for phosphatidylinositol, appear to be involved in the phosphatidylinositol cycle (see following paragraph), and are regulated by a wide range of receptor-directed stimuli (Daniel 1985). The phospholipase C activity in the soluble fraction from lysosomes is unusual in that a wide spectrum of phospholipids is attacked. The enzyme purified from muscle and from platelets hydrolyzes mono-, di-, and tri-phosphoinositides, although the diphosphoinositide appears to be the preferred substrate. All the phospholipases C, except the lysosomal one, appear to require Ca^{2+}, but the pH optimum ranges from 4.5 in the lysosomes to the neutral range for the cytosolic enzymes.

The function of phospholipase C in the phosphatidylinositol cycle appears to be the key to a number of potentially significant physiological stimulators, as outlined by Daniel (1985) (Figure 10.12) and discussed in Chapter 8. In this cycle, phospholipase C degrades all three phosphoinositides, yielding diacylglycerol plus the mono-, di-, and triphosphoinositols. The diacylglycerol can be reconverted to phosphatidylinositol or be degraded by a lipase to yield arachidonate for the arachidonate cascade. The diacylglycerol also stimulates protein kinase C, which is believed to be involved in a series of physiological responses such as cellular differentiation and activation of platelets.

Phospholipase D   Plants are the major source of the phospholipases D. However, a rather specialized Mg^{2+}-requiring phospholipase D has been described in brain and other tissue. The substrate for the brain enzyme is 1-ether-lysophosphatidyl-choline or -ethanolamine, which sug-

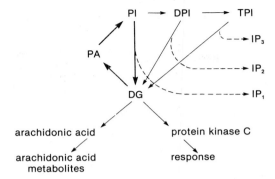

Figure 10.12. Phosphatidylinositol (PI) cycle including diphosphoinositide (DPI) and triphosphoinositide (TPI). Phospholipase C hydrolysis of the compounds yields diglyceride (DG) plus inositol phosphate (IP_1), diphosphate (IP_2), and triphosphate (IP_3). The inositol phosphates may regulate Ca^{2+} metabolism. DG may stimulate the protein kinase C, be hydrolyzed to release arachidonate with subsequent metabolism to the bioactive eicosanoids, or be recycled to PI via phosphatidate (PA). From Daniel (1984).

gests that this enzyme is involved in the catabolism of ether lipids (Wykle, Kraemer, and Schremmer 1980).

The plant phospholipases D, first identified in carrots by Hanahan and Chaikoff (1947), are found in a wide variety of plant tissue—savoy cabbage, Brussels sprouts, and peanut seeds being the most commonly used sources. The enzyme as usually isolated has a molecular weight of about 115,000, which might represent an aggregate, since under the appropriate conditions a minimum molecular size of 20,000 is found. For reasons that are not yet clear, rather high concentrations (20–$100mM$) of Ca^{2+} are required for full activity. Unlike other phospholipases, which are specific for the sn-3 isomer of phospholipids, phospholipases D also act on the sn-1 isomer. The enzymes have a broad substrate specificity.

It is known that the enzyme acts by a phosphatidate exchange with a covalent phosphatidyl-enzyme intermediate. For this reason, the enzyme can catalyze a base-exchange reaction in which alcohols can substitute for water as the phosphatidate acceptor. Although "base exchange" is somewhat misleading, and the term "phosphatidate exchange" is more accurate, this activity has been used in the laboratory for the preparation of a variety of phospholipids. For example,

Phosphatidylcholine + glycerol ⟷ phosphatidylglycerol + choline

The products of the exchange reaction have the same acyl group composition as the starting substrate. Alcohols are better than water as phosphatidate acceptors, since about 1% of an alcohol (for example, glycerol) in water yields an equal mixture of phosphatidic acid and exchanged phospholipid. The presumed phosphatidate intermediate is a thioester, since activity is inhibited by thiol reagents. When lysophospholipids are substrates, the enzyme also forms 1-acyl-*sn*-2,3 phosphoglycerol using the hydroxyl at position 2 as its own phosphatidate acceptor.

1-Acyl-*sn*-2,3-phosphoglycerol

In a reaction analogous to that forming cardiolipin (Chapter 8), the phospholipase D of cauliflower florets produces bis(phosphatidyl)-inositol when phosphatidylinositol is the substrate. This reaction is of interest, since free inositol does not act as a phosphatidate acceptor.

It is curious that plants have a system for the degradation of phospholipids distinct from that commonly associated with other kingdoms. Some work has demonstrated that the activity of the enzyme is associated with growth processes such as germination. It is possible, therefore, that plant phospholipases D are involved in phospholipid turnover during cell division or in the mobilization of phospholipids for degradation as an energy source.

FUTURE DIRECTIONS

The considerable interest shown in phospholipases over the past two decades has provided insight into the mode of action of the extracellular digestive phospholipases. Unfortunately, this information has not yet been available for the cellular regulatory phospholipases, so there is a rather large void in the understanding of these enzymes. Physiologically, very little is known about their regulation other than a few reports on protein(s) that are synthesized in response to steroid administration that blocks phospholipase activity (Daniel 1985). These proteins are named *lipomodulin* or *macrocortin*. Unlike the case with the enzymes that degrade sphingolipids, no disease states are known to exist that involve a genetic defect leading to the deficiency of a phospholipase. Therefore, with the exception of bacterial studies, mutants are not available that might shed insight into the regulation and function of phospholipases. Thus, it is possible that a deficiency of these enzymes is lethal. Clearly, the challenge is to resolve and to define the regulation and functions of the long list of phospholipases described thus far, starting with their structures and properties and the reactions they catalyze. In many respects, this field is similar to the degradation of cellular proteins by proteases both in complexity and in essentiality. The fruits of effort should be equally rewarding.

PROBLEMS

1. A cell culture filtrate was shown to lyse red cells.
   a. You suspect that the active toxin is a phospholipase. Describe the assay you would use to establish its site of attack (that is, positional specificity).
   b. What assay would you use to purify the enzyme from the filtrate? (Activity of the enzyme in the filtrate was 2 μmol product/ min/mL fluid.)

2. How would you establish the substrate (not positional) specificity of this enzyme?
3. How would you characterize the kinetic parameters of this enzyme?
4. This enzyme is present in culture medium yet does not degrade the bacterial cellular membranes. What factors might be responsible for this phenomenon?

BIBLIOGRAPHY

Entries preceded by an asterisk are suggested for further reading.

Bangham, A. D., and Dawson, R. M. C. 1958. Control of lecithinase activity by the electrophoretic charge on its substrate surface. *Nature.* 182:1292–93.

Bokay, A. 1877–78. Über die verdaulichkeit des nucleins und lecithins. *Z. Physiol. Chem.* 1:157–64.

Daniel, L. W. 1985. "Phospholipases." In *Prostaglandins, leukotrienes and cancer*, Vol. 1 of *Basic biochemical processes*, ed. W. E. M. Lands. Hingham, Mass.: Martinus Nijhoff.

*Dawson, R. M. C. 1966. The metabolism of animal phospholipids and their turnover in cell membranes. *Essays Biochem.* 2:69–115.

Delezenne, C., and Fourneau, E. 1914. Constitution du phosphatide hemolysant (lysolecithine) provenant de l'action du venin de cobra sur le vitellus de l'oeuf du poule. *Bull. Soc. Chim.* XV:421–34.

Demel, R. A.; Guerts van Kessel, W. S. M.; Zwaal, R. F. A.; Roelofsen, B.; and van Deenen, L. L. M. 1981. Relation between various phospholipase actions on human red cell membranes and the interfacial phospholipid pressure in monolayers. *Biochim. Biophys. Acta.* 406:97–107.

*Dennis, E. A. 1983. Phospholipases. *Enzymes.* XVI:307–53.

deWinter, J. M.; Vianen, G. M.; and van den Bosch, H. 1982. Purification of rat liver mitochondrial phospholipase A$_2$. *Biochim. Biophys. Acta.* 712:332–41.

Dufton, M. J., and Hider, R. C. 1983. Classification of phospholipase A$_2$ according to sequence-evolutionary and pharmacological implications. *Eur. J. Biochem.* 137:545–51.

Hanahan, D. J., and Chaikoff, I. L. 1947. The phosphorous-containing lipides of the carrot. *J. Biol. Chem.* 168:233–40.

Hendrickson, H. S., and Rank, P. N. 1981. Continuous fluorometric assay of phospholipase A$_2$ with pyrene-labeled lecithin as a substrate. *Anal. Biochem.* 116:553–58.

Jesse, R. L., and Franson, R. C. 1979. Modulation of purified phospholipase A$_2$ activity from human platelets by calcium and indomethacin. *Biochim. Biophys. Acta.* 575:467–70.

Kawasaki, N., and Saito, K. 1973. Purification and some properties of lysophospholipase from *Penicillium notatum*. *Biochim. Biophys. Acta.* 296:426–30.

Kyes, P. 1902. Über die wirkungsweise des cobragiftes. *Klin. Woch. Berlin.* 39:889–90, 918–22.

Macfarlane, M. G., and Knight, B. C. J. G. 1941. The

biochemistry of bacterial toxins—the lecithinase activity of *Cl. welchii* toxins. *Biochem. J.* 35:884–902.

Parce, J. W.; Cunningham, C. C.; and Waite, M. 1978. Mitochondrial phospholipase A_2 activity and mitochondrial aging. *Biochem.* 17:1634–39.

Robinson, M. and Waite, M. 1983. Physical-chemical requirements for the catalysis of substrates by lysosomal phospholipase A_1. *J. Biol. Chem.* 258:14371–78.

Scandella, C. J., and Kornberg, A. 1971. A membrane-bound phospholipase A_1 purified from *E. coli. Biochem.* 10:4447–56.

Sloane-Stanley, G. H. 1953. Anaerobic reactions of phospholipins in brain suspensions. *Biochem. J.* 53:613–19.

*van den Bosch, H. 1980. Intracellular phospholipases A. *Biochim. Biophys. Acta.* 406:97–107.

van den Bosch, H., and Aarsman, A. J. 1979. A review on methods of phospholipase A determination. *Agents and Actions.* 9:382–89.

van'T Hooft, F. M.; van Gent, T.; and van Tol, A. 1981. Turnover and uptake by organs of radioactive serum high-density lipoprotein cholesteryl esters and phospholipids in the rat in vivo. *Biochem. J.* 196:877–85.

*Verger, R., and Pattus, F. 1982. Lipid-protein interactions in monolayers. *Chem. Phys. Lipids.* 30:1–39.

Verger, R.; Mieras, M. C. E.; and deHaas, G. H. 1973. Action of phospholipase A at interfaces. *J. Biol. Chem.* 248:4023–34.

*Verheij, H. M.; Slotboom, A. J.; and deHaas, G. H. 1981. Structure and function of phospholipase A_2. *Rev. Physiol. Biochem. Pharmacol.* 91:91–203.

Wykle, R. L.; Kraemer, W. F.; and Schremmer, J. M. 1980. Specificity of lysophospholipase D. *Biochim. Biophys. Acta.* 619:58–67.

CHAPTER 11

The Eicosanoids:
Prostaglandins,
Thromboxanes, Leukotrienes,
and Hydroxy-Eicosaenoic Acids

William L. Smith
Pierre Borgeat

OVERVIEW OF EICOSANOID METABOLISM

Eicosanoid is a term applicable to any C_{20} fatty acid; a large number of oxygenated eicosanoids are formed biosynthetically from the three most commonly occurring C_{20}, polyunsaturated fatty acids found in animals: arachidonic acid (5-*cis*-, 8-*cis*-, 11-*cis*-, 14-*cis*-eicosatetraenoic acid), 8-*cis*-, 11-*cis*-, 14-*cis*-eicosatrienoic acid, or 5-*cis*-, 8-*cis*-, 11-*cis*-, 14-*cis*-, 17-*cis*-eicosapentaenoic acid. Arachidonic acid is the precursor of the most important eicosanoid family. The biosynthetic interrelationships among arachidonic acid and various oxygenated eicosanoids are diagrammed in Figure 11.1. The pathways for the oxygenation of arachidonate are known collectively as the *arachidonate cascade*. Oxygenated eicosanoids are conveniently considered in two groups. The first group includes the prostaglandins (or prostanoids) and the thromboxanes. The second group includes the hydroxy and hydroperoxy fatty acids and the leukotrienes. Prostaglandins and thromboxanes are referred to as *cyclooxygenase* products because these compounds are formed through the action of a bisdioxygenase originally dubbed the cyclooxygenase. The second group of compounds are often called *lipoxygenase* products, since the initial step in the formation of the hydroxy and hydroperoxy eicosaenoic acids and the leukotrienes is catalyzed by one of a number of related dioxygenases called lipoxygenases.

In an introduction to the first published volume devoted to eicosanoids, Von Euler (1967) notes that the great Swedish chemist Berzelius "commented on the burning taste of seminal fluid." However, the first quantitative measurements of prostaglandin activity

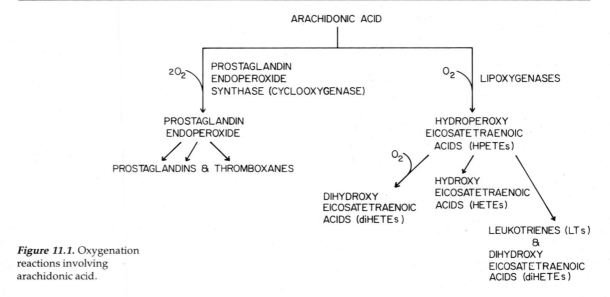

ARACHIDONIC ACID

$2O_2$ — PROSTAGLANDIN ENDOPEROXIDE SYNTHASE (CYCLOOXYGENASE)

O_2 — LIPOXYGENASES

PROSTAGLANDIN ENDOPEROXIDE

HYDROPEROXY EICOSATETRAENOIC ACIDS (HPETEs)

PROSTAGLANDINS & THROMBOXANES

O_2

DIHYDROXY EICOSATETRAENOIC ACIDS (diHETEs)

HYDROXY EICOSATETRAENOIC ACIDS (HETEs)

LEUKOTRIENES (LTs) & DIHYDROXY EICOSATETRAENOIC ACIDS (diHETEs)

Figure 11.1. Oxygenation reactions involving arachidonic acid.

are usually attributed to Von Euler in Sweden and Goldblatt in the United States. In the mid-1930s these workers described the uterine smooth muscle activity of acidified lipid extracts of seminal plasma. The real burst of research in the eicosanoid area occurred after the chemical characterization of the prostaglandins PGE_1 and $PGF_{1\alpha}$ by Bergstrom, Samuelsson, Ryhage, and Sjovall in the late 1950s and early 1960s and the elucidation of the biosynthetic pathway for PGE_2 and $PGE_{2\alpha}$ formation by Samuelsson and Hamberg and by Van Dorp and Nugteren in the mid 1960s. The sophistication and elegance of some of these pioneering studies can be appreciated only by reading the original papers (Hamberg and Samuelsson 1967).

PROSTAGLANDINS AND THROMBOXANES
Introduction

The first part of this chapter focuses on the prostaglandins and thromboxanes. In addressing the metabolism of these compounds, we shall develop the concept that prostanoids are local hormones, or *autocoids*, which act near their sites of synthesis. Aspirin and related nonsteroidal anti-inflammatory compounds exert their major effects by blocking prostaglandin synthesis. We shall discuss the mechanism of action of these drugs as part of a discussion of prostaglandin biosynthesis. Finally, we shall discuss the observation that prostaglandins are rapidly metabolized during passage through the circulatory system, a finding upon which is built the concept of prostaglandins as local hormones. Having presented a synopsis of prostaglandin metabolism and the concept that prostaglandins are local hormones, we shall then discuss two examples of the local action of prostaglandins. These examples are intended to illustrate how prostaglandins may operate physiologically and how prostaglandins may function at the molecular level. At the time of this writing much is known about the chemistry and pharmacology of prostanoids, but major gaps exist in our understanding of how prostaglandin synthesis is regulated and how prostaglandins act physiologically.

Structure and Nomenclature of Prostaglandins and Thromboxanes

The abbreviation PG is used for prostaglandin and Tx is used for thromboxane. All the naturally occurring prostaglandins contain a cyclopentane ring. The letters following the abbreviation PG indicate the nature and position of the oxygen-containing substituents present in the cyclopentane ring. Letters are also used to label thromboxane derivatives (for example, TxA and TxB). The numerical subscript indicates the total number of carbon-carbon double bonds present in the side chains emanating from the cyclopentane ring. The 2-series prostaglandins are formed from arachidonic acid; the structures of the most important prostaglandins derived from arachidonate are shown in Figure 11.2. In general, the 1-series prostaglandins are synthesized from 8,11,14-eicosatrienoic acid, although a major exception is the stable hydrolysis product of PGI_2, which is called 6-keto-$PGF_{1\alpha}$. The 3-series prostaglandins are formed from 5,8,-11,14,17-eicosapentaenoic acid. There are a number of asymmetric centers in all the prostaglandins. The configuration of substituents at asymmetric centers in the cyclopentane ring in naturally occurring prostaglandins is shown in Figure 11.2. The Greek subscript (for example, $PGF_{2\alpha}$) indicates that the hydroxyl group at C-9 lies on the same side of the ring as the carboxyl side chain. Thromboxane A derivatives contain an oxane-oxetane grouping in place of the usual cyclopentane ring (Figure 11.2). TxA_2 is the only TxA found in appreciable quantities in nature. TxA_2 is hydrolyzed rapidly ($t_{\frac{1}{2}} = 30$ s at 37°C in neutral aqueous solution) to a hemiacetal called TxB_2.

Analysis of Prostaglandins and Thromboxanes

Prostaglandins and thromboxanes are extracted efficiently from acidified aqueous solutions by semipolar, nonmiscible organic solvents such as chloroform-methanol, ethyl acetate, or ether. Prostaglandins can be separated from other major lipids by chromatography on silicic acid columns. All the common prostaglandin derivatives generated from arachidonate can be separated in a single run by reversed-phase high-pressure liquid chromatography. Two different solvent systems are required to separate the same set of derivatives by silica gel

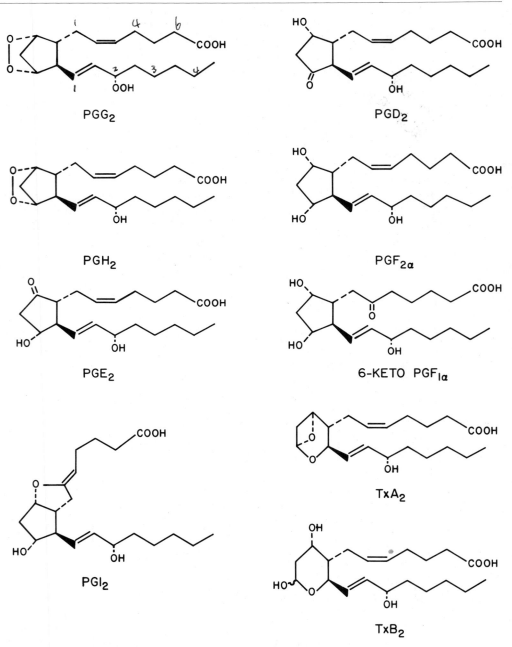

Figure 11.2. Structures of the major prostaglandins and thromboxanes derived from arachidonic acid. Numbering of carbon atoms begins with the carboxyl group.

thin-layer chromatography. High-pressure liquid chromatography is commonly used where a high degree of purity is required—for example, for subsequent gas chromatographic–mass spectrometric analysis. Thin-layer chromatography is faster and less expensive and is commonly used to separate products of enzyme assays or radioimmunoassays. Radioimmunoassays are available commercially for PGE, PGD, PGF_α, 6-keto-$PGF_{1\alpha}$, TxB_2, and several of the common 15-keto- and 13,14-dihydro-15-keto PG catabolites (see next section).

The chemistry of prostaglandins and thromboxanes is complex and will not be considered in detail (see Stehle 1982). For our purposes the most important point is that those derivatives for which radioimmunoassays exist are stable in neutral aqueous or organic solvent solutions for 1–2 d at 4°C. The properties of the less stable PGI_2, TxA_2, PGH_2, and PGG_2 are discussed in the next sections.

PROSTAGLANDIN AND THROMBOXANE BIOSYNTHESIS
Cells and Tissues Which Form Prostaglandins

Prostaglandin synthesis has been documented in crustaceans, insects, fish, amphibians, and mammals. The requisite 20-carbon precursor fatty acids are absent from procaryotes and lower eucaryotes such as yeast. Plants contain 8,11,14-eicosatrienoic acid, but there is no evidence that plants form eicosanoids. Prostaglandins are thought to be found only in animals. Homogenates prepared from each of a variety of different rat organs all form at least small amounts of prostaglandins. However, while virtually all animal organs synthesize prostaglandins, usually only certain cell types within each organ form cyclooxygenase products. Unfortunately, there is no rule to predict which cells will form prostaglandins. One generality is that all smooth muscle and vascular endothelial cells form cyclooxygenase products. Another is that those cell types which form prostaglandins in an organ from one species will be found to form prostaglandins in the same organ in other species. For example, PGE_2 is formed by the collecting tubule in kidneys from all mammals, but prostaglandins are not formed to a major extent by other parts of the renal tubule in any species.

Prostaglandin Biosynthesis: An Overview

Unlike many circulating hormones (such as insulin), eicosanoids are not stored. Moreover, the basal rate of prostaglandin synthesis is low in most cells. Prostaglandins are formed only in bursts in response to certain stimuli. Following the application of a stimulus to a cell, prostaglandins are detected within 10–30 s. Synthesis proceeds for 1–5 min and then stops. Depending on the cell, stimulation of prostaglandin formation increases the rate of cellular prostaglandin synthesis 5- to 50-fold.

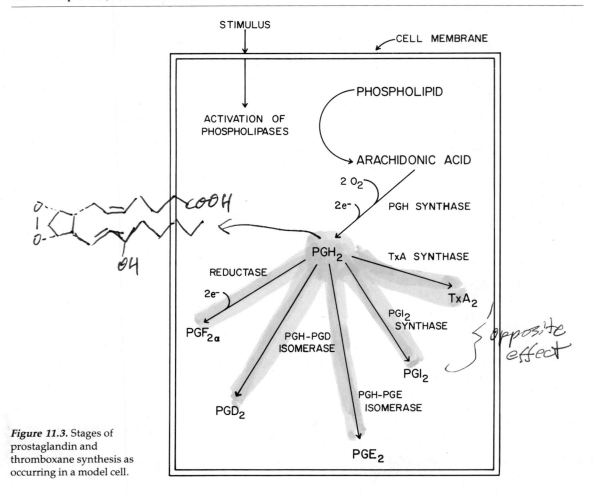

Figure 11.3. Stages of prostaglandin and thromboxane synthesis as occurring in a model cell.

Figure 11.3 illustrates the pathway for the biosynthesis of PGE_2, $PGF_{2\alpha}$, PGD_2, TxA_2, and PGI_2 from arachidonic acid as it might occur in a model cell. Prostaglandin biosynthesis is most easily visualized in three stages: (a) the release of C_{20} precursor fatty acids from phospholipids in response to cellular stimuli, (b) the oxygenation of the fatty acid to yield the prostaglandin endoperoxide derivative PGH, and (c) the cell-specific conversion of PGH to the major biologically active prostanoid forms (for example, PGE_2, $PGF_{2\alpha}$, PGD_2, TxA_2, and PGI_2). Each of these events will be discussed in turn.

Release of Arachidonic Acid from Phospholipids   The essential prerequisite for prostaglandin formation is the availability of precursor fatty acid. In most cells this acid is arachidonic acid. Normally, most arachidonic acid is found esterified at the *sn*-2 position of membrane phosphoglycerides. The concentration of free arachidonic acid in cells is less than 10^{-6} M, which is one to two orders of magnitude below the K_m of PGH synthase for arachidonate (about $5\mu M$). PGH synthase converts free arachidonic acid to the prostaglandin endoperoxides PGG_2 and PGH_2. Cellular stimuli of prostaglandin formation initiate a chain of events which elevate the concentration of free arachidonic acid near the active site of PGH synthase. This, in turn, increases the rate of PGH_2 formation 5- to 50-fold. Thus, the first level of regulation occurs at the stage of the mobilization of arachidonate from phospholipid stores. This process is referred to as *arachidonate release*.

The mechanisms whereby cellular stimuli elevate arachidonic acid levels are incompletely resolved, and current perceptions of arachidonate release are certain to be modified in the future. A brief description of arachidonate release is presented here. Additional information is presented in Chapter 10.

There are two broad classes of stimuli of arachidonic acid release. These are termed *physiological* and *pathological* or, alternatively, *specific* and *nonspecific*. When fatty acids at the *sn*-2 position of phosphoglycerides of cultured cells, platelets, or organs are radiolabeled by preincubation with radioactive polyunsaturated fatty acids, physiological stimuli are found to cause the release of arachidonic acid but not other fatty acids present at the 2-position; in contrast, pathological stimuli cause the release of linoleate, oleate, arachidonate, and other fatty acids from the 2-position in amounts reflecting the proportion of these fatty acids at that position.

Among physiological stimuli of arachidonic acid release are hormones such as angiotensin II, bradykinin, and epinephrine; proteases such as thrombin; and certain antibody-antigen complexes. Each prostaglandin-forming cell has a specific set of receptors and responds only to certain of these stimuli. In contrast, pathological stimuli can affect any cell. Any process or factor causing a generalized effect on cellular membranes resulting in the turnover of phospholipids fits into the category of pathological stimuli. Examples include mechanical damage, ischemia, membrane-active venoms such as mellitin, Ca^{2+} ionophores such as A23187, and tumor promoters such as phorbol esters.

Mechanism of Arachidonate Release in Response to Stimuli

The mechanism of arachidonate release is still the subject of controversy. This process has proven extremely difficult to study. Among the technical problems that have been encountered are the facts that (a) arachidonate release is rapid, usually reaching completion in less than 1 min; (b) only 5–10% of the total arachidonate in the cell is released; (c) all phospholipid classes contain esterified arachidonate, and the metabolism of phospholipids is interdependent; and

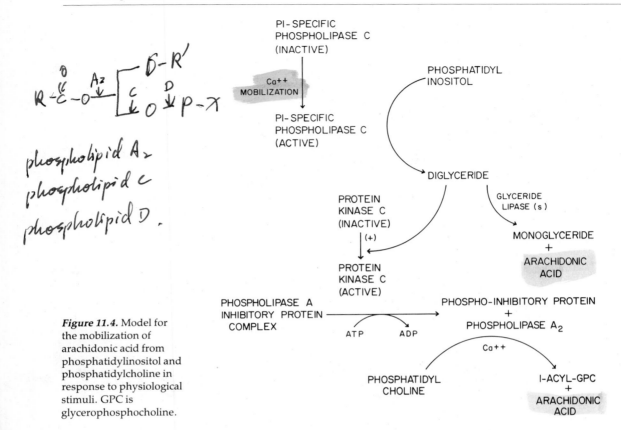

Figure 11.4. Model for the mobilization of arachidonic acid from phosphatidylinositol and phosphatidylcholine in response to physiological stimuli. GPC is glycerophosphocholine.

(d) radiolabeling of phospholipid pools with $[^{14}C]$ or $[^3H]$ arachidonate does not reflect the mass composition in the different pools.

Arachidonate release has been studied most extensively in platelets which form TxA_2 as their major cyclooxygenase product (Neufeld and Majerus 1983; Rittenhouse-Simmons and Deykin 1981). One view of arachidonate release in platelet cells is as follows (Figure 11.4). Arachidonate is derived primarily from two phospholipids, phosphatidylinositol (PI) and phosphatidylcholine. Treatment of platelets with thrombin or collagen causes the formation of diglyceride from phosphatidylinositol via the action of a cytoplasmic, phosphatidylinositol specific phospholipase C. This phospholipase C is probably activated as a result of stimuli-induced Ca^{2+} mobilization. Newly formed diglyceride apparently serves two functions. First, part of the diglyceride is cleaved by glyceride lipase(s) releasing arachidonate. Arachidonate is the major fatty acid at the sn-2 position of platelet phosphatidylinositol. The remaining diglyceride may activate a diglyceride-dependent protein kinase C. This enzyme, in turn, may cause the activation of a phospholipase A_2 which cleaves arachidonate from phosphatidylcholine. The substrate for protein kinase C has not yet been identified, but it may be a phospholipase A_2–inhibitory protein which is rendered inactive by phosphorylation. Certain glucocorticoids with anti-inflammatory activity have been shown to inhibit arachidonate release in some cells by eliciting production of phospholipase A_2 inhibitors. These inhibitory proteins, which all appear to be related immunologically, are variously referred to as macrocortin, lipomodulin, and renocortin.

Conversion of Arachidonic Acid to PGH$_2$; PGH Synthase

The second stage of prostaglandin formation is the conversion of arachidonic acid to PGH_2 (Figure 11.5). Oxygenation is initiated by the stereospecific removal of the hydrogen atom from the 13-pro-*S* position of arachidonate. The first oxygen molecule is thought to be inserted at C-11, since 11-*cis*,14-*cis*-eicosadienoic acid, which lacks a C-8, C-9 double bond and cannot undergo cyclization, can be converted to 11-hydroperoxy-12-*trans*,14-*cis*-eicosadienoic acid by purified PGH synthase. The oxygen atoms at C-9 and C-11 of prostaglandin endoperoxides are derived from the same oxygen molecule. This fact was inferred from experiments in which a precursor acid was incubated in atmospheres containing both $^{18}O^{18}O$

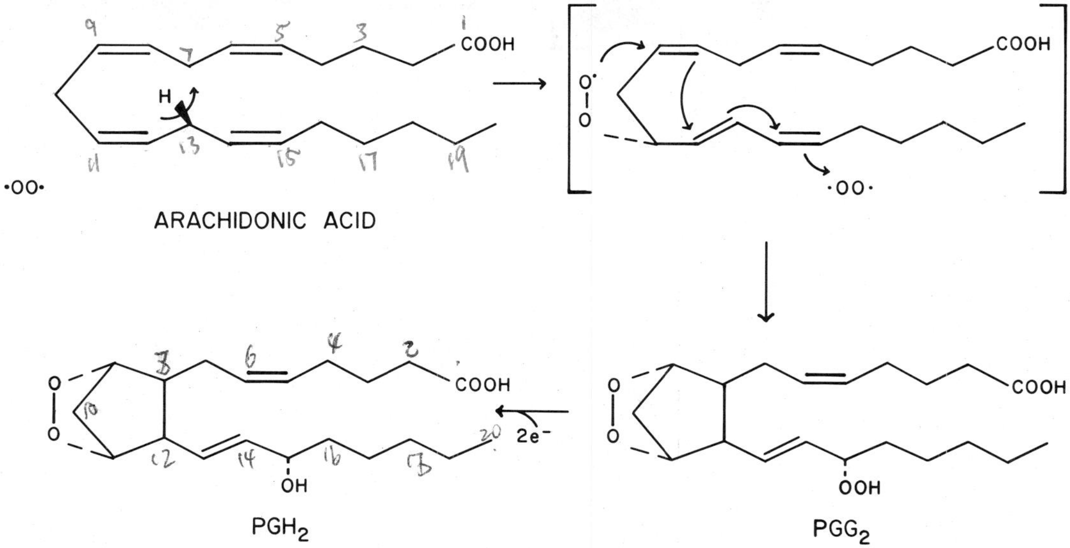

Figure 11.5. Proposed mechanism for the oxygenation of arachidonic acid by PGH synthase.

and $^{16}O^{16}O$ molecules; the prostaglandins that formed contained only ^{18}O atoms or only ^{16}O atoms at both C-9 and C-11, but never a combination of ^{16}O and ^{18}O atoms at these positions.

The amount of prostaglandin formed from the arachidonate released in response to cellular stimuli is governed by the activity of the PGH synthase. This enzyme has been purified and characterized extensively. The conversion of arachidonate to PGH_2 involves two distinct catalytic events (Figure 11.6). The first is a bisdioxygenase reaction in which two molecules of oxygen are inserted into arachidonic acid to yield the endoperoxide PGG_2. This bisdioxygenase activity is called the cyclooxygenase activity. The second step in PGH_2 synthesis is catalyzed by a hydroperoxidase and involves a two-electron reduction of the 15-hydroperoxy group of PGG_2 to yield PGH_2. Cyclooxygenase and hydroperoxidase activities copurify, and a single polypeptide chain with a molecular weight of about 70,000 is found in the isolated enzyme preparation. Furthermore, both activities are precipitated by monoclonal antibodies directed against this polypeptide. Thus, one can conclude that both cyclooxygenase and hydroperoxidase activities reside in the same protein chain. PGH synthase is a hemoprotein. The enzyme is membrane-bound and has been purified following solubilization with nonionic detergent.

The cyclooxygenase component of PGH synthase has three interesting catalytic properties: (a) It requires a hydroperoxide for activity (for example, 15-hydroperoxyarachidonate or PGG_2). (b) It undergoes a self-catalyzed destruction, or suicide, reaction. (c) It is inhibited by nonsteroidal anti-inflammatory drugs, the best known of which is aspirin.

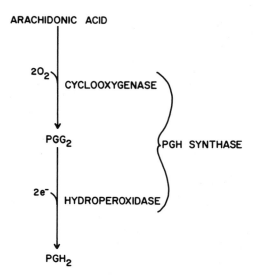

Figure 11.6. Catalytic events catalyzed by PGH synthase.

The hydroperoxide requirement for cyclooxygenase activity was discovered almost 15 years ago during studies designed to trap a putative 11-hydroperoxyl reaction intermediate. Reduced glutathione (GSH) and GSH peroxidase were added to a reaction mixture containing radioactive arachidonic acid and microsomal PGH synthase. It was anticipated that GSH plus GSH peroxidase would reduce any hydroperoxides or hydroperoxyl radicals formed during oxygenation, thereby aborting the reaction after insertion of the first oxygen molecule. Curiously, the combination of GSH and GSH peroxidase prevented any reaction from occurring, but oxygenation did proceed if GSH were removed or if high concentrations of hydroperoxide were added. It is now known that the PGH synthase requires a hydroperoxide activator to catalyze the oxygen insertion reactions. Including GSH and GSH peroxidase in the reaction mixture reduces the ambient hydroperoxides always present in small concentrations in suspensions of polyunsaturated fatty acid substrates. It is believed that the hydroperoxide is first bound to an activator site on the cyclooxygenase and is then cleaved to yield an enzyme-bound hydroperoxyl radical (reaction 1, Figure 11.7). This radical is thought to abstract the 13-pro-*S* hydrogen atom from an incoming

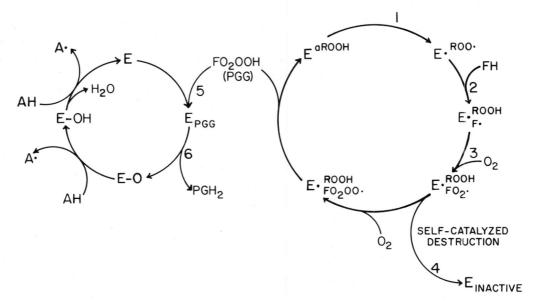

Figure 11.7. Proposed mechanism for the cyclooxygenase and hydroperoxidase reactions catalyzed by PGH synthase (E). R is the alkyl group of the activator hydroperoxide. FH is the fatty acid substrate. The circle on the right diagrams the cyclooxygenase reaction. The hydroperoxidase reactions are illustrated on the left. aROOH is the activator hydroperoxide.

arachidonate molecule (reaction 2, Figure 11.7), permitting reaction with the first oxygen molecule (reaction 3, Figure 11.7). This is an unusual dioxygenase mechanism because it involves activation of the organic substrate (that is, arachidonate) instead of oxygen.

The second unusual feature of the cyclooxygenase activity is its propensity to undergo a self-catalyzed destruction. During the bis-dioxygenase reaction, which appears to involve free-radical intermediates, one of the intermediates can either go on to product or react in such a way as to inactivate the cyclooxygenase (reaction 4, Figure 11.7). Statistically, a cyclooxygenase active site is inactivated once in every 400 substrate turnovers. This inactivation process was first discovered in vitro, but it apparently occurs in vivo as well, since incubation of intact cells, for example, vascular endothelial cells, with arachidonic acid leads to a progressive loss in cyclooxygenase activity. This self-catalyzed destruction may be a crude regulatory mechanism which assures that only a certain amount of prostaglandin endoperoxide is formed from arachidonic acid released from phospholipid stores. The chemical changes that occur in PGH synthase as a result of this destructive event are unknown.

A third interesting characteristic of the cyclooxygenase is its interaction with nonsteroidal anti-inflammatory drugs. In 1971, John Vane and his co-workers published the startling observations that the nonsteroidal anti-inflammatory drugs aspirin and indomethacin inhibit prostaglandin production by several different cells and tissues. Subsequent work has indicated that there is a close parallel between the anti-inflammatory potencies and prostaglandin inhibitory activities of drugs of this type. Thus, the mechanism of action of nonsteroidal anti-inflammatory agents appears to be due to their inhibition of prostaglandin formation.

Nonsteroidal anti-inflammatory drugs exert their effects on the cyclooxygenase component of PGH synthase and prevent the initial oxygen insertion reaction. Each of these drugs competes with arachidonate for the cyclooxygenase active site (reaction 2, Figure 11.7). In addition, aspirin selectively acetylates a serine residue in isolated PGH synthase (Figure 11.8). Presumably this serine is at or near the cyclooxygenase active site. There are no repair mechanisms capable of restoring activity to acetylated PGH synthase. Therefore, cells treated with aspirin to inactivate PGH synthase must synthesize new enzyme protein to restore prostaglandin biosynthetic activity. Aspirin is apparently the only nonsteroidal anti-inflammatory drug which causes covalent modification of the enzyme. Indomethacin also irreversibly inactivates the isolated enzyme in vitro but without covalent modification. Certain nonsteroidal anti-inflammatory drugs act simply as reversible cyclooxygenase inhibitors. Included in this group are common prescription drugs such as Clinoril, Motrin, and

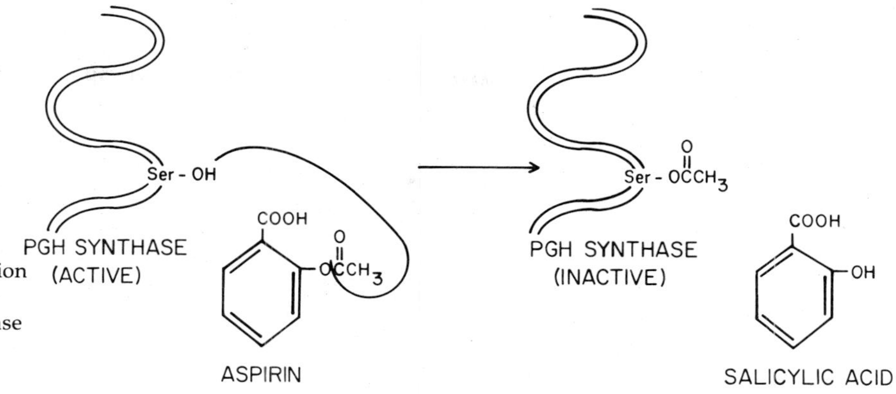

Figure 11.8. Proposed mechanism for the modification and inactivation of the cyclooxygenase component of PGH synthase by acetylsalicylic acid (aspirin).

Naproxen. Like aspirin and indomethacin, these drugs compete with arachidonic acid for binding to the cyclooxygenase active site.

Hydroperoxidase Component of PGH Synthase

The other catalytic activity associated with PGH synthase is a hydroperoxidase. The hydroperoxidase catalyzes a two-electron reduction of PGG_2 to PGH_2 (reactions 5 and 6, Figure 11.7). This reaction appears to involve a sequence of two one-electron reductions analogous to those seen in the reduction of hydrogen peroxide by horseradish peroxidase. The hydroperoxidase actually shows little specificity with respect to the carbon skeleton of the hydroperoxide and will even reduce hydrogen peroxide. The K_m for PGG_2 is approximately $20\mu M$. A number of artificial electron donors, including epinephrine and guaiacol, will function in this reaction. However, recent work suggests that the natural electron donor is NADPH and that electrons are channeled through cytochrome b_5 to the heme group of PGH synthase.

The interrelationships between the cyclooxygenase and hydroperoxidase activities of PGH synthase are not understood. Aspirin and indomethacin inactivate the cyclooxygenase activity without appreciably affecting the hydroperoxidase. What is perplexing is that heme is required for both cyclooxygenase and hydroperoxidase activities, and there appears to be only one heme group per two protein subunits (PGH synthase exists as a dimer in solution). The affinity of hydroperoxide for the activator site on the cyclooxygenase is at least 200 times greater than the affinity of the hydroperoxide for the hydroperoxidase site. In vitro under saturating substrate conditions, considerable PGG_2 formation occurs before reduction of the hydroperoxide begins. However, in vivo when the ratio of arachidonate to PGH synthase is probably low, little PGG_2 accumulates.

Conversion of PGH$_2$ to Prostaglandins and Thromboxanes

The last phase of prostaglandin and thromboxane synthesis is the conversion of PGH$_2$ to what are considered to be the biologically active forms of prostaglandins—PGE$_2$, PGF$_{2\alpha}$, PGD$_2$, TxA$_2$, and PGI$_2$. Although all these products are shown in Figure 11.3, any one cell type forms only one of these derivatives as its major prostaglandin product. For example, platelets form thromboxane A$_2$ (TxA$_2$) almost exclusively, smooth muscle and arterial endothelial cells form mainly prostacyclin (PGI$_2$), and renal collection tubule cells form mostly PGE$_2$. Apparently, each prostaglandin-forming cell possesses an arachidonate release mechanism, PGH synthase, and one of the enzymes which catalyzes the metabolism of PGH$_2$.

Enzymes that Metabolize PGH$_2$

For various reasons, the enzymes that metabolize PGH$_2$ to the active prostaglandins have received little experimental attention. There is evidence that each of these PGH$_2$-metabolizing enzymes is constitutive and present in catalytic excess over PGH synthase. Thus, it is thought that these enzymes play no regulatory role in prostaglandin production other than in determining the disposition of PGH$_2$. Furthermore, besides the usual difficulties associated with fractionating membrane-bound proteins, the assays of PGH$_2$-metabolizing enzymes present major technical difficulties. The two most formidable problems are the necessities (a) of synthesizing and using an unstable substrate PGH$_2$ which at best is never more than 90% pure and (b) of purifying and measuring the products by laborious thin-layer radiochromatography. Studies of the conversion of PGH$_2$ to PGE$_2$ and to PGD$_2$ have proven especially difficult because PGH$_2$ undergoes facile, nonenzymic rearrangement to these classical prostaglandins in aqueous solution and on silica gel thin-layer plates.

PGH–PGD Isomerase

Enzyme activities which catalyze the isomerization of PGH$_2$ to PGD$_2$ are known as PGH–PGD isomerases. PGH–PGD isomerases are the only enzymes which have been purified that catalyze formation of classical prostaglandins (that is, PGF$_{2\alpha}$, PGE$_2$, PGD$_2$). Proteins catalyzing this activity have been isolated from both rat brain and rat spleen. These two proteins are quite different in size. The PGH–PGD isomerase from spleen requires GSH as a cofactor, while the enzyme from brain has no cofactor requirements. It is noteworthy that rat serum albumin can also catalyze the PGH–PGD isomerization reaction at a significant rate. The fact that serum albumins can catalyze the formation of PGD$_2$ from PGH$_2$ raises the question of whether any of the proteins which have been isolated as PGH–PGD isomerases are actually important in forming PGD$_2$ in vivo. To date it has not been conclusively established whether any of the purified PGH–PGD isomerases are present in cells which contain PGH synthase and thus generate the endoperoxide precursor. Another curious factor about purified PGH–PGD isomerase activities is that both enzymes are soluble proteins. All other proteins involved in the metabolism of PGH_2 are membrane bound.

PGH–PGE Isomerase PGH–PGE isomerases catalyze the rearrangement of the endoperoxide group of PGH_2 to yield PGE_2. Membrane preparations from sheep vesicular gland and rabbit renal medulla catalyze a GSH-dependent isomerization of PGH_2 to PGE_2. Attempts to solubilize and fractionate this enzyme have indicated that there are several different proteins capable of catalyzing the reaction. As in the case of PGD_2 formation, this raises the question of whether there is a distinct PGH–PGE isomerase devoted to PGE_2 synthesis or whether the ability to facilitate PGE_2 production from PGH_2 is a property shared by catalytic domains in a number of different proteins.

It has been known for almost 20 years that GSH enhances the rate of production of PGE_2. In fact, GSH appears to be required in these reactions. The mechanism by which GSH participates in PGE_2 synthesis has not been established. However, an attractive possibility is the mechanism shown in Figure 11.9 in which GSH facilitates a 1,2-hydride shift analogous to the hydride shift that occurs in the glyoxylase I reaction.

PGH–PGF$_\alpha$ Reductase The formation of $PGF_{2\alpha}$ from PGH_2 requires two electrons. This is the only commonly recognized transformation of PGH_2 that requires an electron donor. The process(es) by which $PGF_{2\alpha}$ formation occurs in vivo are not understood. No heat-labile activity has ever been demonstrated to facilitate $PGF_{2\alpha}$ production. Moreover, the cofactor for this reaction is unknown. It is known that dithiols in the presence of certain transition metals ions, such as Cu^{2+}, increase the rate of $PGF_{2\alpha}$ synthesis. However, this process is nonenzymic. It is possible that in vivo $PGF_{2\alpha}$ is formed from PGE_2 by reduction of the 9-keto

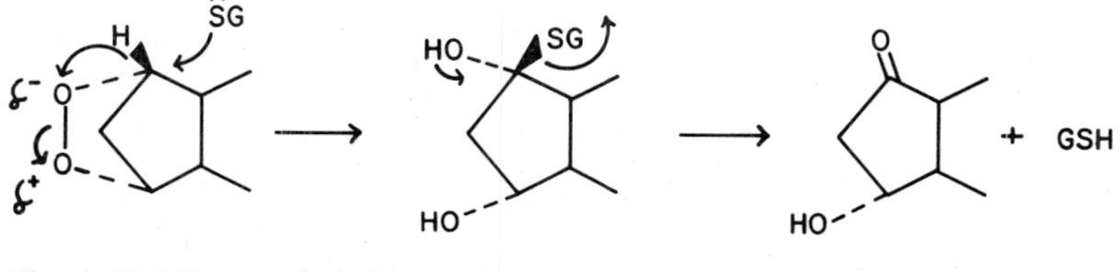

Figure 11.9. Proposed mechanism for the participation of reduced glutathione (GSH) in the conversion of PGH_2 to PGE_2 catalyzed by PGH–PGE isomerase. δ^- and δ^+ refer to partial negative and partial positive charges, respectively.

group and not from PGH_2. For example, PGE_2 infused into rabbit kidney is efficiently converted to $PGF_{2\alpha}$. A 9-keto prostaglandin reductase activity which catalyzes this reaction has been identified, but this protein exhibits other catalytic properties and may not be involved in the putative conversion of PGE_2 to $PGF_{2\alpha}$ in vivo.

There is considerable evidence that $PGF_{2\alpha}$ is a luteolytic factor produced by ovine and guinea pig uterus. Intramuscular injection of $PGF_{2\alpha}$ in cows causes regression of the corpus luteum and subsequent ovulation. $PGF_{2\alpha}$ is sold commercially as Lutylase, a regulator of ovulation in dairy cows. $PGF_{2\alpha}$ is also useful as a midtrimester abortifacient in humans. When injected into humans in a slowly released form, $PGF_{2\alpha}$ functions as a long-acting birth control agent probably by acting to prevent implantation of fertilized ova on the uterine wall.

TxA Synthase

Thromboxane A_2 was discovered in 1973 by Hamberg and Samuelsson in studies on the role of prostaglandins in platelet aggregation. At that time, the structures of the classical prostaglandins were known. It was also known that none of the classical prostaglandins by themselves would promote platelet aggregation but that cyclooxygenase inhibitors would prevent certain types of platelet aggregation. When Hamberg and Samuelsson incubated arachidonic acid with preparations of washed human platelets and analyzed the resulting metabolites, they discovered two cyclooxygenase products (Hamberg and Samuelsson 1974). One of these compounds was TxB_2 and the other was 12-hydroxy-5-cis,8-trans,10-trans-heptadecatrienoic acid (HHT) (Figure 11.10). TxB_2 is a hemiacetal derivative. It has no known important biological activity, but is a stable hydrolysis product of TxA_2. TxA_2 has not yet been isolated and characterized because of its short half-life (about 30 s in aqueous solution at 37°C). The structure of TxA_2 was deduced from the structure of TxB_2 and other derivatives. In fact, some question still remains about whether TxA_2 has the structure shown in Figure 11.2. Questions about its structure arise from the fact that endoperoxide analogues which are structurally different from TxA_2 are TxA_2 agonists. Interestingly, TxA_2 has biological properties identical to rabbit aorta-contracting substance (RCS), an active principle released by lung during anaphylactic shock.

The formation of TxA_2 is catalyzed by thromboxane A (TxA) synthase. This enzyme is present in high concentrations in platelet cells, macrophages, and lung. TxA synthase has been purified from porcine platelets and exhibits an absorption spectrum similar to that of cytochrome P-450. Ullrich and co-workers have suggested that TxA synthase catalyzes a rearrangement of the endoperoxide group by a

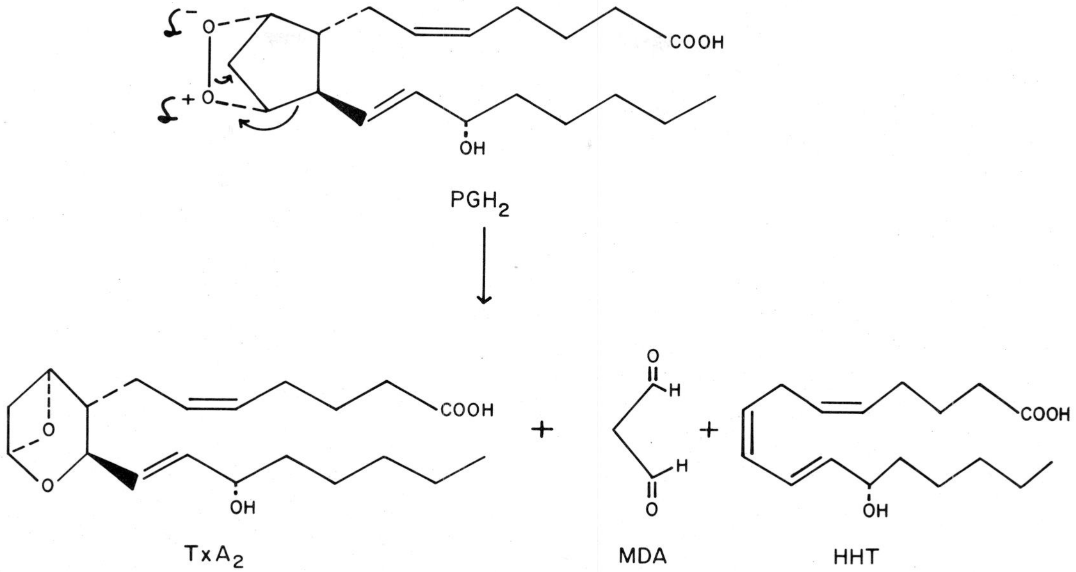

Figure 11.10. Proposed mechanism for the rearrangement of PGH$_2$ to TxA$_2$, malondialdehyde (MDA), and 12-hydroxy-5-*cis*,8-*trans*,10-*trans*-heptadecatrienoic acid (HHT).

mechanism similar to the oxene transfer mechanism of cytochrome P-450. Whatever the initial step, it is clear that rearrangement of PGH$_2$ to TxA$_2$ requires the development of a transient, partial positive charge on the oxygen at C-11 of the endoperoxide (Figure 11.10).

A curious feature of TxA$_2$ production is the coproduction of HHT (plus malondialdehyde) by purified preparations of the enzyme. The ratio of the production of HHT (or related C$_{17}$ hydroxy acids) to the production of TxA derivatives depends on the position of the carbon-carbon double bond in the carboxyl side chain of endoperoxide substrates. For PGH$_2$, the ratio is roughly 1:1. With unnatural endoperoxide substrates having double bonds at positions other than between C-5 and C-6, considerably more C$_{17}$ hydroxy acid production occurs, and TxA derivatives account for only a few percent of the total product. Apparently, the double bond between C-5 and C-6 is used by TxA synthase to position the endoperoxide group at the active site of the enzyme during catalysis. Maintaining proper placement of the endoperoxide is envisioned as favoring formation of TxA derivatives.

Considerable efforts have been expended to develop specific inhibitors of TxA synthase, since abrogation of TxA$_2$ synthesis can prevent certain types of platelet aggregation. The most effective and specific inhibitors are relatively simple pyridine and histidine derivatives.

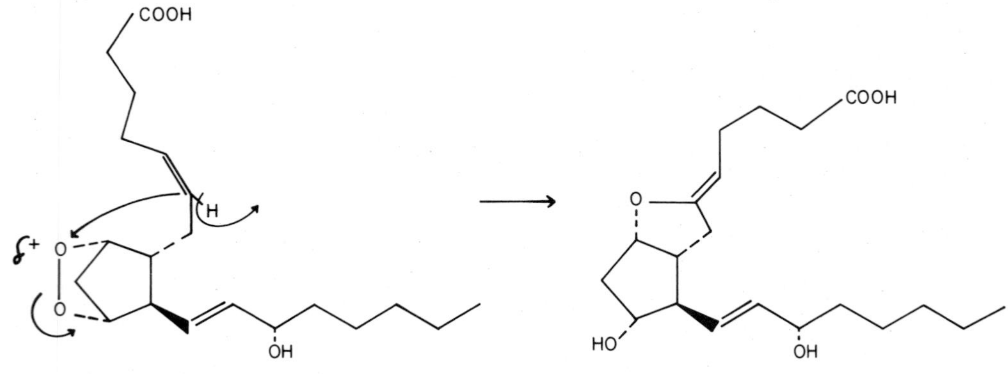

Figure 11.11. Proposed mechanism for the rearrangement of PGH$_2$ to PGI$_2$.

PGI$_2$ Synthase   In contrast with TxA$_2$ formation, synthesis of PGI$_2$ involves transient formation of a positive charge on the oxygen atom at C-9 of PGH$_2$ instead of at C-11 (Figure 11.11). This electropositive oxygen reacts at the electronegative center at C-6. Carbon-oxygen bond formation occurs with concomitant loss of the proton from C-6. In fact, PGI$_2$ synthase, the enzyme that catalyzes the formation of PGI$_2$ from PGH$_2$, can be assayed using PGH$_2$ radiolabeled with ^{3}H at C-6 by measuring the release of tritium. A more common assay involves the extraction of PGI$_2$ into organic solvents following acidification of the aqueous reaction mixture. The enol ether linkage of PGI$_2$ is rapidly hydrolyzed in acid to yield 6-keto-PGF$_{1\alpha}$. 6-Keto-PGF$_{1\alpha}$ is stable and can be analyzed by radioimmunoassay or by radio-thin-layer chromatography.

PGI$_2$ is a vasodilator and antiplatelet aggregatory (antithrombogenic) substance. Diminished production of PGI$_2$ by the vasculature may accompany the development of atherosclerotic plaques. Interest in determining the properties of PGI$_2$ synthase stems from the concept that maintaining PGI$_2$ synthesis is important for optimizing cardiovascular function.

PGI$_2$ synthase has been purified to electrophoretic homogeneity from both bovine and porcine aorta. The isolated protein has a single protein subunit with a molecular weight of 50,000. PGI$_2$ synthase is a hemoprotein which like TxA synthase may be a cytochrome P-450. It has been proposed that the heme group is involved in the heterolytic cleavage of the endoperoxide grouping in a manner analogous to the P-450–catalyzed oxene transfer mechanism. A curious property of the enzyme is its susceptibility to inactivation by lipid hydroperoxides. This inactivation involves a direct interaction of the hydroperoxide with the heme at the active site.

Subcellular Location of Prostaglandin Biosynthesis

Except for PGH–PGD isomerases, prostaglandin biosynthetic enzymes are integral membrane proteins. The membrane site or sites from which arachidonate is mobilized have not been established. However, there appears to be efficient coupling of the release of arachidonate (in response to physiological stimuli) and the conversion of arachidonate to cyclooxygenase products. This suggests that arachidonate is released from membranes which contain PGH synthase activity. Phosphatidylinositol-specific phospholipase C is a soluble protein. Thus, one would expect that this enzyme could operate on phosphatidylinositol molecules on the cytoplasmic surface of a variety of cellular membranes. Phospholipase A_2 and diglyceride lipase activities are membrane-bound. In each case most of the activity is found in the microsomal fraction. Most of the PGH synthase activity of prostaglandin-forming cells is also associated with a microsomal fraction. However, it is clear that PGH synthase is associated with a number of subcellular membranes within the same cell. For example, PGH synthase is present on the nuclear membrane, the endoplasmic reticulum, and the cell surface of smooth muscle cells. In both the endoplasmic reticulum and the plasma membrane, the PGH synthase active site is on the cytosolic surface. The subcellular distribution(s) of the PGH_2-metabolizing enzymes has not been examined extensively. However, in smooth muscle cells, PGI_2 synthase and PGH synthase are associated with the same membrane fractions, which suggests that PGH_2 is generated and isomerized at the same location.

In short, data on the subcellular distribution of prostaglandin biosynthetic enzymes suggests that prostaglandins are generated on the cytosolic surface of several membrane fractions within the same cell, including the endoplasmic reticulum, nuclear membrane, plasma membrane, and possibly the mitochondria. This distribution is quite unusual and difficult to rationalize in the context of the current model for prostaglandin action, which is that prostaglandins are rapidly synthesized and released from cells and then act only on ectoreceptors found on the outside of the plasma membrane.

Prostaglandin Catabolism: Prostaglandins as Local Hormones

Our concept of prostaglandin catabolism has come from studies in which prostanoids have been infused into animals and the disposition of the products determined. In the late 1960s, Vane and Piper infused prostaglandins intravenously into different vascular beds and measured the appearance of products in the arterial blood. They found that PGE_1, PGE_2, and $PGF_{2\alpha}$ were rapidly catabolized to inactive forms and did not survive a single pass through the circulation. The most active site of catabolism was the lung. Studies on the catabolism of prostaglandins by lung tissue and analyses of

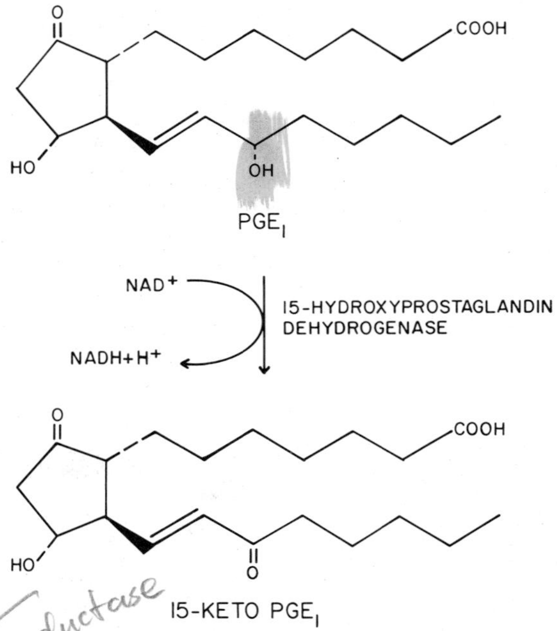

Figure 11.12. Conversion of PGE$_1$ to 15-keto-PGE$_1$ catalyzed by 15-hydroxyprostaglandin dehydrogenase.

prostaglandin metabolites in the urine indicate that the initial step in prostaglandin catabolism is oxidation of the hydroxyl group at C-15 (Figure 11.12). The resulting 15-keto products have less than one-tenth the biological activity of the parent molecules in bioassays on smooth muscle.

The lung contains an active NADH-dependent 15-hydroxy prostaglandin dehydrogenase which catalyzes oxidation of the C-15 hydroxyl group. A related activity is found in kidney. Apparently, only PGE$_2$ and PGF$_{2\alpha}$ are catabolized by lung. Neither PGD$_2$ nor PGI$_2$ is taken up by this organ. However, PGD$_2$- and PGI$_2$-specific 15-hydroxy dehydrogenases are found in kidney. Thus, the first step in the catabolism of PGE$_2$, PGF$_{2\alpha}$, PGD$_2$, and PGI$_2$ is oxidation of the C-15 hydroxyl group, although this reaction occurs in different organs depending on the prostanoid.

The second step in catabolism is reduction of the Δ^{13} double bond. A Δ^{13} reductase activity which utilizes NADPH as the electron donor catalyzes this reaction. The resulting 15-keto-13,14-dihydro prostaglandins are biologically inactive. The major urinary metabolites of most prostaglandins are C$_{16}$(tetranor) dioic acids that result from a combination of β-oxidation and ω-oxidation of the 15-keto-13,14-dihydro prostaglandins. Both β-oxidation and ω-oxidation occur in the liver. In the case of TxA$_2$, the major route of biological inactivation may be nonenzymic hydrolysis to TxB$_2$.

The concentrations of the major active prostaglandin products in blood are $10^{-10}M$ or less. Moreover, infused prostaglandins are rapidly catabolized. The fact that prostaglandins are formed by cells present in virtually all organs suggests that there is no single prostaglandin exocrine gland. All this information is consistent with the now well-established concept that prostaglandins are local hormones, or autocoids, which modify biological events only near their sites of synthesis.

MECHANISM OF ACTION OF PROSTANOIDS
Introduction

There is a consensus that steroid hormones act on target cells to modulate production of specific mRNA molecules. Although there is yet no such paradigm to explain the mechanism of action of prostaglandins, one presumes that there is some process by which all prostaglandins act. Work on the mechanism of prostaglandin action has focused on two events: (a) modulation of adenylate cyclase activity and (b) facilitation of ion transport. To establish that a cyclooxygenase product is important in a physiological process, one must show (a) that prostaglandins are found at the site of the event at concentrations necessary to elicit the event, (b) that inhibitors of prostaglandin formation block the event, and (c) that exogenous prostaglandins mimic the effect. There is no physiological process for which all these criteria are established. Moreover, there are as yet no specific antagonists for prostaglandins, so it is not currently possible to test for antagonism of prostaglandin bioactivity.

Two biological processes which have been widely studied in the context of the mechanism of action of prostaglandins are the interplay between TxA_2 and PGI_2 in platelet aggregation and the inhibition of the hydroosmotic effect of antidiuretic hormone (ADH) by PGE_2 in renal collecting tubules. In considering potential mechanisms of action of prostaglandins, we shall discuss the perceived role of prostaglandins in these two processes.

Thromboxane, Prostacyclin, and Platelet Aggregation

Arteries are lined by endothelial cells, under which are layers of smooth muscle cells (Figure 11.13). A principal function of the endothelium is to regulate the movement of molecules between the blood and the underlying tissue. Vessels can be injured as a consequence of vascular turbulence, noxious chemicals, or physical trauma (Figure 11.13). Injuries result in the loss of endothelial cells, thereby disrupting the normal barrier between blood and tissue. In the arterial circulation, where blood flow is relatively rapid, circulating platelets adhere to the subendothelium at sites of injury and then aggregate. The result is a vascular bandage called a *platelet plug*.

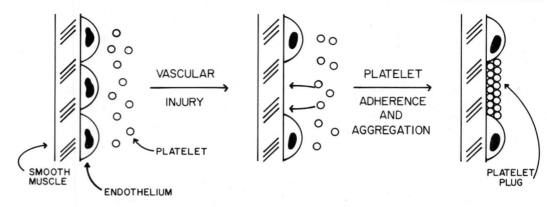

Figure 11.13. Role of platelets in the formation of a hemostatic plug following deendothelialization of an arterial wall as a consequence of a vascular injury.

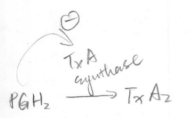

Platelets or thrombocytes are small enucleated cell remnants of precursor blood cells called *megakaryocytes*. There is approximately 1 platelet for every 20 red cells in human blood. The aggregation of platelets in response to stimuli such as subendothelial collagen is accompanied by the formation of TxA_2. Moreover, inhibitors of TxA synthase inhibit aggregation of platelets by PGH_2 or arachidonate, and synthetic TxA derivatives induce platelet aggregation. Thus, TxA_2 is important in promoting irreversible platelet aggregation.

The molecular basis for the action of TxA_2 is unresolved. Mobilization of Ca^{2+} is required for platelet aggregation. There is some evidence that TxA_2 has Ca^{2+}-ionophoretic activity and mediates Ca^{2+} mobilization from intracellular stores. Other studies suggest that cyclooxygenase products of platelets promote the formation of another Ca^{2+} ionophore, phosphatidic acid. Increasing the concentration of unbound Ca^{2+} in platelets prevents increases in platelet cAMP levels which otherwise would prevent aggregation. Thus, TxA_2 may promote platelet aggregation by causing Ca^{2+} mobilization either directly or indirectly.

In early studies on the role of prostanoids in platelet aggregation, it was found that PGE_1 inhibited platelet aggregation and that this effect was due to the elevation of intracellular cAMP. Following the discovery of PGI_2 in 1975, it was found that this prostanoid also activated platelet adenylate cyclase. In fact, PGI_2 and PGE_1 appear to interact with the same receptor on platelet cells. Immobilized PGE_1 inhibits platelet aggregation, which suggests that these prostanoids interact with ectoreceptors coupled to adenylate cyclase. TxA_2 agonists prevent the increase in platelet cAMP levels that occur in response to PGI_2.

The physiological significance of PGI$_2$ in regulating platelet aggregation is still controversial. The concentration of PGI$_2$ in the circulation is too low to activate platelet adenylate cyclase. However, it is argued that the concentration of PGI$_2$ released by the vasculature may become elevated in localized areas of the circulation in response to protein factors released by stimulated platelets. Despite questions about the physiological importance of PGI$_2$ in regulating platelet function, PGI$_2$ derivatives are useful in pharmacological doses as therapeutic agents to prevent platelet aggregation during some types of cardiovascular surgery.

Inhibition of the Hydroosmotic Effect of ADH by PGE$_2$

The final event in the elaboration of urine occurs in the terminal portion of the renal tubule called the *collecting* tubule. Antidiuretic hormone (ADH) interacts with receptors on the blood surface of the collecting tubule, causing the movement of water across the tubule from the hypoosmotic urine to the hyperosmotic interstitium surrounding the blood side of the cell (Figure 11.14). This process is

COLLECTING TUBULE CELL

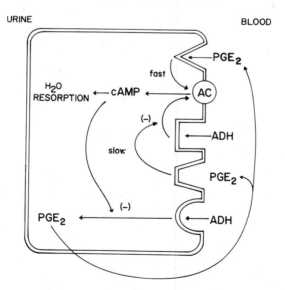

Figure 11.14. Proposed mechanism for the interactions of antidiuretic hormone (ADH) and PGE$_2$ in renal collecting tubule cells. ADH induces cAMP formation, which in turn leads to the resorption of water from the lumen of the tubule. ADH also induces PGE$_2$ synthesis. The newly formed PGE$_2$ has two effects. Low concentrations of PGE$_2$ inhibit ADH-induced cAMP formation (that is, heterologous desensitization). High concentrations of PGE$_2$ activate adenylate cyclase (AC), leading to elevated cAMP and subsequent inhibition of further arachidonic acid release.

the *hydroosmotic effect* of ADH. ADH activates the adenylate cyclase in collecting tubules, thereby increasing cAMP concentrations in these cells. The hydroosmotic effect of ADH is mediated by this increase in cAMP, since cAMP itself will elicit the hydroosmotic effect.

It has been known for 15 years that low concentrations of PGE_2 inhibit the hydroosmotic effect of ADH on collecting tubules. More recent studies have shown that collecting tubule cells have high levels of PGH synthase and that these cells form PGE_2 in response to stimulation with ADH. Pretreatment of collecting tubule cells with very low concentrations of PGE_2 (about $10^{-11}M$) prevents the formation of cAMP that normally occurs in response to ADH.

The biochemical mechanism by which low concentrations of PGE_2 inhibit ADH-induced cAMP formation in the collecting tubule is not fully resolved. However, inhibition may be due to heterologous desensitization of collecting tubule adenylate cyclase by PGE_2. That is, PGE_2 may prevent activation of the adenylate cyclase catalytic unit by the ADH-receptor complex by modifying the properties of one of the GTP binding protein(s) (Figure 11.14). This desensitization observed with collecting tubules may be a general model for the mechanism of action of PGE_2. It should be recalled that a somewhat analogous situation exists in platelets where TxA_2 agonists inhibit the normal accumulation of cAMP in response to PGI_2. Curiously, heterologous desensitization of the adenylate cyclase of collecting tubules is caused by low concentrations of PGE_2; higher concentrations of PGE_2 actually activate adenylate cyclase. This second effect of PGE_2 apparently causes a cAMP-dependent feedback inhibition of arachidonate release and subsequent prostaglandin formation.

HYDROXY- AND HYDROPEROXY- EICOSAENOIC ACIDS AND LEUKOTRIENES Introduction and Overview

Plant enzymes called *lipoxygenases* which catalyze the introduction of oxygen into polyunsaturated fatty acids were discovered almost four decades ago. In fact, studies on the oxygen insertion catalyzed by soybean lipoxygenase were important in elucidating the mechanism of PGH_2 formation (Hamberg and Samuelsson 1967). The discovery of 12S-hydroperoxy-5,8,10,14-eicosatetraenoic acid (12-HPETE) production by platelet cells in 1973 was the first indication that animal tissues also catalyze lipoxygenase reactions (Hamberg and Samuelsson 1974). It is now known that mammalian lipoxygenase activities exist which will catalyze the insertion of oxygen at positions 5, 12, and 15 of various eicosaenoic acids. The immediate products are hydroperoxy fatty acids. In the case of arachidonic acid, the products are hydroperoxy-eicosatetraenoic acids (HPETEs).

HPETEs can subsequently undergo one of three different transformations (Figure 11.15). One reaction is a two-electron reduction of

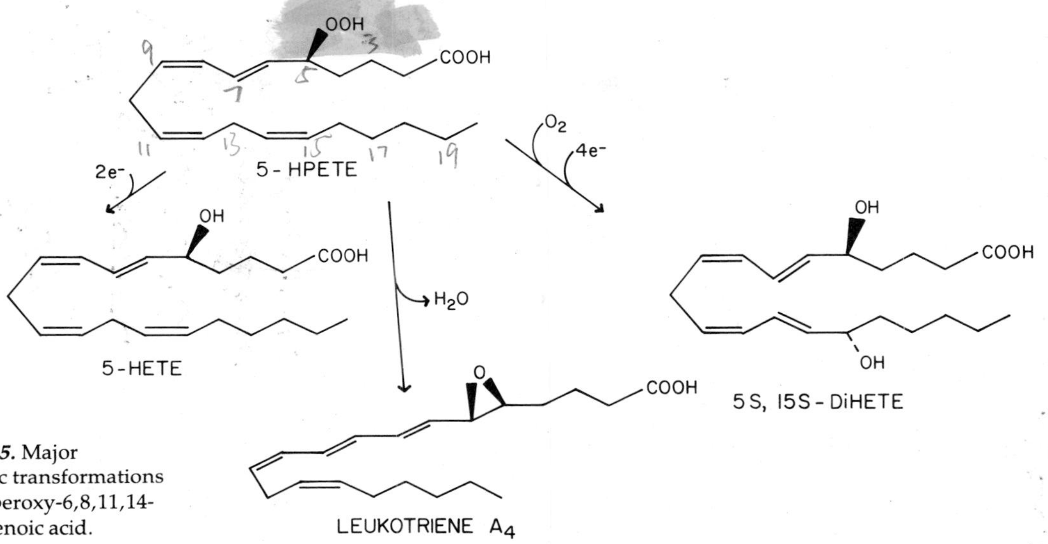

Figure 11.15. Major biosynthetic transformations of 5-hydroperoxy-6,8,11,14-eicosatetraenoic acid.

the hydroperoxy group to an alcohol, yielding the corresponding hydroxy-eicosatetraenoic acids (HETEs). A second type of reaction is lipoxygenation at another position in the aliphatic chain, yielding (after reduction of the two hydroperoxy groups) dihydroxy-eicosatetraenoic acids (diHETEs). Finally, a third type of transformation of HPETEs is a dehydration to produce an epoxy (oxido) fatty acid.

Epoxy fatty acids can undergo three additional transformations (Figure 11.16): (a) nonenzymic hydrolysis to a variety of epimeric diHETEs, (b) stereospecific enzymic hydrolysis to a specific diHETE, or (c) ring opening in the presence of GSH to yield peptide derivatives in which GSH is attached via a sulfoether linkage to the fatty acid. Epoxy eicosatetraenoic acids and the products derived from these epoxy acids contain a conjugated triene unit and are called *leukotrienes* (LT). The term leukotriene denotes the cells (leukocytes) originally recognized to form these products. The term does not apply to diHETEs formed by two successive lipoxygenase reactions, although some of these products do contain a conjugated triene unit.

Some of the peptido-leukotrienes are now recognized as the bioactive components of slow-reacting substance of anaphylaxis (SRS-A). This finding and recent discoveries of the chemotactic and myotropic activities of lipoxygenase products have stimulated interest in these substances in many research fields, particularly in the areas of allergy and inflammation.

It should be recognized that the metabolism of eicosaenoic acids by lipoxygenase pathways has received serious attention only in the past

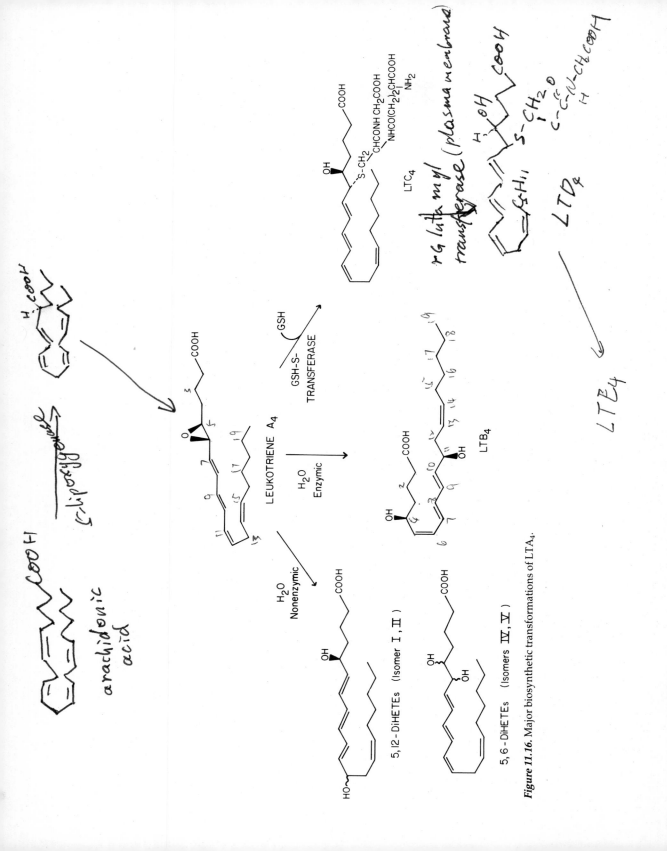

Figure 11.16. Major biosynthetic transformations of LTA₄.

5 years. Considerable progress has been made in identifying the structures of many of these compounds. However, none of the enzymes which catalyze these reactions has been characterized in any detail. Consequently, we know little about the regulation of the biosynthesis of lipoxygenase products. Moreover, although the pharmacological activities of many of these compounds have been determined, the mechanisms of action of lipoxygenase products are not understood.

Our discussion of the metabolism of eicosaenoic acids via lipoxygenase pathways will necessarily focus on the chemical transformations that occur. We will not attempt to present an all-inclusive description of these pathways. Undoubtedly, new metabolites remain to be discovered.

Lipoxygenase Reactions

The synthesis of HPETEs and HETEs is mainly associated with neutrophils, eosinophils, monocytes, mast cells, macrophages, and platelets. Lipoxygenase reactions are also catalyzed by preparations of skin, heart, colonic tissue, lung, spleen, and liver; however, it is possible that the reactions in these tissues are actually catalyzed by tissue macrophages, mast cells, or polymorphonuclear leukocytes (PMNL).

Lipoxygenases catalyze the type of reaction shown in Figure 11.17. This is a simple dioxygenase reaction; there is no net oxidation-reduction of either the fatty acid or the oxygen. A cis-trans conjugated diene is formed in the reaction. All HPETEs and HETEs contain this chromophore, which absorbs strongly at 230–235 nm and is useful in the detection and quantitation of these compounds.

Porcine PMNL are capable of catalyzing the insertion of oxygen into positions 5, 12, and 15 of arachidonic acid. Human PMNL have 5- and 15-lipoxygenase activities, whereas human platelets contain a 12-lipoxygenase. The structures of 5-HETE, 12-HETE, and 15-HETE are shown in Figure 11.18. It is likely that there are several distinct lipoxygenase isozymes with different positional specificities. Only one of these enzymes, the 15-lipoxygenase, has been purified to apparent homogeneity. Ca^{2+} enhances the activity of this enzyme in the impure, but not the purified, form. Similarly, crude preparations of 5-lipoxygenase from PMNL require Ca^{2+} for activity. The 5-lipoxygenase from PMNL appears to be a soluble enzyme. The

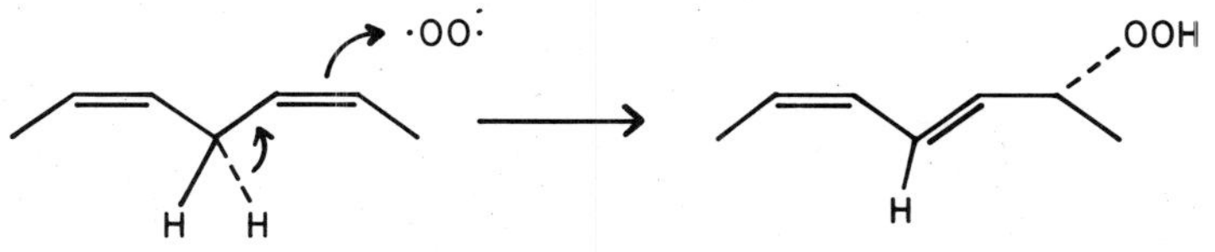

Figure 11.17. Proposed mechanism for the oxygenation reactions catalyzed by lipoxygenases.

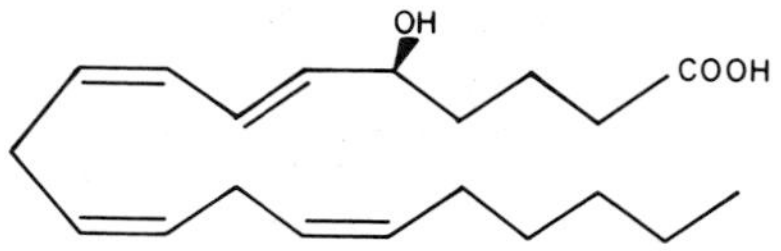

Figure 11.18. Structures of common monohydroxy fatty acids formed through the actions of different mammalian lipoxygenases.

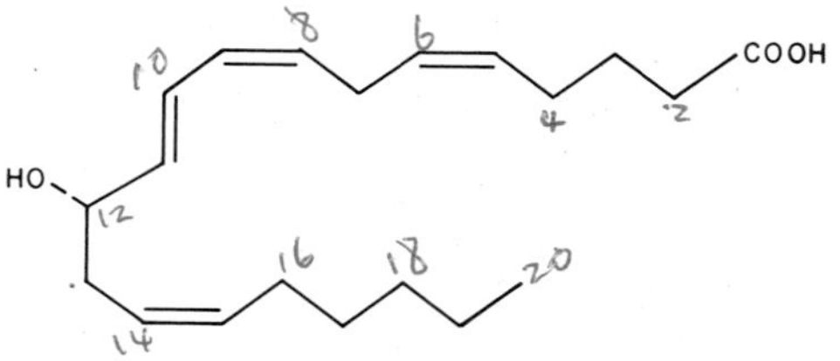

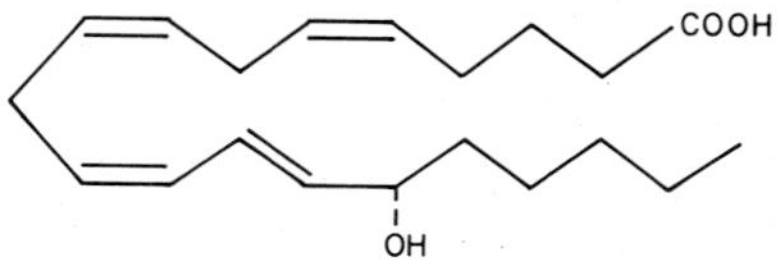

12-lipoxygenase from platelets is present in significant quantities in both soluble and membrane-bound forms. The significance of this apparent bimodal distribution is not understood.

The reaction mechanism of mammalian lipoxygenases has not been studied in detail. However, soybean lipoxygenase, which catalyzes a similar oxygen insertion, has been studied for years because of its importance in the plant seed oil industry. Interestingly, this enzyme has several catalytic properties in common with the cyclooxygenase activity of PGH synthase. For example, soybean lipoxygenase requires a hydroperoxy acid activator and undergoes a self-catalyzed destruction reaction. However, unlike the cyclooxygenase, soybean lipoxygenase does not have a heme group but may contain nonheme iron at its active site. Undoubtedly, the soybean enzyme will serve as a prototype in future studies on mammalian lipoxygenases.

Biosynthesis of Leukotrienes

For ease of nomenclature in discussing lipoxygenase pathways, we have used the *E-Z* system to denote cis (*Z*) and trans (*E*) carbon-carbon double bonds. In human PMNL, arachidonic acid is first

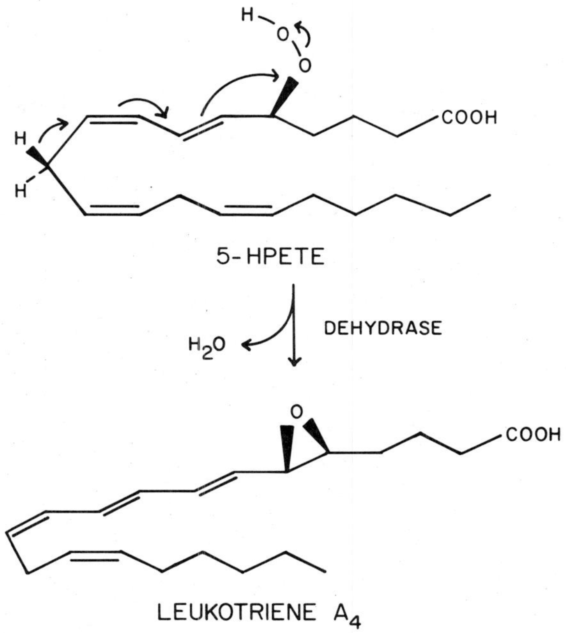

Figure 11.19. Mechanism for the conversion of 5-HPETE to LTA$_4$.

transformed into 5-HPETE ((5S)-5-hydroperoxy-(E,Z,Z,Z)-6,8,11,14-eicosatetraenoic acid) through the action of the 5-lipoxygenase. Intact PMNL rapidly convert 5-HPETE to a combination of 5-HETE ((5S)-5-hydroxy-(E,Z,Z,Z)-6,8,11,14-eicosatetraenoic acid) and leukotriene A$_4$ (LTA$_4$,(5S,6S)-5(6)-oxido-(E,E,Z,Z)-7,9,11,14-eicosatetraenoic acid Figure 11.15)). In LTA$_4$, the letters *LT* indicate leukotriene, the letter *A* indicates the nature and position of the oxygen-containing substituent, and the numerical subscript indicates the number of double bonds in the molecule. This nomenclature is analogous to that used for prostaglandins.

Formation of LTA$_4$ from 5-HPETE is enzymic, but the enzyme has not yet been characterized. LTA$_4$ synthesis involves the stereospecific removal of the 10D(R)-hydrogen atom of 5-HPETE and loss of water (Figure 11.19). The biosynthesis of LTA$_4$ by rabbit peritoneal PMNL incubated with arachidonic acid is maximal within 90 s and is followed by rapid enzymic and nonenzymic conversions into a variety of more-polar compounds.

As expected for an allylic epoxide, LTA$_4$ is highly unstable and undergoes rapid hydrolysis (the $t_{\frac{1}{2}}$ is less than 10 s in aqueous buffer at pH 7.4 and 25°C). LTA$_4$ is hydrolyzed instantaneously at acid pH but can be stabilized under alkaline conditions or upon binding to serum albumin. The formation of LTA$_4$ in biological systems can be demonstrated by trapping with methanol and

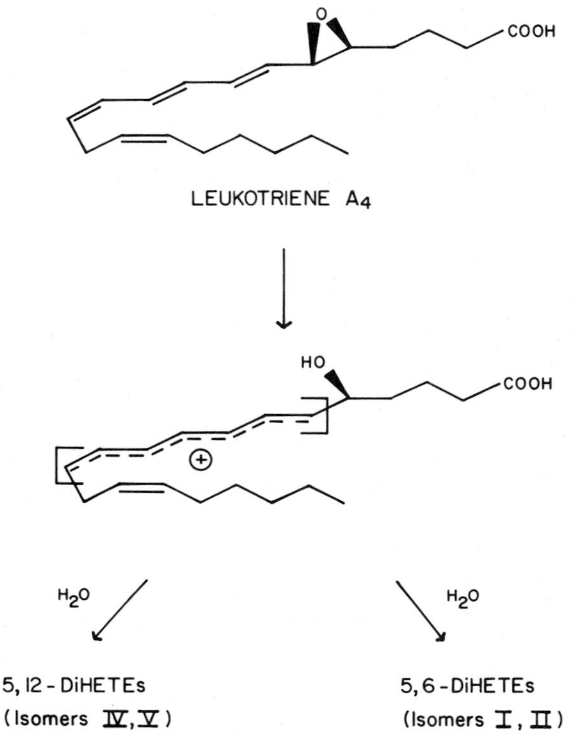

LEUKOTRIENE A$_4$

Figure 11.20. Formation of a carbonium-ion intermediate resulting from opening the 5,6-epoxide ring of LTA$_4$.

5, 12 - DiHETEs

(Isomers IV, V)

5, 6 - DiHETEs

(Isomers I, II)

measuring the stable methanolysis products. Nonenzymic hydrolysis of LTA$_4$ leads to the formation of four diHETEs (two 5,12-diHETEs and two 5,6-diHETEs) (Figure 11.20). The proportion of products depends on the reaction conditions (for example, pH). Each product is derived from a carbonium-ion intermediate produced upon opening of the 5,6-epoxide (Figure 11.20). Studies on the biosynthesis and nonenzymic hydrolysis of LTA$_4$ in the presence of either H$_2$^{18}O or ^{18}O$_2$ indicate that the oxygen atoms at positions C-12 and C-6 of the 5,12-diHETEs and 5,6-diHETEs, respectively, come from water. The oxygen atom at C-5 is from molecular oxygen (via the 5-lipoxygenase reaction). The diHETE products of nonenzymic hydrolysis of LTA$_4$ have little biological activity. However, the identification of these four isomeric dihydroxy acids in initial studies on arachidonate metabolism by the 5-lipoxygenase pathway was the key step in deducing the nature of the precursor, LTA$_4$.

LTA$_4$ undergoes enzymic conversions to various compounds which are important biologically (Figure 11.16). LTB$_4$ ((5S,12R)-5,12-dihydroxy-(Z,E,E,Z)-6,8,10,14-eicosatetraenoic acid) is the product of the enzymic hydrolysis of LTA$_4$. LTB$_4$ is formed in human PMNL incubated with both arachidonic acid and the ionophore A23187. The hydroxyl groups at C-5 and C-12 in LTB$_4$ are derived, respectively, from the epoxide ring of LTA$_4$ and from water.

LTA$_4$ is also metabolized by a glutathione S-transferase (Figure 11.16). LTC$_4$ ((5S,6R)-5-hydroxy-6-S-glutathionyl-(E,E,Z,Z)-7,9,11,14-eicosatetraenoic acid) is the product of the conjugation of glutathione and LTA$_4$. The enzyme involved in this reaction has not yet been characterized, but it is apparently membrane bound. Inhibition of glutathione synthesis in cells or treatment of cells with agents which reduce the levels of intracellular GSH leads to decreased LTC$_4$ synthesis.

A metabolite of LTC$_4$ called LTD$_4$ ((5S,6R)-5-hydroxy-6-S-cysteinylglycyl-(E,E,Z,Z)-7,9,11,14-eicosatetraenoic acid) is formed upon enzymic hydrolysis of the peptide chain of LTC$_4$ and loss of the γ-glutamyl residue (Figure 11.21). The peptidase involved has not yet been characterized, although γ-glutamyl transpeptidase will catalyze

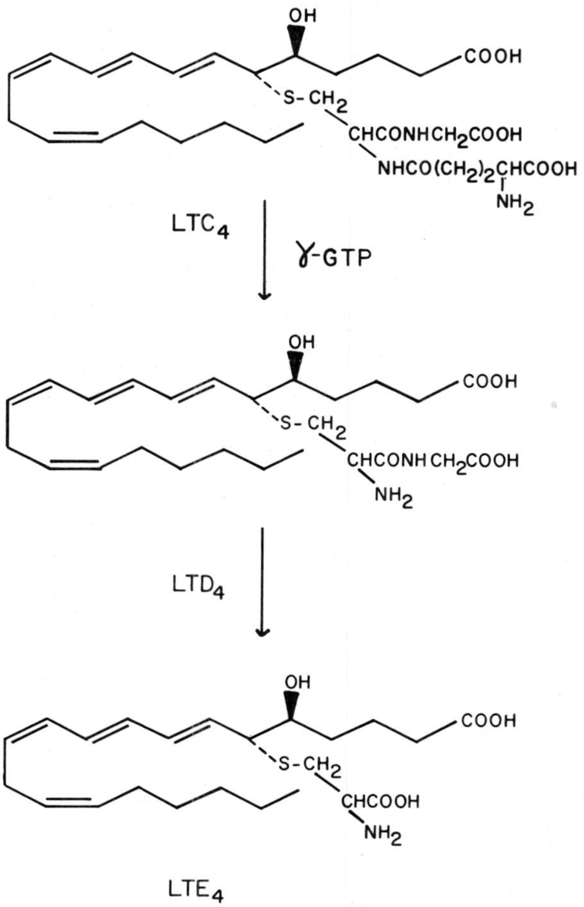

Figure 11.21. Conversion of LTC$_4$ to other peptido-leukotrienes. γ-GTP is γ-glutamyl transpeptidase.

the conversion of LTC_4 to LTD_4. LTD_4 is released from guinea pig lungs following anaphylaxis. LTC_4 and LTD_4 are the major components of SRS-A.

A peptidase-catalyzed cleavage of a glycyl residue from LTD_4 leads to LTE_4 ((5S,6R)-5-hydroxy-6-S-cysteinyl-(E,E,Z,Z)-7,9,11,14-eicosatetraenoic acid) (Figure 11.21). A peptidase activity capable of catalyzing this conversion is present in human plasma and PMNL-specific granules.

A geometric isomer of LTC_4, called Δ^{11}-trans-LTC_4 ((5S,6R)-5-hydroxy-6-S-glutathionyl-(E,E,E,Z)-7,9,11,14-eicosatetraenoic acid), as well as Δ^{11}-trans-LTD_4 and Δ^{11}-trans-LTE_4, which have a trans carbon-carbon double bond at C-11, have been described. The biological importance of these compounds is unknown, but their existence illustrates the variety and complexity of leukotriene chemistry.

Metabolism of 8,11,14-Eicosatrienoic Acid, 5,8,11,14,17-Eicosapentaenoic Acid, and HETEs by the 5-Lipoxygenase Pathway

It was shown in early studies on the metabolism of unsaturated fatty acids that PMNL catalyze the formation of 8(S)-8-hydroxy-(E,Z,Z)-9,11,14-eicosatrienoic acid from exogenous all-cis 8,11,14-eicosatrienoic acid, but that under the same conditions, 5-HETE, rather than the corresponding 8-hydroxy-eicosatetraenoic acid, is produced from arachidonic acid. More recently, the formation of 8-hydroxy-9-S-glutathionyl-10,12,14-eicosatrienoic acid from 8,11,14-eicosatrienoic acid has been shown in vitro. The existence of this peptide derivative indicates the formation of 8-hydroperoxy-9,11,14-eicosatrienoic acid and the subsequent conversion of this hydroperoxy acid to the allylic epoxy acid, 8(9)-oxido-10,12,14-eicosatrienoic acid. The 5-lipoxygenase and other enzymes involved in the synthesis of LTA_4 and LTC_4 from arachidonic acid may be involved in these transformations of 8,11,14-eicosatrienoic acid. 5,8,11,14,17-Eicosapentaenoic acid is also metabolized through the 5-lipoxygenase pathway, yielding LTB_5, LTC_5, LTD_5, and LTE_5.

Studies with a series of synthetic polyunsaturated fatty acids have shown that a pair of cis, methylene-interrupted double bonds at C-5 and C-8 is important for the expression of 5-lipoxygenase activity. The length of the carbon chain and the presence of substituents in other parts of the carbon chain are of secondary importance. For example, both 12-HETE and 15-HETE, which contain double bonds at C-5 and C-8, are oxygenated at C-5 by the action of 5-lipoxygenase. The products are 5S,12S-diHETE and 5S,15S-diHETE. In each case both oxygen atoms are derived from molecular oxygen, unlike the diHETEs formed by nonenzymic or enzymic hydrolysis of LTA_4. The biological importance of these compounds is unknown. The 5S,15S-diHETE is a product of PMNL, and 5S,12S-diHETE can be formed in incubation mixtures containing both platelets and PMNL.

Catabolism of Leukotrienes

Our current knowledge of the catabolism of leukotrienes is fragmentary. LTB_4 undergoes ω-oxidation when incubated with human blood PMNL, leading to the formation of ω-hydroxy-LTB_4 ((5S,12R)5,12,20-trihydroxy-(Z,E,E,Z)-6,8,10,14-eicosatetraenoic acid) and then ω-carboxy-LTB_4. The ω-carboxy-LTB_4 retains little of the biological activity of the parent compound, but the ω-hydroxy-LTB_4 shows either increased or decreased activity depending on the target cells or tissues. It is likely that these ω-oxidation pathways are important in the catabolism and biological inactivation of LTB_4 in vivo.

Recent studies have indicated that there are two pathways for the metabolism of LTC_4, one involving hydrolysis of the glutathionyl group by peptidases, and another involving a peroxidase-mediated oxidation of the sulfur atom. Peptidases present in human plasma, lung, and kidney rapidly convert LTC_4 and LTD_4 into LTE_4. Thus, LTC_4 and LTD_4, like bioactive prostaglandins, appear to have very short half-lives in the circulation. However, it is noteworthy that LTE_4 shows significant biological activity in most systems; therefore, the transformation of LTC_4 and LTD_4 into LTE_4 does not represent an inactivation process. LTC_4, LTD_4, and LTE_4 are also known to undergo oxidative catabolism by intact neutrophils, eosinophils, and by eosinophil peroxidase in reactions requiring H_2O_2 and halide ions. Products formed in the oxidation of the peptido-leukotrienes include the corresponding diastereoisomeric sulfoxides as well as Δ^6-trans-LTB_4, Δ^6-trans-12-epi-LTB_4, and two unknown compounds. These four latter compounds lack appreciable biological activity. Another oxidative mechanism of leukotriene inactivation has been found in myeloperoxidase-deficient PMNL. This inactivation process involves hydroxyl radicals.

Control of Leukotriene Biosynthesis

The formation of lipoxygenase products, unlike the formation of cyclooxygenase products, is not solely determined by the concentration of free arachidonic acid in cells. Indeed, in most systems investigated so far, the 5-lipoxygenase appears to be present in an inactive form in cells and tissues, and the addition of exogenous arachidonic acid usually results in the formation of only small amounts of products (less than 5% of the maximal release obtained upon ionophore A23187 stimulation). Thus, the synthesis of leukotrienes requires both the release of arachidonic acid and the activation of the 5-lipoxygenase. As is the case with cyclooxygenase products, lipoxygenase products are synthesized and released on demand. Immunological stimuli and inflammatory stimuli including phagocytosis and chemotactic peptides induce the formation of leukotrienes by causing arachidonate release and by stimulating the activity of the 5-lipoxygenase. It appears that in cells carrying IgE receptors, the 5-lipoxygenase pathway is activated upon stimulation with antigens.

Some cells form both cyclooxygenase and lipoxygenase products. For example, platelets form 12-HETE and TxA_2 from arachidonic acid, and mouse peritoneal macrophages form PGE_2 and 5-lipoxygenase products. It is not clear how the synthesis of these two types of products is coordinated. However, the arachidonate used by the two pathways may come from different pools. For example, macrophages release prostaglandins but not leukotrienes when treated with soluble stimuli, while insoluble, phagocytotic stimuli (for example, zymosan, bacteria) induce both prostaglandin and leukotriene synthesis.

Interestingly, 15-HETE, a product of the 15-lipoxygenase (which is present in PMNL) is a potent inhibitor of 5-lipoxygenase and platelet 12-lipoxygenase. Furthermore, the platelet lipoxygenase product 12-HPETE was found to activate the synthesis of 5-HETE and LTB_4 in human PMNL. Thus, interactions among lipoxygenase products might be important in regulating leukotriene synthesis.

Analysis of Leukotrienes and Other Lipoxygenase Products

Except for the highly polar peptido-leukotrienes (LTC_4, D_4, and E_4) and the unstable epoxy acid (LTA_4), lipoxygenase products can be extracted from biological media using classical organic solvent extraction procedures. LTC_4, D_4, and E_4 are extracted by adsorption on hydrophobic resins like Amberlite XAD or on octadecylsilylsilica particles.

As noted earlier, the conjugated-diene systems of HPETEs and HETEs are strong chromophores. Leukotrienes exhibit a characteristic spectral triplet with maxima around 260, 270, and 280 nm or 270, 280, and 290 nm (Figure 11.22). The spectral properties of lipoxygenase products have facilitated the use of high-performance liquid chromatography for the analysis of these compounds. Reversed-phase high-pressure liquid chromatography is widely used for the purification and analysis of hydroxy acids and the peptido-lipids LTC_4, LTD_4, and LTE_4. Recently, radioimmunoassays with sensitivities in the 10- to 100-pg range have been developed for LTC_4 and for LTB_4.

Biological Effects of Lipoxygenase Products

It has been recognized for a number of years that there is a connection between arachidonic acid metabolism and the formation of SRS-A. SRS-A was the name coined to describe an active principle released by lung tissue in response to immunological challenge. The main biological effect of SRS-A is contraction of respiratory tract smooth muscles. Studies performed in the mid-1970s by Borgeat and Samuelsson showed that leukocytes would convert arachidonate into 5-HETE as well as several 5,12-diHETE and 5,6-diHETE derivatives in

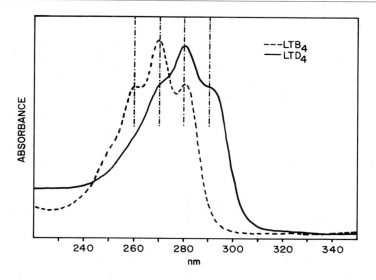

Figure 11.22. Ultraviolet absorption spectra of LTB₄ and LTD₄.

response to treatment with a Ca^{2+} ionophore. Earlier studies had already indicated that the Ca^{2+} ionophore would also elicit production of SRS-A from leukocytes. It now appears that SRS-A activity prepared from lung tissue consists of a mixture of LTC_4, LTD_4, and LTE_4, and it is now clear from pharmacological studies with purified leukotrienes that SRS-A activity can be expressed by LTC_4, LTD_4, and LTE_4 independently. As suggested more than 20 years ago in early studies on SRS-A, there is increasing evidence that leukotrienes are mediators of the bronchoconstriction associated with immediate hypersensitivity reactions such as asthma. These pathological responses to exposure to antigens may be due to increased sensitivity to or excessive production of sulfidopeptide leukotrienes induced by the antigens. Among other biological activities it is now well established that leukotrienes increase vascular permeability in skin and that the same compounds are potent constrictors of skin and lung microvasculature as well as coronary arteries.

Although the impetus for work on the leukotrienes was the challenge of determining the chemical nature of SRS-A, it is clear that another leukotriene with no SRS-A activity, namely, LTB_4, has important roles in nonimmunological inflammatory responses. LTB_4 indeed shows inflammatory properties. It is highly chemotactic toward PMNL, it causes aggregations of these cells, and it causes lysosomal enzyme release in vitro. The chemotactic activity of LTB_4 is also observed in many systems in vivo; interestingly, LTB_4 induces the adherence of PMNL to vascular endothelium and stimulates migration of PMNL into extravascular tissues. This chemotactic activity is not a property of sulfidopeptide leukotrienes.

The mechanism of action of leukotrienes at the molecular level has not been defined. Radioactive leukotrienes have been shown to bind membrane fractions from lung tissue and PMNL. Presumably, these binding activities represent binding to protein receptors. However, neither the properties of the receptors nor the response machinery to which these putative receptors are coupled has been defined.

PROBLEMS

1. When the urinary metabolites of PGI_2 and TxA_2 are measured in humans who have ingested a single 325-mg dose of aspirin, it is found that there is a rapid fall in each of these metabolites during the first 24 h after aspirin ingestion. However, the level of PGI_2 metabolites then returns to normal, while the level of TxA_2 metabolites requires 4–5 d to return to normal levels. Explain.
2. The study of LTA_4 has been facilitated by treatment of LTA_4 with diazomethane to form a methyl ester and by subsequent extraction of the methyl ester of LTA_4 into ether. Discuss the rationale for this extraction procedure.
3. Soybean lipoxygenase has been used as a tool for the structural identification of certain leukotrienes containing a cis methylene-interrupted double-bond system. What is the rationale for the use of this enzyme as a structural probe.
4. $PGF_{2\alpha}$ inhibits ADH-induced cAMP formation in renal collecting tubule cells, but $PGF_{2\alpha}$ does not activate adenylate cyclase. In contrast, PGE_2 both inhibits ADH-induced cAMP formation and (at higher concentrations) activates adenylate cyclase. Provide a possible explanation for these findings.

BIBLIOGRAPHY

Entries preceded by an asterisk are suggested for further reading.

*Frolich, J. C., ed. 1978. *Advances in prostaglandin and thromboxane research*, Vol. 5, New York: Raven Press.

Hamberg, M., and Samuelsson, B. 1967. On the mechanism of the biosynthesis of prostaglandins E_1 and $F_{1\alpha}$. *J. Biol. Chem.* 242:5336–43.

——— 1974. Prostaglandin endoperoxides. Novel transformations of arachidonic acid in human platelets. *Proc. Natl. Acad. Sci. USA* 71:3400–3404.

*Hammarstrom, S. 1983. Leukotrienes. *Ann. Rev. Biochem.* 52:355–77.

*Lands, W. E. M., and Smith, W. L., eds. 1982. *Methods in enzymology*, Vol. 86, New York: Academic Press.

Neufeld, E. J., and Majerus, P. W. 1983. Arachidonate release and phosphatidic acid turnover in stimulated human platelets. *J. Biol. Chem.* 258:2461–67.

*Pace-Asciak, C. R. and Smith, W. L. 1983. "Enzymes in the biosynthesis and catabolism of the eicosanoids: Prostaglandins, thromboxanes, leukotrienes and hydroxy fatty acids." In *The Enzymes*, Vol. XVI, ed. P. D. Boyer. 544–604. New York: Academic Press.

Rittenhouse-Simmons, S., and Deykin, D. 1981. "Release and metabolism of arachidonate in human platelets." In *Platelets in biology and pathology*, 2, ed. J. L. Gordon, 349–71, Amsterdam: Elsevier/North Holland.

*Samuelsson, B.; Goldyne, M.; Granström, E.; Hamberg, M.; Hammarström, S.; and Malmsten, C. 1978. Prostaglandins and thromboxanes. *Ann. Rev. Biochem.* 47:997–1029.

Stehle, R. 1982. Physical chemistry, stability, and handling of prostaglandins E_2, $F_{2\alpha}$, D_2 and I_2: A critical summary. *Methods in Enz.* 86:436–64.

Von Euler, U. S. 1967. Welcoming address. In *Prostaglandins: Nobel symposium 2*, ed. S. Bergstrom and B. Samuelsson, 17–20. New York: Interscience Publishers.

CHAPTER 12

Sphingolipids

Charles C. Sweeley

treatise
√&?

HISTORICAL DEVELOPMENTS Sphingolipids are nearly ubiquitous constituents of membranes in animals, plants, and some lower forms of life. Their initial discovery and partial characterization are attributed to Johann Ludwig Wilhelm Thudichum, a physician scientist in London. In a remarkable treatise on the chemical constitution of the brain (Thudichum 1962), first published in 1874, Thudichum described the systematic fractionation of human and bovine brain lipids and the recovery of several new substances. Among them were three related lipids, which he called *sphingomyelin*, *cerebroside*, and *cerebrosulphatide*. To characterize these lipids Thudichum degraded them with hot barium hydroxide or sulfuric acid and identified the products by their melting points and elemental composition.

$$\text{Cerebroside} \xrightarrow{\text{acid}} \text{fatty acid} + \text{sphingosine} + \text{cerebrose}$$

$$\downarrow \text{base}$$

$$\text{Psychosine} \xrightarrow{\text{acid}} \text{sphingosine} + \text{cerebrose}$$
$$+ \text{ fatty acid}$$

$$\text{Sphingomyelin} \xrightarrow{\text{acid}} \text{sphingosine} + \text{fatty acid} + \text{choline} + \text{phosphate}$$

Some of the substituents of these lipids were themselves novel, such as the extraordinary long-chain fatty acids which Thudichum named *lignoceric* and *cerebronic* acids, and the long-chain aliphatic amine which he found in all these lipids and called *sphingosine*.

The substantial problems involved in the isolation of brain sphingolipids, and the relative neglect of this class of lipids by most biochemists in the early twentieth century, prolonged the time required to establish firmly the complete structures of Thudichum's originally described sphingolipids. The correct chemical structures of sphingosine and related long-chain bases from animal and plant sphingolipids were not reported until the 1940s and 1950s. Although additional types have since been found, the major components of most sphingolipids are 4-sphingenine (sphingosine), 4-D-hydroxysphinganine (phytosphingosine), and sphinganine (dihydrosphingosine).

$$C18 \quad CH_3(CH_2)_{12}CH\!=\!\overset{4}{CH}\!-\!CH\!-\!\overset{2}{CH}\!-\!CH_2OH$$
$$\underset{OH}{|} \quad \underset{NH_2}{|}$$

4-Sphingenine

$$C18 \quad CH_3(CH_2)_{13}\overset{4}{CH}\!-\!CH\!-\!\overset{2}{CH}\!-\!CH_2OH$$
$$\underset{OH}{|} \quad \underset{OH}{|} \quad \underset{NH_2}{|}$$

4-D-Hydroxysphinganine

$$C18 \quad CH_3(CH_2)_{14}CH\!-\!\overset{2}{CH}\!-\!CH_2OH$$
$$\underset{OH}{|} \quad \underset{NH_2}{|}$$

Sphinganine

These bases have the D-erythro configuration at C-2 and C-3; the double bond in 4-sphingenine has the trans (E) configuration. Minor components are usually chain-length homologues or multiply unsaturated forms of these sphingoid bases.

Thudichum suspected that the sugar component (cerebrose) of cerebroside was milk sugar (lactose) but never proved it, and it was about 80 years before the structure of brain cerebroside was fully understood, including the anomeric configuration of the galactosyl residue and the composition of the fatty acid mixture.

A notable difference between sphingolipids and most other lipids is the amide linkage which covalently binds the long-chain bases and fatty acids in a substance referred to as *ceramide*. The complete name of brain cerebroside (Figure 12.1) is galactopyranosyl-(β1-1')-

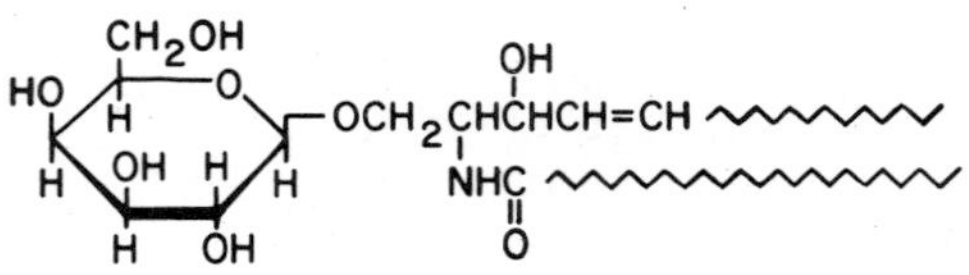

Figure 12.1. Galactosylceramide (cerebroside).

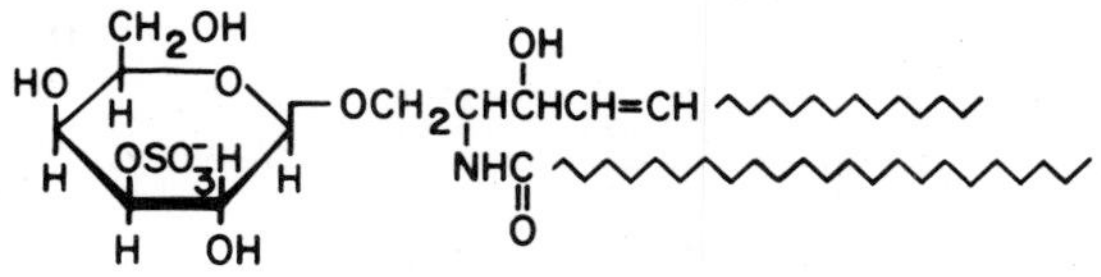

Figure 12.2.
3-Sulfogalactosylceramide
(sulfatide).

ceramide, currently referred to as galactosylceramide and abbreviated GalCer.

Cerebrosulfatide (Figure 12.2) is a sulfate ester form of GalCer. The sulfate group was shown to be on C-3 of the galactosyl residue by comparative resistance of the sugar to periodate oxidation and by permethylation analysis (to be discussed later).

Sphingomyelin is a phosphosphingolipid; it is structurally similar to phosphatidylcholine (Chapter 8) but has ceramide rather than 1,2-diacyl-*sn*-glycerol as the hydrophobic moiety.

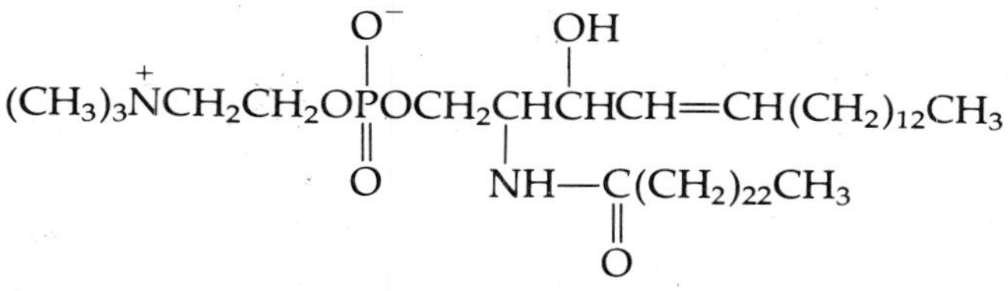

Sphingomyelin

CHEMISTRY AND DISTRIBUTION

The impetus to study the chemistry of sphingolipids and, ultimately, the mechanisms for their biosynthesis and metabolism was the discovery that several rare human diseases could be attributed to the occurrence of abnormal levels of various sphingolipids, resulting in pathological conditions and, in many instances, early death. In the course of studies of brain tissue from patients with Tay-Sachs disease, a new sphingolipid was discovered and named *ganglioside*. This sphingolipid gave a characteristic color reaction with orcinol and *p*-dimethylaminobenzaldehyde, due to a sugar substituent which was called *N-acetylneuraminic acid* (the same sugar was isolated from submaxillary mucins at about the same time and called *sialic acid*). Other kinds of ganglioside were subsequently found in normal brain and in various extraneural tissues and fluids.

Today many different gangliosides are known and are suspected to have fundamental roles in membrane phenomena such as cell-cell recognition, modulation of receptor activity, antigenic specificity, and perhaps even regulation of growth. Gangliosides are acidic glycosphingolipids because of the presence of free carboxyl groups on the sialic acid residues. The detailed chemical structures of the gangliosides from many species and tissues have been elucidated

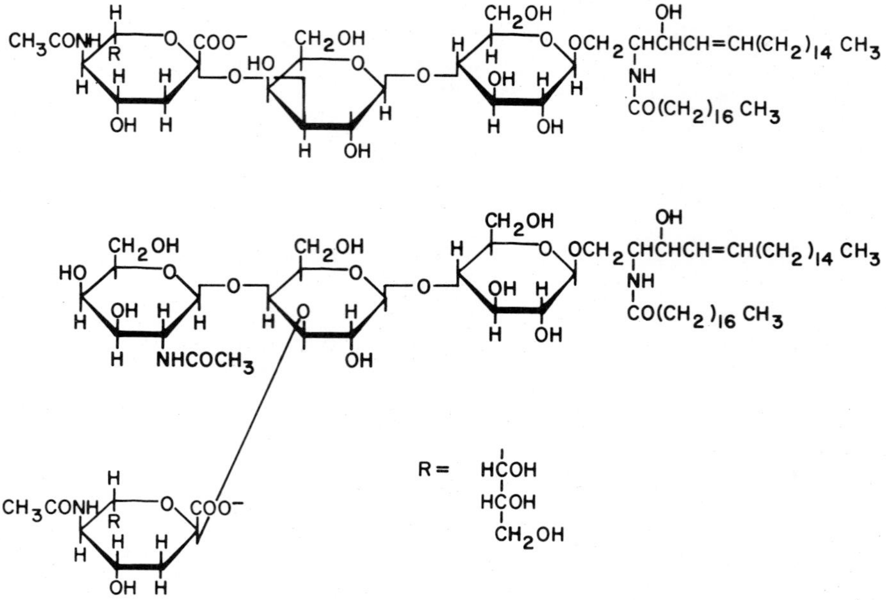

Figure 12.3. Structures of two brain gangliosides: hematoside (GM3) (top) and Tay-Sachs ganglioside (GM2) (bottom).

(Ledeen and Yu 1982). The structures of Tay-Sachs ganglioside (N-acetylgalactosylaminyl-(β1-4)-[N-acetylneuraminosyl-(α2-3)]-galactosyl-(β1-4)-glucosylceramide), which is generally referred to as GM2 ganglioside, and GM3 ganglioside (N-acetylneuraminosyl-(α2-3)-galactosyl-(β1-4)-glucosylceramide), are shown in Figure 12.3. Other gangliosides have more complex carbohydrate chains, with up to 20 or more sugar residues and with up to 5 sialic acid residues.

There is also a large number of carbohydrate-containing sphingolipids which do not possess sialic acid moieties and are therefore generally called *neutral glycosphingolipids* (Sweeley and Siddiqui 1977). GalCer is the simplest type within this broad class. A related cerebroside, GlcCer, containing glucose (Glc) instead of galactose, was first isolated from the spleens of patients with Gaucher's disease. The structure is isomeric with GalCer, the only difference being the configuration at C-4 of the sugar. GlcCer is generally found in extraneural tissues, whereas GalCer is predominant in the white matter of brain, the spinal cord, and, interestingly, kidney. *Globoside* (GalNAc(β1-3)Gal(α1-4)Gal(β1-4)-GlcCer, GbOse$_4$Cer) was originally isolated from erythrocytes. It can also be found in the plasma fraction of blood and in many extraneural tissues. The first evidence of a clear biosynthetic relationship between globoside and some other neutral glycosphingolipids resulted from the finding that kidney from patients with Fabry's disease contains abnormal amounts of a neutral glycosphingolipid shown to be Gal(α1-4)Gal(β1-4)GlcCer-(GbOse$_3$Cer), which differs from globoside only in the absence of a terminal N-acetylgalactosamine (GalNAc) residue.

One of the blood group A–active neutral glycosphingolipids, isolated from human erythrocytes, is illustrative of more-complex carbohydrate chains. The antigenic determinant is the terminal GalNAc(α1-3)[Fuc(α1-2)]Gal structure, which can also be found in other glycosphingolipids and in glycoproteins as well (Fuc = fucose).

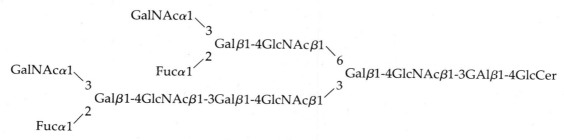

Blood group A glycosphingolipid

The glycosphingolipids from molluscs appear to be very different from those of higher animals. The following typical example shows the presence of two mannose (Man) residues and one xylose (Xyl) in the branched tetrasaccharide chain of a neutral glycosphingolipid.

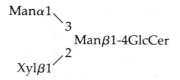

Glycosphingolipid from a mollusc

The glycosphingolipids of plants are usually distinguished by the presence of an inositol-phosphoceramide group. Mannose, glucuronic acid (GlcU), and other sugars may also be present in the carbohydrate chain. It is likely that the plant glycosphingolipids will be as complex in structural diversity as those of higher animals. A typical phytoglycosphingolipid, isolated from wheat flour, is shown in Figure 12.4. This lipid resembles sphingomyelin at its hydrophobic end, where ceramide is linked to phosphate. The carbohydrate chain possesses another acidic site, in the glucuronic acid residue, and may be similar to the gangliosides of animals in some of its functions.

The large number of glycosphingolipids discovered in recent years has caused a problem in the systematic naming of these substances.

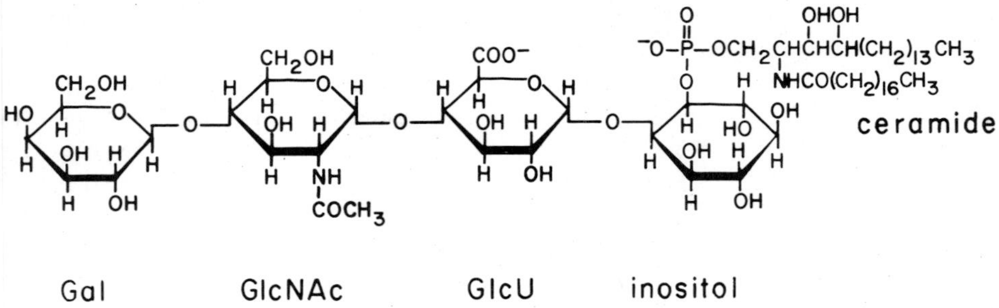

Figure 12.4. The structure of a plant phytoglycosphingolipid, Gal(β1-4)GlcNAc(α1-4)GlcU-(α1-2)-D-inositol-1-O-phosphoceramide.

The semisystematic names used in this chapter were proposed in 1977 by the IUPAC-IUB Commission on Biochemical Nomenclature (Nomenclature 1977). Essential features of the system are the classification of glycosphingolipids into structurally or biosynthetically related families and designation of abbreviations that indicate structural features on individual lipids. Brain gangliosides are abbreviated by the Svennerholm system of nomenclature (Nomenclature 1977), and complex neutral and acidic glycosphingolipids, by an extension of the recommended nomenclature.

Modern instrumental methods have greatly simplified the task of structural analysis and have enabled investigators to characterize glycosphingolipids that are present only in trace quantities in cells. Gas-liquid chromatography, mass spectrometry, and nuclear magnetic resonance spectroscopy provide sufficient information in many cases to elucidate a complete structure, even for complex glycosphingolipids with more than 10 sugar residues. The application of these analytical methods to the isolation and complete chemical characterization of GgOse$_3$Cer, a relatively simple neutral glycosphingolipid of the ganglio series with the structure GalNAc(β1-4)-Gal(β1-4)GlcCer, will illustrate the features of each technique.

Lipids of this kind can be recovered from tissues, fluids, or cultured cells by *Folch extraction*, using 19 volumes of chloroform-methanol (2:1, v/v) per volume of homogenized tissue. More complex (more polar) glycosphingolipids are more effectively extracted with 1:1 or 1:2 chloroform-methanol. After filtration or centrifugation, the mixture is partitioned into two phases by the addition of water. Crude lipids in the lower solvent phase can be fractionated by silicic acid column chromatography or thin-layer chromatography, depending on the amount of tissue, and the glycosphingolipid of interest further purified by thin-layer chromatography. Methanolysis of as little as 50 μg of glycosphingolipid in 0.75N dry methanolic HCl at 80°C and gas-liquid chromatography of the trimethylsilylated methyl

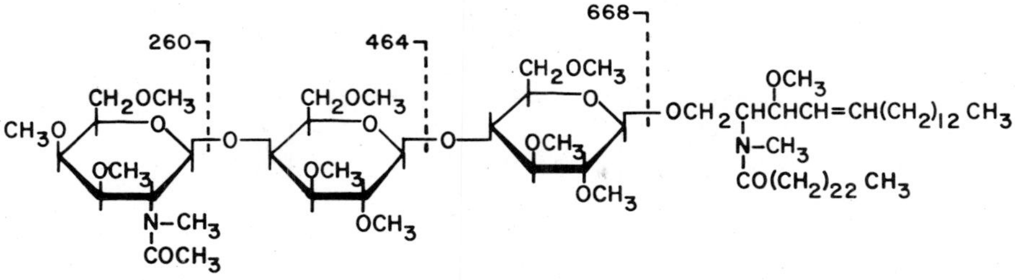

Figure 12.5. Fragment ions in the electron-impact mass spectrum of permethylated and reduced GalNAc(β1-4)Gal(β1-4)GlcCer (GgOse$_3$Cer). Details of the fragmentation pathways of permethylated glycosphingolipids have been published (Karlsson 1980).

glycosides will establish the presence in GgOse$_3$Cer of equimolar amounts of galactosamine, galactose, glucose, 4-sphingenine, and fatty acid.

The positions of attachment of the sugars can be established by stepwise hydrolysis of the glycosphingolipid with exoglycosidases of known specificity or by permethylation analysis. For the latter approach, about 25 μg of the completely permethylated lipid, shown in Figure 12.5, are sufficient to give a series of fragment ions in the electron-impact mass spectrum, indicating the arrangement of the sugar residues but not the positions of the glycosidic linkages nor their anomeric configuration. The arrangement GalNAc-Gal-Glc is determined by the fragment ions at m/z 260, 464, and 668 and a companion series, resulting from the loss of methanol, at m/z 228, 432, and 636. Molecular ions are usually not apparent in the mass spectra of permethyl derivatives.

The procedure for analysis of the glycosidic linkages involves hydrolysis of the permethylated compound followed by reduction of the partially methylated sugars with NaBH$_4$ or NaBD$_4$ and acetylation of the resulting alditols; these derivatives (Figure 12.6) are used to determine the glycosidic linkages by combined gas chromatography–mass spectrometry. As shown in Figure 12.6, fragment ions with charge localization on carbon atoms bearing OCH$_3$ groups occur preferentially (Lindberg and Lonngren 1978). The masses of these ions indicate how many acetyl and methyl groups are present. On this basis, the presence of m/z 158, 161, and 202 in Figure 12.6A point to the location of the GalNAc residue at the terminal (nonreducing) end of the oligosaccharide. The ions at m/z 161 and 233 in Figures 12.6B and 12.6C indicate 1,4-glycosidic linkages to the Gal and Glc residues. Isomeric structures, such as the Gal (Figure 12.6B) and Glc (Figure 12.6C) alditol derivatives, give the same mass spectra, but are usually separated by gas chromatography and can therefore be distinguished by their retention times.

$$
\begin{array}{ccc}
\text{CH}_2\text{OCOCH}_3 & \text{CH}_2\text{OCOCH}_3 & \text{CH}_2\text{OCOCH}_3 \\
158\ \ \text{HCN}\!\!<\!\!\substack{\text{CH}_3\\ \text{COCH}_3} & \text{HCOCH}_3\quad 117 & \text{HCOCH}_3\quad 117 \\
202\ \ \text{CH}_3\text{OCH} & 161\ \ \text{CH}_3\text{OCH}\quad 233 & 161\ \ \text{CH}_3\text{OCH}\quad 233 \\
161\ \ \text{CH}_3\text{OCH} & \text{CH}_3\text{COOCH} & \text{HCOCOCH}_3 \\
\text{HCOCOCH}_3 & \text{HCOCOCH}_3 & \text{HCOCOCH}_3 \\
\text{CH}_2\text{OCH}_3 & \text{CH}_2\text{OCH}_3 & \text{CH}_2\text{OCH}_3 \\
\\
\text{A} & \text{B} & \text{C}
\end{array}
$$

Figure 12.6. Partially methylated alditol acetates derived from permethylated GalNAc(β1-4)Gal(β1-4)GlcCer (Figure 12.5) by hydrolysis, sodium borohydride reduction, and acetylation. Compound A is from the terminal GalNAc residue; compound B is from the internal Gal, substituted in the glycolipid at C-4; and compound C is from the internal Glc, substituted in the glycolipid at C-4.

Only the size (furanose versus pyranose) and conformation of the sugar rings and their anomeric configurations remain to be determined. Proton nuclear magnetic resonance spectroscopy (^{1}H-NMR) yields this information on very small samples. For example, the chemical shifts of the anomeric protons have been established for many different glycosphingolipids (Dabrowski, Hanfland, and Egge 1982). In the solvent dimethylsulfoxide-d_6 (perdeutero-labeled), the anomeric proton region in a 360- or 500-MHz ^{1}H NMR spectrum of underivatized GgOse$_3$Cer would give doublet signals at about 4.10–4.17 ppm (chemical shift from an internal standard) for the H-1 proton of GlcCer, at 4.22–4.27 ppm for the H-1 proton of Gal-β1-4-, and at 4.45–4.49 ppm for the H-1 proton of GalNAc-β1-4-. The observed coupling constants ($J_{1,2}$) of 7.8, 7.8, and 8.4 Hz, respectively, indicate that all three anomeric linkages have the β-configuration. The $^4C_1$ chair conformation can be assumed for all three sugars on the basis of the chemical shifts of these and other protons, as shown in Figure 12.7.

In contrast with the steps given for the analysis of the glycosidic linkages, the determination of the partial structure of GgOse$_3$Cer might be even more readily accomplished by a recent adaptation of

Figure 12.7. High-resolution ^{1}H-NMR spectrum of GalNAc(β1-4)Gal(β1-4)GlcCer (GgOse$_3$Cer). Signals of anomeric protons of GalNAc (III), Gal (II), and Glc (I) are labeled III-1, II-1, and I-1, respectively; the coupling constants of these doublets indicate that all three anomeric linkages have the β-configuration. Analysis of the remainder of the ^{1}H-NMR spectrum was facilitated by a technique called two-dimensional spin-echo correlation spectroscopy (SECSY); details about how the peaks were correlated with structure (involving the graphic display at the bottom) have been published (Prestegard et al. 1982). Permission to reproduce this figure was kindly provided by the authors and publisher.

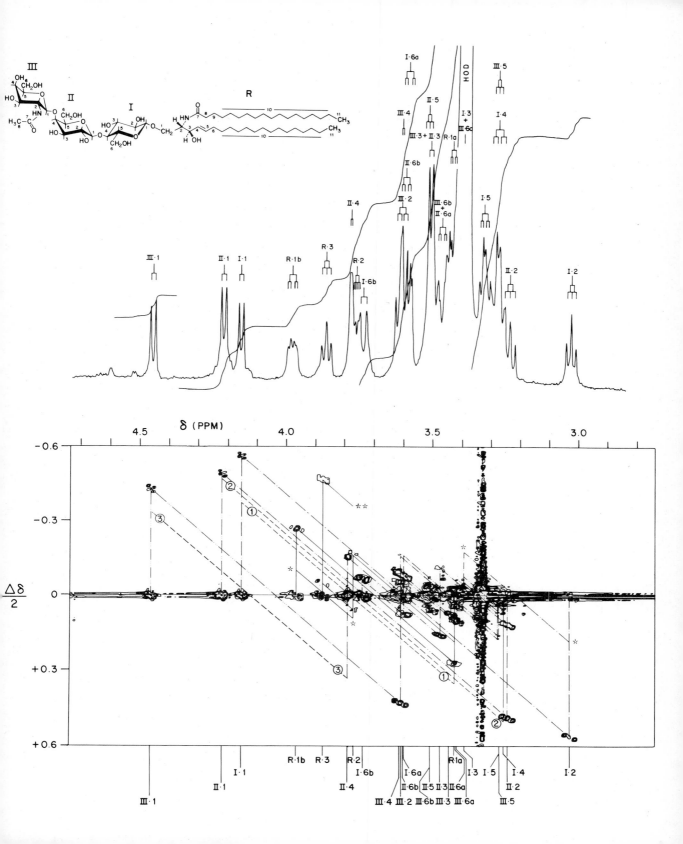

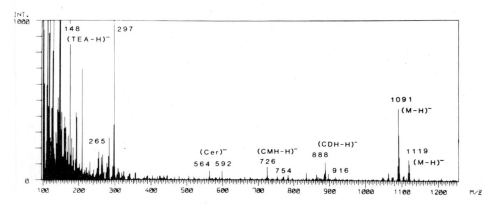

Figure 12.8. Negative-ion fast-atom bombardment (FAB) mass spectrum of underivatized GalNAc(β1-4)Gal(β1-4)GlcCer (GgOse$_3$Cer). Permission to reproduce this figure was kindly provided by the authors and publisher (Arita et al. 1983).

mass spectrometry called fast-atom bombardment (FAB) mass spectrometry (Arita et al. 1983). Figure 12.8 shows the negative-ion mass spectrum of the underivatized glycosphingolipid, ionized from a matrix of triethanolamine and tetramethylurea by a highly energetic beam of xenon atoms. The molecular weight is determined from ions at m/z 1091 and 1119, which result from the loss of a proton and negative-charge retention. The occurrence of fragment ions from the ceramide residue at m/z 564 and 592 indicates that the long-chain base consists of approximately equal proportions of 4-sphingenine (18:1) and eicosa-4-sphingenine (20:1) linked to stearic acid, which explains the two (M-H)$^-$ ions that are 28 mass units apart. The difference between the ion pair at m/z 564 and 592 (Figure 12.9) and the next prominent set at m/z 726 and 754 (Δ = 162 mass units) indicates that a hexose residue is attached to the ceramide. The difference to the next pair at m/z 888 and 916 (Δ = 162 mass units) indicates attachment of a second hexose residue, and the increase in mass to the molecular ion region, (M-H)$^-$, at m/z 1091 and 1119

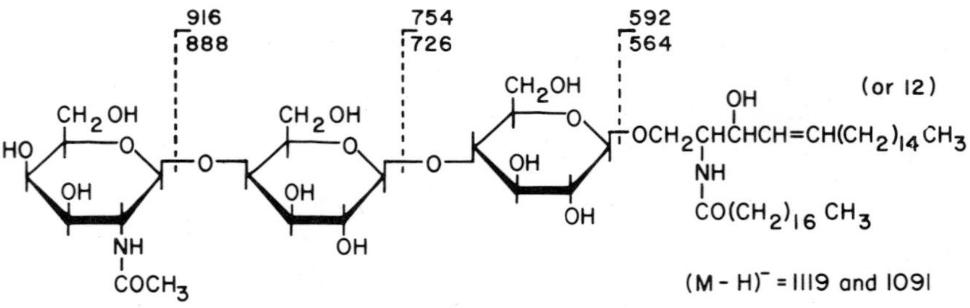

Figure 12.9. Fragment ions in the negative-ion fast-atom bombardment (FAB) mass spectrum of GgOse$_3$Cer.

(Δ = 203 mass units) indicates a terminal *N*-acetylhexosamine. FAB mass spectrometry affords molecular weight and sequence information of moderately complex glycosphingolipids, is quicker than permethylation analysis, and consumes less sample.

SUBCELLULAR LOCALIZATION

Sphingolipids are amphipathic substances (Figure 12.10). The water solubility of various sphingolipids ranges from nearly complete insolubility (sphingomyelin and galactosylceramide) to the opposite extreme, in which complex neutral glycosphingolipids and gangliosides are ''soluble'' in the form of aggregates or micelles. In cells the sphingolipids occur predominantly in bilayer lipid membranes. The highest concentration is in the plasma membrane of most eucaryotic cells, where there may be an asymmetric distribution, especially of the glycosphingolipids that appear to be in the outer leaflet of the bilayer.

Since the sphingolipids are synthesized in the subcellular compartments where membranes are assembled, the endoplasmic reticulum,

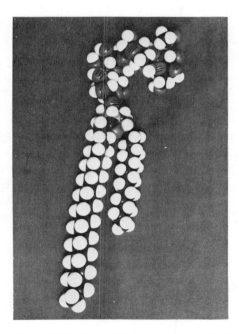

Figure 12.10. Space-filling molecular model of globoside (GbOse$_4$Cer), illustrating a cleft (bay area) in the carbohydrate chain and different acyl chain lengths of 4-sphingenine and lignoceric acid groups of the hydrophobic ceramide moiety. Yamakawa has described early work on the isolation and characterization of globoside from erythrocytes (Yamakawa 1982).

and Golgi apparatus, low levels would be expected to occur in the membranes of these organelles as well. Experimental studies of the lipid composition of different cell membranes support this view. Nothing is known, however, about the distribution of various types of sphingolipid on the cytosolic and luminal aspects of these bilayer membranes. Some predictions can be made from a biosynthetic standpoint (see next section), but they have not been experimentally confirmed.

BIOSYNTHESIS OF SPHINGOLIPIDS

Enzymes involved in the synthesis of sphingolipids are membrane-bound or integral membrane proteins. As such, they are hydrophobic and often very unstable in a membrane-free environment. This factor and the requirement for detergents to solubilize most of the substrates have complicated studies of the steps involved and the kinetics of individual reactions in cell-free systems. Nevertheless, the general pathways of sphingolipid biosynthesis are relatively well understood, and the enzymes catalyzing many of the individual steps have been at least partially characterized (Kishimoto 1983). In the remainder of this chapter, emphasis will be on the relationship of the synthetic pathways to the problem of translocation of sphingolipids from the site of their synthesis to the plasma membrane. Additionally, several aspects of regulation, about which relatively little is known, will be considered.

Sphingosine and Related Long-Chain Bases

The process of sphingolipid biosynthesis begins in the microsomes (endoplasmic reticulum), where the sphingoid bases are assembled. Early in vivo studies with various isotope-labeled compounds provided good evidence that these long-chain bases are synthesized from palmitic acid and serine. It was later shown in liver homogenates and microsomal fractions of the yeast *Hansenula ciferrii* that the first step involves a pyridoxal phosphate–requiring condensation of palmitoyl-CoA and serine with loss of the carboxyl group of the serine. The product of this reaction, 3-oxosphinganine, is a precursor of the

$$CH_3(CH_2)_{14}-\overset{\overset{\displaystyle O}{\|}}{C}-CH-CH_2OH + CoASH + CO_2$$

3-Oxosphinganine

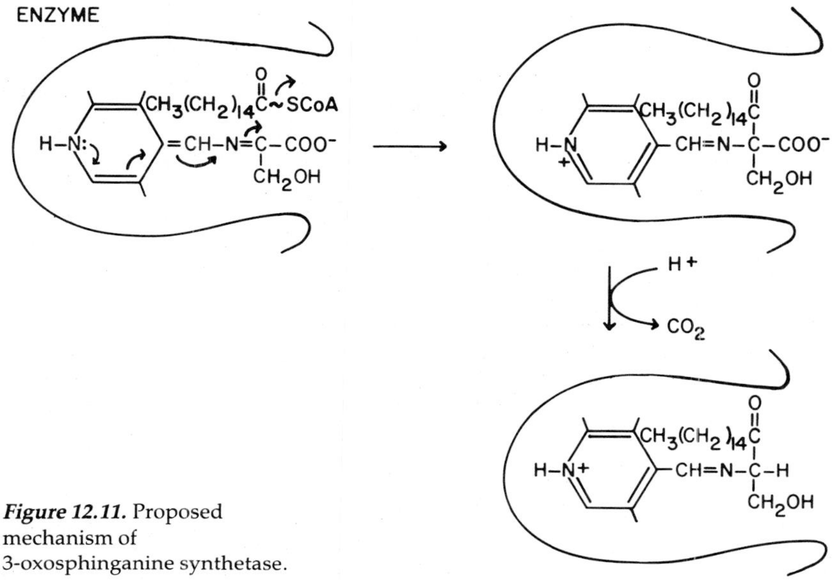

Figure 12.11. Proposed mechanism of 3-oxosphinganine synthetase.

three most common C_{18} bases, sphinganine, 4-sphingenine, and 4-hydroxysphinganine. The mechanism of the synthetase-catalyzed reaction was clarified by the finding that the α-hydrogen of serine is eliminated in the reaction and replaced by a proton from water. This step might therefore be pictured as shown in Figure 12.11, where serine has reacted with the pyridoxal phosphate moiety prior to the condensation step, and the α-hydrogen is lost during formation of the active imine form of the complex. The chirality (S) of the asymmetric carbon atom of serine is retained in the 3-oxosphinganine. L-Cycloserine has been shown to be a potent inhibitor of the 3-oxosphinganine synthetase from *Bacteroides levii* as well as from a crude brain microsomal preparation.

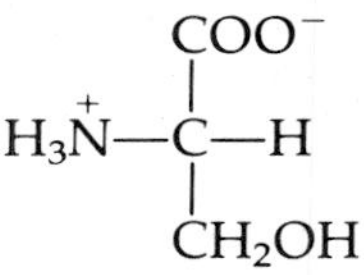

L-Serine 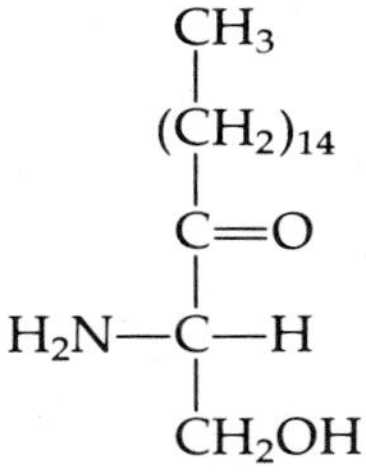   (2S)-3-Oxosphinganine

The synthesis of sphinganine and other long-chain bases from 3-oxosphinganine requires NADPH and a reductase which is probably closely associated with the synthetase in the membrane. Only the D (2S) isomer of 3-oxosphinganine is a substrate for this enzyme, which involves transfer of the α-hydrogen of the reduced pyridine nucleotide. The product, sphinganine, has the 2S,3R configuration, as shown.

$CH_3(CH_2)_{14}$C ... CH_2OH / C / H, NH_2 + H$^+$ + NADPH ⟶ $CH_3(CH_2)_{14}$ ... CH_2OH

(2S)-3-Oxosphinganine   **Sphinganine**

Most sphingolipids contain predominantly sphingoid bases with a trans double bond, as in 4-sphingenine. This base can be formed from sphinganine by a dehydrogenation catalyzed by a microsomal flavo-protein (fp). The mechanism of the reaction has not been studied, and the substrate specificity is uncertain. There is evidence that the double bond might be inserted into the aliphatic chain prior to synthesis of the 3-oxo intermediate in some cases; other studies have suggested that ceramide is the preferred substrate.

$CH_3(CH_2)_{12}$ ... CH_2 ... CH_2 ... CH_2OH / NH_2 →(fp) $CH_3(CH_2)_{12}$ ... CH_2OH

Sphinganine   **4-Sphingenine**

Chain-length homologues of sphinganine and 4-sphingenine, such as a C_{14} sphingoid base and the C_{20} sphingoid base of brain sphingolipids, are assumed to be synthesized in an analogous manner. However, the synthesis of 4-hydroxysphinganine involves an additional enzyme; the mechanism of the reaction has been studied in some detail in yeast. The major pathway for the introduction of the hydroxyl group at C-4 of 4-hydroxysphinganine is by hydroxylation of sphinganine. There is displacement of the pro-R hydrogen at C-4 of the substrate, with retention of configuration in the product, and the reaction involves molecular oxygen.

$CH_3(CH_2)_{13}$ ... CH_2OH / S, R / NH_2 →(O_2) $CH_3(CH_2)_{13}$ ... CH_2OH / OH, NH_2

Sphinganine   **4-Hydroxysphinganine**

Two laboratories have shown that a small incorporation of ^{18}O from $H_2^{18}O$ occurs consistently in yeast. They have suggested that there is an alternate (minor) mechanism for the hydroxylation of sphinganine that could involve a hydratase-catalyzed addition of water to 4-sphingenine, but they have not published details of the reaction. Efforts to clarify the mechanism of 4-hydroxysphinganine synthesis are hampered by the failure to retain any enzyme activity in cell-free homogenates of the yeast.

Ceramide   Four kinds of fatty acid occur in sphingolipids, namely, (1) the saturated long-chain fatty acids that are the products of the fatty acid synthase complex of most organisms: myristic (14:0), palmitic (16:0), and stearic (18:0) acids; (2) very long chain fatty acids formed by a microsomal chain-elongation system, such as behenic (22:0) and lignoceric (24:0) acids; (3) monoenoic fatty acids, such as oleic (18:1(n–9)), and nervonic (24:1(n–9)) acids; and (4) α-hydroxy very long chain fatty acids, such as cerebronic (h24:0) and oxynervonic (h24:1) acids. Although linoleic acid (18:2) is found as a minor component of some sphingolipids, polyunsaturated fatty acids are conspicuously absent from these lipids. The very long chain fatty acids and their α-hydroxy derivatives appear to occur exclusively in sphingolipids, suggesting that there are mechanisms of compartmentalizing the pools of these fatty acids, or that chain elongation and α-hydroxylation occur after the regular chain length fatty acids have been introduced into a sphingolipid or some specialized precursor. Further consideration of this aspect of sphingolipid biosynthesis is beyond the scope of this chapter. Additional details on chain elongation and desaturation can be found in Chapter 6. There is also an excellent review of sphingoid base and ceramide biosynthesis (Kishimoto 1983).

Several different pathways have been proposed for the biosynthesis of ceramide. The two most-studied reactions involve the transfer of either free fatty acid or an acyl-CoA intermediate to the sphingoid base (shown on p. 376). The synthesis from free fatty acids is catalyzed by a hydrolase and is the reverse of the catabolism of ceramide, whereas the second reaction is catalyzed by an acyltransferase. The reaction involving free fatty acids has been demonstrated in vitro and may be of biological significance.

Most of the work on the mechanism of ceramide synthesis has been carried out with rat brain microsomal fractions. Studies of the acyltransferase activity of crude microsomal fractions indicate that there are at least four different enzymes, each with a different specificity for the chain length of the fatty acid and the presence or absence of an α-hydroxyl group. The tissue distribution of these

$$\text{RCOOH} + \text{CH}_3(\text{CH}_2)_{14}\underset{\underset{\text{OH}}{|}}{\text{CH}}-\underset{\underset{\text{NH}_2}{|}}{\text{CH}}-\text{CH}_2\text{OH} \longrightarrow \text{CH}_3(\text{CH}_2)_{14}\underset{\underset{\text{OH}}{|}}{\text{CH}}-\underset{\underset{\underset{\underset{R}{|}}{\underset{C=O}{|}}}{\underset{\text{NH}}{|}}}{\text{CH}}-\text{CH}_2\text{OH} + \text{H}_2\text{O}$$

Fatty acid

Sphinganine

Ceramide

$$\text{RCO—SCoA} + \text{CH}_3(\text{CH}_2)_{14}\underset{\underset{\text{OH}}{|}}{\text{CH}}-\underset{\underset{\text{NH}_2}{|}}{\text{CH}}-\text{CH}_2\text{OH} \longrightarrow$$

Fatty acyl-CoA

Sphinganine

$$\text{CH}_3(\text{CH}_2)_{14}\underset{\underset{\text{OH}}{|}}{\text{CH}}-\underset{\underset{\underset{\underset{R}{|}}{\underset{C=O}{|}}}{\underset{\text{NH}}{|}}}{\text{CH}}-\text{CH}_2\text{OH} + \text{CoASH}$$

Ceramide

enzymes is probably different, since the levels of glycosphingolipids containing α-hydroxy fatty acids vary greatly from one tissue to another. Curiously, there does not seem to be an absolute specificity for the long-chain base, as both erythro and threo isomers can be incorporated into ceramide. However, glycosphingolipids that contain α-hydroxy fatty acids generally have 4-hydroxysphinganine as a major sphingoid component.

There are also odd-chain-length fatty acids in most sphingolipids, particularly the C_{23} homologue. These constituents are assumed to be formed by the oxidative decarboxylation of α-hydroxy fatty acids, but the reaction has not been studied in detail. In this connection, there is also some evidence for a second CoA-independent pathway of ceramide synthesis, involving an unknown activated form of fatty acids that may also be a pivotal intermediate in α-hydroxylation.

The enzymes involved in the synthesis of long-chain bases and ceramide can be assumed to be localized with their active sites on the cytosolic aspect of the endoplasmic reticulum. There is as yet no direct evidence in support of this assumption, but the limited permeation of palmitoyl-CoA into the microsomal lumen and the fact that the enzymes of sphingoid base and ceramide synthesis utilize acyl-CoA as a substrate argue strongly for this kind of topography. This is an important consideration, since subsequent steps in the synthesis of glycosphingolipids and the asymmetric localization of these substances on the extracellular aspect of the plasma membrane infer transfer of an unknown intermediate (perhaps ceramide) from the cytosolic to the luminal side of the endoplasmic reticulum or Golgi apparatus.

Phosphosphingolipids   Crude membrane fractions from liver or brain catalyze the transfer of phosphorylcholine from CDP-choline to ceramide. This reaction is analogous to the synthesis of phosphatidylcholine (Chapter 8).

$$\text{Ceramide + CDP-choline} \longrightarrow \text{sphingomyelin + CMP}$$

The exclusive stereospecificity of this enzyme for ceramide containing the threo isomer of 4-sphingenine (the natural isomer is erythro) suggests that this may not be the principal mechanism involved in the biosynthesis of sphingomyelin. An alternative pathway is supported by several groups who have shown that phosphatidylcholine can also be a donor of the phosphorylcholine moiety of sphingomyelin.

$$\text{Ceramide + phosphatidylcholine} \longrightarrow \text{sphingomyelin + acylglycerol}$$

The exact mechanism of the transfer reaction is not known, but ceramide containing the naturally predominant erythro form of the sphingoid base is the preferred substrate. Pulse-chase studies with tritium-labeled choline support this pathway. There is a precursor-product relationship between phosphatidylcholine and sphingomyelin. Phosphatidylcholine is labeled initially, after which the choline label disappears from phosphatidylcholine and is incorporated into sphingomyelin.

The spatial relationship in the membrane of ceramide:phosphatidylcholine phosphocholinetransferase and CDP-choline:diacylglycerol phosphocholinetransferase is not known. However, a previous step in the synthesis of phosphatidylcholine is catalyzed by the enzyme CTP:phosphocholine cytidylyltransferase, which is present in both cytosol and endoplasmic reticulum but is active only when bound to the membrane (Chapter 8). Since this is the regulated step of phosphatidylcholine synthesis, it would be interesting to know the extent to which sphingomyelin and phosphatidylcholine are coordinately regulated by the activity of the cytidylyltransferase.

Nothing is known about the biosynthesis of plant phosphosphingolipids. These glycolipids contain phosphorylceramide, as does sphingomyelin, and the pathway for the incorporation of phosphate might mimic the transfer of phosphocholine from phosphatidylcholine to sphingomyelin. In the case of plants, this pathway would involve the initial synthesis of phosphatidylinositol from CDP-diacylglycerol and inositol (Chapter 8) followed by the transfer of phosphoinositol to ceramide. This pathway has also been proposed for the phosphoinositol-containing sphingolipids of yeast.

Subsequent steps in the plant (the incorporation of mannose and glucuronic acid) would presumably be similar to the reactions of mammalian glycosphingolipid synthesis (see next section).

Sulfoglyco-sphingolipids

Sulfoglycosphingolipids are relatively more abundant in the central nervous system, kidney, gastrointestinal tract, and testis than in other mammalian tissues. The synthesis of 3'-sulfogalactosyl-ceramide (sulfatide) involves sulfation of GalCer by an activated sulfate donor, 3'-phosphoadenosine-5'-phosphosulfate (PAPS), which contains a mixed anhydride of phosphoric and sulfuric acids.

$$\text{GalCer} + \text{PAPS} \longrightarrow 3\text{-SO}_3\text{-GalCer} + 3'\text{-P-AMP}$$

The reaction is catalyzed by galactosylceramide sulfotransferase, the level of which appears to be more important than that of PAPS in regulation of sulfatide biosynthesis. Sulfate transfer is preceded by receptor-mediated translocation of PAPS from the cytosol across the Golgi membrane, a process that can be inhibited by 3'-P-AMP and palmitoyl-CoA, both of which have a diphosphorylated ribose moiety similar to that found in PAPS (Capasso and Hirschberg 1984a). Interestingly, transport of PAPS across the Golgi membrane is also inhibited by atractyloside, a disulfated C_{30} glucoside from *Atractylis gummifera* L. that has strychnine-like toxic properties and is a known inhibitor of the ATP–ADP translocase of mitochondria. The synthesis of more complex sulfoglycosphingolipids occurs by similar reactions.

Neutral Glycosphingolipids

Galactosylceramide and glucosylceramide are synthesized by the transfer of galactose or glucose, respectively, from the appropriate sugar nucleotide to ceramide. The reaction is catalyzed by a glycosyl-transferase and requires a divalent cation (Mn^{2+} is usually most effective). The sugar nucleotides are synthesized in the cytosol from uridine triphosphate (UTP) and hexose-1-monophosphate, as illustrated in Figure 12.12 for the glucose derivative. Uridine diphosphate glucose (UDP-Glc) was discovered by L. Leloir as an activated glucose intermediate in the biosynthesis of glycogen (he received the Nobel Prize in 1970 for this work). The formation of UDP-Glc is catalyzed by UDP-Glc pyrophosphorylase and is typical of the reactions involved in the synthesis of other sugar nucleotides such as UDP-Gal, UDP-GlcNAc, and GDP-Man. Glucosylceramide synthesis requires the direct transfer of the glucose residue from UDP-Glc to ceramide, with an inversion of the configuration of the glycosidic bond (α to β). This reaction (Figure 12.13), catalyzed by UDP-Glc:ceramide glucosyltransferase, is a model for the biosynthetic steps involved in the synthesis of more complex glycosphingolipids.

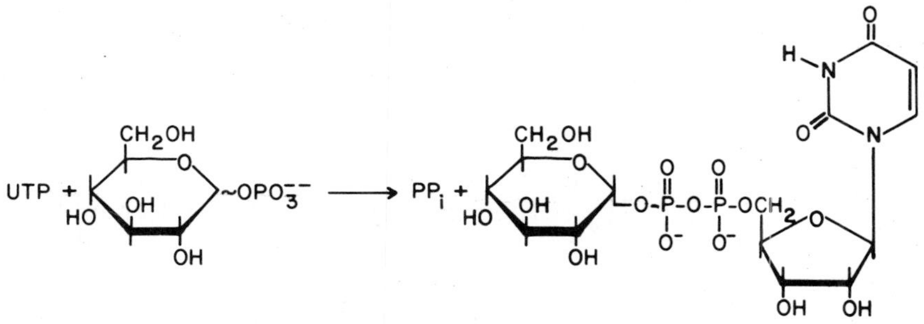

Figure 12.12. Biosynthesis of uridine diphosphate glucose (UDP-Glc).

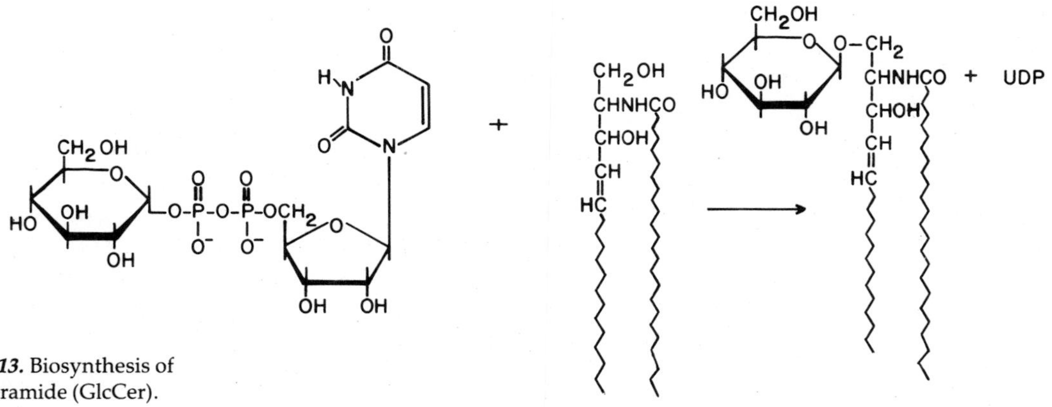

Figure 12.13. Biosynthesis of glucosylceramide (GlcCer).

In all cases, there is a direct transfer of a single sugar from the appropriate sugar nucleotide to the nonreducing end of the growing carbohydrate chain attached to ceramide.

Glucosylceramide and galactosylceramide are synthesized by crude microsomal fractions. Both the Golgi apparatus and endoplasmic reticulum have been reported to contain transferase activity, and the enzyme for galactosylceramide synthesis has also been found in myelin. In brain, microsomes also catalyze the transfer of galactose from UDP-Gal to a sphingoid base to form psychosine (galactosyl-(β1-1')-sphinganine), which is subsequently converted to GalCer by a condensation reaction with acyl-CoA. The relative importance of the two pathways for GalCer synthesis remains to be clarified.

The translocation of newly synthesized galactosylceramide from the microsomal membrane (endoplasmic reticulum) to myelin may involve a cytosolic carrier protein, which has been isolated from brain and other organs such as liver and spleen. Topologically, the galactosylceramide is presumably localized on the cytosolic aspect of the microsomal lipid bilayer membrane in order to be accessible to the transfer protein. This suggests that the active site of the UDP-Gal:ceramide galactosyltransferase is also cytosolic in its active conformation. Thus, it seems likely that the entire set of enzymes involved in the synthesis of sphingoid bases, ceramides, and cerebrosides is localized on the cytosolic side of the endoplasmic reticulum, perhaps as a functional multienzyme complex. Definitive experiments to prove the orientation of these enzymes have not been reported.

The fatty acids of galactosylceramide and glucosylceramide are quite different in composition. The galactosylceramide of myelin contains a substantial amount of α-hydroxy fatty acids, the total proportion of which varies at different stages of development (myelination). On the other hand, glucosylceramide contains little, if any, α-hydroxy fatty acid, which may reflect the different cellular localizations of the two cerebrosides in the brain—galactosylceramide is made in oligodendroglia (cells responsible for myelin synthesis), whereas glucosylceramide is predominantly of neuronal origin. It has also been proposed that the fatty acid chain length and presence or absence of α-hydroxyl groups may be involved as determinants of specificity for various glycosyltransferases.

Lactosylceramide (Gal(β1-4)GlcCer) is a common precursor of nearly all the neutral glycosphingolipids and gangliosides of higher animals. The enzyme that catalyzes the synthesis of lactosylceramide is UDP-Gal:glucosylceramide galactosyltransferase. It is believed to be localized in Golgi membranes although the activity in rat liver Golgi-enriched fractions is very low.

$$\text{UDPGal} + \text{GlcCer} \xrightarrow{\text{Mn}^{2+}} \text{UDP} + \text{Gal}(\beta\text{1-4})\text{GlcCer}$$
$$\text{(LacCer)}$$

The kinetics of the reaction are complicated by the requirement for a detergent to solubilize the glucosylceramide and to stabilize the enzyme in membrane-solubilized preparations. Much remains to be learned about the regulation of lactosylceramide synthesis; such studies will be dependent upon the successful purification of the enzyme and its reconstitution in a unilamellar vesicle.

Subsequent steps catalyzed by glycosyltransferases lead to a large number of neutral glycosphingolipids which can be grouped into several pathways. These glycosphingolipids are synthesized by a strictly ordered sequence of transfers of sugars from sugar nucleotides to glycolipid acceptors. The synthesis of a pentaglycosylceramide (Gb5b) with Forssman-positive antigenic activity and a blood group B–active hexaglycosylceramide (Lc6c) relates two groups of glycosphingolipids, called the *globo* (Gb) and *lacto* (Lc) series, respectively, according to their precursor-product relationships. These pathways are illustrated in Figures 12.14 and 12.15. The enzymes have all been partially characterized, primarily with respect to substrate specificity,

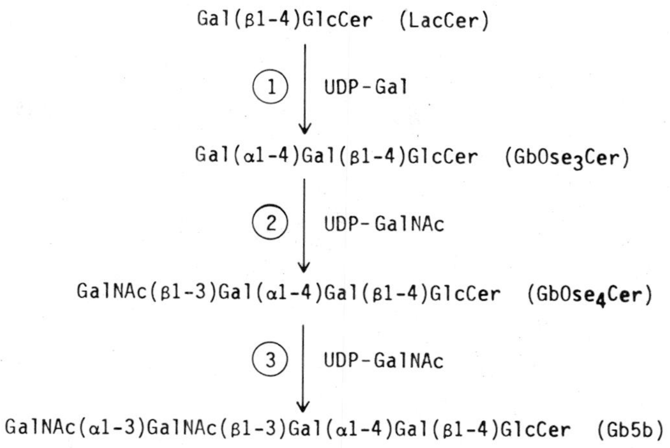

Gal(β1-4)GlcCer (LacCer)
1 UDP-Gal
Gal(α1-4)Gal(β1-4)GlcCer (GbOse3Cer)
2 UDP-GalNAc
GalNAc(β1-3)Gal(α1-4)Gal(β1-4)GlcCer (GbOse4Cer)
3 UDP-GalNAc
GalNAc(α1-3)GalNAc(β1-3)Gal(α1-4)Gal(β1-4)GlcCer (Gb5b)

Figure 12.14. Biosynthesis of Forssman-active pentaglycosylceramide (Gb5b) from lactosylceramide (LacCer), catalyzed by stepwise additions of galactose (Gal) and *N*-acetylgalactosamine (GalNAc) with membrane-bound galactosyltransferase (1) and *N*-acetylgalactosaminyltransferases (2 and 3).

Gal(β1-4)GlcCer (LacCer)
1 UDP-GlcNAc
GlcNAc(β1-3)Gal(β1-4)GlcCer (Lc3a)
2 UDP-Gal
Gal(β1-4)GlcNAc(β1-3)Gal(β1-4)GlcCer (Lc4a)
3 UDP-Gal
4 GDP-Fuc
Gal(α1-3)Gal(β1-4)GlcNAc(β1-3)Gal(β1-4)GlcCer
(Lc5c)
Gal(β1-4)GlcNAc(β1-3)Gal(β1-4)GlcCer
Fuc(α1-2)
(Lc5e)
GDP-Fuc 5
6 UDP-Gal
Gal(α1-3)Gal(β1-4)GlcNAc(β1-3)Gal(β1-4)GlcCer (Lc6c)
Fuc(α1-2)

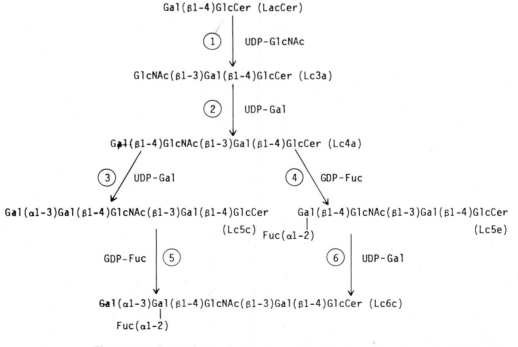

Figure 12.15. Biosynthesis of a blood group B–active hexaglycosylceramide (Lc6c) of the lacto series from lactosylceramide (LacCer) by stepwise additions of *N*-acetylglucosamine (GlcNAc), galactose (Gal), and fucose (Fuc) with *N*-acetylglucosaminyltransferase (1), galactosyltransferases (2, 3, and 6), and fucosyltransferases (4 and 5).

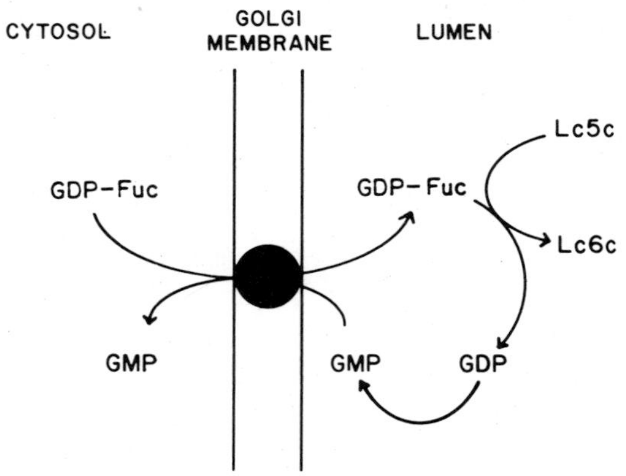

Figure 12.16. Receptor-mediated transfer of a sugar nucleotide from the cytosol of the lumen of Golgi cisternae by an antiport mechanism, with cotransport from lumen to cytosol of the nucleoside monophosphate.

pH optimum, metal-ion requirements, and optimal conditions of detergent solubilization (Basu and Basu 1982). They are localized on the luminal aspect of Golgi membranes and may be organized into multiglycosyltransferase complexes. Sugar nucleotides are transported across the Golgi membrane from the cytosol by an antiport mechanism, with cotransport of a nucleoside monophosphate from the lumen to the cytosol, as shown in Figure 12.16 (Capasso and Hirschberg 1984b).

The existence of several galactosyltransferases with different glycosphingolipid substrate specificities and/or product structures has been experimentally demonstrated (Table 12.1) (Basu and Basu). At least three different galactosyltransferases have been solubilized from an embryonic chicken brain membrane-bound fraction. The enzymes involved in the conversion of Lc3a to Lc4a (Table 12.1, a β1-4 galactosyltransferase) and GM2 to GM1 ganglioside (Table 12.1, a

Table 12.1. Some Specific Galactosyltransferases of Glycosphingolipid Biosynthesis

Substrate	Product	Linkage
GlcCer	LacCer	β1-4
LacCer	Gb3a	α1-4
LacCer	Gb3b	α1-3
Lc3a	Lc4a	β1-4
Lc4a	Lc5c	α1-3
GM2	GM1	β1-3

β1-3 galactosyltransferase) have been resolved by ion-exchange chromatography and affinity chromatography. Some glycosyltransferases are active with either glycosphingolipid or glycoprotein as substrates; the extent to which there are unique enzymes for the two kinds of glycoconjugates is not known.

Investigators in this field must use rigorous methods to characterize the products formed in vitro from labeled sugar nucleotides. Perhaps the best method involves incubation with a glycolipid substrate that is radiolabeled in the sugar moiety at the nonreducing end, permethylation of the product isolated by thin-layer chromatography (autoradiography detection), and mass spectrometry of the labeled, partially methylated alditol acetate produced after hydrolysis and derivatization (see previous section). This method, which allows very small amounts of product to be detected and isolated, also provides direct chemical evidence for the linkage position of the newly added glycose unit (from a known sugar nucleotide) and therefore establishes the structure of the product.

Another important, but generally overlooked, feature of these glycosyltransferase reactions involves a stereochemical consideration. It is commonly accepted that transfer from a sugar nucleotide occurs with inversion of anomeric configuration. Since some of the products have α-glycosidic linkages and some have β-linkages, it must be assumed that double inversions are involved in some cases, unless the sugar nucleotides can have either configuration. For example, the Forssman antigen Gb5b (Figure 12.14) contains Gal(α1-4)- and Gal(β1-4)- as well as GalNAc(α1-3)- and GalNAc(β1-3)-glycosidic linkages. Since UDP-Gal and UDP-GalNAc are assumed to exist exclusively in the α-configuration, double inversions must be involved in the syntheses of Gal(α1-4) and GalNAc(α1-3) linkages. It will be interesting to learn whether an intermediate acceptor is involved in the double inversion, which could be the enzyme itself (Ping-Pong mechanism) or some other membrane-bound substance. In another area of metabolism involving glycosyltransferases of Golgi apparatus—the biosynthesis of asparagine-linked carbohydrate chains of glycoproteins—the synthesis of the dolichol pyrophosphate–linked oligosaccharide chains involves a double inversion where mannose is transferred from GDP-α-Man to dolichol phosphate (with inversion to Man-β-O-P-dolichol) and thence (with a second inversion) to the growing oligosaccharide chain to yield Man(α1-2) linkages.

Gangliosides   More than 60 gangliosides have been isolated and characterized. Most of them fall into two general structural types, called the *ganglio* (Gg) and *lacto* (Lc) series. They are synthesized by the same general

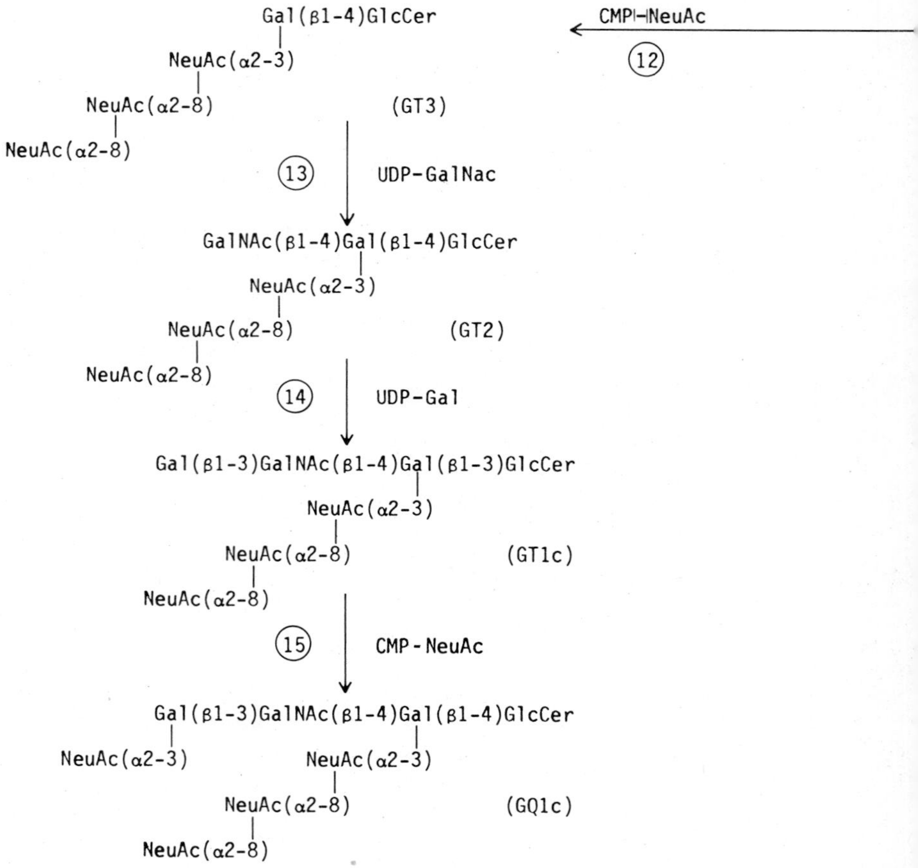

```
                              Gal(β1-4)GlcCer                    CMPI-NeuAc
                           NeuAc(α2-3)                    ◄────────────────────
                        NeuAc(α2-8)          (GT3)            (12)
                     NeuAc(α2-8)
                                         │
                                    (13) │  UDP-GalNac
                                         ▼
                           GalNAc(β1-4)Gal(β1-4)GlcCer
                              NeuAc(α2-3)
                           NeuAc(α2-8)          (GT2)
                        NeuAc(α2-8)
                                    (14) │  UDP-Gal
                                         ▼
                        Gal(β1-3)GalNAc(β1-4)Gal(β1-3)GlcCer
                              NeuAc(α2-3)
                           NeuAc(α2-8)          (GT1c)
                        NeuAc(α2-8)
                                         │
                                    (15) │  CMP - NeuAc
                                         ▼
                        Gal(β1-3)GalNAc(β1-4)Gal(β1-4)GlcCer
                   NeuAc(α2-3)              NeuAc(α2-3)
                              NeuAc(α2-8)          (GQ1c)
                           NeuAc(α2-8)
```

Figure 12.17. Biosynthesis of gangliosides of the ganglio series from lactosylceramide (LacCer) by the stepwise additions of galactose (Gal), *N*-acetylgalactosamine (GalNAc), and sialic acid (NeuAc) with membrane-bound galactosyltransferases (4, 8, and 14), *N*-acetylgalactosaminyltransferases (3, 7, and 13), and sialyltransferases (1, 2, 5, 6, 9, 10, 11, 12, and 15).

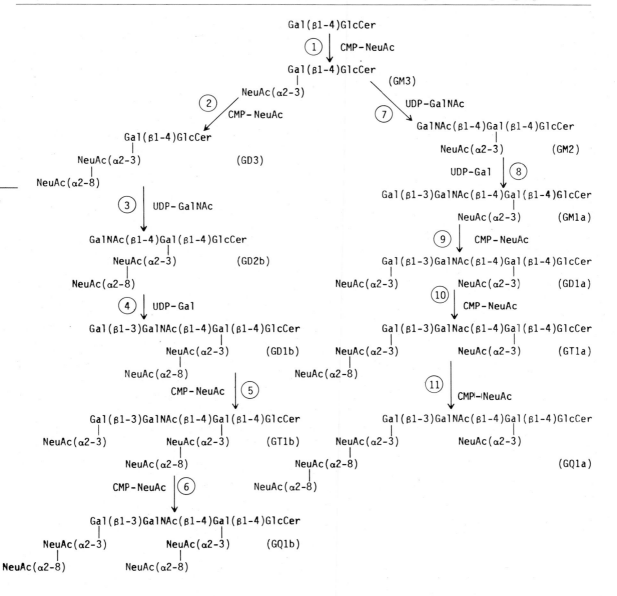

mechanism as the neutral glycosphingolipids, that is, the stepwise transfer of sugars from sugar nucleotides to the nonreducing end of the oligoglycosylceramide substrate. The major pathway for the synthesis of brain gangliosides of the ganglio series is illustrated in Figure 12.17. This scheme was originally proposed in 1966 and still represents the most likely sequence of individual glycosyltransferase steps. It must be emphasized, however, that not all the reactions have been demonstrated with cell-free extracts, and the large number of steps affords the possibility of alternate pathways. The glycosyltransferases of reactions 1, 2, 7, 8, and 9 have also been found in rat liver Golgi membranes.

While GQ1b is the major tetrasialoganglioside in the brains of higher vertebrates, GQ1c is the major tetrasialoganglioside of fish brain. The pathway on the left in Figure 12.17, involving GT3, GT2, and GT1c, may be unique in fishes, since these simpler gangliosides are primarily found in fish brain but not elsewhere.

The pathways of ganglioside biosynthesis in the lacto series have not yet been formulated in detail. On the basis of what is known about the occurrence and structures of these gangliosides, and in vitro demonstration of some of the enzyme-catalyzed steps, a logical pathway (Figure 12.18) can be predicted which is analogous to the synthesis of gangliosides of the ganglio series. Fucose-containing gangliosides are shown here as the products of neutral fucoglycosphingolipid precursors, but the order of addition of sialic and fucose residues has not been firmly established. Fucose-containing gangliosides of the ganglio series are also known; it is assumed that they are synthesized by variations in the pathway shown in Figure 12.18.

Many of the gangliosides contain modified sialic acids, of which there are at least two different kinds. The most common modification is the presence of an N-glycolyl group (RNH—$COCH_2OH$) instead of the N-acetyl group. Interestingly, the hydroxyl group is introduced into the sialic acid group of glycosphingolipids by hydroxylation of the N-acetyl residue. Another modified form, which is probably of considerable biological importance, is acetylated at one or more of the hydroxyl groups of sialic acid. Two such gangliosides are 4-O-acetyl-N-glycolylneuraminosyl-(α2-3)-galactosyl-(β1-4)-glucosylceramide and a 9-O-acetyl-N-acetylneuraminosyl derivative of the tetrasialoganglioside GQ1b, whose structure is shown.

$$\text{Gal}(\beta\text{1-3})\text{GalNAc}(\beta\text{1-4})\text{Gal}(\beta\text{1-4})\text{GlcCer}$$

Gal(β1-3)GalNAc(β1-4)Gal(β1-4)GlcCer
| |
NeuAc(α2-3)   NeuAc(α2-3)
| |
NeuAc(α2-8)   9-O-Ac-NeuAc(α2-8)

GQ1b

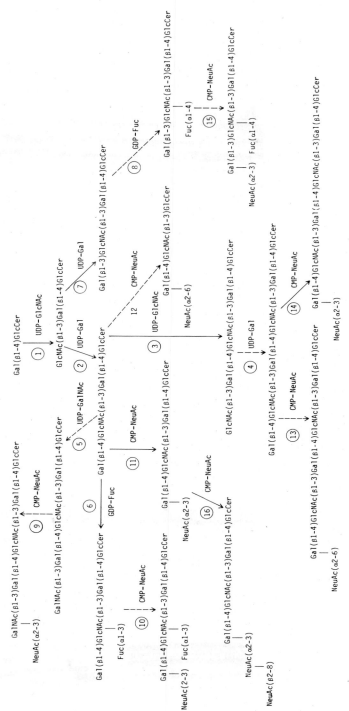

Figure 12.18. Biosynthesis of gangliosides of the lacto series by the stepwise additions of galactose (Gal), N-acetylglucosamine (GlcNAc), N-acetylgalactosamine (GalNAc), fucose (Fuc), and sialic acid (NeuAc) with membrane-bound galactosyltransferases (2, 4, and 7), N-acetylglucosaminyltransferases (1 and 3), fucosyltransferases (6 and 8), N-acetylgalactosaminyltransferase (5), and sialyltransferases (9, 10, 11, 12, 13, 14, 15, and 16). Dotted arrows indicate reactions that have not been demonstrated in vitro.

LYSOSOMAL METABOLISM OF SPHINGOLIPIDS

Turnover of plasma membrane and subcellular membrane constituents as well as sphingolipids of extracellular origin involves specific lysosomal hydrolases. The investigation of these enzymes was greatly stimulated by the discovery that several heritable diseases could be attributed to abnormal lysosomal sphingolipid catabolism. This section is organized in a manner that emphasizes the normal pathways of sphingomyelin and glycosphingolipid metabolism and some of the important lysosomal storage diseases where the biochemistry is relatively well understood (Stanbury et al. 1983).

Sphingomyelin

Sphingomyelin can be completely degraded in the lysosome to ceramide and phosphorylcholine, as shown.

$$CH_3(CH_2)_{12}CH{=}CH{-}\underset{\underset{\displaystyle CH_3(CH_2)_{22}CO}{|}}{\underset{\displaystyle NH}{|}}{\underset{\displaystyle OH}{|}}CH{-}CH{-}CH_2O{-}\overset{\displaystyle O}{\overset{\|}{\underset{\underset{\displaystyle O^-}{|}}{P}}}{-}OCH_2CH_2\overset{+}{N}(CH_3)_3$$

Sphingomyelin $\Big|\ H_2O$

$$CH_3(CH_2)_{12}CH{=}CH{-}\underset{\underset{\displaystyle CH_3(CH_2)_{22}CO}{|}}{\underset{\displaystyle NH}{|}}{\underset{\displaystyle OH}{|}}CH{-}CH{-}CH_2OH\ +\ {}^-O{-}\overset{\displaystyle O}{\overset{\|}{\underset{\underset{\displaystyle O^-}{|}}{P}}}{-}OCH_2CH_2\overset{+}{N}(CH_3)_3$$

Ceramide   **Phosphorylcholine**

The reaction is catalyzed by *sphingomyelinase*, a hydrolytic enzyme that has an acidic pH optimum. No cofactors are involved, but the reaction kinetics can be affected by other phospholipids, synthetic detergents, and cholesterol when studied in cell-free systems. Sphingomyelinase has a preferred stereospecificity for the D-erythro form of sphingoid bases. A nonlysosomal form of sphingomyelinase has also been described; it has a pH optimum near neutrality and requires magnesium ions for maximal activity.

Patients with *Niemann-Pick disease* are deficient in lysosomal sphingomyelinase activity and have large lipid-laden reticuloendothelial cells (foam cells) scattered throughout the spleen, bone marrow, lymph nodes, liver, and lungs. This is a relatively rare panethnic disorder of genetic origin, named after the two medical scientists who first described the symptoms early in the twentieth

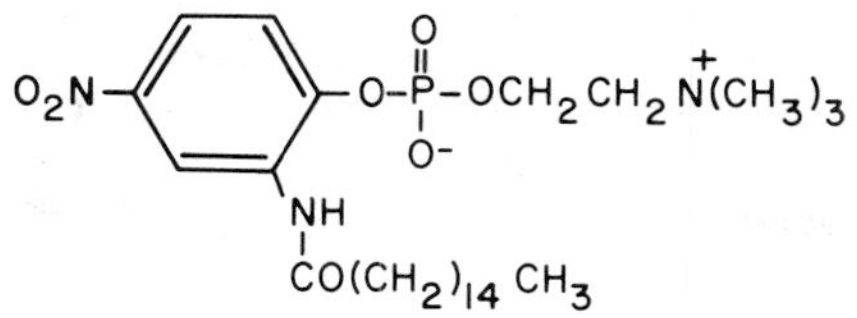

Figure 12.19. Artificial chromogenic substrate used to assay sphingomyelinase activity.

$$O_2N-\underset{\underset{CO(CH_2)_{14}\ CH_3}{\overset{|}{NH}}}{\bigcirc}-O-\overset{\overset{O}{\parallel}}{\underset{\underset{O^-}{|}}{P}}-OCH_2CH_2\overset{+}{N}(CH_3)_3$$

century. Chemical analyses reveal that sphingomyelin is the major substance accumulated in the abnormal cells of these patients, and it is assumed that the characteristic hepatosplenomegaly, osteoporosis, and central nervous system damage stem from the sphingomyelinase deficiency. Unesterified cholesterol is accumulated along with sphingomyelin in the foam cells of some cases of Niemann-Pick disease, which complicates enzymatic assays because cholesterol inhibits sphingomyelinase.

There are three forms of Niemann-Pick disease, known as types A, B, and C. They appear to be variants with different mutations of the structural gene for sphingomyelinase. It has been proposed that the mutations involved in types A and B affect the active site of sphingomyelinase, whereas type C may involve the structural alteration of a domain where sphingomyelinase interacts with an allosteric activator (a heat-stable protein from human spleen or liver).

An interesting chromogenic substrate has been designed to simplify the assays of sphingomyelinase activity for detection of heterozygotes and for prenatal diagnosis of Niemann-Pick disease. This substrate (Figure 12.19) has the advantage of forming a yellow nitrophenol product that is much more water-soluble than sphingomyelin and provides a high sensitivity in the colorimetric determination.

Catabolism of Ceramide

Ceramide is the ultimate sphingolipid product of lysosomal sphingomyelin and glycosphingolipid metabolism. It can be degraded to long-chain bases, such as 4-sphingenine and sphinganine, and long-chain fatty acids by an acid hydrolase called *ceramidase*. The enzyme

$$\text{Ceramide} + H_2O \longrightarrow \text{long-chain base} + \text{fatty acid}$$

has a pH optimum of 4.8, requires no cofactors, and is somewhat inhibited by the products of the reaction. It also catalyzes the reverse reaction in cell-free homogenates, a synthetic potential which has unknown significance to the cell, since ceramide is normally synthesized from acyl-CoA. There are neutral and alkaline ceramidases in some tissues such as cerebellum and small intestine; they also catalyze both the breakdown and synthesis of ceramide.

A deficiency of acid ceramidase has been consistently observed in patients with *Farber's lipogranulomatosis*. This rare disorder is believed to be inherited in an autosomal recessive mode. It is characterized by less than 5% of normal ceramidase activity, variably impaired mental function, granulomatous lesions in the skin, joints, and larynx of most patients, and heart and lung involvement. Massive amounts of ceramide occur in swollen cytoplasmic vacuoles of lysosomal origin in the affected tissues. The disease is usually fatal in early childhood.

Sphingoid bases released from ceramide are further degraded to a C_{16} aldehyde and phosphoethanolamine, as shown.

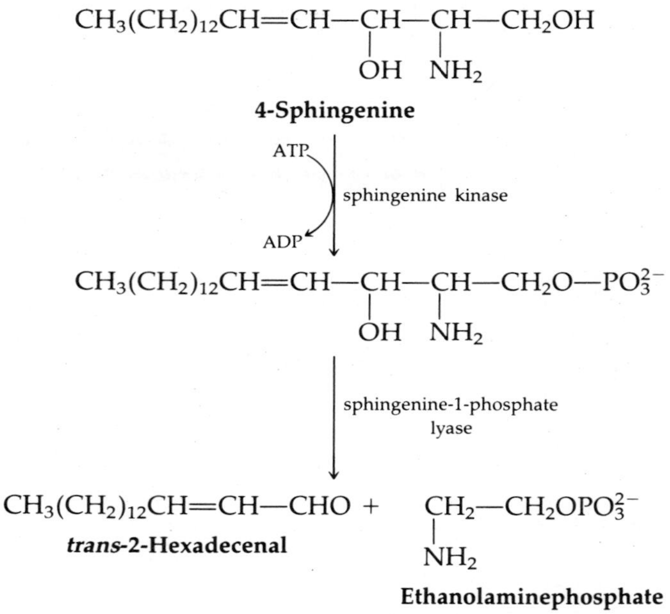

$$CH_3(CH_2)_{12}CH\!\!=\!\!CH\!-\!\underset{\underset{OH}{|}}{CH}\!-\!\underset{\underset{NH_2}{|}}{CH}\!-\!CH_2OH$$

4-Sphingenine

ATP — sphingenine kinase — ADP

$$CH_3(CH_2)_{12}CH\!\!=\!\!CH\!-\!\underset{\underset{OH}{|}}{CH}\!-\!\underset{\underset{NH_2}{|}}{CH}\!-\!CH_2O\!-\!PO_3^{2-}$$

sphingenine-1-phosphate
lyase

$$CH_3(CH_2)_{12}CH\!\!=\!\!CH\!-\!CHO \;+\; \underset{\underset{NH_2}{|}}{CH_2}\!-\!CH_2OPO_3^{2-}$$

trans-2-Hexadecenal

Ethanolaminephosphate

Cleavage of the long-chain base requires phosphorylation of the terminal primary hydroxyl group, followed by lyase-catalyzed scission between C-2 and C-3 to yield phosphoethanolamine and palmitaldehyde (from sphinganine) or *trans*-2-hexadecenal (from 4-sphingenine). The phosphoethanolamine might be utilized for the biosynthesis of glycerophospholipids (Chapter 8).

Lysosomal Glycosidases   Neutral and acidic glycosphingolipids are catabolized by the stepwise removal of sugars from their nonreducing ends by a family of lysosomal exoglycosidases. The glycosidases are acid hydrolases and the products are free sugars. The catabolism of a trisialogan-

NeuAc(α2-8)NeuAc(α2-3)Gal(β1-3)GalNAc(β1-4)Gal(β1-4)GlcCer   (GT1a)
           NeuAc(α2-3)

NEURAMINIDASE I

NeuAc(α2-3)Gal(β1-3)GalNAc(β1-4)Gal(β1-4)GlcCer   (GD1a)
       NeuAc(α2-3)

NEURAMINIDASE II

Gal(β1-3)GalNAc(β1-4)Gal(β1-4)GlcCer   (GM1)
     NeuAc(α2-3)

β-GALACTOSIDASE I

GalNAc(β1-4)Gal(β1-4)GlcCer   (GM2)
   NeuAc(α2-3)

β-N-ACETYLHEXOSAMINIDASE I
(Hex A)

Gal(β1-4)GlcCer   (GM3)
NeuAc(α2-3)

NEURAMINIDASE III

Gal(β1-4)GlcCer   (LacCer)

β-GALACTOSIDASE II

GlcCer

β-GLUCOSIDASE I

Cer

Figure 12.20. Lysosomal catabolism of a trisialoganglioside (GT1a), catalyzed by specific exoglycosidases (Dawson 1978).

glioside (GT1a) to ceramide requires seven sequential exoglycosidase-catalyzed reactions, as shown in Figure 12.20. A somewhat different complement of enzymes is required for the catabolism of the blood group A- and B-active antigens of the lacto series (Lc6e and Lc6c, respectively) and globoside (GbOse$_4$Cer), where nine reactions are involved (Figure 12.21).

The substrate specificities of the glycosidases involved in these pathways are not entirely understood. Neuraminidase preparations from *Vibrio cholerae* and *Vibrio pneumoniae* cleave NeuAc(α2-3)Gal-, NeuAc(α2-8)NeuAc-, and NeuAc(α2-6)Gal- linkages of gangliosides but are unable to hydrolyze the internal NeuAc of GM1a and GM2 gangliosides, probably because of steric hindrance. The literature contains conflicting results about the specificity of mammalian neuraminidases, and the question of whether neuraminidases I, II, and III (Figure 12.20) are identical must be deferred.

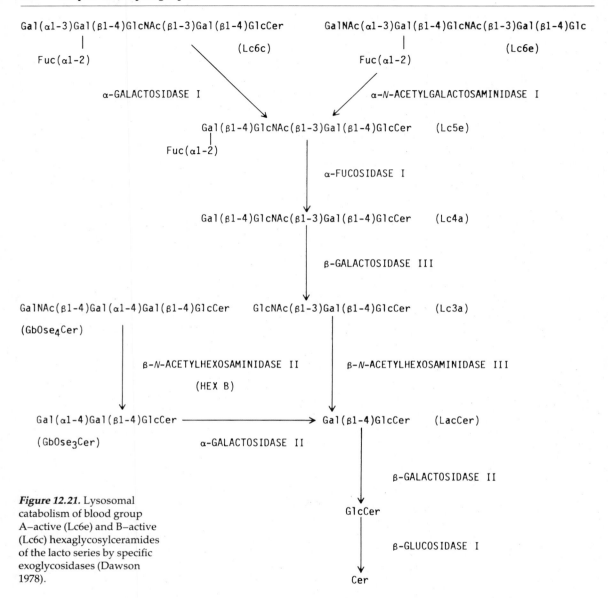

Figure 12.21. Lysosomal catabolism of blood group A–active (Lc6e) and B–active (Lc6c) hexaglycosylceramides of the lacto series by specific exoglycosidases (Dawson 1978).

The identity of α-galactosidases I and II in Figure 12.21 can be inferred from the finding that globotriaosylceramide (GbOse$_3$Cer) and blood group B–active glycosphingolipids such as Lc6c accumulate in the cytosolic vacuoles of kidney and vascular endothelium of patients with an X-linked genetic disorder called *Fabry's disease*. The abnormal cellular levels of these lipids and galabiosylceramide, Gal(α1-4)GalCer, are explained by the deficiency of one of two forms of α-galactosidase activity, called α-galactosidase A (α-galactosidases I and II in Figure 12.21). This enzyme catalyzes the hydrolysis of GbOse$_3$Cer in vitro in the presence of a detergent such as sodium cholate or sodium taurocholate, or when a heat-stable glycoprotein activator from liver is added to the incubation mixtures.

Another form of α-galactosidase activity (α-galactosidase B) is normal in patients with Fabry's disease. Although α-galactosidase B can cleave GbOse$_3$Cer in vitro, it apparently has no activity with this substrate in cells. This enzyme probably functions as an α-N-acetylgalactosaminidase in vivo, since it hydrolyzes Forssman antigen (Gb5b) and the blood group A–active antigen (Lc6e) shown in Figure 12.21.

Both α-galactosidases A and B catalyze the hydrolysis of a fluorogenic substrate, 4-methylumbelliferyl-α-D-galactopyranoside (Figure 12.22), but the α-galactosidase B activity with this substrate can be inhibited by $50 mM$ N-acetylgalactosamine, whereas α-galactosidase A cannot. The product, 4-methylumbelliferone, is highly fluorescent at 450 nm when excited at 365 nm at a pH greater than 11.0.

Tay-Sachs disease is probably the best-known sphingolipid storage disease. In the infantile form of the disease, large amounts of GM2 ganglioside accumulate in the brain due to the nearly absolute deficiency of β-N-acetylhexosaminidase I (Hex A in Figure 12.20). The incidence of this disorder, which occurs primarily in Jews of Ashkenazi origin, has been dramatically reduced in the United States by a mass screening for heterozygotes and by prenatal diagnosis in the case when both parents are carriers.

The activities of Hex A and another form of β-N-acetylhexosaminidase (Hex B in Figure 12.21) are both deficient in patients with *Sandhoff disease*. Although both globoside (GbOse$_4$Cer) and GM2 ganglioside accumulate in the brain and visceral tissues, respectively, of these patients, the clinical manifestations are nearly identical with those of Tay-Sachs disease. If it were not for the fact that Sandhoff disease does not afflict a particular population group, the different biochemical mechanism might not have been so avidly pursued.

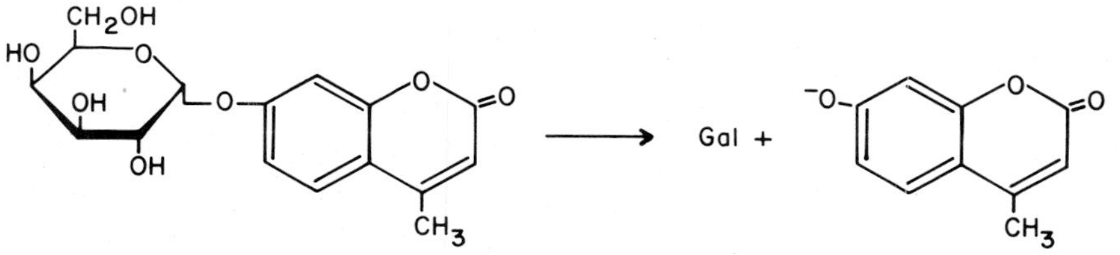

Figure 12.22. Artificial fluorogenic substrate used to assay α-galactosidase A and α-galactosidase B activities. Differential assays can be made in the presence and absence of N-acetylgalactosamine, an inhibitor of α-galactosidase B.

Tay-Sachs and Sandhoff diseases, as well as other forms of *GM2 gangliosidosis*, can be understood in terms of a molecular genetic model involving three separate gene products, two of which are polypeptide subunits of Hex A or Hex B and one of which is an activator protein. The Hex A enzyme consists of two nonidentical subunits, an α-chain coded at a locus on chromosome 15 and a β-chain coded at a locus on chromosome 5. Recent work suggests that the active form of Hex A contains two α-chains, one β-chain, and one activator protein. Mutations which affect the α-locus on chromosome 15 are responsible for the several forms of Tay-Sachs disease that can be characterized by a deficiency of Hex A. A variant form of Tay-Sachs disease, in which the activity of Hex A in vitro is normal with chromogenic substrates, has been attributed to a mutation at the activator locus.

The other form of β-*N*-acetylhexosaminidase, Hex B, is a tetrameric homopolymer of β-chains which appears to be active without an activator protein. Since the β-chain is a component of both Hex A and Hex B, mutations at the β locus on chromosome 5 will affect the activity of both enzymes, which is observed in patients with Sandhoff disease. Thus, the similarity of clinical manifestations can be understood, as well as the disposition of the disorder among different population (ethnic) groups.

Another group of genetic storage diseases results from the deficiency of β-galactosidase. There are at least four isozymes of β-galactosidase in normal tissues. Their specificities toward different glycoconjugates (glycosphingolipids and glycoproteins) containing β-linked galactose residues have not yet been clearly defined. Patients with the autosomal recessive disorder called *generalized gangliosidosis* are deficient in three of the isozymes, which results in the accumulation of GM1 ganglioside and several galactose-containing oligosaccharides derived from glycoproteins.

Other genetic diseases that result from the deficiency of a lysosomal glycosidase are defined in Table 12.2 according to the accumulated substrate and the missing enzyme activity. When the lists are compared with the pathways of glycosphingolipid catabolism in Figures 12.20 and 12.21, it is clear that additional inherited disorders may be found that are related to these substances. Such disorders may be fatal prior to birth, however, and would therefore escape detection unless abortus tissues were routinely assayed.

The lysosomal enzymes of sphingolipid metabolism are relatively minor gene products in most cells. As a result, sufficient quantities of most of the enzymes have not been available for primary sequence analyses. Currently, intensive efforts are being directed toward cloning the structural genes of these enzymes, which will allow their polypeptide sequences to be determined from DNA sequences. The

Table 12.2. Sphingolipids Accumulated in Various Disorders of Lysosomal
Enzymes

Inherited disease	Lysosomal enzyme	Accumulated substrate
Farber's lipogranulomatosis	Acid ceramidase	Ceramide
Niemann-Pick disease	Sphingomyelinase	Sphingomyelin
Gaucher's disease	β-Glucosidase	Glucosylceramide
Krabbe's globoid cell leukodystrophy	β-Galactosidase	Galactosylceramide
Metachromatic leukodystrophy	Arylsulfatase A	3-Sulfogalactosylceramide
Fabry's disease	α-Galactosidase A	Globotriaosylceramide (GbOse$_3$Cer)
GM1 gangliosidosis	β-Galactosidase	GM1 Ganglioside
Tay-Sachs disease	β-N-acetylhexosaminidase A	GM2 Ganglioside
Sandhoff disease	β-N-acetylhexosaminidases A and B	GM2 Ganglioside and globoside (GbOse$_4$Cer)

molecular basis of how mutations lead to each of the sphingolipid
storage diseases will then become reasonable questions.

REGULATION OF SPHINGOLIPID METABOLISM

Insight into the molecular mechanisms by which sphingolipid metabolism is regulated comes from several different areas of research
including hematopoiesis, chemical carcinogenesis, changes in the
organization of cell-surface glycosphingolipids with oncogenic
transformation, embryogenesis, and the effects of tumor promoters
and "antipromoters." Embryonic development and cellular differentiation are accompanied by programmed changes of cell-surface
carbohydrate structure. It can be assumed that these changes are
associated with functional characteristics of membranes such as
antigenic specificity, adhesion, receptor specificity, and perhaps
even the regulation of cell growth.

Developmental Changes in Sphingolipid Metabolism

Several glycoconjugates on the surface of human erythrocytes express specific antigenic determinants, such as the blood group I and
i activities (Hakomori 1981). Structural analyses and immunochemical
binding studies with specific monoclonal antibodies have revealed
that the expression of I and i antigens on human erythrocytes is
developmentally regulated. The i phenotype, associated with linear
glycosphingolipid chains of N-acetyllactosamine, is the predominant
antigen on fetal erythrocytes, while the I phenotype, associated with
branched carbohydrate chains, is the predominant antigen of adult

erythrocytes. The i and I phenotypes can be pictured structurally as shown in Figure 12.23 ($\bullet$ = Gal, $\bigcirc$ = GlcNAc, and $\square$ = Glc). Other I and i antigens, with and without sialic acid chains and two or more branch points, have recently been reported. The only novel feature of the branched (I) structures is the presence of one or more GlcNAc(β1-6)-Gal- linkages. Therefore, it can be assumed that at least one structural gene, for a specific β-N-acetylglucosaminyltransferase, is repressed in fetal erythropoiesis but becomes actively expressed during erythropoiesis at a later stage of development. Occasionally, this gene is not expressed at all, and i antigen is the predominant form on adult erythrocytes.

Regeneration of the epithelium of small intestine is another example of cell-surface structural changes related to differentiation. Glucosylceramide is the major neutral glycosphingolipid in neonatal rat intestine. During postnatal growth the content of globotriaosylceramide (GbOse$_3$Cer) increases dramatically along with GM3 ganglioside, and the fatty acid composition of these glycosphingolipids changes to favor α-hydroxylated species. A striking aspect of this work is the observation that not only are there age-dependent changes but also there are changes in glycosphingolipid metabolism during crypt to villus cell differentiation (Bouhours and Bouhours 1983).

Cytodifferentiation of nucleated myotubes in vitro involves aggregation and fusion of mononucleated myoblasts, both in primary cultures and established cell lines. A rat myoblast cell line (L6) can synthesize gangliosides of the ganglio series, including GM1 and GD1a. The concentration of GD1a in the plasma membranes increases about threefold just prior to fusion of myoblasts, then returns to basal levels in myotubes. Mutant myoblasts that cannot synthesize GM1 and GD1a are unable to fuse and differentiate, suggesting the possibility that these gangliosides may be involved in the alteration of membrane components that accompany fusion (Whatley et al. 1976).

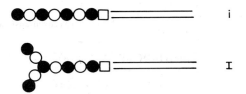

Figure 12.23. The structural basis of I and i antigenic specificity is illustrated by the glycosphingolipids Gal(β1-4)GlcNAc(β1-3)Gal(β1-4)GlcNAc(β1-3)Gal(β1-4)-GlcNAc(β1-3)Gal(β1-4)GlcCer (i specificity) and Gal(β1-4)GlcNAc(β1-6)[Gal(β1-4)Glc-NAc(β1-3)]Gal(β1-4)GlcNAc(β1-3)Gal(β1-4)GlcNAc(β1-3)Gal(β1-4)GlcCer(I specificity). The branch point is a galactose (Gal) residue substituted at β1-3 and β1-6 with N-acetylglucosamine residues. $\bullet$ = Gal, $\bigcirc$ = GlcNAc, $\square$ = Glc.

Oncogenic Transformation

Oncogenic transformation induces changes in the composition of cell-surface glycosphingolipids, which Hakomori has associated with a blocked synthetic step and accumulation of the appropriate precursor in some tumors, and the appearance of new glycosphingolipids in others (Hakomori and Kannagi 1983). In either case, an altered complement of cell-surface structures can have an effect on such properties as contact inhibition of growth (which is lacking in transformed cells), adhesion, antigenic specificity, and tumorigenicity. Some specific tumor markers and the transformed cells with which they have been associated are given in Table 12.3. The mechanisms by which glycosphingolipid profiles on the cell surface are affected by oncogenic transformation are not entirely clear, but transforming-gene activation or repression of certain glycosyltransferases is almost certain to be involved.

The activation of a specific rat liver fucosyltransferase has been associated with oncogenic transformation resulting from chemical carcinogenesis by N-2-acetylaminofluorene. This carcinogen induced hepatoma in rats when fed along with a tumor promoter over a prolonged period. Two fucose-containing gangliosides were found in the hepatoma cells and in precancerous tissue but not in normal rat liver. The biosynthesis of these fucose-containing gangliosides from GM1 has been related to the activity of a specific fucosyltransferase that is absent in normal liver, and a galactosyltransferase that is present in both hepatoma and normal liver. All three gangliosides involved are of the ganglio series. The final product (Sgm7b) has blood group B activity and is unusual because the blood group B determinant is usually associated with glycosphingolipids of the lacto series. The intermediate product (Sgm6g), while not present in normal rat liver, has been found in rat erythrocytes and in bovine brain, thyroid, and liver.

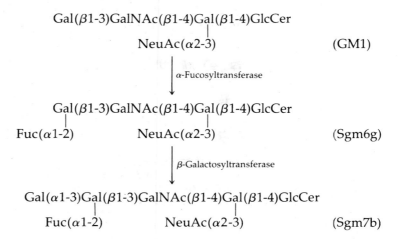

Table 12.3. Structures of Some Tumor-Associated Glycosphingolipid Antigens

Trivial name	Structure	Source
Paragloboside (Lc4a)	Gal(β1-4)GlcNAc(β1-3)Gal(β1-4)GlcCer	Hamsters infected with polyoma virus–transformed NIL cells
Asialo GM2 (Gg3a)	GalNAc(β1-4)Gal(β1-4)GlcCer	Tumors of mice inoculated with Kirsten murine sarcoma virus–transformed 3T3 cells or with L5178 mouse lymphoma cells
SIm6c	NeuAc(α2-3)Gal(β1-3)GlcNAc(β1-3)Gal(β1-4)GlcCer | Fuc(1-4)	Human colorectal and pancreas adenocarcinomas
GD3	NeuAc(α2-8)NeuAc(α2-3)Gal(β1-4)GlcCer	Human melanoma cells
GbOse3Cer	Gal(α1-4)Gal(β1-4)GlcCer	Burkitt's lymphoma cells
Forssman antigen (Gb5b)	GalNAc(α1-3)GalNAc(β1-3)Gal(α1-4)Gal(β1-4)GlcCer	Gastric and colorectal adenocarcinomas of normally Forssman-negative tissues

Note: Monoclonal antibodies against these glycosphingolipids have been produced by hybridoma techniques, using tumor tissue or crude glycolipid fractions as immunogens. Chemical analyses and/or quantitative immunochemical methods have shown that the glycolipids accumulate in these tumors.

Other Aspects of Regulation

Cultured cells have been used to study the mechanisms involved in the regulation of glycosphingolipid metabolism. Tumor promoters of the phorbol ester type, butyrate, and retinoic acid all impinge on the control mechanisms of growth and differentiation and have been found to affect glycosphingolipid metabolism in cultured cells. Phorbol esters of the greatest tumor-promoting activity, such as 12-O-myristoyl-phorbol-13-acetate (TPA), increase the saturation density of cultured cells. Butyrate and retinoic acid, on the other hand, make transformed cells less tumorigenic, so their growth is more readily arrested in dense cultures.

TPA has multiple biochemical effects, including the induction of ornithine decarboxylase and phosphodiesterase activities. Recent work has implicated TPA as a mimic of membrane phosphatidylinositol-derived diacyl-sn-glycerol in the stimulation of the calcium-phospholipid-dependent protein kinase (C-kinase) (Hakomori and Kannagi 1983). If this is true, it seems certain that the tumor-promoting activity of TPA will be related to the phosphorylation of certain proteins in the plasma membrane or elsewhere in the cell. The effect of TPA on the metabolism of gangliosides is not yet entirely clear. The sialyltransferase involved in GM3 ganglioside synthesis (Figure 12.17), the first committed step of ganglioside biosynthesis, is substantially increased following TPA treatment of Chinese hamster V79 cells, and the amount of GM3 is increased by TPA treatment of these cells and of HL-60 cells, whereas the de novo synthesis of a more complex glycosphingolipid, trisialoganglioside, is decreased by TPA treatment of a promoter-sensitive cell line. There is no evidence that glycosyltransferases in the Golgi apparatus are phosphorylated as a result of the TPA treatment. It is more likely that an intracellular messenger of TPA, perhaps a phosphorylated protein, affects the transcription of numerous genes, including those of some of the glycosyltransferases.

Butyrate and retinoic acid inhibit some of the effects of tumor promoters and have been referred to as tumor anti promoters. The activity of the sialyltransferase that catalyzes the synthesis of GM3 ganglioside (Figure 12.17) is increased by both of these substances. Butyrate reversibly blocks transformed cells in the early G1 stage of the cell cycle after about 12 to 18 h of treatment, during which time the sialyltransferase activity increases approximately 10-fold along with the levels of GM3 ganglioside and other less prominent plasma membrane gangliosides. Upon removal of the butyrate, the sialyltransferase activity decreases, and the cells resume growth in synchrony, entering the S-phase about 8 h later. The effect of retinoic acid on this sialyltransferase activity is shown in Figure 12.24. All the cells tested had greater sialyltransferase activity following treatment with retinoic acid, the most dramatic increases being in the sarcoma

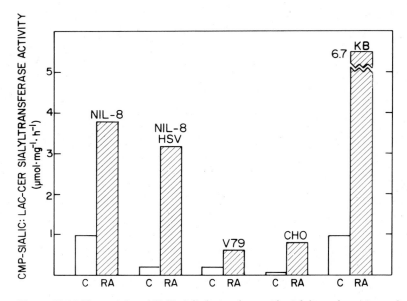

Figure 12.24. The activity of CMP-sialic:lactosylceramide sialyltransferase in crude cell-free membrane preparations from control (C) and retinoic acid–treated (RA) normal hamster cells (NIL-8), sarcoma virus–transformed hamster cells (NIL-8HSV), Chinese hamster lung fibroblasts (V79), Chinese hamster ovary cells (CHO), and human larynx epithelial carcinoma cells (KB). Cells were treated with medium containing 10 μg per mL ($1.6 \times 10^{-6}M$) of retinoic acid (Moskal, Lockney, and Sweeley, unpublished results).

virus–transformed hamster cells (NIL-8 HSV) and a human carcinoma cell line (KB). Since both butyrate and retinoic acid can be transported to the nucleus, where they perturb the histone fraction, it seems likely that the mechanism of their effect on glycosphingolipid metabolism may be related to the transcription of certain glycosyltransferase genes.

Posttranslational covalent modification of key enzymes involved in the biosynthesis of glycosphingolipids is a logical mechanism for the regulation of various pathways, such as those shown in Figures 12.14, 12.15, 12.17, and 12.18. In fact, the Golgi apparatus contains protein kinase activity, but the function of protein phosphorylation has not been determined. The activity of the partially purified sialyltransferase for the first step in the biosynthesis of most gangliosides (reaction 1, Figure 12.17) can be modulated by a cAMP-dependent protein kinase (increased activity) and by an alkaline phosphatase (decreased activity). Similarly, phosphorylation-dephosphorylation has been implicated in the regulation of the N-acetylgalactosaminyltransferase which catalyzes the conversion of GM3 ganglioside to GM2 ganglioside (reaction 7, Figure 12.17), the first biosynthetic step in the ganglio series. The activity of this GalNAc transferase in a mouse neuroblastoma clonal cell line and a neuroblastoma-glioma hybrid cell line was inhibited by substances that down-regulate cAMP synthesis, such as β-endorphin, enkephalins, and opiates. The results to date, however, only suggest that phosphorylation of some glycosyltransferases of the Golgi membrane may be a mechanism of regulation.

FUNCTIONAL ROLES OF SPHINGOLIPIDS IN MEMBRANES

The sphingolipids may have specialized structural functions in certain membranes such as the myelin sheath (sphingomyelin) and more general functions in the stabilization of plasma membranes (Chapter 2). The relatively high concentrations of 3-sulfogalactosylceramide in the collecting tubules of kidney and in the salt-forming gland of ducks have suggested to some investigators that this glycosphingolipid may be involved in the transport of metal ions across the bilayer membranes in these specialized organs.

The biological effects of cholera toxin result from several cell-membrane events, the first of which requires GM1 ganglioside in the outer leaflet of the plasma bilayer membrane. Binding of the five B-subunits of cholera toxin to GM1 ganglioside is an obligatory step which allows penetration of the catalytic A-subunit to the site of a second messenger system. Specifically, the A-subunit activates adenyl cyclase by enzymatically converting a GTP-binding protein associated with adenyl cyclase to an ADP-ribosylated form. Thus, the ultimate biological effects of cholera toxin in the intestine—intense fluid secretion and subsequent diarrhea—are triggered by the binding of this toxic bacterial protein to the GM1 of intestinal mucosal epithelial cells.

Other toxins may similarly bind to the glycosphingolipids of the plasma membrane. Bacteria and viruses may also have specific glycosphingolipid binding sites on cells. For example, pyelonephritic *Escherichia coli* seem to recognize and bind to $GbOse_3Cer$ on the surface of epithelial cells of the gastrointestinal tract, suggesting that this glycosphingolipid may have an important role in the establishment of some *E. coli* infections.

The appearance and disappearance of some glycosphingolipids during early stages of embryonic and fetal development and differentiation suggests their involvement in stage-specific recognition phenomena (Hakomori and Kannagi 1983). Although definitive roles in the complex events of development are lacking, the two functions most often proposed are cell recognition and adhesion, and modulation of the activities of growth-factor receptors and other cell-surface receptors. Recent studies have shown that glycosphingolipids exogenously added to cell-culture medium are incorporated into the plasma membranes and modify growth behavior. The effects were somewhat structure-specific; that is, different glycosphingolipids modulated growth in preferential ways. Apparently, the glycosphingolipids are not themselves receptors, but rather they affect the function of receptor proteins in a manner that remains to be elucidated.

PROBLEMS

1. What reaction of sphingolipid biosynthesis requires pyridoxal phosphate?
2. Explain what is meant by an antiport mechanism of transport of sugar nucleotide across the Golgi membrane.
3. What constituent of most plant glycosphingolipids distinguishes them from those of higher animals?
4. What would be the product of the overnight treatment of GD1a ganglioside with a neuraminidase from *Vibrio cholerae*?
5. Suppose you have isolated a new glycosphingolipid from bovine spleen and you have determined its partial structure to be

    Gal(?1-3)Gal(?1-2)Gal(?1-4)GlcCer
    IV   III   II   I

    a. Describe two methods that could be used to determine the anomeric configurations of the three galactose residues.
    b. Assuming that you obtained the glycosidic linkages by the Lindberg technique of permethylation analysis, what ions did you find in the mass spectra of the partially methylated alditol acetates from Gal III and Gal II that allowed the assignment of linkage positions?

c. If you found the linkage between Gal IV and Gal III to have the α-configuration, would you report that you have discovered a new blood group B–active glycosphingolipid?
6. Why must detergent be added to Golgi membrane preparations to demonstrate the in vitro (cell-free) transfer of radiolabeled sulfate from PAPS to *exogenously* supplied galactosylceramide, added in the form of a mixed micelle with phospholipid and cholesterol?
7. Which glycosyltransferase catalyzes the committed step in the incorporation of radiolabeled lactosylceramide into a pathway leading to the Forssman-active glycosphingolipid called Gb5b, rather than into blood group–active glycosphingolipids of the lacto series?
8. What partially degraded products of GQ1b and GQ1a would you expect to find in abnormally large quantities in the brain tissue from a patient with generalized gangliosidosis?
9. What mechanism(s) of regulation of glycosphingolipid biosynthesis (transcriptional, posttranslational, substrate availability, and the like) would you wish to study to understand how viral transformation effects cell-surface glycosphingolipid changes in cultured cells?

BIBLIOGRAPHY

Entries preceded by an asterisk are suggested for further reading.

Ariga, T.; Murata, T.; Oshima, M.; Maezawa, M.; and Miyatake, T. 1980. Characterization of glycosphingolipids by direct inlet chemical ionization mass spectrometry. *J. Lipid. Res.* 21:879–87.
Arita, M.; Iwamori, M.; Higuchi, T.; and Nagui, Y.

1983. Negative ion fast atom bombardment mass spectrometry of gangliosides and asialogangliosides: A useful method for the structural elucidation of gangliosides and related neutral glycosphingolipids. *J. Biochem.* 94:249–56.
*Basu, S., and Basu, M. 1982. "Expression of glycosphingolipid glycosyltransferases in development and transformation." In *The glycoconjugates*,

vol. 3, ed. M. I. Horowitz, 265–84. New York: Academic Press.

Bouhours, D., and Bouhours, J-F. 1983. Developmental changes of hematoside of rat small intestine. *J. Biol. Chem.* 258:299–304.

Capasso, J. M., and Hirschberg, C. B. 1984a. Effect of atractylosides, palmitoyl-CoA and anion transport inhibitors on translocation of nucleotide sugars and nucleotide sulfate into Golgi vesicles. *J. Biol. Chem.* 259:4263–66.

———1984b. Mechanisms of glycosylation and sulfation in the Golgi apparatus: Evidence for nucleotide sugar/nucleoside monophosphate and nucleotide sulfate/nucleoside monophosphate antiports in the Golgi apparatus membrane. *Proc. Natl. Acad. Sci. USA.* 81:7051–55.

Dabrowski, J.; Hanfland, P.; and Egge, H. 1982. "Analysis of glycosphingolipids by high resolution proton nuclear magnetic resonance spectroscopy." In *Methods in enzymology*, ed. V. Ginsburg, 69–86. New York: Academic Press.

Dawson, G. 1978. "Glycolipid catabolism." In *The glycoconjugates*, vol. 2, ed. M. I. Horowitz and W. Pigman, 287–336. New York: Academic Press.

*Hakomori, S. 1981. Blood group ABH and Ii antigens of human erythrocytes: Chemistry, polymorphism, and their developmental change. *Seminars in Hematology* 18:39–62.

*Hakomori, S., and Kannagi, R. 1983. Glycosphingolipids as tumor-associated and differentiation markers. *J. Nat. Cancer Inst.* 71:231–51.

Karlsson, K.-A. 1980. Structural fingerprinting of gangliosides and other glycoconjugates by mass spectrometry. *Adv. Exp. Med. Biol.* 125:47–61.

*Kishimoto, Y. 1983. "Sphingolipid formation." In *The enzymes.* 3rd ed. vol. 16, ed. P. D. Boyer, 358–408. New York: Academic Press.

Ledeen, R. W., and Yu, R. K. 1982. "Gangliosides: Structure, isolation, and analysis." In *Methods in enzymology*, vol. 83, ed. V. Ginsburg, 139–90. New York: Academic Press.

*Li, Y.-T., and Li, S.-C. 1982. "Biosynthesis and catabolism of glycosphingolipids." In *Advances in carbohydrate chemistry and biochemistry*, vol. 40, ed.

R. S. Tipson and D. Horton, 235–86. New York: Academic Press.

Lindberg, B., and Lonngren, J. 1978. "Methylation analysis of complex carbohydrates: General procedure and application for sequence analysis." In *Methods in enzymology*, vol. 50, ed. V. Ginsburg, 3–31. New York: Academic Press.

The nomenclature of lipids: Recommendations. 1977. *Eur. J. Biochem.* 79:11–21; *Lipids* 12:455–68.

Prestegard, J. H.; Koerner, T. A. W., Jr.; Demou, P. C.; and Yu, R. K. 1982. Complete analysis of oligosaccharide primary structure using two-dimensional high-field proton NMR. *J. Am. Chem. Soc.* 104:4993–95.

Sonnino, S.; Ghidoni, R.; Chigorno, V.; and Tettamanti, G. 1982. "Chemistry of gangliosides carrying O-acetylated sialic acid." In *New vistas in glycolipid research*, ed. A. Makita, S. Handa, T. Taketomi, and Y. Nagai, 55–69. New York: Plenum.

Stanbury, J. B.; Wyngaarden, J. B.; Fredrickson, D. S.; Goldstein, J. L.; and Brown, M. S. 1983. *The metabolic basis of inherited disease.* 5th ed. New York: McGraw-Hill.

Sugimoto, Y.; Whitman, M.; Cantley, L. C.; and Erickson, R. L. 1984. Evidence that the Rous sarcoma virus transforming gene product phosphorylates phosphatidylinositol and diacylglycerol. *Proc. Natl. Acad. Sci. USA* 81:2117–21.

Sweeley, C. C., and Siddiqui, B. 1977. "Chemistry of mammalian glycolipids." In *The glycoconjugates*, vol. 1, ed. M. I. Horowitz and W. Pigman, 459–540. New York: Academic Press.

Thudichum, J. L. W. 1962. *A treatise on the chemical constitution of the brain.* Hamden, Conn.: Archon Books.

Whatley, R.; Ng, K-C.; Rogers, J.; McMurray, W. C.; and Sanwal, B. D. 1976. Developmental changes in gangliosides during myogenesis of a rat myoblast cell line and its drug resistant variants. *Biochem. Biophys. Res. Comm.* 70:180–85.

Yamakawa, T. 1982. "Glycolipids and I: Past, present, and future." In *New vistas in glycolipid research*, ed. A. Makita, S. Handa, T. Taketomi, and Y. Nagai, 1–13. New York: Plenum.

CHAPTER 13

Metabolism of Cholesterol and Lipoproteins

Christopher J. Fielding
Phoebe E. Fielding

HISTORICAL DEVELOPMENTS

Cholesterol is found in all mammalian cells. It is largely confined to the plasma membrane, where its functions are still incompletely understood (Chapters 1 and 2). Nevertheless, the cholesterol content of cell membranes is tightly regulated, and cells which divide retain all the enzymes necessary for cholesterogenesis. It is possible to show modulation of some cell-surface enzyme and receptor functions when membrane cholesterol content is varied. Physically, cholesterol in membranes increases their viscosity and decreases their permeability to small water-soluble molecules, such as monosaccharides. In studies of the effects obtained with analogues of cholesterol, the important structural factors which regulate the permeability effects of cholesterol (such as the β-orientation of the 3-position hydroxyl group, and the planar conformation of the A and B rings) are those which affect the orientation of the sterol at the lipid-water interface.

The early history of the discovery of cholesterol and its structure is contained in the monumental volume by Myant (1981). This chapter deals mainly with its synthesis, metabolism in plasma, and irreversible catabolism. Most of the data in this chapter relate to cholesterol metabolism in mammalian, often human, systems for the good reason that it is only in these systems that our present knowledge is sufficient to give an integrated overall view of the birth, life, and death of the sterol nucleus.

The synthesis of cholesterol from its major physiological precursor (acetyl-CoA) proceeds by a sequence of more than 30 enzyme-catalyzed reactions. The details of this sequence were determined largely between the years 1945 and 1970. This research, by Bloch, Cornforth, Lynen, Popjak, Rudney, and others, represents one of the monuments of modern biochemistry. Several of the reactions in the sequence are among the most complex known. More recently, work by Gaylor and others has led to an essentially complete understanding of the reaction sequence between lanosterol and cholesterol. Most of the enzymes of cholesterogenesis are membrane bound, and our knowledge of the proteins that catalyze the synthetic steps is in many cases fragmentary. The proposal by Rudney (1959) that the conversion of hydroxymethylglutaryl-CoA to mevalonate by hydroxymethylglutaryl-CoA reductase might be the rate-limiting step of the entire reaction sequence has since received much support. The realization that this enzyme regulates cholesterol synthesis has led to an unprecedented focus of attention on its structure and mode of action and, recently, to the isolation of the enzyme protein and the sequencing of its gene.

In studies on the regulation of cholesterol synthesis by extracellular sources of cholesterol (specifically the plasma lipoproteins), the pioneering work of Bailey and Rothblat (1961–68) correctly identified the specific roles of low- and high-density lipoproteins in the uptake and efflux of cholesterol. The demonstration by Goldstein and Brown in 1973 that the action of low-density lipoproteins (LDL) can be mediated by a cell-surface receptor has further advanced our knowledge of the mechanisms by which cells can control their cholesterol content. Since that time, at least three other receptors with different lipoprotein specificities have been identified in mammalian cells. It has also become clear that nonreceptor thermodynamic processes, such as the movement of free cholesterol down its concentration gradient, can result in lipid transfer and may be important in maintaining cell cholesterol content.

While it was naturally recognized from the first that pathways increasing cell cholesterol (de novo synthesis, interiorization of lipoprotein cholesterol) must be balanced by others mediating the transfer of cholesterol out of cells to their medium, the major factors involved have been identified only recently. Glomset in 1968 proposed that the plasma enzyme lecithin:cholesterol acyltransferase (LCAT), an enzyme that esterifies free cholesterol, might play a key role in regulating cholesterol efflux from cells. Much work has since confirmed this view. The paradox that most of the cholesteryl ester generated by this transferase was recovered in lipoproteins that were not substrates for LCAT was solved in 1975 by Zilversmit and colleagues, who first found in plasma a transfer protein that catalyzed the movement of cholesteryl esters between lipoprotein particles. It now appears that all the major changes in cholesterol distribution that occur within plasma can be explained by the action of LCAT and transfer proteins.

For practical purposes the catabolism of cholesterol is limited to the liver and to glands producing steroid hormones. The concept that the first catabolic step in the liver (the formation of 7-α-hydroxycholesterol from cholesterol) is rate limiting was proposed by Danielsson in 1960 and has been confirmed by much subsequent research. Further work by Mosbach and others has identified the major pathways for the formation of the principal polyhydroxylated degradation products of cholesterol. The major problem encountered in this area has been the multiplicity of bile acids and intermediates found in greater or lesser amounts in the bile of different mammalian species and the presence of plausible alternate routes for their synthesis.

The drive behind much of the research on cholesterol metabolism has come from an appreciation of the significance of cholesterol in *atherogenesis*, the pathological deposition of lipids, particularly cholesterol and its esters, in the vascular bed. The original epidemiological findings reported by Kannel and colleagues from the data of the Framingham study implicated plasma cholesterol as a major independent risk factor for atherosclerosis. These findings have been confirmed in a number of subsequent reports, which showed the particular significance of low-density lipoprotein. However, it has been extremely difficult to correlate the data on the cell biology and biochemistry of cholesterol with pathological findings from the arterial wall. It appears that no current hypothesis adequately explains the incidence, focal distribution, structure, and composition of human atherosclerotic lesions. This ought to encourage more research on the biology of cholesterol. Our purpose in this chapter is to introduce students to cholesterol and lipoprotein metabolism at an advanced level. At the same time, we shall look critically at available data and indicate, where possible, areas in which important gaps in our knowledge remain.

BIOSYNTHESIS OF CHOLESTEROL

The synthetic pathway is conveniently considered in four sections: the synthesis of mevalonic acid from acetyl-CoA, including the rate-limiting synthesis of mevalonic acid from hydroxymethylglutaryl-CoA (HMG-CoA); the synthesis of squalene from mevalonic acid by the sequential condensation of isoprene units; the cyclization of squalene to lanosterol; and, finally, the series of demethylation and isomerization reactions that convert lanosterol to cholesterol.

Acetyl-CoA to Mevalonate

The precursor pool of cholesterol synthesized in mammalian cells is the acetyl-CoA present in the cytosol. Such acetate units are derived, for example, from the catabolism of long-chain fatty acids in the mitochondria (Chapters 4 and 5). The acetyl-CoA pool is in rapid equilibrium with intracellular and extracellular acetate, and rates of cholesterogenesis, particularly in cultured-cell systems, are often measured from the rate of incorporation of 1- or [2-^{14}C]acetate into cholesterol. The biosynthetic sequence involves a series of condensations leading to the incorporation of 12 carbons from the C-1 position of acetate and 15 carbons from the C-2 position of acetate into the C_{27} skeleton of cholesterol. This stoichiometry is taken into account in the calculation of cholesterol synthesis rates.

The early steps in the reaction sequence leading to cholesterol are well known. The enzymes involved have been isolated or highly purified, usually from liver or yeast (Figure 13.1). The formation of acetoacetyl-CoA from two acetyl units is catalyzed by a specific thiolase. The reaction is driven to completion by the subsequent condensation of the reaction product with another acetyl-CoA unit, which forms hydroxymethylglutaryl-CoA (HMG-CoA). The mechanism of this reaction (catalyzed by HMG-CoA synthase) has been

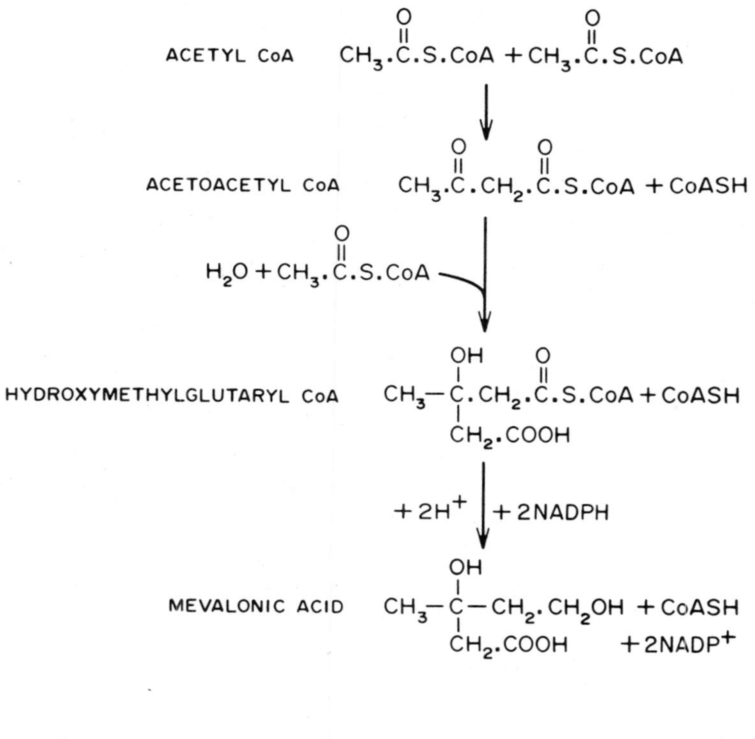

Figure 13.1. Cholesterol synthesis: reactions from acetyl-CoA to mevalonic acid.

studied in detail. Acetyl-CoA is bound to the enzyme first, and the HMG-CoA is released before free CoA. Both thiolase and HMG-CoA synthase are present in the cytosol. An earlier concept, that cholesterol was synthesized at least in part from the condensation of malonyl-CoA with successive acetyl-CoA units, is no longer supported.

HMG-CoA is converted to mevalonic acid by HMG-CoA reductase in the presence of NADPH. The reductase is a microsomal enzyme whose activity is generally rate limiting in the reaction sequence leading to cholesterol. The enzyme protein has been recently isolated from cultured chinese hamster ovary cells in which the gene for reductase had been amplified by selection in the presence of an inhibitor of sterol synthesis. It is a transmembrane protein (with a molecular weight of 97,092) which is glycosylated and from which products of lower molecular weight, also catalytically active, have been isolated. It is possible that proteolysis of the reductase is part of a natural pathway by means of which this enzyme protein is rapidly turned over ($t_{\frac{1}{2}}$ is about 3 h under unstimulated conditions).

As would be expected from its rate-limiting role in cholesterogenesis, the activity of HMG-CoA reductase is highly regulated. Its activity in liver varies at least 10-fold over a well-defined diurnal cycle. It is also regulated by the level of cellular cholesterol. Two kinds of regulatory mechanisms have been proposed. The first involves the reversible phosphorylation of reductase mediated by a phosphatase-kinase cascade similar to that identified for several other microsomal enzymes (Figure 13.2).

It is clear from an analysis of highly purified HMG-CoA reductase that when the enzyme is phosphorylated there is a loss of catalytic activity. Under several conditions it has been shown that catalytic activity changes in parallel with the extent of phosphorylation of the enzyme protein. For example, when reductase activity is inhibited by mevalonic acid or glucagon, the amount of phosphorylated enzyme increases (Beg and Brewer 1982). Under other circumstances, large changes in the activity of the reductase have been reported which were not mediated by changes in phosphorylation. For example, when reductase was inhibited by oxygenated sterols such as 25-hydroxycholesterol, the amount of enzyme protein decreased (Tanaka et al. 1983). This is an active area of continuing research.

It had been suggested earlier that reductase could be regulated directly by cholesterol. It now appears clear that pure cholesterol does not regulate reductase and that the regulatory effects ascribed to cholesterol earlier are mediated by oxygenated cholesterol derivatives (Schroepfer 1981). A wide variety of cholesterol oxidation products, including those which can be generated during intracellular metabolism, inhibit cholesterogenesis, probably by lowering the level of

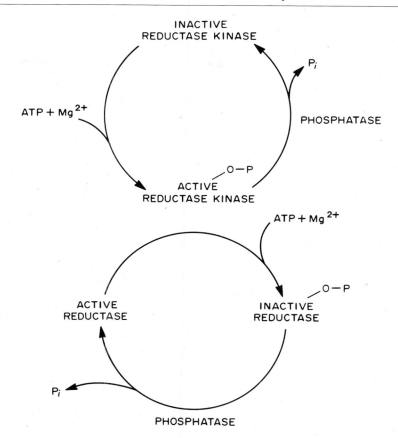

INACTIVE
REDUCTASE KINASE

P_i

$ATP + Mg^{2+}$

PHOSPHATASE

O—P

ACTIVE
REDUCTASE KINASE

$ATP + Mg^{2+}$

O—P

ACTIVE
REDUCTASE

INACTIVE
REDUCTASE

P_i

PHOSPHATASE

Figure 13.2. Regulation of HMG-CoA reductase activity by reversible phosphorylation-dephosphorylation. P_i, inorganic phosphate; reductase, HMG-CoA reductase. Modified from Ingebritsen et al. (1981) with permission.

HMG-CoA reductase. HMG-CoA reductase is also inhibited by several structural analogues of HMG-CoA that have been isolated from microorganisms. Two of these metabolites (compactin and mevinolin) have been tested as inhibitors of cholesterol synthesis in cultured cells and in vivo.

Mevalonate to Squalene

Mevalonate is the precursor of dolichols, quinones, and other compounds derived from isoprene units. Isopentenyl pyrophosphate is derived from mevalonate by phosphorylation and loss of a CO_2 unit; then six units of isopentenyl pyrophosphate are condensed to give presqualene pyrophosphate, which is converted in the presence of NADPH to squalene, with the loss of pyrophosphate (Figures 13.3 and 13.4).

The conversion of mevalonic acid to its pyrophosphate is achieved by the sequential addition of phosphate groups catalyzed by separate kinases. The subsequent formation of isopentenyl pyrophosphate

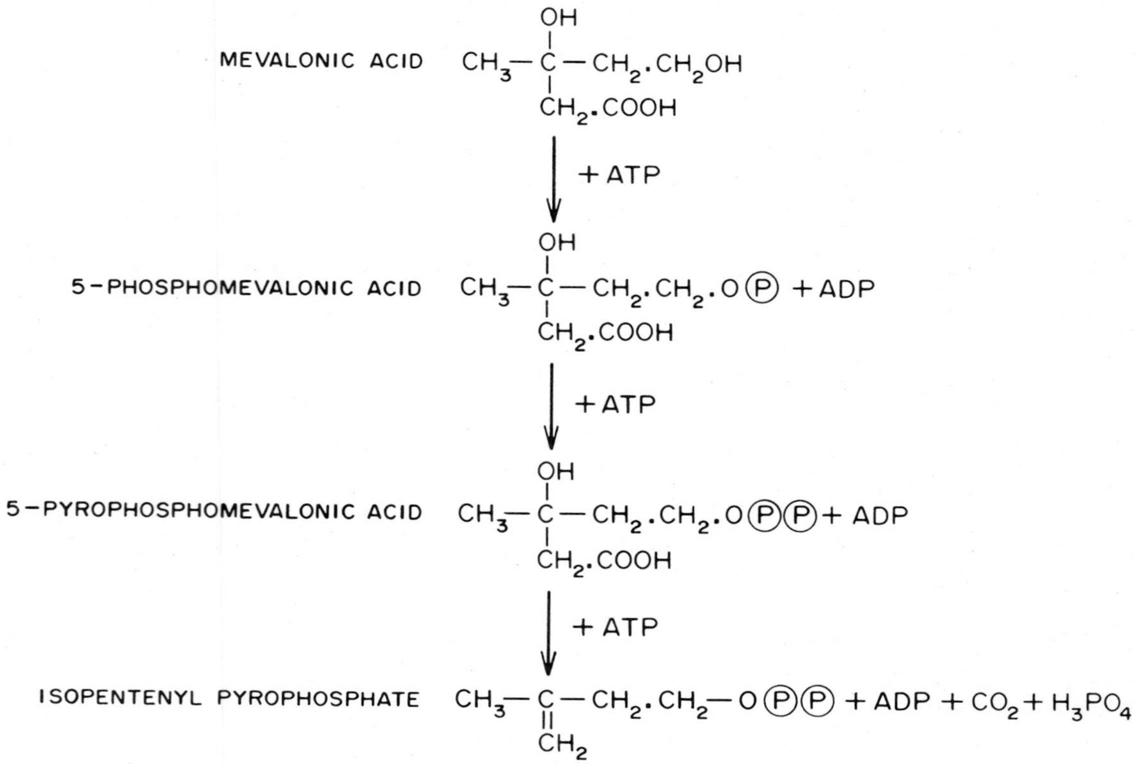

MEVALONIC ACID $CH_3-\underset{\underset{CH_2 \cdot COOH}{|}}{\overset{\overset{OH}{|}}{C}}-CH_2 \cdot CH_2OH$

$\downarrow + ATP$

5-PHOSPHOMEVALONIC ACID $CH_3-\underset{\underset{CH_2 \cdot COOH}{|}}{\overset{\overset{OH}{|}}{C}}-CH_2 \cdot CH_2 \cdot O \textcircled{P} + ADP$

$\downarrow + ATP$

5-PYROPHOSPHOMEVALONIC ACID $CH_3-\underset{\underset{CH_2 \cdot COOH}{|}}{\overset{\overset{OH}{|}}{C}}-CH_2 \cdot CH_2 \cdot O \textcircled{P}\textcircled{P} + ADP$

$\downarrow + ATP$

ISOPENTENYL PYROPHOSPHATE $CH_3-\underset{\underset{CH_2}{||}}{C}-CH_2 \cdot CH_2-O \textcircled{P}\textcircled{P} + ADP + CO_2 + H_3PO_4$

Figure 13.3. Cholesterol synthesis: reactions from mevalonic acid to isopentenyl pyrophosphate.

proceeds with the formation of CO_2, ADP, and inorganic phosphate. While the stereochemistry of these reactions is known, the enzyme proteins involved are not yet fully characterized. The condensation of dimethylallyl and isopentenyl pyrophosphates is probably achieved by the same enzyme complex that also catalyzes the formation of farnesyl pyrophosphate by subsequent condensation with another isopentenyl unit. The enzyme involved (prenyl transferase) has been isolated. It is likely that the formation of presqualene pyrophosphate from the condensation of two farnesyl pyrophosphate units, and its subsequent reduction to squalene, are both catalyzed by the same complex (squalene synthase), but this complex has not been isolated.

An alternative pathway of mevalonate metabolism is the *trans-methylglutaconate shunt*, by which dimethylallyl pyrophosphate is degraded via dimethylacrylyl-CoA and HMG-CoA to acetate units. This pathway is particularly active in kidney tissue; it has been

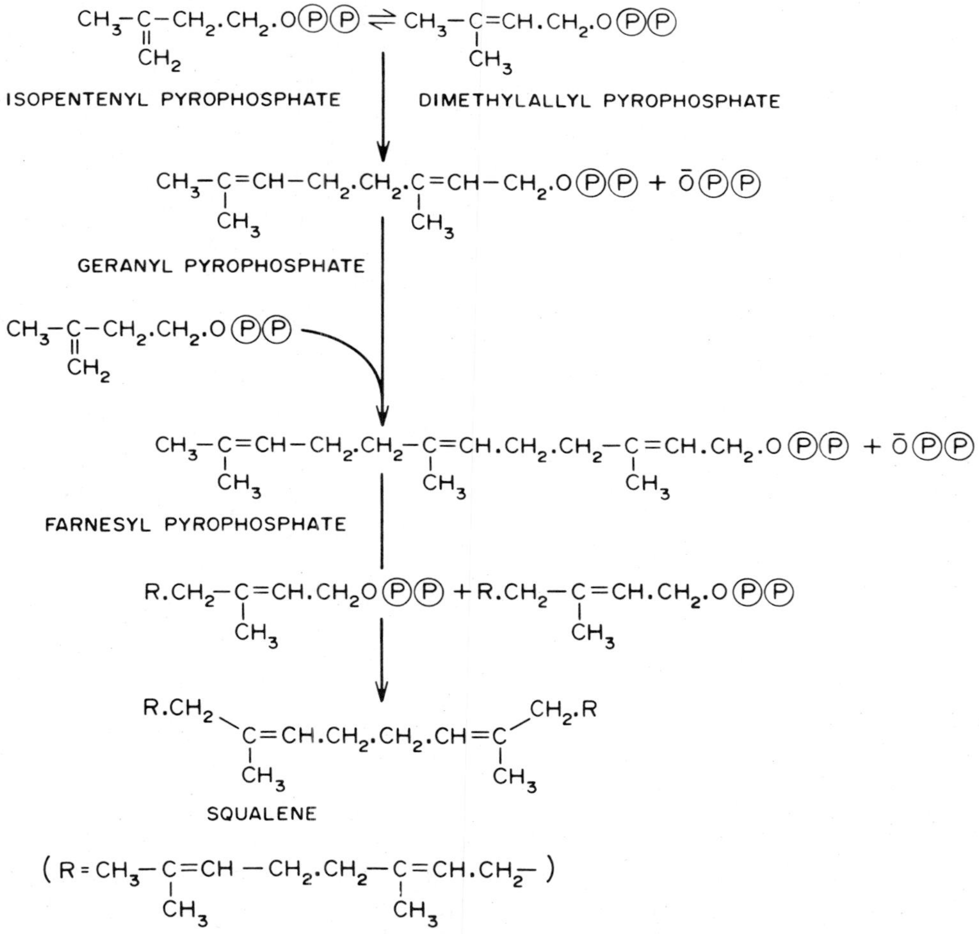

$$CH_3-\underset{\underset{CH_2}{\|}}{C}-CH_2.CH_2.O\,\textcircled{P}\textcircled{P} \rightleftharpoons CH_3-\underset{\underset{CH_3}{|}}{C}=CH.CH_2.O\,\textcircled{P}\textcircled{P}$$

ISOPENTENYL PYROPHOSPHATE   DIMETHYLALLYL PYROPHOSPHATE

$$CH_3-\underset{\underset{CH_3}{|}}{C}=CH-CH_2.CH_2.\underset{\underset{CH_3}{|}}{C}=CH-CH_2.O\,\textcircled{P}\textcircled{P} + \bar{O}\,\textcircled{P}\textcircled{P}$$

GERANYL PYROPHOSPHATE

$$CH_3-\underset{\underset{CH_2}{\|}}{C}-CH_2.CH_2.O\,\textcircled{P}\textcircled{P}$$

$$CH_3-\underset{\underset{CH_3}{|}}{C}=CH-CH_2.CH_2-\underset{\underset{CH_3}{|}}{C}=CH.CH_2.CH_2-\underset{\underset{CH_3}{|}}{C}=CH.CH_2.O\,\textcircled{P}\textcircled{P} + \bar{O}\,\textcircled{P}\textcircled{P}$$

FARNESYL PYROPHOSPHATE

$$R.CH_2-\underset{\underset{CH_3}{|}}{C}=CH.CH_2O\,\textcircled{P}\textcircled{P} + R.CH_2-\underset{\underset{CH_3}{|}}{C}=CH.CH_2.O\,\textcircled{P}\textcircled{P}$$

$$\underset{\underset{CH_3}{|}}{\overset{R.CH_2}{\diagdown}}C=CH.CH_2.CH_2.CH=\underset{\underset{CH_3}{|}}{\overset{CH_2.R}{\diagup}}C$$

SQUALENE

$$\left(R = CH_3-\underset{\underset{CH_3}{|}}{C}=CH - CH_2.CH_2-\underset{\underset{CH_3}{|}}{C}=CH.CH_2- \right)$$

Figure 13.4. Cholesterol synthesis: reactions from isopentenyl pyrophosphate to squalene.

estimated that 7–12% of mevalonate in humans may be catabolized via the shunt (Fogelman, Edmond, and Popjak 1975).

Squalene to Lanosterol   Squalene is converted to squalene-2,3-oxide (Figure 13.5) by a microsomal mixed-function oxidase (squalene epoxidase) in the presence of NADPH, FAD, and O_2 (there is no requirement for cytochrome P_{450} in this reaction). For full activity, the enzyme also requires the presence of phospholipids and a specific carrier protein (Chapter 1). The conversation of squalene 2,3-oxide to lanosterol is catalyzed by the microsomal 2,3-oxidosqualene:lanosterol cyclase. This reaction,

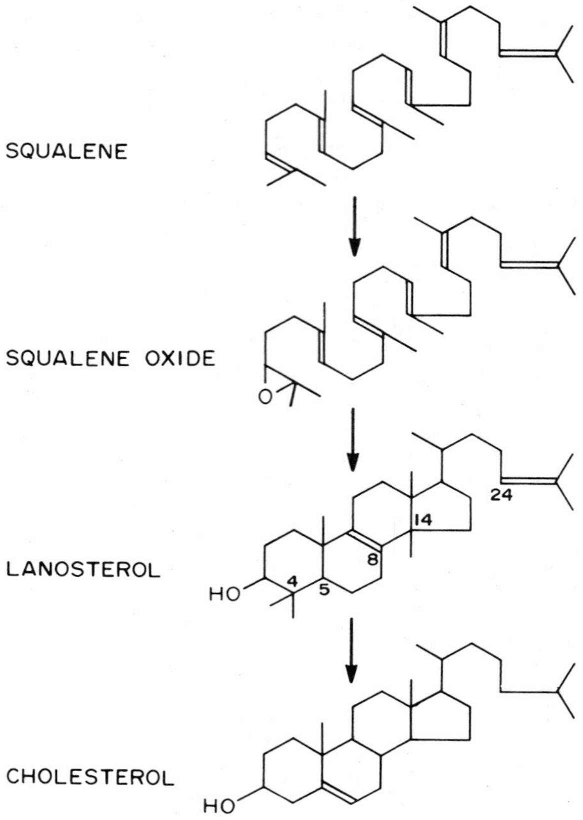

Figure 13.5. Cholesterol synthesis: reactions from squalene to cholesterol.

SQUALENE

SQUALENE OXIDE

LANOSTEROL

CHOLESTEROL

which has been described as the most complex known enzyme-catalyzed reaction, depends on the presence of a surprisingly small (molecular weight of 90,000) protein.

Lanosterol to Cholesterol

The conversion of lanosterol to cholesterol involves a 19-step reaction sequence catalysed by microsomal enzymes (Gaylor 1981). The synthesis of cholesterol proceeds sequentially as follows:

1. demethylation of lanosterol at C-14 with liberation of the methyl group as formic acid
2. two successive demethylations at C-4, involving three microsomal enzymes and the release of two equivalents of CO_2
3. reduction of the C-24 double bond by a specific reductase
4. rearrangement of the 8-position double bond to position 7 by the microsomal 8-isomerase
5. formation of the 5-position double bond
6. reduction of the 7-position double bond by a specific reductase

Table 13.1. Cytosolic Sterol Carrier Proteins

Synonyms	Molecular weight	Isoelectric point	Specificity
Z-protein; fatty acid binding protein (FABP); sterol carrier protein (SCP)	14,184	7.0	Methyl sterol oxidase; also modifies partition of acyl-CoA between membranes
Nonspecific lipid transfer protein; sterol carrier protein 2 (SCP$_2$)	12,300	9.0	7-Dehydrocholesterol-Δ^7-reductase; also acyl-CoA:cholesterol acyltransferase; the exchange of phosphatidylethanolamine; and net transfer of unesterified cholesterol
Soluble protein activator	47,000	?	Squalene epoxidase

Sources: The data were taken from Billheimer and Gaylor (1980), Ferguson and Bloch (1977), Gordon et al. (1983), Grinstead et al. (1983), and Trzaskos and Gaylor (1983).

Sterol Carrier Proteins The earlier literature contains many reports of cytosolic protein factors, usually of low molecular weight, that stimulate the enzymatic conversion of intermediates of cholesterol synthesis. The biological function of such factors within the cell has been difficult to determine, since many of the intermediates of cholesterol synthesis are relatively insoluble in aqueous media. In vitro stimulation of a reaction by a sterol carrier protein may not have significance in the cell. In addition, the carrier proteins when highly purified appear to aggregate readily, and there has been considerable uncertainty concerning the identity of factors purified in different laboratories. Two carrier proteins can now be unambiguously recognized (Table 13.1); a third (soluble protein activator of squalene epoxidase) appears to be a distinct protein. Several other "candidate" carrier proteins have been described.

Measurement of Cholesterol Synthetic Rates In many experiments, particularly those involving cells in culture, all that is required is a valid method to measure the effects of some external factor, for example, the cholesterol contained in extracellular lipoprotein, on the relative rate of endogenous cholesterol synthesis. In other instances, for example, those determining the significance of contributions of different tissues to total cholesterogenesis in an animal, absolute synthesis rates are required. None of the available methods gives unambiguous, absolute synthesis rates. For that reason, synthesis rates are often determined in the same study by two or more methods which rely on different assumptions about intracellular metabolite pools.

Acetate labeled in the 1- or 2-position carbon atom has been extensively used to measure relative synthesis rates. The advantage of this method is that acetate enters the synthetic sequence before the rate-limiting step. Its major disadvantage is that while acetate (unlike acetyl-CoA) readily crosses cell membranes, it is acetyl-CoA and not acetate that is the true precursor for sterol synthesis. Under some conditions, the activity of the enzyme which converts acetate into acetyl-CoA (the cytosolic acetate thiokinase) might be rate limiting; under many conditions the pool of labeled acetyl-CoA might not be in equilibrium with the acetyl-CoA derived from glucose or from the β-oxidation of long-chain fatty acids.

One way to circumvent this problem is to use $[1\text{-}^{14}C]$octanoic acid as a precursor, which at suitably high concentrations represents the major donor of acetyl-CoA units for cholesterol synthesis, although as with labeled acetate, it is difficult to be sure that all acetyl-CoA pools have the same specific activity. In those tissues synthesizing ketone bodies, the specific activity of acetyl-CoA in the mitochondrial compartment can be determined from the radioactivity in acetoacetate.

A different approach is to use tritiated water as precursor for locally synthesized cholesterol. The problem here is that while the origin of hydrogen atoms in cholesterol is known, some of these hydrogens are derived from sources such as NADPH, not water. It is difficult to demonstrate that hydrogen atoms in such pools are equilibrated.

Relative rates of cholesterol synthesis among various tissues have been determined in different mammalian species, using several of these techniques. In all species tested, the liver played a major role in the synthesis of cholesterol, while the intestine was also important. These tissues secrete cholesterol in plasma-lipoprotein form, so their dominant role in synthesis was not unexpected. However, the same studies showed that a major proportion (25–50%) of total-body cholesterol synthesis takes place in other tissues, including peripheral (nonhepatic) tissues in which cholesterol synthesis had been assumed almost fully suppressed by cholesterol uptake from lipoproteins by lipoprotein receptors in cell membranes. This is an important finding because it emphasizes the necessity of pathways that return cholesterol from the peripheral tissues (which do not themselves degrade cholesterol to bile acids) to the liver (which does).

A potential artifact in studies of cholesterol synthesis in cells is that under some conditions, it is not cholesterol itself but C_{27} precursors and other intermediates that accumulate. The most extreme examples come from the use of lipoprotein-deficient serum (produced by centrifugation from native serum) to stimulate cholesterogenesis in cultured cells or tissues. As first shown in 1975, the major sterol products, for example, from $[2\text{-}^{14}C]$acetate, were intermediates between lanosterol and cholesterol. The following were the products when vascular smooth muscle cells were incubated with lipoprotein-deficient serum (Assmann, Brown, and Mahley 1975):

	Radioactivity incorporated
Cholesta-5,24-dien-3-ol (desmosterol)	33%
Lanosterol	35%
Dihydrolanosterol	21%
Cholesterol	5%

That these products were not the result of either incubation period or other factors relating to cell-culture conditions in general was shown by other studies in which identical levels of HMG-CoA reductase were induced in leukocytes, either by a mitogen (which stimulated cholesterogenesis by the need for cholesterol for new membrane synthesis) or by lipoprotein-deficient serum (which induced cholesterogenesis pharmacologically by maximizing cholesterol efflux into the medium). Under the same conditions, in the dividing cells the major sterol product was cholesterol, while in the cells exposed to lipoprotein-deficient serum the major product was lathosterol (5-cholest-7-en-3-ol). In many of the published studies on the regulation of "cholesterol" synthesis, it is likely that little cholesterol was really synthesized. Single-dimension thin-layer chromatography, used in many studies, or digitonide precipitation are insufficient to separate the later intermediates of cholesterogenesis, whereas gas-liquid chromatography, high-pressure liquid chromatography, and/or mass spectrometry can provide unambiguous identification. This difference between assumed and actual products is important in studies of the regulation of cholesterol synthesis in cells.

Finally, other methods not involving the use of labeled precursors have been developed to estimate whole-body cholesterol synthesis in humans. These studies assume the existence of a steady state in the rates of cholesterol uptake, synthesis, and degradation, and for this reason strictly apply only to the adult (nongrowing) subject under metabolic ward conditions. Nevertheless, important information on synthesis rates in normal subjects and in patients with various disorders of cholesterol metabolism has been obtained. In the steady state, *synthesis* can be defined as the difference between the dietary intake of cholesterol and the combined excretion rates of cholesterol and its degradation products. The amounts of these components can be measured by gas-liquid chromatography. Whole-body synthesis rates in normal humans on a cholesterol-free diet have been found to vary from 689 to 1080 mg/d. Obviously, such a method cannot estimate the contributions of individual tissues to total-body cholesterol synthesis.

An alternative method for determining whole-body cholesterol synthesis relies on the observation that detectable levels (about

50nM) of mevalonate are present in plasma. Since mevalonate is the product of the rate-limiting step of cholesterogenesis (from HMG-CoA by the activity of HMG-CoA reductase), its concentration might be expected to reflect reductase activity and hence the synthesis rate of cholesterol. Plasma mevalonate concentrations were measured along with whole-body synthesis rates in normal subjects and in those where cholesterol synthesis was increased by treatment with bile acid sequestering agents or decreased by cholesterol feeding. Under these conditions, plasma mevalonate was proportional to synthesis rate. The method should be applied to other situations (for example, in diabetes mellitus), where studies of whole-body cholesterol synthesis indicate an increased synthetic rate.

REGULATION OF CHOLESTEROL SYNTHESIS

Rates of cellular cholesterol synthesis measured in cultured cells, isolated organs or organ slices, or in the whole organism vary over a manifold range in response to cholesterol demand. Under physiological conditions, cholesterol is needed mainly in cell division (for growth or to replace desquamated cells), where it is required for the synthesis of new cell membranes; or the replacement of cholesterol metabolized to steroid hormones in the adrenals or gonads, or catabolized in bile acids in the liver.

Experimentally, sterol synthesis can be stimulated by exposure of cells to media that contain little or no cholesterol (for example, those containing lipoprotein-deficient serum) and high concentrations of cholesterol acceptors (particularly lysolecithin and the dissociated apoproteins of plasma lipoproteins). Under this condition (a common manipulation), many cells in culture lose to these acceptors in the medium as much as one-third of total cell cholesterol, the major part of which is plasma membrane cholesterol, over a 24-h exposure period. As a result, total sterol synthesis in these cells is increased many times, as is the cellular level of HMG-CoA reductase activity.

However, there is little evidence to suggest that under normal physiological conditions membrane cholesterol depletion is an important factor in regulating cholesterogenesis or reductase activity. Incubation experiments with cells and physiological fluids suggest that cell membranes are normally in close thermodynamic equilibrium with their extracellular medium. Consequently, the necessary rates of compensatory cholesterogenesis are probably always much lower than those maximal rates induced and studied in cholesterol-depleting media. This distinction is important because although only one reaction in a coordinated sequence (the slowest) is rate limiting under given experimental conditions, the rate-limiting reaction need not be the same when overall synthesis rates are high and low.

**Cholesterol Balance
in the Cell**

The following pathways maintain cholesterol balance within the cell (Figure 13.6):

1. uptake of intact lipoproteins via receptor pathways (for example, the LDL receptor)
2. net uptake of free cholesterol from lipoproteins by lipid transfer (for example, from cholesterol-rich lipoproteins)
3. cholesterogenesis
4. efflux of cholesterol (promoted particularly by high-density lipoprotein (HDL) outside the cell)
5. esterification of cholesterol, promoted by acyl-CoA:cholesterol acyltransferase (ACAT)
6. hydrolysis of cholesteryl esters, promoted by neutral cholesterol esterase.

Different experimental manipulations can be designed to affect the relative contributions of the different pathways of Figure 13.6 to overall cellular cholesterol balance (Figure 13.7). For example, incubation of cells in lipoprotein-deficient serum in the absence of LDL

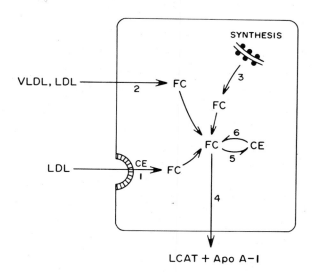

Figure 13.6. Regulation of cellular cholesterol content. In the simplified example shown, a single receptor species (for example, the apo B,E receptor of fibroblasts) catalyzes the endocytosis of LDL, and free cholesterol is transferred into the cell from VLDL and LDL. FC, free cholesterol; CE, cholesteryl esters; 1, receptor-mediated endocytosis and lysosomal cholesteryl ester hydrolase activity; 2, transmembrane free-cholesterol transfer; 3, de novo cholesterogenesis; 4, cholesterol efflux catalyzed by the plasma cholesterol acceptor system; 5, 6, reversible esterification of free cholesterol by acyl-CoA:cholesterol acyl transferase (ACAT).

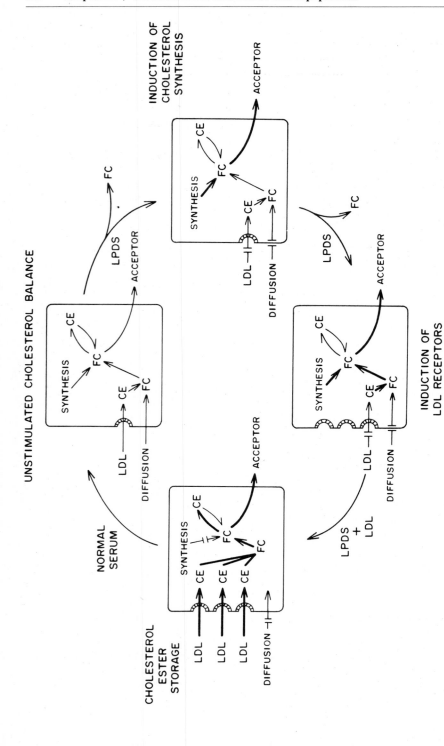

Figure 13.7. Modifications of cellular cholesterol homeostasis and manipulation of receptor number inducible in vitro by sterol depletion (with lipoprotein-deficient serum, LPDS) or sterol loading (with low-density lipoprotein, LDL). Upper panel, normal cholesterol balance; right panel, cholesterol synthesis induced by the loss of free cholesterol from the cell into lipoprotein-deficient serum; lower panel, up-regulation of LDL receptors as loss of free cholesterol to the medium continues; left panel, addition of isolated LDL to the up-regulated, sterol-depleted cell activates acyl-CoA:cholesterol acyltransferase (ACAT) and suppresses cholesterogenesis; top panel; reestablishment of normal cholesterol balance in the unstressed cell; native serum provides a balance of influx and efflux in the presence of moderate cholesterogenesis and down-regulated LDL receptors.

blocks (1) and (2) and promotes maximal cholesterogenesis (3) (right panel of Figure 13.7). Also, incubation of cells in lipoprotein-deficient serum in the presence of LDL stimulates (1) and blocks (3) (left panel of Figure 13.7). Incubation of cells with native (uncentrifuged) HDL in the absence of LCAT (which esterifies cholesterol in the medium and thereby maintains the outward cholesterol gradient from cells to the medium) inhibits (4). Finally, in cells lacking functional LDL receptors, endocytosis (1) is blocked while uptake of free cholesterol (2) is unchanged, relative to normal cells.

It has been argued that the uptake and efflux of free cholesterol by cells simply represent a nonproductive molecular exchange process at the cell surface. This is unlikely to be the case because influx of free cholesterol from plasma is almost entirely from low- and very-low-density lipoproteins (VLDL), while efflux is almost entirely into HDL. Also, free cholesterol interiorized via direct transfer to cellular plasma membranes (2) is esterified and stored as effectively as is cholesterol interiorized via endocytosis (1). This suggests that a single substrate pool of interiorized and newly synthesized cholesterol reacts with ACAT.

The LDL Receptor Hypothesis for the Regulation of Cholesterol Synthesis

In the LDL receptor hypothesis, cholesterol (carried by a plasma lipoprotein, LDL) regulates cholesterogenesis at the level of the rate-limiting reaction, HMG-CoA reductase (Goldstein and Brown 1977). The model has three important features.

First, cholesterol is delivered to specific cell-surface receptors whose ligand is the protein moiety of LDL (apolipoprotein B). Cholesterol delivered as LDL is considered more effective on a molar basis for inhibition of HMG-CoA reductase than is cholesterol delivered in other forms—for example, other lipoproteins that do not react with the LDL receptor—or dispersions of pure cholesterol. The LDL is taken into the cell by endocytosis and delivered to lysosomes. In this organelle, the cholesteryl esters are hydrolyzed to yield fatty acids and free cholesterol.

Second, this free cholesterol transfers from the lysosomes to the rough endoplasmic reticulum, where it has the following regulatory effects: (a) it inhibits HMG-CoA reductase, either by modulating the enzyme activity in the microsomal membrane or by affecting the translation rate of mRNA for reductase; (b) because the inhibition of mevalonate synthesis is not complete, there is a concomitant stimulation of the incorporation of mevalonate into nonsterol pathways such as that leading to ubiquinone; and (c) the free cholesterol activates acyl CoA:cholesterol acyltransferase (ACAT) and thus stimulates the esterification of "excess" cholesterol. Such stimulation has been suggested as a model for atherogenesis.

Third, the interiorized cholesterol also down-regulates the number and expression of receptors at the cell surface, thus inhibiting the uptake of further cholesterol.

It is unquestionable that on many cell surfaces there are LDL receptors that facilitate the degradation of LDL cholesterol (and that of several other lipoprotein classes). However, it is much harder to show that the LDL receptor pathway is rate limiting for cholesterogenesis in vivo under nonpharmacological conditions, or in cultured cells that are in normal cholesterol balance. In fact, the following experimental results seem to argue strongly against this hypothesis:

1. Whole-body cholesterol synthesis was in general no higher in patients with homozygous *familial hypercholesterolemia* (an inherited disease with partial or complete deficiency of LDL receptors) than in normal subjects.
2. Freshly isolated leukocytes or follicle cells from the same patients had similar LDL binding and cholesterol synthesis rates as normal cells.
3. When cultured cells were depleted of cholesterol in lipoprotein-deficient serum so that the cholesterol content of the cell membranes was progressively reduced, induction of HMG-CoA reductase preceded the induction of lipoprotein receptors.
4. Cells in culture incubated in the absence of LDL did not increase their receptor number, contrary to the reaction of the same cells in sterol-depleting media.

These observations concur and suggest that the LDL receptor mechanism is not the rate-limiting factor in determining cholesterol synthesis rates either in cultured cells at rest or under the conditions that characterize normal human metabolism. This does not mean, of course, that considerable amounts of LDL are not degraded via the LDL receptor pathway, only that the rate of entry into the cell of cholesterol from LDL is not normally the rate-limiting factor in cholesterogenesis.

The balance of experimental evidence, summarized from the work of different laboratories, instead suggests that cellular cholesterol content is normally maintained by local cholesterol synthesis to balance any transient inequities between molecular (nonreceptor) pathways of cholesterol influx and efflux (Stange and Deitschy 1983).

Multivalent Feedback Hypothesis and Oxygenated Sterols

As part of a broader concept, it has been proposed that HMG-CoA reductase shows *multivalent feedback regulation* (Brown and Goldstein 1980). In this hypothesis, cholesterol not only blocks the activity of reductase but also blocks a synthetic step subsequent to the formation of farnesyl pyrophosphate, with the result that alternate products of mevalonate (for example, ubiquinone) accumulate. A variety of important metabolic functions, such as the regulation of cell growth and of isoprenoid synthesis, have been suggested as the reason why the cell blocks cholesterol synthesis at a later step as well as at the level of HMG-CoA reductase. However, the following recent findings are contrary to the multivalent theory as originally formulated.

The radioactive compound that was shown to accumulate after inhibition of reductase was identified as ubiquinone but has since been shown to be squalene-2,3:22,23-dioxide (Sexton et al. 1983). Squalene dioxide is also formed when the conversion of 2,3-oxido-squalene to lanosterol is inhibited by the reagent U18666A. The squalene dioxide subsequently cyclizes to sterols such as 24,25-oxidolanosterol and 24(S),25-epoxycholesterol. These, like 25-hydroxycholesterol, are powerful inhibitors of HMG-CoA reductase. Research in several laboratories has shown the major site of this inhibition to be at the level of translation of mRNA for HMG-CoA reductase.

In the presence of U18666A, where LDL cholesterol no longer inhibits cholesterol synthesis, the degradation of LDL apolipoprotein (apo B) is unchanged.

Highly purified cholesterol delivered to cells does not inhibit reductase, whereas oxysterols are potent repressors.

Oxysterols mimic all the effects of LDL in cholesterol metabolism: the suppression of LDL receptor activity; the stimulation of ACAT; and the suppression of other enzymes of cholesterol synthesis as well as HMG-CoA reductase.

Finally, a specific oxysterol carrier protein has been identified in the cytosol. The affinities of oxysterols for this protein reflect their relative ability to suppress reductase. The activity of a large number of oxysterols for inhibition of reductase is proportional to their affinity for the oxysterol carrier protein.

The balance of evidence, recently reviewed (Gibbons 1983), seems against the simple concept that LDL, and its cholesterol, under physiological conditions directly regulate cell cholesterol metabolism or, indirectly, the metabolism of other isoprenoids.

These data suggest a modified reaction scheme in which oxysterols, not cholesterol itself, are the intermediary inhibitors of cholesterogenesis. In this scheme (Figure 13.8), sterols derived from squalene-2,3:22,23-dioxide would be the intracellular messengers that alter reductase activity. Oxidosterols derived from lipoprotein cholesterol would also act via the same oxysterol carrier protein to modulate the effects of extracellular cholesterol.

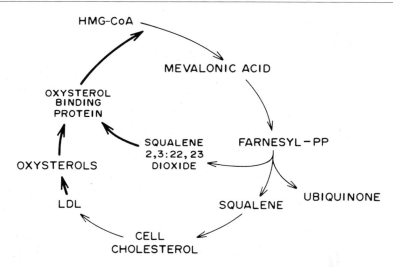

Figure 13.8. Regulation of cholesterogenesis by external or internal oxysterols via the activity of oxysterol binding protein. This scheme summarizes the results of recent data (see text), suggesting that a common signal derived from either external or internal oxysterols mediates HMG-CoA reductase activity.

Finally, there is an important technical problem in working with LDL and cells in tissue culture. Oxygenated sterols may be derived either by intracellular sterol epoxidases or by autoxidation of cholesterol in air. Trace amounts of naturally formed oxysterols are probably present in plasma and tissues, but certainly much larger amounts of oxysterols are present in isolated lipoproteins. Among these, cholesterol in LDL that has been isolated from plasma, is the most susceptible to autoxidation. Indeed, the level of oxidation products in LDL prepared by standard techniques, such as those used in the LDL receptor studies, is sufficient to dissociate endothelial cells from their substratum. Such effects are clearly due to a non-physiological autoxidation and are reversible by the inclusion of the natural antioxidants. The level of oxidized sterols present in isolated LDL is evidently many times that in unfractionated plasma and could be sufficient to account for the observed inhibition of cholesterogenesis by LDL cholesterol. It is possible that much of the reported effect of LDL cholesterol on HMG-CoA reductase was produced by autoxidation products generated during isolation of the lipoprotein.

An additional effect of sterol epoxides is their inhibition of cholesterol 7-α-hydroxylase, which catalyzes the initial and rate-limiting step of the catabolism of cholesterol to bile acids. However, the regulatory effects of this reaction are limited to hepatocytes, where cholesterol is catabolized to bile acids.

Regulation of Cellular Cholesteryl Ester Synthesis

In addition to the effects on HMG-CoA reductase, a second regulatory step associated with the uptake of cholesterol (probably mediated by an oxysterol derivative) is the formation of intracellular cholesteryl ester synthesized by acyl-CoA:cholesterol acyltransferase (ACAT):

$$\text{Acyl-CoA} + \text{cholesterol} \longrightarrow \text{cholesteryl ester} + \text{CoASH}$$

This microsomal enzyme is present in most cells and its activity under normal conditions is probably rate limiting for the synthesis of cholesteryl esters, since progesterone (an inhibitor of ACAT activity) decreases the rate of incorporation of labeled fatty acids into cholesteryl esters in intact cells such as macrophages. The activity of ACAT is typically increased 10- to 20-fold in cells in which the cholesterol content is rapidly increasing above a stable baseline. This occurs in fibroblasts that have previously been depleted of cholesterol by incubation in lipoprotein-deficient serum and then supplied with LDL. The rate of uptake of cholesteryl esters via induced LDL receptors temporarily exceeds the rate of efflux, and ACAT is induced (fibroblasts do not store large amounts of cholesteryl esters under normal conditions). ACAT is also induced in macrophages under conditions where uptake of cholesterol from chemically modified lipoproteins (by a distinct *scavenger* receptor) exceeds efflux. It is not yet clear whether the increased activity results from activation of preformed enzyme or increased availability of substrate, an increased rate of synthesis, or decreased rate of degradation.

ACAT has been only partially purified. There is some evidence that ACAT, like HMG-CoA reductase, might be regulated via its phosphorylation state (inactive when dephosphorylated, active when phosphorylated). However, the data are not yet conclusive. In tissues that secrete plasma lipoproteins (liver, intestine), ACAT is normally active in synthesizing cholesteryl esters, which are secreted in lipoprotein form via the Golgi apparatus, as discussed later. It appears that all mammals possess intestinal ACAT activity, so chylomicrons secreted from the intestine are relatively low in free cholesterol and have a cholesteryl ester content proportional to the rate of uptake of cholesterol from the intestinal lumen. In some species (including humans) the level of ACAT in the liver is very low. It is likely, then, that human hepatic lipoproteins contain little if any cholesteryl esters when secreted.

An important recent finding is that the synthesis of cholesteryl esters in cells, catalyzed by ACAT, is in equilibrium with their hydrolysis by a neutral, nonlysosomal cholesterol esterase (Brown, Ho, and Goldstein 1980). This esterase has been unambiguously identified in macrophages, although not yet characterized biochemically, and its presence can be deduced in other cells. In macrophages, the neutral esterase catalyzes the turnover of cellular cholesteryl ester stores with a half-life of about 12 h. In the presence of an active pathway for the efflux of cholesterol excess cholesteryl ester stores are rapidly depleted, and the cell returns to a baseline cholesterol content in equilibrium with its medium.

THE PLASMA LIPOPROTEINS

Cholesterol, both free and esterified, is secreted in lipoprotein form from intestinal mucosal cells into the mesenteric lymph ducts and thence into the plasma (Chapter 7), and from hepatocytes directly into the plasma. These lipoproteins are multimolecular complexes which are classified on the basis of their lipid composition or their protein composition. While these classifications generally overlap, they do not always, which has historically accounted for a great deal of confusion, in both data and nomenclature.

The lipoproteins are most commonly classified on the basis of their flotation density (Table 13.2). They are classically separated by preparative ultracentrifugation. There are several practical single-spin gradient centrifugation systems suitable for small volumes of plasma. Larger amounts of lipoproteins are prepared by preparative centrifugation. For metabolic or biochemical studies, lipoproteins are usually centrifuged several times to remove trace components of other lipoprotein classes or adsorbed nonlipoprotein proteins. There are a number of caveats that should accompany any formal classification of plasma lipoproteins.

1. The lipoprotein classes represent a continuum of related species, rather than complexes of fixed composition.
2. The functional properties of lipoproteins are usually determined more by their protein composition than by their lipid composition, since the apoproteins include cofactors and inhibitors of several of the reactions of plasma lipid metabolism.

Table 13.2. Human Plasma Lipoproteins Classified by Density

Density range (g/mL)	Lipoprotein fraction	Synonyms
$d < 1.006$	Very low density lipoprotein	VLDL
$1.006 < d < 1.019$	Intermediate-density lipoprotein	IDL
1.006 or $1.019 < d < 1.063$	Low-density lipoprotein	LDL, β-lipoprotein
$1.063 < d < 1.21$	High-density lipoprotein	HDL, α-lipoprotein
$1.063 < d < 1.12$	High-density lipoprotein$_2$	HDL$_2$
$1.12 < d < 1.21$	High-density lipoprotein$_3$	HDL$_3$
$1.21 < d < 1.25$	Very high density lipoprotein	VHDL

3. The lipids and proteins of lipoproteins in native plasma are not in equilibrium but in a dynamic nonsteady state, which reflects the fact that newly synthesized lipoproteins are continually being secreted into the plasma compartment, and particles that have been transformed by reactions in the plasma are constantly being removed. Even in plasma rapidly obtained from freshly isolated blood and subjected to only mild fractionation processes, there may be significant differences between circulating lipoprotein complexes and those studied in vitro.

4. Ultracentrifugation causes major changes in the metabolic properties of some native lipoproteins. It is therefore essential that the composition and properties of lipoproteins obtained by centrifugation be confirmed by an independent isolation procedure. For example, the multiple discrete species of HDL detectable by the direct electrophoresis of native plasma are converted to a broad homogeneous band by centrifugation, probably representing a fundamental rearrangement of lipids and HDL proteins under the influence of shearing and ionic-strength forces. Other fractionation methods need to be validated on an individual basis.

5. There is no absolute distinction in binding properties between "genuine" lipoprotein apoproteins and some "contaminating" plasma proteins; rather, there is a range of affinities covering stronger and weaker associations of proteins with the surface lipids of the lipoprotein particles. For example, some accepted apolipoproteins (such as apo A-1) are readily removed by fractionation and probably exist in an equilibrium between lipoprotein-associated and soluble fractions. It has been suggested that apo A-1 may be removed from plasma for catabolism as its water-soluble species. On the other hand, albumin, usually considered a contaminant of VLDL, plays an important metabolic role as acceptor of triglyceride-derived fatty acids. However, it is difficult to remove albumin completely from VLDL without the dissociation of "genuine" apolipoproteins such as apo E.

6. All the plasma lipoprotein complexes contain mixtures of more- and less-polar lipids (Table 13.3). The major lipids of the plasma lipoproteins are cholesteryl esters, triglycerides, cholesterol, and phospholipids. It was initially believed that proteins, free cholesterol, and phospholipids together made up a monomolecular film of "surface," organized around a spherical "core" of nonpolar lipids (triacylglycerol and cholesteryl esters). More recent experiments have modified this simple picture. Direct chemical analysis of surface and core elements shows that free and esterified cholesterol and

Table 13.3. Composition of the Major Human Plasma Lipoprotein Classes

Lipoprotein	Isolation method	Protein	Free cholesterol	Phospholipid	Cholesteryl ester	Triacylglycerol
				(w%)		
VLDL*	Affinity chromatography	10.4	5.8	15.2	13.9	53.4
IDL[†]	Density gradient centrifugation	17.8	6.5	21.7	22.5	31.4
LDL*	Affinity chromatography	25.0	8.6	20.9	41.9	3.5
HDL_2[‡]	Rate zonal centrifugation	42.6	5.2	30.1	20.3	2.2
HDL_3[‡]	Rate zonal centrifugation	54.9	2.6	25.0	16.1	1.4
VHDL[§]	Preparative ultracentrifugation	62.4	0.3	28.0	3.2	4.6

* The data are from Fielding et al. (1984).
[†] IDL = LDL_1; the data are from Lee and Downs (1982).
[‡] The data are from Groot et al. (1982).
[§] The data are from Alaupovic et al. (1966).

triacylglycerol are distributed between surface and core (Miller and Small 1983). Nuclear magnetic resonance studies indicate that, at least in the case of free cholesterol, the equilibrium between surface and core is very rapid relative to the rate of metabolic reactions in plasma (Lund-Katz and Phillips 1984). The significant solubility of some polar lipids (for example, free cholesterol itself) and apoproteins in aqueous media means that any isolation procedure inevitably modifies by dissociation the native composition of both surface and core.

Lipoprotein Complexes Present in Plasma

The distribution of the major individual apolipoproteins and their molecular weights are given in Table 13.4. While a few of the apolipoproteins have well-defined metabolic roles, the roles of others are more controversial, and some have no known functions at all. The identifications of mutant deficiencies of individual apoproteins have been particularly helpful, because functions deduced from the activities of isolated apoproteins reincorporated with synthetic lipid dispersions can be confirmed with native plasma lipoproteins. For example, apo C-2 activates triacylglycerol hydrolysis by lipoprotein lipase. Genetic apo C-2 deficiency is associated with hypertriglyceridemia similar to that present when the lipase itself is absent.

Most plasma lipoproteins contain more than a single apolipoprotein species. Some information has been obtained on the apoprotein composition of lipoproteins in native plasma by *immunoaffinity chromatography*. In this technique, plasma is passed sequentially through columns that contain, immobilized on a solid support, antibodies to individual apolipoproteins. Analysis of the mixture of apoproteins retained at each step provides information on the com-

Table 13.4. Apolipoproteins of the Human Plasma Lipoproteins

Apolipoprotein	Molecular weight	Lipoprotein distribution
Apo A-1	28,331	HDL
Apo A-2	17,380	HDL
Apo B-48	200,000	Chylomicrons
Apo B-100	350,000	VLDL, LDL
Apo C-1	7,000	HDL, VLDL
Apo C-2	8,837	Chylomicrons, VLDL, HDL
Apo C-3	8,751	Chylomicrons, VLDL, HDL
Apo D	32,500	HDL
Apo E	34,145	Chylomicrons, VLDL, HDL

Note: Several reported minor apolipoproteins, of unknown properties and functions, are not included. Further research will be required before their significance as lipid-binding proteins in plasma is established.

plexes present. The technique does not appear to modify the association of apolipoproteins in native lipoprotein complexes. The results of some of these experiments are shown in Table 13.5. Interestingly, some apoprotein combinations are not present. For example, in spite of the relatively high level (300 μg/mL) of apo A-2 in plasma, none of this apoprotein appears by itself in lipoprotein form (that is, it is always associated with other apolipoproteins). In the same way, complexes containing both of the minor HDL apolipoproteins D and E are absent from normal plasma. In spite of the high levels of both apo A-1 and A-2 in plasma, these apoproteins are not found in appreciable concentration in plasma on the same particles as apo B, the major protein of VLDL and LDL. Other apoproteins (for example apo E) appear to distribute freely among all the major lipoprotein classes, although the distribution is affected by the lipid composition of the lipoproteins; apo E appears to associate particularly with cholesterol-rich lipoproteins. The physical basis of these distributions is not understood. Nevertheless, it is clear that the native plasma lipoproteins represent a complex nonequilibrium dynamic system whose components reflect both the biophysical properties of both lipids and proteins as well as the effects of recent metabolic transformations.

Table 13.5. Lipoprotein Complexes Present in and Absent from Normal Human Plasma

Apolipoprotein	Complexes present	Complexes absent
A-1	A-1 only;	
	A-1:A-2; A-1:D; A-1:E	A-1:B
	A-1:A-2:D; A-1:A-2:E	A-1:B:D; A-1:B:E
A-2		A-2 only;
	A-1:A-2;	A-2:B; A-2:D; A-2:E
	A-1:A-2:D; A-1:A-2:E	A-2:B:D; A-2:B:E; A-2:D:E
B	B	
	B:E	A-1:B; A-2:B; B:D; B:E;
D		D only;
	A-1:D;	A-2:D; B:D; D:E;
	A-1:A-2:D	A-2:D:E; A-2:B:D;
E		E only;
	A-1:E; B:E	A-2:E; D:E
	A-1:A-2:E	A-1:D:E; A-2:D:E

Sources: The data are from Fielding and Fielding (1980), Fielding et al. (1982), Fielding et al. (1984); and Castro and Fielding (1984).
Notes: The complexes represent those detectable or undetectable by sequential immunoaffinity chromatography of normolipidemic human plasma. There appear to be no data at present on the distribution of apo proteins C-1, C-2, or C-3, in terms of apoprotein associations. Apo B in this table represents the major apo B species of fasting plasma, apo B-100.

SECRETION OF PLASMA LIPOPROTEINS
Secretion by the Liver of Lipoproteins Containing Apo B

The liver is a major site for the secretion of lipoproteins containing apo B. Recent research with human and many other animal plasmas indicates that apo B is secreted by the liver largely or completely as its high molecular weight (apo B-100) form (Kane 1983). The particles secreted that contain apo B are triacylglycerol rich and contain smaller amounts of phospholipid, free cholesterol, and (in species containing sufficient levels of hepatic ACAT) cholesteryl esters. Primary hepatic lipoprotein particles fall mostly within the VLDL density range. They are rapidly converted after secretion into plasma to a composition typical of plasma VLDL, mostly by the addition of further apolipoproteins from plasma. The newly secreted particles are usually termed *nascent VLDL*. The apo B of nascent VLDL is synthesized in the rough endoplasmic reticulum, where the enzymes of phospholipid synthesis and ACAT are also present. On the other hand, the enzymes of triacylglycerol synthesis have been reported to be in the smooth endoplasmic reticulum.

On these grounds it has been proposed that a pronascent VLDL consisting of protein, phospholipid, and cholesteryl ester is first synthesized, then combined with a preformed triglyceride droplet. It is hard to see how this type of combination would be physically favorable if the VLDL contained any significant proportion of cholesteryl esters, because of the almost complete insolubility of this lipid in the aqueous phase. Additionally, as the VLDL synthesis rate is systematically varied, the syntheses of triacylglycerol and cholesteryl ester are well coupled, while synthesis of apo B is poorly coupled to both. The cholesteryl ester to triacylglycerol ratio of intracellular lipid droplets in rat liver after cholesterol feeding is the same as that of the secreted VLDL, which also suggests a common precursor pool of less-polar lipids. Overall, it appears more likely that cholesteryl ester, synthesized in the rough endoplasmic reticulum, is transferred to the smooth endoplasmic reticulum (perhaps with the assistance of a transfer protein, such as is present in plasma) and that this complex is stabilized by the incorporation of apo B and phospholipid (Figure 13.9).

There appears to be no direct evidence for the point of addition of free cholesterol to VLDL. But the synthesis of cholesterol in the endoplasmic reticulum, the solubility of cholesterol in phospholipids, and the presence of higher-than-normal levels of free cholesterol in VLDL when (as in diabetes) whole-body cholesterol synthesis is increased make it likely that free cholesterol is added along with apo B and phospholipid. However, there are kinetic data suggestive of a preferential recycling of cholesterol from receptor-interiorized LDL into nascent VLDL (see later section on cholesterol degradation). Since LDL contains high free cholesterol levels relative to VLDL, and since there is a proximity in the cell between the degradative

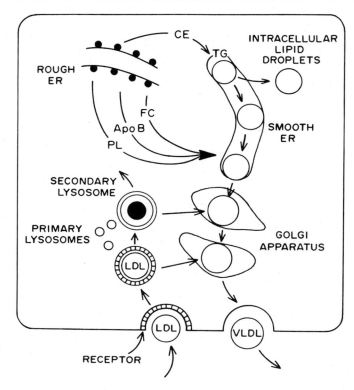

Figure 13.9. Synthesis of nascent VLDL in the hepatocyte and its possible relationship to the recycling of free cholesterol (FC) within the cell predicted by recent data on the sources of free cholesterol used for bile acid synthesis. PL, phospholipid; CE, cholesteryl ester; TG, triacylglycerol. Arrows between lysosomes and Golgi indicate the potential pathway for FC transfer.

(secondary lysosome) and synthetic (Golgi) organelles (Hornick et al. 1984), it is possible that free cholesterol also diffuses down a concentration gradient via a countercurrent distribution within the hepatocyte to VLDL.

The triacyglycerol-rich lipoproteins secreted by the liver also contain smaller amounts of non–apo B apoproteins (for example, apo E and the C proteins), which are found associated with them in preparations of Golgi bodies. It is likely that these apoproteins are added at the same point as apo B. Lipoproteins containing apo B as the only apoprotein are physically stable, however, so these other proteins probably play no important intracellular role.

Secretion by the Intestine of Lipoproteins Containing Apo B

The second source of triacylglycerol-rich lipoproteins is the intestine. Dietary cholesterol is adsorbed by the mucosa, reesterified in large part from the same pool of unesterified fatty acids from which triacylglycerol is derived, then secreted as cholesteryl ester into the intestinal lymph in a complex with the smaller form of apo B (apo B-48), together with phospholipids and small amounts of free cholesterol. Lipoproteins of this origin are usually termed *chylomi-*

crons. Although some of the particles fall within the same size range as hepatic VLDL, the term *intestinal VLDL* for the smaller sizes of particles within the continuous distribution present in lymph, has little to recommend it. All mammalian species investigated have significant levels of ACAT in intestinal mucosal cells. During cholesterol feeding, the level of cholesteryl ester in chylomicrons increases, although whether this results from induction of ACAT synthesis or activiation of preformed enzyme is uncertain. Chylomicrons also contain low levels of free cholesterol. Its concentration does not increase in cholesterol feeding, and it may originate in large part from intestinal cholesterogenesis, rather than dietary cholesterol.

The fatty acid moieties of the phospholipids of chylomicrons are derived mainly from fatty acids synthesized endogenously, since in animals fed pure fatty acids, the fatty acid compositions of triacylglycerol and cholesteryl esters approach the fatty acid composition of the diet, while that of phospholipids is little changed. These data fit well with the same model that was proposed for hepatic VLDL synthesis. Apo B is required for the secretion of chylomicrons, as it is for hepatic VLDL, although the synthesis of apo B and of neutral lipids is not well coordinated. There is evidence that apo B is stored within the mucosal cell for rapid release in triacylglycerol-associated form following fat absorption (Figure 13.10).

One difference between the triacylglycerol-rich lipoproteins of intestine and liver lies in the apolipoproteins other than apo B which are associated with them during secretion. All the apoproteins may be synthesized by both tissues to some extent. However, in humans a major site of apo A-1 synthesis is the intestine, but the liver also plays a significant role. Apoproteins E, C-2, and C-3 are derived mainly from the liver and possibly other sites.

Other Sources of Plasma Apolipoproteins

It has recently become clear that several other tissues that do not secrete triacylglycerol-rich lipoproteins nevertheless synthesize and secrete lipoprotein apoproteins. Studies in different species on the mRNA of macrophages, kidney cells, and some brain cells indicate that these cells can secrete apo E. The secretion of apo E by macrophages is increased during cholesterol feeding; this was originally considered a specific response of the cell to the cholesterol loading that occurs under these conditions. However, further investigation has shown that this increased synthesis of apo E during cholesterol feeding reflects a general stimulation of protein synthesis and secretion by the macrophage. It is unlikely that the apo E of macrophages plays any local role in the promotion of cellular cholesterol efflux.

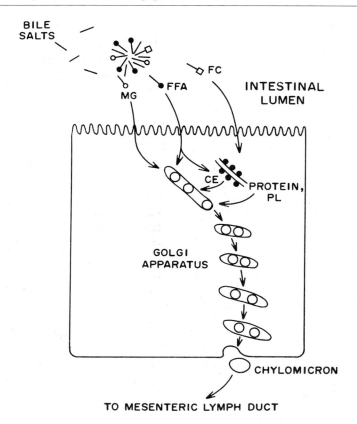

Figure 13.10. Synthesis of nascent chylomicrons in the intestinal mucosal cell. The figure incorporates the evidence suggesting cholesteryl ester (CE) transfer within the cell from the rough to the smooth endoplasmic reticulum (see text). Triacylglycerols are resynthesized from the absorbed monoacylglycerols and fatty acids. MG, monoacylglycerol; FFA, free fatty acid; PL, phospholipid; FC, unesterified cholesterol.

Apo A-1 is secreted by the liver (and by a variety of tissues in chickens) in a process distinct from the secretion of VLDL. The apo A-1 appears in the plasma space as a discoidal lipoprotein with a density within the HDL range. It has not yet been determined at which stage the lipid (largely phospholipid and unesterified cholesterol) is added to apo A-1 in such lipoprotein discs. Because of the spontaneous association and ready dissociation of these lipids and proteins, the complex may represent a form synthesized during the later stages of secretion.

Some general points should be kept in mind about the significance of apoproteins in lipid transport from cells to extracellular fluids such as plasma.

Apo B, which is insoluble in aqueous solution, probably serves a stabilizing function during lipoprotein synthesis and secretion. When apo B synthesis is inhibited or genetically deficient, there is no secretion of chylomicrons or VLDL. The other apoproteins, however, are appreciably soluble in aqueous media but associate spontaneously with phospholipid surfaces and readily exchange among them, probably via the aqueous phase. In a sense, it is not important whether these apoproteins are secreted as apolipoproteins or as lipoproteins because they reach equilibrium with available lipid surfaces as soon as they are secreted.

Even though a protein such as apo E may represent a significant proportion of total protein synthesized by the macrophage, this may be unimportant in terms of whole-body apo E production if the proportion of whole-body protein synthesis by the macrophage is small, as is the case. This argument could be countered if it could be shown that the apoprotein acts in local lipid transport. However, it has not been possible to show that apo E plays any special role in the net transport of cholesterol out of cells. On the other hand, apo A-1, which is present in relatively high concentrations in both plasma and intercellular fluids, is far more active in the removal of cholesterol from cells. Apo E-HDL is less active than other HDL fractions in promoting cholesterol efflux.

Perhaps only in the case of lipoproteins containing apo B is the secretion of apoprotein necessarily associated with the local net secretion of lipids. It may be relevant that the apo B-containing lipoproteins are responsible for the secretion of the major nonpolar lipid product, triacylglycerol. The transport of more-polar lipids into the extracellular fluid probably occurs via the action of transfer proteins or through the aqueous medium to an apoprotein or lipid-apoprotein complex such as that formed from apo A-1 or apo E. Such apoproteins can act as acceptors of lipids diffusing through the aqueous phase. Except for apo B, apoproteins may not have local functions in relation to their tissue of origin. Their physiological roles are in the plasma as cofactors, transfer proteins, or receptor ligands.

The enzymes and transfer proteins involved in modification of circulating lipoproteins are themselves secreted into the plasma. Lecithin:cholesterol acyltransferase (LCAT), which catalyzes plasma cholesteryl ester synthesis, originates in the liver. Cholesteryl ester transfer protein, which catalyzes the distribution of these esters among plasma lipoproteins, has the same source. Lipoprotein lipase, which catalyzes triacylglycerol hydrolysis on the endothelial surface, is derived locally from the adjacent parenchymal cells. Hepatic lipase, which may play a role in VLDL remodeling, is derived from liver. The origin of the phospholipid transfer protein of plasma has not been reported. Nor is it known whether these enzymes and transfer proteins are secreted in association with lipids.

DYNAMICS OF CHOLESTEROL TRANSPORT AND METABOLISM IN THE CIRCULATION

Cells that do not secrete into their medium cholesterol or its esters in lipoprotein form nevertheless can lose free cholesterol from their cell membranes. This section concerns the pathway by which such cholesterol entering the plasma is returned to the liver for catabolism.

Free cholesterol in plasma is derived from those cells in which the net rate of cholesterogenesis is greater than that required for the synthesis of new membranes. Since peripheral cholesterogenesis is not completely suppressed, and since many tissues contain few dividing cells, the majority of cells are in positive cholesterol balance. If the accumulation in cells of free and esterified cholesterol is to be prevented, such excess must be recycled through the plasma to the liver, where it can be degraded. The mechanism by which those tissues not secreting lipoprotein particles return cholesterol to the liver has been called *reverse cholesterol transport*. However, as we shall see, there probably is no important difference in the physical processes involved in the uptake or efflux of nonlipoprotein cholesterol. Because cholesterol is partly soluble in aqueous media, its net transport can be explained largely in terms of the chemical potential of cholesterol in donor and recipient surfaces, and reverse cholesterol transport may itself be reversed under changing physiological conditions. *Cholesterol net transport* is proposed as a better term for describing directional mass transport of the kind that occurs between cell membranes and their media.

Two points need to be kept in mind. First, the use of cholesterol-loaded or cholesterol-depleted cells or media can cause cholesterol to move either from or into the cells. While such maneuvers are useful for uncovering available biochemical pathways, the net movement of cholesterol so demonstrated may have little relevance to physiological conditions if the gradients imposed (that is, concentration differences between cells and medium) are much greater than, or reversed from, those occurring naturally among plasma, extracellular fluids, individual cells, and tissues. The evidence suggests that the cholesterol of cell membranes and normal physiological fluids is in fairly close equilibrium. This equilibrium may be disturbed temporarily by changes in plasma lipoprotein composition, for example, as occurs in postprandial lipemia.

Second, the turnover time of some lipoprotein species—for example, chylomicrons and VLDL—is of the same order as that of the diffusion of cholesterol between lipid surfaces. Accordingly, the cholesterol of such lipoproteins will not normally reach equilibrium with that of the other lipoproteins (LDL, HDL) in plasma.

These two points indicate the experimental difficulty of studying free cholesterol transfers in plasma. However, the magnitude of such transfers makes them the major part of the total movement of cholesterol between nonhepatic cells and plasma (see next section).

Thermodynamics of Cholesterol Flux and Net Transport

The flux of cholesterol between cell membranes and extracellular fluids can easily be measured with cells preincubated to equilibrium with radiolabeled cholesterol. Different studies agree that the flux of cholesterol between cell membranes and medium lipid surfaces is rapid, typically 5–10% of total cellular sterol per hour at 37°C. This finding, together with the observation that the rate of cholesterol flux would be the same at low plasma or medium lipid concentrations, has led a number of investigators to conclude that cholesterol must be transported between cells and their medium by mechanisms that involve either nonspecific (collision) or specific (receptor-mediated) contact between cell membranes and acceptor lipid surfaces.

However, thermodynamic considerations argue against this view and in favor of the idea that net movement of cholesterol depends mainly on the chemical potential of cholesterol in donor and recipient surfaces. Although cholesterol is only slightly soluble in aqueous media (with a critical micelle concentration of about $3 \times 10^{-8}M$), this solubility appears to be sufficient to account for a substantial proportion of the observed flux rates between cells and plasma lipoproteins. Additionally, flux rates calculated from theoretical considerations of diffusion rates (Table 13.6), as well as flux rates among synthetic lipid vesicles containing no protein receptors, are very similar to those found for the reaction between living cells and lipoproteins in native plasma. Very similar values are also obtained for the activation energy associated with efflux in these three systems. While such data do not rule out some contribution of collision (whether between lipoprotein particles, or lipoproteins and cell membranes) to total efflux, they do indicate that in many cases it must be relatively small. It appears that the major part of cholesterol efflux is due to a rate-limiting desorption step and diffusion to the acceptor lipoprotein through the aqueous medium (Figure 13.11).

Table 13.6. Thermodynamic Constants for Free-cholesterol Flux in Cell Membranes and Model Systems

Diffusion coefficient (J)	(mol/cm²/s)	Activation energy (G)	(kcal/mol)
J_{theor}	2×10^{-15}	G_{theor}	14
$J_{cells \rightarrow plasma}$	1×10^{-14}	$G_{cells \rightarrow plasma}$	11
$J_{vesicle \rightarrow vesicle}$	4×10^{-15}	$G_{vesicle \rightarrow vesicle}$	8–16

Notes: J_{theor} is the diffusion rate calculated from Fick's Law ($J = Dc/\chi$), where D is the lateral diffusion coefficient of free cholesterol (8×10^{-5} cm²/s), c is the monomeric saturating concentration of cholesterol (3×10^{-11} mol/cm³), and χ is the thickness of the unstirred water layer (0.03 cm) for a continuous cell surface. $J_{cells \rightarrow plasma}$ was determined for normal fibroblasts in normal human plasma (Fielding and Moser 1982); $J_{vesicle \rightarrow vesicle}$ is for synthetic lecithin cholesterol liposomes (McLean and Phillips 1981). G_{theor} was determined as $G = RT\log_e x_w f_w$, where x_w is the aqueous mole fraction and f_w the activity coefficient (Tanford 1980, modified as described by McLean and Phillips 1984). $G_{cells \rightarrow plasma}$ is for normal fibroblasts (Fielding and Moser 1982) and $G_{vesicle \rightarrow vesicle}$ is from McLean and Phillips (1984). Reprinted from Fielding (1984) by permission.

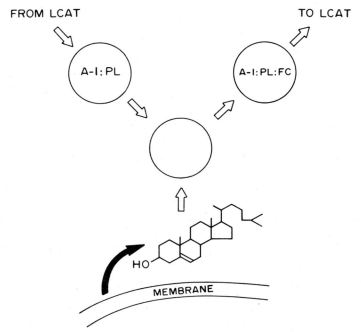

FROM LCAT   TO LCAT

A-I:PL   A-I:PL:FC

MEMBRANE

Figure 13.11.
Diffusion-dependent transfer
of free cholesterol from the
cell surface to a lipoprotein
acceptor containing apo A-1
in the aqueous phase. PL,
phospholipid; FC,
unesterified cholesterol.
Modified from Phillips et al.
(1980), by permission.

For cholesterol efflux from cells to be significant in terms of net transport, it must be greater than the simultaneous influx of cholesterol from the medium to the cells. For maintenance of net transport, such a difference (efflux greater than influx) must be coupled either to an efficient continuous removal of the cholesterol acceptor from the medium, and its replacement by new acceptors, or to a chemical reaction that modifies cholesterol and thereby continuously reduces its concentration in the medium.

Cholesterol Efflux and LCAT

The evidence is very strong that the net transport of cholesterol from nonhepatic cells into the circulation is maintained by a chemical reaction, the esterification of cholesterol by LCAT (Dobiasova 1983). When the activity of LCAT in plasma is blocked by an inhibitor such as the sulfhydryl reagent dithiobis(2-nitrobenzoic acid) (DTNB), net transport of cholesterol from cultured cells to medium ceases. Similar results have been reported with DTNB in rats in vivo. Furthermore, LCAT-dependent net transport rates of cholesterol parallel efflux rates in a variety of cell types (including all the major cell types of the vascular bed) over a wide range of medium acceptor concentrations. Finally, in patients with a genetic deficiency of LCAT, free cholesterol accumulates in the tissues, while LCAT-deficient plasma is ineffective at promoting the net transport of cholesterol from cell membranes.

Several studies have investigated whether or not LCAT is bound to a particular lipoprotein species. Early data showed cholesteryl esters in all the major plasma lipoprotein fractions. The absence of almost the entire plasma cholesteryl ester in congenital LCAT deficiency indicated that cholesteryl esters in human plasma are derived from LCAT activity. Experiments with ultracentrifugally isolated lipoproteins indicated that lipoproteins with apo A-1 are the only effective direct substrate for LCAT in vitro. These results raise two more specific questions, however: (1) Is LCAT associated with HDL in general or with a specific subfraction of the many HDL species present in plasma? (2) Does LCAT receive its substrate (cholesterol) only (if at all) from HDL, or also by transfer from other lipoproteins? Recent research has provided considerable information on these points.

It was shown that a protein of HDL (apo A-1) is the cofactor of LCAT activity in plasma. Studies since that time have indicated that only a small part of total HDL is the primary substrate for LCAT. The relevant findings are the following: First, there are many more (about 500-fold) HDL particles in plasma than molecules of LCAT. Kinetic data on cholesterol turnover in human plasma supports the concept that a small pool of HDL, rapidly turning over, provides the majority of cholesteryl esters in plasma lipoproteins. Second, immunoaffinity chromatography studies by two laboratories have indicated that the subfraction of HDL containing only apo A-1 (and lacking apo E) is the major acceptor of cholesterol from cell membranes. The removal of apo A-2 (the second major apoprotein of HDL, present in 90–95% of HDL particles) has little effect on efflux, net transport, or LCAT activity. Third, small HDL particles containing only apo A-1 accumulate in the plasma of patients with LCAT deficiency. Finally, LCAT itself is found in plasma not as a free protein but as a complex with apo A-1 and usually LCAT is also associated with a minor HDL apoprotein, apo D, forming a unique, small subfraction of total HDL. Taken together, these data strongly indicate that a subfraction of HDL containing apo A-1 is the major acceptor of cholesterol from cell membranes, and they suggest that this species preferentially reacts with LCAT to maintain a chemical potential gradient (effective concentration difference) of free cholesterol between the cell membrane and the medium (Figure 13.12).

A smaller proportion of cholesterol efflux is to albumin. The molar concentration of albumin in plasma is about 40-fold greater than that of total apo A-1 (and about 600-fold greater than that of the fraction of apo A-1 most active in cholesterol net transport). Since free cholesterol has no detectable physical interaction with albumin, there is probably a complex of lysolecithin (or, less likely, unesterified fatty acids) bound to albumin, which is the true acceptor of cholesterol in this reaction. Efflux to albumin, unlike the efflux promoted by apo

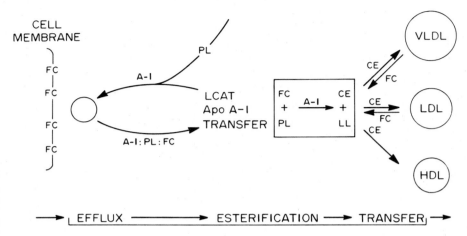

Figure 13.12. The coupling of efflux, esterification, and transfer in maintaining the free-cholesterol potential gradient between cell membranes and plasma in the presence of lecithin:cholesterol acyltransferase (LCAT) activity. FC, free cholesterol; PL, phospholipid; CE, cholesteryl ester; LL, lysolecithin.

A-1, is not coupled effectively to net transport. Removal of albumin from plasma by affinity chromatography does not reduce LCAT-dependent net transport. In human analbuminemia (a genetic disorder in which albumin is absent or present at an extremely low concentration), the whole of cholesterol efflux is apo A-1 dependent and can be linked to net transport via LCAT.

A complicating factor of plasma cholesterol metabolism is that not only cell-membrane cholesterol but also the cholesterol of plasma lipoproteins contributes to the total cholesterol esterified by LCAT. When LCAT is inhibited in normal human plasma, there is neither net transport of cholesterol from cell membranes to plasma nor net transport among the major plasma lipoprotein classes. However, if LCAT activity is present, free cholesterol is drawn from both cell membranes and plasma lipoproteins, but when equal *masses* of cell and plasma lipoprotein cholesterol are present, up to 80% of free cholesterol used for esterification by LCAT is derived from cell membranes (Fielding and Fielding 1982).

Unexpectedly, individual cell types differ severalfold in their efficiency in coupling net transport from cell membranes to esterification; vascular endothelial cells are particularly effective in limiting their cholesterol content by promoting cholesterol efflux from their membranes that is coupled to LCAT activity. It may be relevant that these cells not only are exposed to the full concentration of plasma lipoprotein cholesterol, but also retain a significant rate of unsuppressed cholesterogenesis in the presence of lipoproteins.

There are two possible explanations for the efficiency of cholesterol transport from cells. One possibility is that the small HDL containing apo A-1 (which carries cholesterol from cell membranes) may be particularly effective as a substrate for LCAT, compared with the bulk of HDL or VLDL and LDL. A second possibility is that the unstirred water layer at the plasma membrane surface, which is the effective diffusion barrier for molecules leaving or entering cell membranes, may be a less effective barrier for the small HDL than for the larger HDL which make up the bulk of this lipoprotein fraction. Rapidly dividing cells have a much greater efflux (and net transport) rate for cholesterol than do confluent, nondividing cells. The former have been shown to have an unstirred boundary layer that is thinner by an order of magnitude. The relative importance of these alternative explanations has not been distinguished in research to date. What is clear is that the LCAT reaction cannot be realistically modeled in the absence of cell membranes as a major substrate donor.

Red blood cells are relatively inactive (compared with vascular cells) as donors of cholesterol for LCAT-mediated esterification, and blood platelets are virtually inactive when the plasma membranes of vascular cells are present. While these differences make sense physiologically, their biophysical basis is not yet known.

The contributions of individual plasma lipoproteins to the total plasma cholesterol used as a substrate for LCAT can be determined by one of the rapid fractionation procedures (for example, precipitation with dextran sulfate and Mg^{2+}, or affinity chromatography on immobilized heparin or specific antibodies to individual apolipoproteins). Such studies show clearly that the major lipoprotein contributors of free cholesterol for the LCAT reaction are VLDL and LDL. HDL contributes little, if at all. In some disease states where the free-cholesterol content of VLDL and LDL is increased (such as diabetes), essentially all the free cholesterol for LCAT comes from these sources. However, in these cases the cholesterol of plasma lipoproteins is not at equilibrium and flows spontaneously (that is, in the absence of LCAT activity) from VLDL and LDL to HDL (Fielding 1984). While the rate of cholesteryl ester synthesis by LCAT does not appear to depend on the source of free cholesterol, the origin of the free cholesterol used depends on a variety of factors whose effects are likely to be very important in maintaining cellular cholesterol content. It is clear that in normal plasma, conditions are such that cholesterol is transported very effectively from vascular cell membranes to LCAT and there esterified. Several factors, four of which are outlined here, may contribute to the inefficient transport of cholesterol in disease states where the balance of cholesterol between cells and plasma is disturbed and cholesterol accumulates in cells.

1. Low LCAT activity will reduce cholesterol net transport from cell membranes, since normally net cholesterol transport is numerically nearly equal to total LCAT activity.
2. Since plasma lipoprotein cholesterol competes with cell-membrane cholesterol as a substrate for LCAT, an increased plasma free cholesterol per se will adversely affect the competitive position of cell-membrane cholesterol. Interestingly, the ratio of cell-membrane cholesterol to plasma cholesterol will be an inverse function of the diameter of the blood vessel because it reflects the surface-to-volume ratio. Other things being equal, it may be relatively more difficult for LCAT to esterify free cholesterol from the vascular bed of large blood vessels than of small vessels and capillaries.
3. Since net transport of cholesterol is nearly equal to total efflux, and since efflux is reduced as the thickness of the diffusion barrier (the unstirred water layer at the surface of the cell) is increased, it will be more difficult for LCAT to catalyze cholesterol net transport from surfaces where mixing is inefficient than from regions where mixing is more complete.
4. Decreased HDL levels may play an important role in reducing efflux and net transport rates, insofar as they may reflect a decreased concentration of the small HDL containing only apo A-1, which appears to catalyze such efflux. Interestingly, in non-insulin-dependent diabetic plasma, where total HDL levels are reduced only about one-third, apo A-1-dependent efflux (and cholesterol net transport) is almost undetectable, so the small HDL fraction may be particularly sensitive to metabolic changes.

Cholesteryl Ester Transfer in Plasma

The cholesteryl ester synthesized by LCAT is not retained at the site of synthesis in those animal species whose plasma contains significant levels of VLDL and LDL. In human plasma, for example, these lipoproteins, which contain no apo A-1, are not significant substrates for LCAT, yet they contain about 80% of total cholesteryl ester, all derived from the LCAT reaction; therefore, there must be a mechanism for the transfer of cholesteryl ester in human plasma. Unlike free cholesterol, cholesteryl ester has an insignificant solubility in aqueous media, and so it will not spontaneously diffuse among plasma lipoprotein particles.

Work from several laboratories has identified the presence of at least two transfer proteins in plasma which catalyze the transfer of poorly diffusible lipids among lipoprotein particles (Table 13.7). The essential role of such a transfer reaction is emphasized by the inhibition of LCAT by its product, cholesteryl ester, which would

Table 13.7. Lipid Transfer Proteins in Normal Human Plasma

Transfer protein	Specificity	Activity ($\mu mol/mL/h$)	Probable function
TP-1	CE, TG > PL	0.05–0.10	Transfer of LCAT-derived CE to VLDL, LDL; associated with reverse transfer of TG to HDL from VLDL and probably LDL
TP-2	PL (PC > Sph)	1.5–2.0	Transfer of PL from chylomicrons and VLDL to HDL

Notes: Activity refers to the rate with the first-named lipid. The abbreviations are CE, cholesteryl ester; TG, triacylglycerol; PL, phospholipid; PC, phosphatidylcholine; Sph, sphingomyelin.
Sources: The activity for CE transfer is from Fielding et al. (1983) and Albers et al. (1984); the activity for phospholipid transfer (TP-2) is from Tall, Abreu and Shuman (1983).

otherwise accumulate at its site of synthesis. The mechanism of action of the transfer proteins is presumably to increase the effective solubility of cholesteryl esters in aqueous media, such as plasma. One of these transfer systems (TP-1) catalyzes the exchange of both triacylglycerol and cholesteryl ester. There is no agreement at present about whether this is a single activity carrying both lipids or the activity of two similar proteins.

The following points can be made about cholesteryl ester transfer. First, even in normal plasma, cholesteryl ester is not at its equilibrium distribution among the major plasma lipoprotein species. That is, even when LCAT activity is inhibited, there is a spontaneous net transport of cholesteryl esters from HDL to VLDL and LDL. Whether different HDL species show different activities in cholesteryl ester transfer is not known. Second, in freshly isolated whole plasma, cholesteryl ester transfer is accompanied by an approximately equimolar back-transfer of triacylglycerol from VLDL and LDL to HDL. Third, while transfer of cholesteryl ester to VLDL (which is substantially more triacylglycerol-rich than LDL) is faster than to LDL, most cholesteryl ester transfer in normal whole plasma is to LDL, whose concentration is many times greater than VLDL. Since transfer of cholesteryl ester in whole plasma continues even at 4°C and in the presence of inhibitors of LCAT, it is likely that the cholesteryl ester and triacylglycerol concentrations in lipoproteins separated by relatively slow fractionation procedures, such as centrifugation, do not correspond to those in native plasma. In fact, comparison of rapid separation methods with centrifugation suggests that in centrifuged VLDL the content of cholesteryl ester may be higher than present in VLDL in the original native plasma. The important point in this discussion is that in vivo VLDL do not contain the maximum content of cholesteryl esters, at least in normal plasma.

Plasma also contains a distinct phospholipid transfer protein (TP-2 in Table 13.7) which has little, if any, affinity for cholesteryl esters. The levels of transfer of phosphatidylcholine reported in whole plasma would be sufficient to maintain an equilibrium among lipoprotein species. Hence, phosphatidylcholine degraded by LCAT could be readily replenished. The activity of this transfer protein would also be sufficient to prevent the accumulation of phospholipid at the surface of VLDL and chylomicrons during lipolysis of the triacylglycerol of these particles. Presumably such phospholipid is transferred by the phospholipid transfer protein to the other lipoprotein classes.

Triacylglycerol Metabolism in Plasma; Lipoprotein Lipase

Further discussion of the transfer of cholesteryl esters to VLDL and LDL requires a brief consideration of plasma triacylglycerol metabolism. Plasma cholesterol levels differ little on a day-to-day basis, even after the ingestion of a cholesterol-rich meal. On the other hand, plasma triacylglycerol levels, as VLDL, increase markedly on the consumption of carbohydrate, and circulating levels of triacylglycerols in chylomicrons increase in response to dietary fat. As we shall see, these changes in circulating triacylglycerol levels affect the distribution of free and esterified cholesterol in plasma, and the generation of cholesteryl esters.

The initial step of catabolism of triacylglycerol-rich lipoproteins (both dietary chylomicrons and hepatic VLDL) involves the hydrolysis of lipoprotein triacylglycerol by lipoprotein lipase (LPL) at the endothelial surface of the vascular bed (Cryer 1983). Hydrolysis of about 80% of triacylglycerol occurs in the extrahepatic tissues. It has been shown by several laboratories that LPL, synthesized in parenchymal cells (particularly adipocytes and smooth muscle cells), is bound to specific sites on the endothelial surface. Probably the lipase is bound ionically at the luminal end of cell-surface glycosaminoglycans containing heparin or heparan sulfate. The enzyme has a high affinity for these sulfated polysaccharides. Antibodies to LPL completely block plasma triacylglycerol turnover in vivo in chickens. In addition, human subjects with LPL deficiency have a high and persistent hypertriglyceridemia in response to dietary fat. There can be little doubt that LPL plays an important role in the initial processing of triacylglycerol-rich lipoproteins.

Competition studies in which VLDL and chylomicrons carry triacylglycerol with different radioactive labels show that these particles compete for LPL at the vascular surface of the perfused rat heart. An individual triacylglycerol-rich lipoprotein might remain adsorbed at the LPL binding site during the whole of its catabolism. However, the appearance within the circulation of particles of intermediate

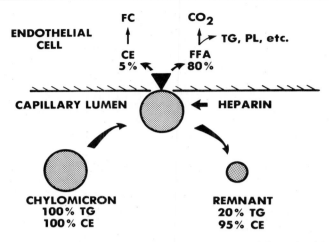

Figure 13.13. Uptake of cholesteryl ester from triacylglycerol-rich lipoproteins in the presence of lipoprotein lipase (LPL) activity. The quantitative data are taken from Fielding (1978) and Fielding and Fielding (1983). TG, triacylglycerol; CE, cholesteryl ester; FC, free cholesterol; PL, phospholipid; FFA, free fatty acids. Heparin releases LPL into plasma.

size (between VLDL and LDL) indicates that there are multiple adsorption-desorption events during catabolism of lipoprotein triacylglycerol by LPL (Figure 13.13). For its activity, LPL requires apo C-2, which is transferred to the surface of the newly secreted chylomicron or VLDL particle from HDL, to which it returns after triacylglycerol depletion by LPL (Figure 13.14). The end point of LPL activity is probably determined by loss of apo C-2 from the surface of triacylglycerol-rich particles. The triacylglycerol mostly present in the central hydrophobic core of VLDL and chylomicrons, is in equilibrium with a small proportion of surface triacylglycerol, which is probably the true substrate of LPL. The same population of surface triacylglycerol is available for exchange with cholesteryl esters by the transfer pathway discussed in the previous section.

Interrelationship of Cholesterol, Cholesteryl Esters, and Triacylglycerol Metabolism

The presence of plasma triacylglycerol has two consequences with regard to plasma cholesterol metabolism. First, in postprandial plasma containing increased chylomicron triacylglycerol, or in the presence of isolated chylomicrons, LCAT activity is increased about twofold. This increase represents an activation of enzyme already present, since increased esterification of cholesteryl is demonstrated in vitro in the presence of chylomicrons with a fixed volume of plasma containing a constant amount of enzyme protein. There are several possible mechanisms for this activation. The most likely

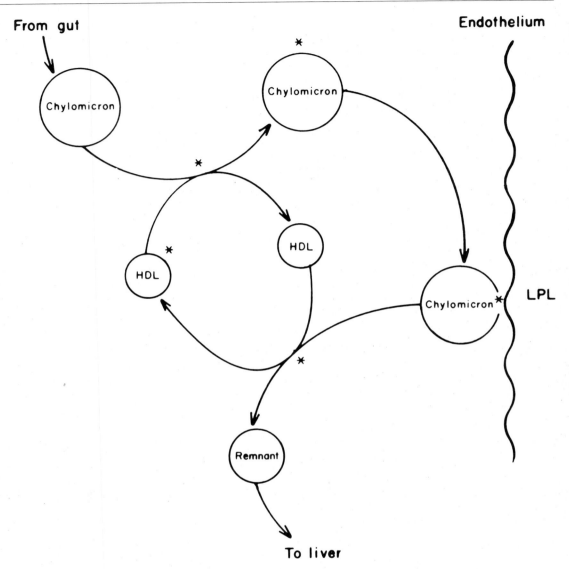

Figure 13.14. Recycling of LPL activator apolipoprotein (apo C-2, indicated by *) between LPL-substrate lipoproteins (for example, chylomicrons) and HDL. The movement of apo C-2 to the chylomicron surface occurs spontaneously as the nascent lipoprotein enters the plasma; loss of apo C-2 occurs continuously during lipolysis and is essentially complete when remnant lipoprotein is formed after loss of about 80% of the initial triacylglycerol content. Reprinted by permission from Fielding and Havel (1977).

explanation for the activation of LCAT by chylomicrons is that triacylglycerol-rich lipoproteins are better acceptors of cholesteryl esters transferred from the LCAT reaction than are LDL. Because transfer of cholesteryl ester is coupled to esterification, an increased rate of esterification would follow the increased rate of transfer.

Second, the transfer of cholesteryl esters from triacylglycerol-rich lipoproteins to cell membranes is also stimulated during LPL-mediated triacylglycerol hydrolysis. It has been shown with several cell types that even in the absence of LPL activity, cholesteryl ester is transferred from triacylglycerol-rich lipoproteins across the plasma membrane to the interior of the cell. This process has the characteristics of a receptor-mediated pathway. It is saturable, specific for triacylglycerol-rich lipoproteins (and not competitive with LDL), and the interiorized cholesteryl ester efficiently suppresses cholesterogenesis. However, uptake of cholesteryl ester does not involve endocytosis of the intact lipoprotein (perhaps understandably, in view of its large size) or degradation of the lipoprotein protein moiety. Such uptake is stimulated about 30-fold in the presence of LPL. For cells such as confluent endothelial cells (which interiorize LDL poorly although they have LDL receptors), cholesteryl ester transfer is probably the major regulatory mechanism of cholesterogenesis (Figure 13.13).

In the isolated perfused heart, it has been shown that the uptake of cholesteryl ester from VLDL and chylomicrons is almost totally inhibited when LPL is removed from the endothelial surface by perfusion with heparin. Two possible mechanisms can be suggested. First, the rapid removal of triacylglycerol from the lipoprotein surface may result in higher cholesteryl ester concentrations, at least transiently, that set up a concentration gradient of cholesteryl ester between the opposed lipoprotein and cell-membrane surfaces. (The lipoprotein particle and the cell membrane are contiguous because of the binding of the triacylglycerol-rich lipoprotein to the membrane.) Second, products of lipolysis by LPL (for example, unesterified fatty acids, lysolecithin, or monoglycerides) may increase the local solubility of cholesteryl ester in the aqueous medium, again resulting in increased transfer to the cell membrane. These possibilities have not been distinguished experimentally.

Cholesteryl Esters and Chylomicrons

Chylomicrons, converted in nonhepatic tissues to their end-product remnant composition by triacylglycerol hydrolysis are cleared quantitatively by the liver (Windler, Chao, and Havel 1980). The remnants do not appear within the LDL density range. The turnover rate of chylomicrons is normally rapid (the $t_{\frac{1}{2}}$ is 3–5 min). Although when chylomicrons are secreted they are relatively cholesterol poor, their half-life in normal plasma is so short that it is unlikely that a significant mass of free cholesterol is transferred to their surfaces during recirculation. Likewise, little cholesteryl ester would be transferred over this time period. The lipid composition of chylomicrons, and the remnant lipoproteins derived from them, is consistent with the concept that there is little interaction of the cholesteryl ester pool in chylomicrons with that in other lipoproteins. Supporting this conclusion is the observation that retinyl ester, secreted into the intestinal lymph, is a useful marker for dietary triacylglycerol-rich lipoprotein particles in the circulation, in spite of the reactivity of retinyl ester as a substrate for the cholesteryl ester transfer protein.

Cholesteryl Esters and VLDL

The fate of VLDL cholesteryl ester is more complex. First, a variable but substantial proportion of VLDL apo B appears in the LDL fraction, where it is the major or only apoprotein on the particle. Second, the circulation time of VLDL is much longer than that of chylomicrons (the $t_{\frac{1}{2}}$ is about 90 min in humans). Third, because of the low level of hepatic ACAT in humans, the major part of cholesteryl ester of VLDL and LDL is derived from the LCAT reaction. There have been several computations of the flux of cholesteryl ester which might occur during the conversion of VLDL to LDL. Unfortunately, these have been based on the composition of VLDL recovered after centrifugation under conditions where substantial (and in this context, artifactual) transfer of cholesteryl ester from HDL during fractionation would be anticipated.

There is evidence from studies in vivo with synthetic lipoprotein complexes that free cholesterol can have a pronounced inhibitory effect on LPL activity as molar ratios of free cholesterol to phospholipid exceed about 0.3. In the case of chylomicrons, this ratio is generally lower (about 0.2), and because of the rapid turnover of these particles, the ratio would not be much modified by diffusion from other lipoprotein particles with higher surface free-cholesterol contents. The rate of LPL activity with chylomicrons is similar to that with synthetic triacylglycerol-phospholipid dispersions containing no free cholesterol. On the other hand, the molar ratio of free cholesterol to phospholipid of VLDL is higher (about 0.6). This may be the result either of a higher concentration of free cholesterol in nascent VLDL than in nascent chylomicrons or from transfer of cholesterol down the concentration gradient from lipoprotein such as LDL, whose molar ratio of free cholesterol to phospholipid is 0.8–0.9. Thus, the rate of LPL activity with VLDL of similar size and apoprotein composition to chylomicrons is only about one-half that with chylomicrons.

Cholesteryl Esters and HDL

In normal human plasma, about one-quarter of the cholesteryl ester generated by the LCAT activity is retained in particles containing apo A-1 (that is, HDL) instead of being transferred to lipoprotein particles containing apo B (VLDL and LDL). The evidence summarized earlier suggests that lipoproteins containing only apo A-1 are the best substrates for LCAT. However, the majority of HDL contains other apolipoprotein in addition to apo A-1 (especially apo A-2), and such HDL (perhaps 95% of total HDL particles) contains the majority of HDL cholesteryl ester. Two possible pathways can be suggested for the enrichment of the bulk of HDL with cholesteryl ester.

1. As apo A-1–HDL becomes enriched with cholesteryl ester, apo A-2 transfers to the particles, which become progressively poorer substrates for LCAT. In this model, species of HDL that contain larger amounts of cholesteryl esters (that is, HDL_2 with a density of 1.063–1.125 g/mL), arise by transfer of apo A-2 to HDL that contain apo A-1. On the other hand, smaller HDL (HDL_3 with a density of 1.125–1.21 g/mL) that have a lower content of cholesteryl esters have a ratio of apo A-1 to apo A-2 that is similar to that of HDL_2.
2. Apo A-1–HDL is the primary acceptor of cholesterol from cell membranes, while HDL containing apo A-1 or apo A-2 might be formed separately either from de novo secretion or as a result of lipolysis of chylomicrons or VLDL.

These alternatives cannot be readily distinguished by the usual isotope methods, since all the components of HDL—lipid and protein—readily equilibrate among the particles. At present we know little or nothing about the features of HDL which make it an attractive acceptor of cholesteryl esters, whether derived by transfer from other lipoproteins or generated in situ by direct LCAT activity. Such ignorance covers an important area of the biochemical pathology of cholesterol. In a number of situations—for example, in some genetic hypercholesterolemias—transfer of cholesteryl esters from HDL to VLDL and LDL is almost completely inhibited, while total LCAT activity is normal, and the HDL receives cholesteryl esters at several times the usual rate.

In summary, the end products of plasma lipoprotein metabolism are the following:

1. chylomicron remnants, derived from chylomicrons by LPL activity
2. VLDL remnants, representing that fraction of VLDL catabolized by LPL and removed as such, without conversion to LDL

3. LDL, generated from VLDL remnants by further remodeling that most importantly involves loss of other apolipoproteins and an additional loss of triacylglycerol

4. HDL, containing cholesterol derived from cell membranes or from other lipoproteins secreted by liver or intestine, transformed by LCAT, and remodeled by the addition of apolipoproteins other than apo A-1, which is already present

REMOVAL OF LIPOPROTEINS FROM PLASMA

In the preceding section, the factors influencing the cholesterol composition of mature circulating lipoproteins were discussed. It is clear that some lipids and apolipoproteins are associated relatively loosely in their lipoprotein form. Examples are apo A-1 among proteins, and cholesterol among lipids. The evidence suggests that both are in equilibrium with a sufficient aqueous concentration so that the transfer of such components occurs among lipoproteins and cell membranes by simple diffusion. On the other hand, the transfer of cholesteryl ester through the aqueous phase (in the absence of a catalyst such as a transfer protein) is likely to be very slow because of its low solubility in water. Nevertheless, all lipoprotein lipids and all lipoprotein proteins (with the possible exception of apo B) have the potential to be cleared as individual molecules down their concentration gradients.

These processes take place simultaneously with different mechanisms from the uptake of intact lipoproteins. The earlier discussion indicates that diffusion is a major pathway for removal of free cholesterol, while the uptake of intact lipoproteins, by one of several mechanisms, is the major pathway for the poorly soluble cholesteryl esters. Some cholesteryl ester is also transferred from lipoproteins across cell membranes; it is most likely that this requires surface-to-surface contact between donor and acceptor.

It is very likely that free-cholesterol concentration gradients between lipoproteins and cell membranes can play an important role in the uptake of lipoprotein cholesterol by cells. This is supported by the following observations.

First, the rate of flux of free cholesterol across cell membranes is equivalent to the maximum rate of receptor-mediated uptake of LDL total cholesterol in fully up-regulated cells. For example cholesterol flux (molecular transfer) in normal fibroblasts is 1.5 μg of cholesterol per milligram of cell protein per hour at 37°C and LDL endocytosis in fully up-regulated normal fibroblasts is 600 ng of apo B per milligram of cell protein per hour at 37°C [that is, $(0.6 \times 40/20) = 1.2$ μg cholesterol per milligram of cell protein per hour]. Uptake via the LDL pathway will be at least 10-fold less relative to free-cholesterol transfer when (as under physiological conditions) these receptors are not up-regulated.

Second, all the groups of humans at increased risk for atherosclerosis (pathological deposition of cholesterol in tissues, especially in the vascular wall) who have been studied in detail have a positive cholesterol net flux from plasma lipoproteins to cell membranes (Fielding 1984).

Among processes for the uptake of intact lipoproteins, *fluid-phase, nonspecific endocytosis* (pinocytosis) is slow, typically 100–200 nL/10^6 cells per hour for vascular cells. The uptake of individual lipoproteins by this pathway is obviously proportional to their concentration in plasma, and minor in relation to other pathways.

A second nonspecific adsorption mechanism is *adsorptive endocytosis*, which involves binding of lipoproteins to the cell surface, probably to phospholipids, via hydrophobic bonds. The lipoprotein, along with other adsorbed proteins, is interiorized intact and degraded as the membrane components are recycled.

Finally, lipoproteins can be cleared intact via specific lipoprotein receptors. The first of these to be recognized as such was the fibroblast receptor shown to recognize apo B (and subsequently, apo E). The number of distinct lipoprotein receptors identified continues to increase, and at this point at least four are recognized (Table 13.8).

Receptor-Mediated Uptake of Lipoproteins

It should be recognized that the term "receptor" used in the description of lipoprotein catabolism is used more loosely than in the case of hormone receptors (for example, those for adrenalin or insulin) from which the terminology and methods of investigation of the lipoprotein receptors were derived. Unlike the hormone receptors, the ligand that binds to the lipoprotein receptor is the lipoprotein apoprotein, whereas the biological effect is modulated not by the ligand but by the lipoprotein lipid (cholesterol or, most likely, an oxysterol derivative of cholesterol). For this reason apo B itself, or complexes of apo B with lipids other than cholesterol, do not regulate receptor number, while suspensions of free cholesterol, or cholesterol-rich lipoproteins which bypass the apo B,E receptor, have all the regulatory functions of LDL. It has been stated that cholesterol entering the cell via a specific receptor is uniquely potent in regulation, but there is no convincing evidence that this is the case.

In the lipoprotein field it has become customary to distinguish between *high-affinity* and *low-affinity* receptors. The terms relate to the physiological concentration of ligand. A high-affinity receptor is one whose dissociation constant is low relative to the circulating level of ligand. Accordingly, such receptors are always occupied by bound

Table 13.8. Cell Receptors Reacting with Native and Modified Plasma Lipoproteins

Receptor	Location	Up/down regulateable	Regulated by	Ligand	K_d in vitro (M)	Circulating concentration of ligand (M)	Receptors/cell
Apo B,E (LDL receptor)	Liver, fibroblast, endothelium, adrenal, smooth muscle, macrophage	Yes	Oxysterols	Apo E >> apo B	2×10^{-10} 3×10^{-9}	1.5×10^{-6} 1.5×10^{-6}	27,700 (apo E) 99,600 (apo B)
Apo E	Liver	No	—	Apo E	2×10^{-10}	1.5×10^{-6}	?
Scavenger (modified LDL receptor)	Macrophage; endothelial cell	No	—	High molecular weight positive charge (polyinosinic acid > dextran sulfate > chemically modified LDL > chemically modified albumin)	—*	—†	?
β-VLDL	Macrophage; endothelial cell	Yes	Oxysterols	Apo B-48?	10–20 g/mL β-VLDL cholesterol	—‡	?

* Half-saturation is about 5 µg/mL acetylated LDL, 50% displaced by 1 µg/mL polyinosinic acid; the major physiological ligand of this receptor is not known.

† Ligand is undetectable in plasma.

‡ β-VLDL a cholesteryl ester-rich lipoprotein, recovered in the VLDL density range, is undetectable in normal plasma; a recent report identifies reactivity of the endothelial β-VLDL receptor with normal chylomicrons.

ligand, and the rate of uptake of ligand depends on the number of receptors, not the concentration of ligand. A low-affinity receptor can be defined as one whose dissociation constant is high relative to the circulating concentration of ligand. Accordingly, a significant proportion of receptor sites are unoccupied except at high ligand concentration, and the rate of uptake of ligand depends on the concentration of ligand, as well as on the number of receptors. Many receptor effects involve functions of intermediate affinity. It is important that the high affinity of a receptor for ligand does not necessarily reflect any significance in regulation. Since the receptor is fully occupied at a low concentration of ligand, a higher concentration of ligand would have no additional effect.

Catabolism of Chylomicrons and Chylomicron Remnants

Chylomicrons newly secreted from the intestinal lymph can be cleared as intact particles by the liver, but only at a very slow rate in the absence of preliminary catabolism by lipoprotein lipase. It is likely that this uptake involves the phagocytic (Kupffer) cells of the liver. Chylomicrons do not penetrate the vascular wall and are not cleared intact by the vascular endothelium. However, chylomicron cholesteryl ester is cleared by endothelial cells by a saturable high-affinity pathway that involves uptake of cholesteryl esters without interiorization or degradation of chylomicron protein. This chylomicron binding site has been detected both in perfused organs (heart, adipose tissue) and in isolated cultured endothelial cells. The number and properties of these chylomicron binding sites are very similar to those of the LPL binding site. LPL is not made by endothelial cells but is synthesized and released by adjacent cells and then binds to the surface of endothelial cells. Therefore, the lipase is undetectable in these cells in culture.

It seems most likely that LPL and chylomicrons each bind independently to the same site in both native and cultured endothelial cells. Both triacylglycerol and cholesteryl ester are interiorized by these cells, even in the absence of LPL, and are subsequently metabolized intracellularly. The cholesterol derived from chylomicron cholesteryl ester effectively suppresses cholesterogenesis in endothelial cells. Uptake of cholesteryl ester in the presence of LPL (about 1.5 μg per milligram of protein per hour) can be compared with the uptake of cholesterol by endothelial cells via the LDL receptor under the same conditions (0.15 μg per milligram of protein per hour). The endothelial chylomicron receptor mechanism may play a role in the regulation of cholesterogenesis in these specialized cells. Since this mechanism of lipid internalization is also operative with VLDL (chylomicrons and VLDL are competitive at the chylomicron site), this endothelial high-affinity site would ensure a fairly constant delivery of sterol to the cells, despite the transient high concentration of chylomicrons postprandially. It is important to bear in mind, however, that only a small proportion of dietary cholesterol is retained in the peripheral endothelium in this way. The major part returns to the liver via a receptor process.

The finding that chylomicron remnants are cleared by the isolated perfused liver by a specific, high-affinity receptor indicated the mechanism by which the bulk of remnant cholesterol is directed to liver cells. This receptor has been characterized both in isolated perfused liver and in partially purified membrane fractions (about 10% plasma membranes and 90% microsomal membranes) of isolated hepatocytes. In the liver, the apparent Michaelis (binding) constant for chylomicron remnants is about 20 μg cholesterol/mL, and the rate of uptake of cholesterol from chylomicron remnants is 0.16 mg/g liver/10 min (Sherrill and Dietschy 1978). Chylomicron remnants are formed by depletion of about 80% of the initial triacylglycerol content of the chylomicron, but most of their cholesteryl ester is retained (Higgins and Fielding 1975). The amount of phospholipid is reduced in proportion to the decreased surface area of the remnant particle. Remnants contain a higher molar ratio of free cholesterol to phospholipid than do newly secreted chylomicrons (0.6–0.8 versus 0.2). The circulation time of the remnant is sufficiently short that little cholesterol is likely to be gained or lost during circulation. These remnants retain all their apo B-48 and the majority of the apo E adsorbed from other lipoproteins during passage of the particle from lymph to plasma. They have lost the major part of the apo C proteins, including apo C-2, the cofactor of LPL, the loss of which probably effectively terminates lipolysis.

There is evidence that the ligand for the chylomicron remnant receptor is apo E (Windler, Chao, and Havel 1980). The apo B of dietary particles (apo B-48) is not a ligand (nor does apo B-48 appear to react with the B,E receptor). The poor reactivity of intact chylomicrons with the remnant receptor is probably due to an interference mediated by the apo C proteins, which are strongly represented in the intact chylomicron. The same receptor binds a species of HDL present in cholesterol-fed animals (HDL$_c$), which contains apo E as its major or only apolipoprotein. Although this receptor mediates endocytosis, and hence the delivery of remnant lipids to the hepatocyte, many investigators have not found effective regulation of hepatic cholesterogenesis from the action of the remnant receptor. This has been interpreted as reflecting the special role of the apo B,E receptor, also present on hepatocytes, as the major regulator of cholesterogenesis.

However, there is a simpler explanation. In experiments on regulation of cholesterol synthesis by chylomicrons, it is usual to prepare chylomicrons from animals fed very high levels of fatty acids or triglycerides. The chylomicrons obtained (and hence their remnants) have very low levels of cholesterol, perhaps an order of magnitude lower than those circulating physiologically. As a result, higher than normal levels of fatty acids are delivered to the liver per unit of cholesterol. These fatty acids stimulate VLDL synthesis and secretion and thus create demand for cholesterol synthesis. This sequence of events would nullify the inhibitory effects of remnants on cholesterol synthesis under these unique circumstances. The relative importance of the apo B,E receptor and the remnant receptor in the regulation of cholesterogenesis remains to be investigated under carefully defined conditions where cholesterol delivery (influx) and cholesterol secretion (efflux) have been matched.

Catabolism of VLDL and VLDL Remnants

VLDL has a structure and composition similar to that of chylomicrons—indeed, sizes and densities partially overlap, although VLDL is on average smaller. The major difference between the two types of particles is that the apo B species of VLDL (apo B-100) reacts with the apo B,E receptor also present on hepatocytes, whereas chylomicron remnants do not. Since VLDL does not pass the endothelial barrier, and since as discussed, the apo B,E receptor in the intact endothelial layer is inactive, there is little nonhepatic catabolism of lipid of intact VLDL or VLDL remnants except for triacylglycerol. Some VLDL cholesteryl ester is cleared by endothelial cells by the saturable high affinity mechanism described in the previous section for the same lipid in chylomicrons. However, as for chylomicrons, the major part of VLDL cholesteryl ester is returned to the liver.

In animals which contain appreciable levels of ACAT in liver, there is probably little addition of cholesteryl ester by transfer from LCAT in the periphery. In humans, however, where ACAT activity in the liver is very low, and cholesteryl ester transfer activity in plasma is relatively high, there is the potential for cholesterol that has been synthesized in tissues other than liver to be transferred back to the liver for catabolism along with the VLDL remnants. It is hard to estimate the importance of this pathway (compared with that mediated by LDL) because information is not yet available on the esterified- and free-cholesterol composition of human nascent VLDL. However, the concentration of cholesterol in VLDL and LDL is estimated to be 5 and 100 mg/dL respectively. In addition, if half the VLDL cholesterol were removed in remnant form (the remainder would be converted to LDL) and the circulating half-lives for VLDL and LDL were 2 h and 48 h, respectively, the ratio of cholesterol removed by liver as VLDL remnants (the ratio of VLDL to LDL) compared with that removed as LDL would be $(5/100 \times 1/2 \times 48/2)$, or $60/100$. Overall, it seems likely that VLDL remnant removal by the apo B,E receptor of liver plays a significant role in cholesterol balance in plasma and therefore in peripheral cells.

VLDL remnants, like the equivalent chylomicron derivatives, lose the major part of their apo C proteins during lipolysis. It is at present uncertain whether, or to what extent, VLDL remnant particles could be removed by the apo E receptor that is reported to clear chylomicron remnants. Studies with rabbits genetically deficient in apo B,E receptors show that VLDL, as well as LDL, accumulates in these animals. It has been suggested that the explanation for this increase in plasma VLDL is that the apo B,E receptor (absent in these animals) normally removes such particles. However, since VLDL remnants competitively inhibit LPL activity, alternatively it is possible that the accumulation of VLDL in plasma of the genetically deficient rabbits is the result of an inhibition of lipolysis. It may also be relevant that humans who are genetically deficient in apo B,E receptors do not seem to accumulate VLDL or VLDL remnants and thus must have an alternative mechanism (perhaps via the apo E receptor) for removal of these lipoprotein particles.

These data indicate that the major catabolic pathway for VLDL cholesteryl ester is endocytosis via the hepatic apo B,E (and possibly apo E) receptors.

In some pathological human plasma samples the VLDL molar ratio of free cholesterol to phospholipid is increased compared with that in normal plasma. Under these conditions, free cholesterol transfers down its concentration gradient to cell membranes and to HDL. As discussed in the previous paragraph, the transfer of cholesterol to cell membranes causes the accumulation of cholesterol, and subsequently its ester, within the cell. Free cholesterol on VLDL (and LDL) also competes with cell-derived cholesterol in supplying substrate to plasma LCAT.

Catabolism of LDL   Human LDL contains apo B-100 as the major or only apolipoprotein. Consequently, LDL reacts with the apo B,E receptor, and its cholesterol is interiorized along with other components of the lipoprotein (Figure 13.6). Esterified cholesterol is hydrolyzed in the lysosomes, and the free cholesterol, or its oxygenated derivative, is available for regulation of cholesterogenesis. In many animals (such as the rat) little VLDL is converted to LDL. A consensus of recent work suggests that in normocholesterolemic humans only a part of VLDL enters the LDL density range (Kesaniemi, Vega, and Grundy 1982). The data suggest that in VLDL about one-third of the cholesteryl ester derived from LCAT activity is cleared as LDL. The factors which contribute to the conversion of VLDL to LDL are poorly understood. It is relatively easy to demonstrate the pathways of lipolysis and transfer by which VLDL *can* be converted to LDL, but much harder to show which of the several plausible routes are most significant under physiological conditions.

In contrast to these outstanding problems of physiology, the LDL (apo B,E) receptor as a protein is well defined. It has been isolated, and its structure and synthesis have been the subject of detailed studies. It is a glycoprotein with a molecular weight of 120,000, whose apparent molecular weight after posttranslational processing is 163,000. Turnover studies with puromycin (an inhibitor of protein synthesis) or monensin (an inhibitor of microfilament formation) indicate that each receptor molecule recycles repeatedly within the cytoplasm, catalyzing the internalization of about five LDL particles per hour. In the presence of LDL, the maximal number of receptors externalized down-regulates to about 10–20% of the maximal inducible with lipoprotein-deficient serum, a finding consistent with the earlier data on down-regulation expressed as LDL binding. Internalization of LDL receptors requires their localization laterally into *coated pits*, which are clathrin-coated invaginations that are interiorized as vesicles along with the receptors and LDL during recycling (Goldstein, Anderson, and Brown 1979). In cells that cannot organize receptors into coated pits (such as contact-inhibited endothelial cells or certain mutants of other cells), LDL binds normally but is not internalized, a mechanism that protects the cell from receptor-mediated influx of cholesterol.

It has been shown by both competition studies and studies with cells genetically deficient in LDL receptors that apo E binds to the apo B,E receptor with an affinity about 23-fold greater than that for apo B.

Significance and Properties of the Apo B,E Receptor in vivo

It has been established without question that a cell-surface receptor can catalyze the internalization of LDL containing apo B. In vitro this pathway has the properties of a high-affinity receptor for endocytosis. Under appropriate experimental manipulation (for example, the presence of lipoprotein-deficient serum), it can be shown that the apo B,E receptor is the major regulator of the cholesterol content of cultured cells such as fibroblasts or vascular smooth muscle cells. In humans, when the receptor is genetically deficient (as in familial hypercholesterolemia), hypercholesterolemia develops with resulting atherosclerosis. It has been concluded that LDL receptors regulate nonhepatic cholesterogenesis and that in their absence cholesterol synthesis is unsuppressed (Goldstein and Brown 1977).

As a result, cells oversynthesize and accumulate cholesterol, and atherosclerosis results. However, countering this reasoning are some experimental findings that challenge this concept of the pre-eminent role of the apo B,E receptor in the regulation of cholesterol metabolism.

Studies of LDL catabolism in vivo suggest that the rate of synthesis of LDL, not the catabolic rates, determines the circulating level of LDL (Kesaniemi and Grundy 1982). Furthermore, whole body choles-terol synthesis is unrelated to circulating LDL levels in normals, and is similar in receptor-deficient and normal subjects (Figure 13.15).

The fractional catabolic rate of LDL in vivo is the fraction of the total pool of LDL that turns over per unit time. Thus, the total amount of LDL degraded per unit time is the product of the fractional catabolic rate and the total pool of LDL. It has been demonstrated (Goldstein and Brown, 1977; Bilheimer, Stone and Grundy, 1979) that the fractional catabolic rate determined in vivo for LDL is propor-tional to the maximal number of apo B,E receptors that can be expressed in vitro. Moreover, in cell culture when the apo B,E receptors are fully saturated with LDL, the amount of LDL degraded by this pathway is independent of the concentration of LDL. Since the level for saturation of these receptors by LDL is well below the concentration of LDL seen in plasma, we would expect that the total amount of LDL degraded in vivo would be directly proportional to the number of receptors. However, the total amount of LDL de-graded is *not* proportional to the number of apo B,E receptors.

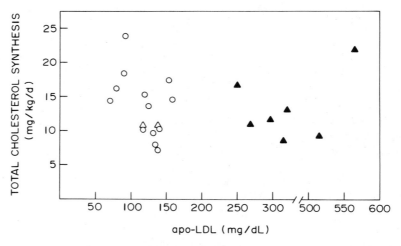

Figure 13.15. Relationship between whole-body cholesterol synthesis and plasma LDL protein concentration. Adapted from Kesaniemi and Grundy (1982). The absence of detectable effect on whole body cholesterol synthesis by the loss of the LDL receptor in homozygous familial hypercholesterolemic (FH) subjects (▲) or half of the LDL receptors in heterozygous FH subjects (△). Normal controls (o).

Rather, the finding that the apo B,E receptor number is proportional to the fractional catabolic rate is more consistent with the concept that the apo B,E receptors are not saturated in vivo.

Large changes in whole-body and tissue cholesterol synthesis in experimental animals are not accompanied by any appreciable change in receptor number (Stange and Dietschy 1984).

Humans with congenital absence of apo B (abetalipoproteinemia) do not deposit cholesterol in their tissues nor do they significantly oversynthesize cholesterol, particularly when the secondary mal-adsorption of biliary cholesterol is taken into account (Herbert et al. 1983). If the uptake of LDL by the apo B,E receptor were a significant pathway in the regulation of cholesterol synthesis, the absence of LDL or the absence of the LDL receptor should have a profound effect on cholesterol synthesis. It has been argued that apo E in abetalipoproteinemia, substitutes for apo B in suppressing choles-terol synthesis. In this case, congenital apo E deficiency should have the same effect on cholesterol deposition as apo B deficiency, since apo B could substitute for apo E. However, this is not the case: In apo E deficiency there is massive accumulation of cholesterol in the tissues (Ghiselli et al. 1981). This suggests that apo B and apo E have different regulatory activities in peripheral tissues in vivo and that apo E has the greater significance.

These data suggest that LDL catabolism occurs with very different kinetics and specificity in the intact animal than in tissue culture. It appears that LDL removal via the receptor-dependent pathway in vivo occurs largely by a low-affinity process not distinguishable kinetically from the low-affinity binding seen in receptor-deficient subjects. Other data (discussed in the next paragraph) indicate clearly that a significant proportion of total LDL degradation occurs via apo B,E receptors. The apo B,E receptor, which acts as a high-affinity receptor in vitro, must have the properties of a low-affinity receptor in vivo. This is readily understandable in view of the effects of diffusion on affinity; even a modest diffusion barrier produces a significant decrease in affinity. A substantial diffusion barrier, as occurs in a multilayer tissue such as the vascular bed, would certainly result in an increase in apparent Michaelis constant of a high-affinity receptor.

The increase in apparent Michaelis constant mediated by a diffu-sion barrier would not be expected to have any selective effect on the competitive position of apo B and apo E at the receptor. The molar concentrations of apo E and apo B in normal plasma are similar (about 1.5 μM). Hence, the occupancy of the apo B,E receptors by the two ligands will be in proportion to their affinities. Lipoproteins contain-ing both apo B and apo E (for example, VLDL remnants) probably bind through apo E, as do lipoproteins containing apo E but not apo B (for example, a fraction of HDL). The ratio of the affinities of apo E as HDL_c and apo B (as LDL), respectively, for the receptor is about 23:1 (Pitas et al. 1979). These considerations indicate that LDL containing apo B may be at a disadvantage for catabolism compared to VLDL and remnants containing apo B as well as one or more apo E molecules. The uptake of apo E will also be favored by its three-fold more rapid binding to the receptor.

It is possible that apo E may exchange rapidly between lipoproteins, and that LDL must be modified so that it will bind apo E *before* it can be effectively removed from the circulation; the conversion of VLDL remnant to LDL would therefore be reversible. This possibility, illustrated in Figure 13.16, would account both for the removal of LDL despite the much greater affinity of the apo B,E receptor for apo E. It would also account for the lack of effect of apo B deficiency or cholesterol deposition (but severe consequences of apo E deficiency) observed in human subjects. A considerable and unexplained proportion of total plasma apo E in any case is present in the LDL density range. An alternative explanation is that unmodified LDL is largely removed via nonspecific pathways, while LDL containing apo E would be removed through the apo B,E receptor. These alternatives cannot be distinguished at present, because there are few data on the turnover of apo E in intact animals.

It is quite clear that the apo B,E receptor cannot have the same properties and significance in vivo as have been ascribed to it on the basis of the extensive in vitro studies. The receptor studies described so far do not strongly support the concept that this pathway plays the key role in the regulation of in vivo cholesterol metabolism. The physiological data suggest that LDL removal in vivo occurs largely by

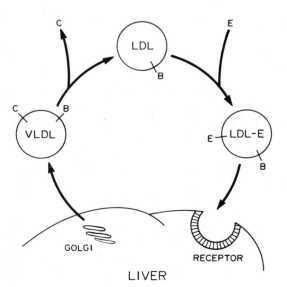

Figure 13.16. A hypothesis for explanation of the removal of LDL from plasma via the apo B,E receptor: the reversible association of apo E with LDL containing only apo B.

low-affinity (receptor and nonreceptor) processes not distinguishable kinetically from nonspecific binding in vitro, and possibly as an apo B,E complex rather than as LDL (apo B); and that cholesterol fluxes modulating local cholesterogenesis play the major role in regulating cellular cholesterol metabolism, particularly in extrahepatic tissues (Stange and Dietschy 1983).

Contributions of Receptor and Nonreceptor Pathways to LDL Catabolism

When the positive charge (for example on lysine and arginine groups) of apo B in LDL is chemically modified, the interaction of LDL with the specific apo B,E receptor is strongly inhibited or abolished. However, both native and chemically modified LDL show similar rates of uptake and degradation with receptor-deficient cells (Shepherd et al. 1979). These rates are similar in turn to the rate of the receptor-independent (nonspecific) degradation of LDL seen in normal cells. These findings have been applied to studies of the contributions of apo B,E receptor and nonreceptor pathways to total apo B degradation in vivo. In such studies, trace amounts of ^{125}I-labeled native LDL (*a*) and ^{131}I-labeled chemically modified LDL (*b*) are injected, into an animal and the relative rates of clearance determined. The difference in rates (*a* − *b*) is considered to represent the receptor-mediated uptake of LDL.

There are two assumptions in this method. The first is that while [^{125}I]LDL has been well validated as a tracer for whole-body LDL turnover, there is no independent evidence that a population of LDL exists in plasma with the properties of chemically modified LDL. The [^{131}I]LDL is a tracer with no defined pool. Moreover, the comparison of the behavior of native and chemically derivatized LDL in tissue culture supports the idea that [^{125}I]LDL is removed by a high affinity process and ^{131}I-modified LDL is removed by a low affinity process. It is not clear that this can be extrapolated to an in vivo situation. Indeed, as we have just seen, it is possible that both receptor-mediated and receptor-independent clearance have similar kinetics in vivo, while the kinetics of each is clearly different in tissue culture.

Second, fibroblasts have been shown to contain only apo B,E receptors among that number identified in different cell lines. In the cultured-cell assay for native and chemically modified LDL, the label therefore discriminates between binding to apo B,E receptors and nonspecific or low-affinity binding that may represent adsorptive endocytosis. In vivo, on the other hand, chemically modified LDL also binds to a specific receptor site of macrophages (see the next paragraph). Several different pathways, some receptor mediated, therefore contribute to the total non–apo B,E degradation of LDL. These considerations suggest caution in the quantitative interpretation of the isotope data used to model receptor and nonreceptor degradation of LDL, in view of the numbers of different receptors, and their uncertain kinetics, in vivo. Nevertheless, it seems clear that a considerable part, and perhaps the major proportion, of LDL is degraded by apo B,E receptors, whatever the regulatory significance of this pathway.

Studies in experimental animals on the tissue distribution of LDL degradative activity indicate that the liver and those glands converting plasma cholesterol to steroid hormones are the most important quantitatively. These data support the concept that the apo B,E receptor pathway contributes substantially to the catabolism of LDL cholesterol but that this may be unrelated to regulation of cholesterogenesis, particularly in extrahepatic tissues.

Degradation of Chemically Modified LDL

Macrophages express a specific scavenger receptor site reactive with LDL in which apo B has been chemically modified. This site is reactive with some other modified proteins (for example, maleylated albumin), but the affinity for modified LDL is much greater. A similar site has been identified on cultured endothelial cells. While the early studies of this pathway were carried out with chemically modified LDL, more recent work has been directed toward identifying physiological processes by which native LDL could become modified in vivo and hence be catabolized via the scavenger receptor for modified LDL, rather than by the apo B,E receptor for native LDL.

Two such physiological modifications of native LDL have been proposed. Blood platelets produce appreciable amounts of malonyldialdehyde, which reacts with lysine residues of LDL under mild conditions to produce a derivatized LDL recognized by the scavenger receptor. Derivatization of 16% of total apo B lysine residues is sufficient for recognition by this receptor. The physiological significance of this modification has been questioned because a high concentration of malonyldialdehyde (about $1mM$) would be required, based on results obtained in vitro. However, the possibility remains that sufficiently high concentrations of malonyldialdehyde could be produced locally during platelet aggregation.

A second physiological modification of LDL suggested as a mediator of LDL catabolism by macrophages is nonenzymatic glucosylation. Glucosylated LDL interacts poorly with the apo B,E receptor, but is probably taken up by a low affinity process. Such glucosylation is proportional to circulating glucose concentration and hence is more marked in the LDL of diabetics. The apo B of normal LDL has 1–2% of available lysines derivatized in this manner, while in diabetics it has typically been reported that 3–5% of the lysine residues of apo B are glycosylated. While there is no question that extensively glucosylated LDL, generated by prolonged incubation or chemical modification in vitro, has a reduced interaction with apo B,E receptors in vitro, it is not yet clear that the extent of glucosylation occurring in vivo is sufficient to have a marked direct effect in determining the metabolic fate of LDL. An interesting recent finding is that LDL extracted from blood vessel walls, were modified so that they would react with scavenger receptors. The nature of the modification is not yet known.

In summary, at least three pathways are therefore available for the catabolism of LDL: the apo B,E receptors, present on hepatocytes and most nonhepatic cells; nonspecific or low-affinity uptake and degradation; and the scavenger pathway, reactive with modified LDL. There appear to be no estimates of the relative significance of pathways other than the apo B,E receptor in vivo in normal or in pathological LDL metabolism.

Catabolism of β-VLDL

β-VLDL are cholesteryl ester–rich lipoproteins recovered in the VLDL density range, which react with a specific receptor on macrophages, and endothelial cells. Macrophages do not bind or react with normal VLDL. This receptor catalyzes the uptake of β-VLDL by endocytosis: the VLDL cholesteryl ester store is hydrolyzed by lysosomal cholesterol esterase and the cholesterol is reesterified by ACAT. Macrophages store large amounts of cholesteryl esters when β-VLDL concentration is high. Unlike the apo B,E receptor, the β-VLDL receptor is present on the cell surface without induction by lipoprotein-deficient serum, although the number of receptors is somewhat down-regulated by continued exposure to β-VLDL. In addition to the lysosomal cholesterol esterase, the macrophage also contains a neutral cholesterol esterase which hydrolyzes stored cholesteryl esters. Thus, in cells grown in a cholesterol-depleting medium (for example, one containing centrifuged HDL) the stored cholesteryl ester is rapidly emptied out of the cell (the $t_{\frac{1}{2}}$ is about 24 h). Recent studies have shown that β-VLDL (which contain apo B-48 and therefore are derived from dietary particles) have a greater ability to promote cholesteryl ester storage in macrophages than do particles of similar lipid composition containing apo B-100. However, the ligand for the β-VLDL receptor has not yet been identified. β-VLDL contain high concentrations of apo E as well as apo B-48.

Catabolism of HDL

Because of the relative mobility of all components of the HDL particle, it has been particularly difficult to determine its major catabolic pathways by the usual isotopic methods. Nevertheless, data on the disposition of both apoprotein and lipids are now available.

In unspecialized cultured cells such as fibroblasts, HDL protein is bound, internalized, and degraded at rates that are similar to those for fluid and adsorptive endocytosis combined, which indicates that

HDL do not bind to apo B,E receptor sites. A number of laboratories have reported for HDL a significant dissociation of lipid and protein catabolism (Gwynne and Hess 1980; Glass et al. 1983). For example, uptake of cholesteryl ester (or its nondegradable analogue, cholesteryl ether) from HDL is 5- to 40-fold greater than that of protein in hepatocytes and gonadal and adrenal cells. The phenomenon has been particularly well characterized in adrenal cells. Binding of HDL is saturable (with an apparent K_m of 230 μg apo HDL/mL). Uptake of cholesterol was determined in terms of synthesis of steroid hormones from HDL cholesterol, and degradation of ^{125}I-labeled protein by [^{125}I]peptide generation. Uptake of lipid in all cases exceeded uptake of protein more than 10-fold. In contrast to adrenal cells, the kidney has been reported to clear apo A-1 more rapidly than lipid. This may represent clearance of apo A-1 in nonlipid or poorly lipidated form via renal filtration mechanisms.

THE FURTHER METABOLISM OF CHOLESTEROL

Total-body cholesterol loss (balanced by synthesis and dietary intake under steady-state conditions) is about 800 mg/d in the normal human adult. Of this, about 80 mg/d are lost as undegraded cholesterol via the skin. About 50 mg/d is converted to steroid hormones. Most of the remainder is lost in the feces as bile salts (about 400 mg/d) or neutral sterols. In the steady state this latter rate is also the bile acid synthesis rate. The rate of secretion of bile salts into the bile is of course many times greater (15–20 g/d) because of the extensive recirculation of bile salts via the enterohepatic circulation.

Regulation of Bile Acid Synthesis

The rate-limiting step of bile acid synthesis is the action of 7-α-hydroxylase with cholesterol in the smooth endoplasmic reticulum of the hepatocytes (Figure 13.17) (Danielsson and Sjovall 1975). No significant rate of direct hydroxylation of cholesteryl esters has been reported. It is not clear whether the regulation of the 7-α-hydroxylase is mediated by changes in the level of enzyme protein or by changes in activation state. The concept that 7-hydroxylation commits cholesterol to bile acid synthesis seems sound. However, while the later intermediates for synthesis of the common bile acids are well known, there is little agreement on the regulation of the steps that produce these intermediates. For example, 12-hydroxylation commits the sterol nucleus to the synthesis of cholic rather than chenodeoxycholic acid (Figure 13.17). However, little or no evidence is available to show whether or not the level of 12-α-hydroxylase activity is rate limiting for cholic acid synthesis in humans. The mechanism of side-chain cleavage, and its regulation, are also poorly

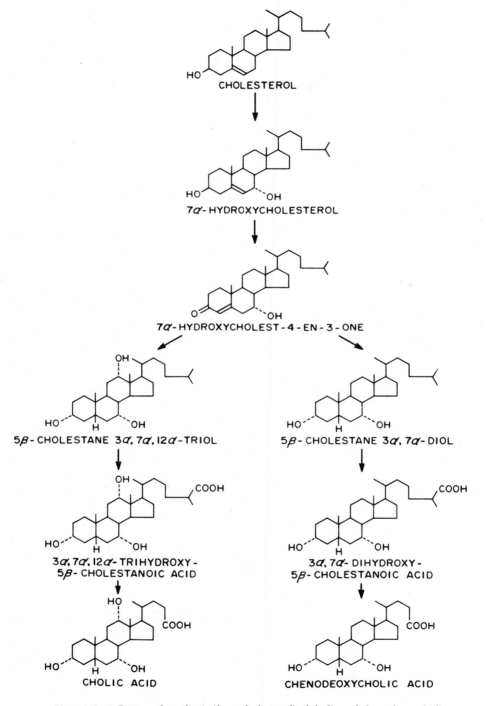

Figure 13.17. Routes of synthesis (from cholesterol) of cholic and chenodeoxycholic acids, the major bile acids of human bile.

understood. Studies of 26-hydroxylase activity (the enzyme that hydroxylates the 26-position before cleavage of the side-chain) indicate that this enzyme has no regulatory importance. On the basis of studies of patients with cerebrotendinous xanthomatosis (a genetic deficiency of 26-hydroxylase activity), the accumulation of 25-hydroxylated derivatives in bile suggests alternate pathways of side-chain degradation (Swell et al. 1980).

Source of Cholesterol for Bile Acid Synthesis

There are two alternative sources of cholesterol as substrate: cholesterol newly synthesized in the adjacent hepatic rough endoplasmic reticulum, and the free and esterified cholesterol of the plasma lipoproteins, interiorized via the different pathways already discussed. Early studies with rodents indicated that newly synthesized cholesterol provided the major substrate for bile acid synthesis. On the contrary, recent whole-body turnover studies with human subjects report that the plasma cholesterol pool provides the major part (about two-thirds) of cholesterol substrate (primarily free cholesterol of lipoproteins) for 7-α-hydroxylase activity (Schwartz et al. 1982). In this model, free cholesterol entering the hepatocyte is partitioned into a bile acid precursor pool. The proportion of this pool that is hydroxylated (and hence the ratio of bile acids to free cholesterol in the bile) depends on the activity of 7-α-hydroxylase (Figure 13.18).

While the tracer methodology and analysis of these studies are complex, the main findings are consistent with other data on the roles of cholesterol of HDL and LDL cholesterol in plasma and cellular cholesterol metabolism, as discussed previously. In human subjects, free cholesterol from HDL (rather than that from LDL) is preferentially incorporated into bile acids. It has been argued that since free cholesterol of HDL is the preferred substrate for plasma LCAT, it could not at the same time provide the mass of free cholesterol required for the secretion of biliary free cholesterol and bile acids. However, while cholesterol esterification by LCAT occurs on particles recovered in the HDL density range (the specialized subfraction of HDL), the free cholesterol used in the esterification is drawn from VLDL and LDL, not from HDL. It has also been argued that the rate of uptake of free cholesterol from plasma HDL, which is required for bile acid synthesis, is greater than the known rate of irreversible degradation of [^{125}I]HDL by the liver. However, it is largely free cholesterol, not cholesteryl ester, that is used for bile acid synthesis, and free cholesterol can be cleared at a rate in excess of the degradation rate of HDL protein. Therefore, it appears that cholesterol from HDL is the major precursor of bile acids and biliary free cholesterol.

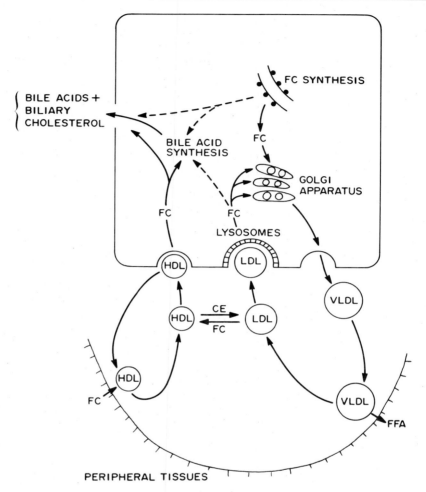

Figure 13.18. Routes of transfer and uptake of free and esterified cholesterol from lipoprotein in the liver associated with bile acid and biliary cholesterol secretion. The pathways shown are consistent with the in vivo kinetic data in normal humans (Schwartz et al. 1982) and with the metabolism and transfer of cholesteryl esters in human plasma (Fielding 1984). Dashed lines represent pathways of lesser importance in human metabolism. However, it should be noted that in rodents, unlike humans, newly synthesized hepatic cholesterol appears to be the major source of biliary cholesterol and bile acids. FC, free cholesterol; CE, cholesteryl ester; FFA, free fatty acids.

The liver also interiorizes large amounts of free and esterified cholesterol from VLDL remnants and from LDL by the apo B,E and apo E receptors present and from the cumulative contribution of other nonspecific or low-affinity processes. Since the synthesis of bile acids represents the only significant degradative pathway for cholesterol in the liver, the cholesterol of newly secreted human

VLDL and hence LDL must be the free cholesterol (and free and esterified cholesterol in animals such as rabbits and rats) recycled into nascent lipoproteins (Figure 13.9), together with the product of local cholesterogenesis. This same cholesterol also contributes in glandular tissues to the cholesterol used for steroid hormone production.

Balance studies indicate that the rate of cholesterol secretion and esterification is two- to threefold that of lipoprotein degradation (measured from studies of lipoprotein protein). There must therefore be extensive recycling of cholesterol between plasma and tissues prior to its irreversible catabolism or excretion. Overall, the data seem most consistent with the concept (Figure 13.18) that a major part of free cholesterol cleared by the liver as VLDL and LDL recycles peripherally, while cholesterol cleared as HDL is directed preferentially to catabolism.

ATHEROSCLEROSIS

Atherosclerosis is a common cause of morbidity and mortality in industrialized nations. This disease involves the accumulation of lipids, particularly free and esterified cholesterol, in and between cells within the vascular bed. The involvement of plasma cholesterol metabolism in atherogenesis is supported most convincingly by epidemiological studies. An increased level of plasma cholesterol (particularly LDL cholesterol) is an independent risk factor for this disease. (It should be pointed out, though, that many who suffer from atherosclerosis have normal plasma and LDL cholesterol levels.) There is also data that HDL "protects" against the deleterious effects of LDL in plasma. No convincing biochemical explanation of this finding has yet been made, although the physical states of the lipids in this disease have been studied in detail.

Structure of Atherosclerotic Lesions

The focal deposits of lipids in the vessel wall are usually classified as follows.

1. *Fatty streaks* are clusters of cells (often called foam cells and mostly macrophages) filled with large lipid droplets, typically in the superficial (luminal) layers. The lipid consists mostly of cholesteryl esters, particularly cholesteryl oleate, and the molar ratio of free cholesterol to phospholipid is also often high (up to 2.0).
2. *Gelatinous plaques* are accumulations of extracellular lipids in the form of fine droplets extending deep through the vascular bed. This lipid includes a high proportion of lipoproteins (particularly LDL) and the high proportion of cholesteryl linoleate typical of plasma LDL.

3. *Fibrous plaques* are similar in lipid content to gelatinous plaques but with a luminal cap of fibrous debris and cells, mostly derived from vascular smooth muscle cells.

Fatty streaks are found in high proportion in the vessels of animals such as rabbits fed high-cholesterol diets. Human lesions are more typically gelatinous and fibrous plaques. While it is generally agreed that fibrous plaques derive from gelatinous plaques, there is no agreement among pathologists that fatty streaks are the precursors of the other deeper and more complicated lesions.

There have been many attempts to relate molecular events of cellular and plasma cholesterol metabolism to atherogenesis, using data from rare genetic diseases or animal models. Our present state of knowledge concerning normal cholesterol metabolism does not convincingly explain atherogenesis in the vast majority of humans who suffer from this disease. Any hypothesis must deal with at least the following considerations: (a) the focal nature of lesions—for example, their prevalence at branch points of vessels—and (b) the slow development of lesions. While fatty streaks in animals can be induced within a few weeks by cholesterol feeding, human lesions such as gelatinous and fibrous plaques are believed to develop over years, if not decades.

Hypotheses for Atherogenesis

Endothelial Damage and Atherosclerosis. The rate of filtration of LDL through the normal vessel wall is low and can probably be explained in terms of the known rate of vacuolar transport of large solutes across the endothelial barrier. Plasma LDL is in equilibrium with the LDL of intercellular fluid and lymph. The LDL concentration in lymph is about 10% of plasma concentration, or 50–100 μg apo B/mL). In areas where the vessel wall shows lipid deposition, the rate of filtration not only of LDL but also of other plasma proteins is increased. The level of LDL in the wall may reach a concentration several times that of plasma (Smith and Ashall, 1983). Some process other than increased permeability must cause this accumulation.

Viruses have been suggested as promoters of endothelial damage. There is no doubt that in some avian models of atherosclerosis, herpes virus can induce endothelial damage and subsequent atherosclerosis. Data demonstrating that some arterial plaques in humans were derived from single cells also suggest a viral origin. However, there is no firm evidence yet that the major proportion of human plaques are virus induced.

Other investigators have pointed out that specific patterns of blood flow may lead to repeated local shearing of endothelial cells. Two factors support this concept. First, high blood pressure is an independent risk factor for atherosclerosis. Second, local rheological forces in the vessels provide an explanation for the focal generation of plaques. Endothelial damage would lead to the attachment of platelets to the exposed basement membrane and, possibly, multiplication of vascular smooth muscle cells if the endothelial layer were not immediately regenerated. While a change in filtration properties does not directly explain the involvement of free cholesterol in the lesion, especially under a cap of proliferating smooth muscle cells, it could easily lead to the movement of excess LDL across the vessel wall.

In a related hypothesis, cholesterol-enriched lipoproteins are themselves the initiating factors of endothelial damage. Endothelial cells, when contact inhibited as in vivo, have little LDL receptor activity but can take up cholesterol along a concentration gradient from plasma lipoproteins. In addition, if cholesterol-loaded endothelial cells were subsequently loosened from their substratum, the tissue would be locally released from constraints imposed by contact inhibition, and active uptake of cholesterol through the receptors would take place as well (Figure 13.19).

Receptor theories of atherosclerosis. As discussed in the preceding section, four major classes of lipoprotein receptors have been identified (Table 13.8): the apo B,E receptor of most nucleated cells, the

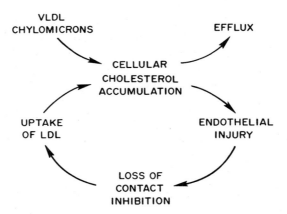

Figure 13.19. A possible mechanism for the accumulation of cholesterol and its esters within the vascular bed. Free cholesterol, interiorized along its concentration gradient in endothelial cells, causes local loss of cells (compatible with the experimental data given for cholesterol-fed primates; see Ross and Harker (1976)) and further cholesterol loading via up-regulated apo B,E receptors. Exposure of the substratum and vascular smooth muscle layers to platelets may lead to infiltration of LDL, cell proliferation, and lipid accumulation.

scavenger receptor for chemically modified LDL on macrophages and endothelial cells, the apo E receptor clearing chylomicron (and perhaps VLDL) remnants and cholesterol-rich HDL (HDL$_c$), and the receptor for β-VLDL.

The apo B,E receptor was first believed to be involved in atherogenesis because those who have receptor deficiency have high circulating levels of LDL and develop early atherosclerosis. It was proposed that this receptor plays a key role in preventing oversynthesis of cholesterol by cells. In the absence of the receptor, cholesterogenesis would be uncontrolled and the cells would fill up with cholesterol and assume the typical foam-cell appearance of the fatty streak. As discussed previously, it has been hard to prove that the apo B,E receptor regulates cholesterogenesis in the intact animal. In addition, this theory does not account for the focal distribution of lesions nor their long-term development. Since the apo B,E receptor is not present in macrophages, the theory does not explain the accumulation of lipid in these cells nor the presence of filtered extravascular LDL in gelatinous or fibrous plaques. It is true, however, that homozygotes lacking the apo B,E receptor have increased LDL levels and die from atherosclerosis at an early age. This may be an indirect effect of secondary changes in lipoprotein concentrations or composition, both altered in atherosclerosis.

The scavenger receptor seems better suited to playing a role, presumably protective, in atherogenesis. Since this receptor, unlike the apo B,E receptor, is not readily down-regulated, the macrophage can continue to accumulate cholesterol as long as there is a supply of chemically modified cholesterol. The recent finding that filtered LDL in the vessel wall is interiorized by the scavenger receptor indicates a possible physiological role for this receptor in atherogenesis. However, macrophages rapidly unload their cholesterol in the presence of HDL, and the maintenance of the lipid load requires that the macrophage remains in an environment where the cholesterol gradient is positive into the cell. The role of the macrophage is most likely a secondary and potentially beneficial one in removing foreign lipoproteins from the vessel wall. There is no evidence that the macrophages gather up lipids from the plasma and deposit them focally in the vessel wall.

The β-VLDL receptor has come into prominence because it is specific for the unusual, cholesterol-rich VLDL that is prominent in the plasma of cholesterol-fed animals. β-VLDL is also a minor (although possibly important) component of plasma lipoprotein in some humans who have an increased risk of atherosclerosis. It is unclear whether β-VLDL is filtered through the blood vessel wall or whether macrophages with matured β-VLDL receptors are present in the plasma. This receptor would be expected to have a beneficial effect if it assisted clearing of cholesterol into scavenger cells. It has not been shown that either scavenger receptors or β-VLDL receptors are absent or poorly functional in those persons with atherosclerosis.

The apo E receptor. In the human genetic disease *dysbetalipoproteinemia*, apo E has an abnormal amino acid sequence and is not recognized by the apo E receptor. As a result, there is a prolonged recirculation of remnant lipoproteins. Those with dysbetalipoproteinemia are at greater risk for atherosclerosis and have increased circulating levels of both cholesterol and triacylglycerol. However, the same genetic mutation is present in many people with a normal plasma cholesterol level who are not at risk for atherosclerosis. Thus, dysbetalipoproteinemia probably results from common inheritance of the gene for an abnormal apo E protein and for a hyperlipidemia. Those with dysbetalipoproteinemia have normal apo B,E receptor activity.

FUTURE DIRECTIONS   It is likely that most of the major metabolic pathways of cellular and plasma cholesterol metabolism have been identified. However, very little is known about the regulation of most of these steps, or of the rate-limiting steps of different metabolic sequences under physiological conditions.

Enough is now known about cholesterol metabolism in humans to establish that it is very different from that in many other mammals, particularly in rats and mice, which are accordingly poor research models for human metabolism in this particular area. The cholesterol metabolism of several primate models (such as cynomolgus and rhesus monkeys) seems to have many similarities to that in humans. The primate model would allow the integration of cultured-cell, organ, and whole-body experimental physiology, which is not possible in human studies. It is by such integrated studies that the next major advances in this complex area are likely to be achieved.

PROBLEMS

1. Epidemiological data indicate that those with high levels of HDL in the plasma are less likely to develop atherosclerosis. Could this be due to the ability of HDL to promote cholesterol efflux from cells?

2. It has been proposed that LDL, acting through the apo B,E receptor, regulates cholesterol synthesis in vivo. What is the evidence for and against this concept?

3. What is the evidence that cholesterol does not regulate its own synthesis in cells?

4. In dysbetalipoproteinemia, where the sequence of apo E is abnormal, intermediate lipoproteins containing both apo B-48 and apo B-100 accumulate in the plasma. What role does this suggest for hepatic apo E receptor?

5. The movement of free cholesterol between cells and a purified plasma lipoprotein occurs by a

specific, high-affinity saturable process. How would you show whether or not a cell-surface receptor was involved?

6. Mutant cultured cells showed only a very low level of apo B,E receptors in lipoprotein-deficient serum. However, inhibition of cholesterol synthesis with mevinolin significantly increased the number of receptors expressed. What is a possible mechanism?

7. Cholesteryl esters are rapidly synthesized from cholesterol and recycled back to free cholesterol in macrophages. A reversible cholesterol esterase has been described in liver. How would you determine
   (a) whether this enzyme was functioning in the macrophages and
   (b) whether synthesis or hydrolysis of cholesteryl esters was rate limiting under given experimental conditions?

8. There appears to be no relationship in humans between the level of 12-α-hydroxylase in liver and the ratio of cholic and chenodeoxycholic acids secreted into the bile. What are reasonable alternative mechanisms that would control the ratio of these two bile acids?

9. (a) In congenital LCAT deficiency, a small HDL containing only apo A-1 becomes an important component of total HDL. Why?
   (b) LCAT-deficient plasma is added to a culture of normal fibroblasts. What will be the *net* flux of free cholesterol under these conditions?

10. What would you expect to happen to plasma mevalonate levels when a normal person goes on a vegetarian diet?

BIBLIOGRAPHY

Most references are to recent reviews. However, in the case of recent developments where no adequate review is available, the original research papers are cited. Entries preceded by an asterisk are suggested for further reading.

Alaupovic, P.; Sanbar, S. S.; Furman, R. H.; Sullivan, M. L.; and Walraven, S. L. 1966. Studies on the composition and structure of serum lipoproteins. Isolation and characterization of very high density lipoproteins of human serum. *Biochemistry* 5:4044–53.

Albers, J. J.; Tollefson, J. H.; Chen, C. H.; and Steinmetz, A. 1984. Isolation and characterization of human plasma lipid transfer proteins. *Arteriosclerosis* 4:49–58.

Assmann, G.; Brown, B. G.; and Mahley, R. W. 1975. Regulation of 3-hydroxy-3-methylglutaryl coenzyme A reductase activity in cultured swine aortic smooth muscle cells by plasma lipoproteins. *Biochemistry* 14:3996–4002.

*Beg, Z. H., and Brewer, H. B. 1982. Modulation of rat liver 3-hydroxy-3-methylglutaryl-CoA reductase activity by reversible phosphorylation. *Fed. Proc.* 41:2634–38.

Billheimer, J. T., and Gaylor, J. L. 1980. Cytosolic modulators of activities of microsomal enzymes of cholesterol biosynthesis. *J. Biol. Chem.* 255:8128–35.

Bilheimer, D. W.; Stone, N. J.; and Grundy, S. M. 1979. Metabolic studies in familial hypercholesterolemia: Evidence in a gene-dosage effect in vivo. *J. Clin. Invest.* 64:524–33.

Björkhem, I.; Eriksson, M.; and Einarsson, K. 1983. Evidence for a lack of regulatory importance of the 12α-hydroxylase in formation of bile acids in man: An in vivo study. *J. Lipid Res.* 24:1451–56.

*Brown, M. S., and Goldstein, J. L. 1980. Multivalent feedback regulation of HMG CoA reductase, a control mechanism coordinating isoprenoid synthesis and cell growth. *J. Lipid Res.* 21:505–17.

Brown, M. S.; Ho, Y. K.; and Goldstein, J. L. 1980. The cholesteryl ester cycle in macrophage foam cells. Continual hydrolysis and reesterification of cytoplasmic cholesteryl esters. *J. Biol. Chem.* 255:9344–52.

Castro, G. R., and Fielding, C. J. 1984. Evidence of the distribution of apolipoprotein E between lipoprotein classes in human normocholesterolemic plasma and for the origin of an associated apolipoprotein E (Lp-E). *J. Lipid Res.* 25:58–67.

*Cryer, A. 1983. "Lipoprotein lipase–endothelial interactions." In *Biochemical interactions at the endothelium*, ed. A. Cryer, 245–74. Amsterdam: Elsevier.

*Danielsson, H., and Sjovall, J. 1975. Bile acid metabolism. *Ann. Rev. Biochem.* 44:233–53.

*Dobiasova, M. 1983. Lecithin:cholesterol acyltransferase and the regulation of endogenous cholesterol transport. *Adv. Lipid Res.* 20:107–94.

Ferguson, J. B., and Bloch, K. 1977. Purification and properties of a soluble protein activator of rat liver squalene epoxidase. *J. Biol. Chem.* 252:5381–85.

Fielding, C. J. 1978. Metabolism of cholesterol-rich chylomicrons. Mechanism of binding and uptake of cholesteryl esters by the vascular bed of the perfused rat heart. *J. Clin. Invest.* 62:141–51.

*_____. 1984. The origin and properties of free cholesterol potential gradients in plasma, and their relation to atherogenesis. *J. Lipid Res.*, 25:1624–28.

Fielding, C. J., and Fielding, P. E. 1982. Cholesterol transport between cells and body fluids. Role of plasma lipoproteins and the plasma cholesterol esterification system. *Med. Clinics N. Am.* 66:363–73.

*_____. 1983. "Lipoprotein binding at the endothelial surface." In *Biochemical interactions at the endothelium*, ed. A. Cryer, 275–99. Amsterdam: Elsevier.

Fielding, C. J., and Havel, R. J. 1977. Lipoprotein lipase. *Arch. Pathol.* 101:225–29.

Fielding, C. J., and Moser, K. 1982. Evidence for the separation of albumin- and apo A-1-dependent mechanisms of cholesterol efflux from cultured fibroblasts into human plasma. *J. Biol. Chem.* 257:10955–60.

Fielding, C. J.; Frohlich, J.; Moser, K.; and Fielding, P. E. 1982. Promotion of sterol efflux and net transport by apolipoprotein E in lecithin: cholesterol acyltransferase deficiency. *Metabolism* 31:1023–28.

Fielding, C. J.; Reaven, G. M.; Liu, G.; and Fielding, P. E. 1984. Increased free cholesterol in plasma low and very low density lipoproteins in non-insulin-dependent diabetes mellitus: Its role

in the inhibition of cholesterol ester transfer. *Proc. Natl. Acad. Sci. USA* 81, 2512–16.

Fielding, P. E., and Fielding, C. J. 1980. A cholesteryl ester transfer complex in human plasma. *Proc. Natl. Acad. Sci. USA* 77:3327–30.

Fielding, P. E.; Fielding, C. J.; Havel, R. J.; Kane, J. P.; and Tun, P. 1983. Cholesterol net transport, esterification and transfer in human hyperlipidemic plasma. *J. Clin. Invest.* 71:449–60.

Fogelman, A. M.; Edmond, J.; and Popjak, G. 1975. Metabolism of mevalonate in rats and man not leading to sterols. *J. Biol. Chem.* 250:1771–75.

*Gaylor, J. L. 1981. "Formation of sterols in animals." In *Biosynthesis of isoprenoid compounds*, vol. 1, ed. J. R. Porter and S. L. Spurgeon, 481–543. New York: Wiley.

*Gibbons, G. F. 1983. The role of oxysterols in the regulation of cholesterol biosynthesis. *Biochem. Soc. Trans.* 11:649–51.

Glass, C.; Pittman, R. C.; Weinstein, D. B.; and Steinberg, D. 1983. Dissociation of tissue uptake of cholesterol ester from that of apoprotein A-1 of rat plasma high density lipoproteins: Selective delivery of cholesterol ester to liver, adrenal, and gonad. *Proc. Natl. Acad. Sci. USA* 80:5435–39.

Ghiselli, G.; Schaefer, E.; Gascon, P.; and Brewer, H. B. 1981. Type III hyperlipoproteinemia associated with apolipoprotein E deficiency. *Science* 214:1239–41.

*Goldstein, J. L.; Anderson, R. G. W.; and Brown, M. S. 1979. Coated pits, coated vesicles, and receptor-mediated endocytosis. *Nature* 279:679–85.

*Goldstein, J. L.; and Brown, M. S. 1977. The low density lipoprotein pathway and its relation to atherosclerosis. *Ann. Rev. Biochem.* 46:897–930.

Gordon J. I.; Alpers, D. H.; Ockner, R. K.; and Strauss, A. W. 1983. The nucleotide sequence of rat liver fatty acid binding protein mRNA. *J. Biol. Chem.* 258:3356–63.

Grinstead, G. F.; Trzaskos, J. M.; Billheimer, J. T.; and Gaylor, J. L. 1983. Cytosolic modulators of activities of microsomal enzymes of cholesterol biosynthesis. Effects of acyl-CoA inhibition and cytosolic Z-protein. *Biochim. Biophys. Acta* 751:41–51.

Groot, P. H. E.; Scheek, L. M.; Havekes, L.; van Noort, W. L.; and van't Hooft, F. M. 1982. A one-step separation of human serum high density lipoproteins 2 and 3 by rate-zonal density gradient ultracentrifugation in a swinging bucket rotor. *J. Lipid. Res.* 23:1342–53.

Gwynne, J. T., and Hess, B. 1980. The role of high density lipoproteins in rat adrenal cholesterol metabolism and steroidogenesis. *J. Biol. Chem.* 255:10875–83.

*Herbert, P. N.; Assmann, G., Gotto, A. M.; and Fredrickson, D. S. 1983. "Familial lipoprotein deficiency: Abetalipoproteinemia, hypobetalipoproteinemia, and Tangier disease." In *The metabolic basis of inherited disease*, ed. J. B. Stanburg, J. B. Wyngaarden, D. S. Fredrickson, J. L. Goldstein, and M. S. Brown, 589–621. New York: McGraw-Hill.

Higgins, J. M., and Fielding, C. J. 1975. Lipoprotein lipase. Mechanism of formation of triglyceride-rich remnant particles from very low density lipoproteins and chylomicrons. *Biochemistry* 14:2288–93.

Hornick, C. A.; Jones, A. L.; Renaud, G.; Hradek, G.; and Havel, R. J. 1984. Effect of chloroquine on low-density lipoprotein catabolic pathway in rat hepatocytes. *Am. J. Physiol.* 246:G187–G194.

Ingebritsen, T. S.; Parker, R. A.; and Gibson, D. M. 1981. Regulation of liver hydroxymethylglutaryl-CoA reductase by a bicyclic phosphorylation system. *J. Biol. Chem.* 256:1138–44.

*Kane, J. P. 1983. Apoliprotein B: Structural and metabolic heterogeneity. *Ann. Rev. Physiol.* 45:637–50.

*Kesaniemi, Y. A., and Grundy, S. M. 1982. Significance of low density lipoprotein production in the regulation of plasma cholesterol level in man. *J. Clin. Invest.* 70:13–22.

Kesaniemi, Y. A.; Vega, G. L.; and Grundy, S. M. 1982. "Kinetics of apolipoprotein B in normal and hyperlipidemic man: Review of current data." In *Lipoprotein kinetics and modelling*, ed. M. Berman, S. M. Grundy, and B. V. Howard, 181–215. New York: Academic Press.

Lee, D. M., and Downs, D. 1982. A quick and large-scale density gradient subfractionation method for low density lipoproteins. *J. Lipid Res.* 23:14–27.

Lund-Katz, S., and Phillips, M. C. 1984. Packing of cholesterol molecules in human high density lipoproteins. *Biochemistry* 23:1130–38.

McLean, L. R., and Phillips, M. C. 1981. Mechanism of cholesterol and phosphatidylcholine exchange or transfer between unilamellar vesicles. *Biochemistry* 20:2893–2900.

McLean, L. R., and Phillips, M. C. 1984. Kinetics of phosphatidylcholine and lysophosphatidylcholine exchange between unilamellar vesicles. *Biochemistry* 23:4624–30.

Miller, K. W., and Small, D. M. 1983. Surface-to-core and interparticle equilibrium distributions of triglyceride-rich lipoprotein lipids. *J. Biol. Chem.* 258:13772–84.

*Myant, N. B. 1981. *The biology of cholesterol and related steroids*. London: Heinemann.

Oram, J. F.; Brinton, E. A.; and Bierman, E. L. 1983. Regulation of high density lipoprotein receptor activity in cultured human skin fibroblasts and human arterial smooth muscle cells. *J. Clin. Invest.* 72:1611–21.

Phillips, M. C.; McLean, L. R.; Stoudt, G. W.; and Rothblat, G. H. 1980. Mechanism of cholesterol efflux from cells. *Atherosclerosis* 36:409–22.

Pitas, R. E.; Innerarity, T. L.; Arnold, K. S.; and Mahley, R. W. 1979. Rate and equilibrium constants in binding of apo E HDL$_c$ (a cholesterol-induced lipoprotein) and low density lipoproteins to human fibroblasts: Evidence of multiple receptor binding of apo E HDL$_c$. *Proc. Natl. Acad. Sci. USA* 76:2311–15.

*Ross, R., and Harker, L. 1976. Hyperlipidemia and atherosclerosis. *Science* 193:1094–1100.

*Schroepfer, G. J. 1981. Sterol biosynthesis. *Ann. Rev. Biochem.* 50:585–621.

Schwartz, C. C.; Berman, M.; Halloran, L. G.; Swell, L.; and Vlahcevic, Z. R. 1982. "Cholesterol disposal in man: Special role of HDL free cholesterol." In *Lipoprotein kinetics and modelling*, ed. M. Berman, S. M. Grundy, and B. V. Howard, 337–49. New York: Academic Press.

Sexton, R. C.; Panini, S. R.; Azran, F.; and Rudney, H. 1983. Effects of 3β-[2-(diethylamino)ethoxy]-androst-5-en-17-one on the synthesis of cholesterol and ubiquinone in rat epithelial cell cultures. *Biochemistry* 22:5687–92.

Shepherd, J.; Bicker, S.; Lorimer, A. R.; and Packard, C. J. 1979. Receptor-mediated low density lipoprotein catabolism in man. *J. Lipid Res.* 20:999–1006.

Sherrill, B. C., and Dietschy, J. M. 1978. Characterization of the sinusoidal transport process responsible for uptake of chylomicrons by the liver. *J. Biol. Chem.* 253:1859–67.

Smith, E. B., and Ashall, C. 1983. Low density lipoprotein concentration in interstitial fluid from human atherosclerotic lesions: Relation to theories of endothelial damage and lipoprotein binding. *Biochim. Biophys. Acta* 754:249–57.

*Stange, E. F., and Dietschy, J. M. 1983. Cholesterol synthesis and low density lipoprotein uptake are regulated independently in rat small intestinal epithelium. *Proc. Natl. Acad. Sci. USA* 80:5739–43.

———. 1984. Age-related decreases in tissue sterol acquisition are mediated by changes in cholesterol synthesis and low density lipoprotein uptake in the rat. *J. Lipid Res.* 25:703–13.

Swell, L.; Gustafsson, J.; Schwartz, C. C.; Halloran, L. G.; Danielssen, H.; and Vlahcevic, Z. R. 1980. An in vivo evaluation of the quantitative significance of several potential pathways to cholic

and chenodeoxycholic acids from cholesterol in man. *J. Lipid Res.* 21:455–66.

Tall, A. R.; Abreu, E.; and Shuman, J. 1983. Separation of a plasma phospholipid transfer protein from cholesterol ester/phospholipid exchange protein. *J. Biol. Chem.* 258:2174–80.

Tanaka, R. D.; Edwards, P. A.; Lan, S.-F.; and Fogelman, A. M. 1983. Regulation of 3-hydroxy-3-methylglutaryl coenzyme A reductase activity in avian myeloblasts. Mode of action of 25-hydroxycholesterol. *J. Biol. Chem.* 258:13331–39.

Trzaskos, J. M., and Gaylor, J. L. 1983. Cytosolic modulators of activities of microsomal enzymes of cholesterol biosynthesis. Purification and characterization of a non-specific lipid-transfer protein. *Biochim. Biophys. Acta* 751:52–65.

Windler, E.; Chao, Y-s.; and Havel, R. J. 1980. Determinants of hepatic uptake of triglyceride-rich lipoproteins and their remnants in the rat. *J. Biol. Chem.* 255:5475–80.

CHAPTER 14

Lipid Assembly into Cell Membranes

Dennis R. Voelker

THE DIVERSITY OF LIPIDS

As discussed more fully in Chapter 2, phospholipids are amphipathic molecules with distinct hydrophilic and hydrophobic domains. In monomeric form the hydrophobic portion of a single molecule has the capacity to order a significant number of water molecules around it. The preferred state of water molecules, however, is not to exist in ordered arrays around the hydrophobic domains of lipid molecules but rather to exist in a more disordered physical state. Water's energetically favored state of maximum entropy leads to the segregation of hydrophobic domains away from water molecules. This interaction between water molecules and hydrophobic molecules has been most lucidly addressed by Tanford (1973) and colleagues; they have coined the term *hydrophobic effect* to define this interaction.

The consequences of the hydrophobic effect and the amphipathic nature of phospholipids are profound with respect to membrane assembly. The hydrophobic portions of phospholipids are excluded from interaction with the bulk aqueous solvent by the hydrophobic effect. However, the hydrophilic portions of phospholipids interact with the bulk aqueous solvent in a manner similar to that of simple salts, and this interaction is the preferred one of this portion of the molecule with water. The net result of the interaction of water with simple mixtures of phospholipids, or in most cases with single phospholipid species, is the formation of two fundamental structures: phospholipid micelles and phospholipid bilayers. In most model membrane systems and virtually all biological membrane systems, the phospholipid bilayer is the observed structure. The important point is that phospholipid bilayers form spontaneously in aqueous solvents as a consequence of the chemical interactions between water and phospholipids. The primary chemical driving force behind this membrane (phospholipid bilayer) formation is the entropy of water.

In addition to phospholipids. other lipid species contribute to the composition of cell membranes. These include primarily cholesterol and the glycosphingolipids. Cholesterol alone is incapable of forming membrane structures, but in the presence of 35 mol% phospholipid it readily associates with and stabilizes bilayer structures. Glycosphingolipids can vary remarkably in their ability to form micelles or bilayers or to exist as free monomers, depending upon the amount of carbohydrate attached to the hydrophobic ceramide moiety.

A cursory survey of the lipid composition of bacteria, fungi, lower animal eucaryotes, and higher animal eucaryotes provides the interesting observation that within each organism or cell there is a substantial amount of lipid diversity. In an excellent review of the genetics of phospholipid synthesis, Raetz (1982) has noted that in an organism such as *Escherichia coli* there exist as many as 100 chemically distinct phospholipids, while in eucaryotes there can be as many as 1000 distinct phospholipids. What is the significance of this diversity? The answer to that question is still for the most part unknown. A reasonable percentage of the diversity is probably accounted for by the variation in the degree of unsaturation of the fatty acid moieties, which ultimately translates into the fluidity of the bilayer. But the functions subserved by lipid diversity (exclusive of the components that contribute to fluidity) remain an enigma. The best one can do is to rely on one of the most fundamental maxims of biochemistry: Differences in structure translate into differences in function. One fact (too often overlooked by biochemists and cell biologists alike), however, remains clear. The chemical evolution of membrane lipids has not been in the direction of simplification to one species of phospholipid and its variant fatty acids. Although this may seem like such an obvious fact, it serves to emphasize the point that the lipid components of membranes probably function as much more than an inert two-dimensional matrix for protein diffusion.

In addition to the multiplicity of chemically distinct lipid species that occur within a procaryotic or eucaryotic cell, there is another level of complexity—the asymmetric distribution of the lipids across the plane of the bilayer. Two archetypical examples of membrane lipid asymmetry are the red blood cell membrane (Verkleij et al. 1973), and the cytoplasmic membrane of *Bacillus megaterium* (Rothman and Kennedy 1977). The data in Figure 14.1 demonstrate that in the red cell membrane the outer leaflet of the lipid bilayer is composed primarily of sphingomyelin and phosphatidylcholine, and

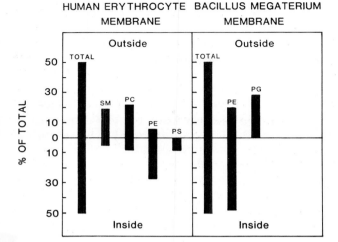

Figure 14.1. The asymmetric distribution of phospholipids in the human red blood cell and *Bacillus megaterium*. The abbreviations are SM, sphingomyelin; PC, phosphatidylcholine; PE, phosphatidylethanolamine; PS, phosphatidylserine; and PG, phosphatidylglycerol.

the inner leaflet of the bilayer contains phosphatidylserine and phosphatidylethanolamine, with lesser amounts of phosphatidylcholine and sphingomyelin. The location of the other major lipid component of the red blood cell membrane, cholesterol, has not been established with certainty. In the procaryote *B. megaterium* the distribution of phosphatidylethanolamine has been shown to be asymmetric, with 30% of this lipid present on the outer leaflet of the bilayer and 70%, on the inner leaflet. Phosphatidylethanolamine comprises about 70%, and phosphatidylglycerol about 30%, of the total phospholipid. Thus, nearly all the phosphatidylglycerol is in the outer leaflet of the bilayer. The location of the trace (about 1%) component cardiolipin is not yet known.

Yet another level of complexity is found to exist in cells that possess multiple membrane systems. In the Gram-positive bacteria there is essentially only one membrane system, the cytoplasmic membrane. In the Gram-negative bacteria there is both an inner and outer membrane system. In photosynthetic bacteria such as *Rhodopseudomonas sphaeroides* there are specialized membranes associated with the photopigments. In eucaryotes there are numerous membrane systems, the best characterized being endoplasmic reticulum, Golgi membranes, and plasma, mitochondrial, lysosomal, and nuclear membranes. Several of these membrane systems have dramatically different lipid compositions; these differences raise a variety of interesting questions: How are the different lipid compositions of different organelles established? How are these differences maintained? Are the different lipid compositions essential for organelle function? These questions remain unanswered and challenge genete-cists, biochemists, and cell biologists to design clever experiments and exploit new systems for their resolution.

INTRAMEMBRANE ASSEMBLY OF LIPIDS
Asymmetric Synthesis of Lipids

The primary consideration in the genesis of any biological membrane is the location of the synthetic apparatus that manufactures the subunits of the membrane and its relationship to the final distribution of its products. The work of Bell and colleagues (1971) has provided substantial evidence that in *E. coli* the synthesis of phospholipids occurs at the inner (cytoplasmic) membrane. Although not unequivo-cally established, it appears likely that most of the synthetic enzymes have their active sites on the cytoplasmic surface of the inner membrane. Such an orientation allows free access of water-soluble substrates and reaction products to the cytosol. Experiments con-ducted by Rothman and Kennedy (1979) demonstrate convincingly that in *B. megaterium*, newly synthesized phosphatidylethanolamine appears first on the cytoplasmic surface of the bilayer.

In eucaryotic systems a better-characterized pattern of synthetic asymmetry has emerged—in particular, the topology of the enzymes of phospholipid synthesis in rat liver microsomal membranes. The findings of Bell and co-workers (1981) have indicated that the active sites of the enzymes fatty acyl-CoA ligase, glycerolphosphate acyl-transferase, lysophosphatidic acid acyltransferase, diacylglycerol cholinephosphotransferase, diacylglycerol ethanolaminephospho-transferase, phosphatidylserine synthase, and phosphatidylinositol synthase are located on the cytosolic face of the membrane. Thus, in both procaryotic and eucaryotic systems it appears that the site of synthesis of the bulk of cellular phospholipid is the cytosolic side of the membrane. This asymmetric localization of synthetic enzymes strongly implicates transbilayer movement of phospholipids as an important event in membrane assembly. The topic of transbilayer movement of phospholipids has received considerable attention. In the next two sections two very different lines of experimental inves-tigation regarding transbilayer movement of phospholipids will be summarized.

Transbilayer Movement of Phospholipids: Lessons from Model Membranes

Model membranes made from either a single species or simple mixtures of phospholipids have been useful tools for studying the transbilayer movement of phospholipids. The most convenient physical form for these membranes has been as small, single-walled, bilayer vesicles commonly referred to as *liposomes* (Chapter 2). A simple consideration of the events that occur in the transbilayer movement of a zwitterionic molecule such as phosphatidylcholine

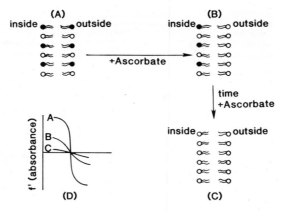

Figure 14.2. Measurement of the transbilayer movement of spin-labeled phosphatidylcholine. (A) Liposomes containing spin-labeled lipid (denoted in black) in both the inner and outer leaflets of the bilayer. (B) The ESR signal from lipid in the outer leaflet is eradicated by treatment with ascorbate. (C) With time the signal from the lipid in the inner leaflet disappears as the lipid moves from the inner to the outer leaflet and reacts with ascorbate. The $t_{\frac{1}{2}}$ for this process was observed to be 6.5 h (Kornberg and McConnell 1971). (D) The ESR spectra observed for conditions A, B, and C. The amplitude of the spectra is proportional to the amount of spin-labeled lipid present.

suggest that there are at least two energetically unfavorable events that must occur. The first is desolvation of the molecule and the second is movement of the charged portion of the lipid through the hydrophobic portion of the bilayer.

Direct experiments to examine the transbilayer movement of phospholipids were conducted by Kornberg and McConnell (1971), who made use of spin-labeled (paramagnetic) analogues of phosphatidylcholine (Figure 14.2). In these analogues the choline moiety of the phospholipids was replaced with the tempocholine probe, N,N-dimethyl-N-(1'-oxyl-2',2',6',6'-tetramethyl-4'-piperidyl)-ethanolamine:

$$HOCH_2CH_2-\underset{\underset{CH_3}{|}}{\overset{\overset{CH_3}{|}}{N_+}}-\text{(piperidyl ring)}-N-O$$

Tempocholine

Liposomes containing 3–5 mol% spin label exhibited a distinct paramagnetic spectrum the amplitude of which was directly proportional to the amount of spin-labeled lipid present. The spectrum was derived from molecules residing on both halves of the bilayer.

The liposomes were treated with ascorbate to reduce chemically the spin-labeled molecules on the outer half of the bilayer, leaving only the spectrum derived from the molecules on the inner half of the bilayer. This ESR spectrum was monitored as a function of time. The amplitude of the ESR signal from the residual spin-labeled probe was observed to decline with a $t_{\frac{1}{2}}$ of 6.5 h. The decline indicated a slow rate of transbilayer movement of phospholipids from the inner to the outer leaflet of the bilayer.

A second approach to measurement of transbilayer movement of phospholipids made use of phospholipid exchange protein (Rothman and Dawidowicz 1975). (These proteins are discussed in more detail on pages 490–493.) In these experiments small unilamellar vesicles containing radiolabeled phosphatidylcholine were incubated with excess phospholipid exchange proteins and excess acceptor membranes. Between 60 and 70% of the phospholipid was removed from the vesicles with a $t_{\frac{1}{2}}$ of 7–10 min. In vesicles of the size used in these experiments, the outer leaflet of the phospholipid bilayer contains 60–70% of the total phospholipid. The remaining 30% of the labeled phospholipid was not efficiently exchanged out of the vesicles. Estimates of the $t_{\frac{1}{2}}$ for the exchange of the residual 30% of the radiolabel were 11–15d. This experiment is summarized in Figure 14.3. These results clearly indicated an almost nonexistent rate of transbilayer movement of phospholipid.

A third experimental approach (Roseman, Litman, and Thompson 1975) to the problem was chemical modification of phosphatidyletha-nolamine in liposomes containing this lipid and phosphatidylcholine. In these experiments small unilamellar vesicles containing 10 mol% phosphatidylethanolamine were made. The vesicles were treated with isethionylacetimidate, a reagent that chemically modifies the phosphatidylethanolamine molecules only on the outer leaflet of the bilayer. The vesicles were next treated with trinitrobenzenesul-fonate, a reagent that also chemically modifies phosphatidyletha-

Figure 14.3. Measurement of the transbilayer movement of radiolabeled phosphatidyl-choline using phospholipid exchange protein. (A) Liposomes containing radiolabeled phosphatidylcholine (denoted in black) were made by sonication and incubated with a high concentration of phospholipid exchange protein and excess acceptor membranes. (B) The results obtained indicated that only the lipid in the outer leaflet of the membrane was removed by the exchange protein. The $t_{\frac{1}{2}}$ for movement of the lipid from the inner to the outer leaflet was determined to be 11–15 d (Rothman and Dawidowicz 1975).

nolamine molecules (on the outside of the bilayer) but which will not react with those molecules that have previously been modified with isethionylacetimidate.

$$RNH_2 \;+\; {}^-O_3S-\underset{NO_2}{\overset{NO_2}{\bigcirc}}-NO_2 \longrightarrow$$

Phosphatidylethanolamine   **Trinitrobenzene sulfonate**

$$HSO_3^- \;+\; R\overset{H}{\underset{N}{}}-\underset{NO_2}{\overset{NO_2}{\bigcirc}}-NO_2$$

N-Trinitrophenylphosphatidylethanolamine

$$RNH_2 \;+\; H_3\overset{+}{C}-\overset{\overset{NH_2}{\|}}{C}-OCH_2CH_2SO_3^- \longrightarrow$$

Phosphatidylethanolamine   **Isethionylacetimidate**

$$HOCH_2CH_2SO_3^- \;+\; R\overset{H}{\underset{N}{}}-\overset{\overset{+NH_2}{\|}}{C}-CH_3$$

N-Acetimidoylphosphatidylethanolamine

When phosphatidylethanolamine reacts with trinitrobenzenesulfonate, the reaction product, N-trinitrophenylphosphatidylethanolamine, absorbs light at 410 nm and thus can be quantitated spectrophotometrically. The only phosphatidylethanolamine molecules available for reactions with trinitrobenzenesulfonate are those that did not originally react with isethionylacetimidate (that is, those on the inner leaflet of the bilayer). However, these molecules of unreacted phosphatidylethanolamine can react with trinitrobenzenesulfonate only if they are translocated across the bilayer. The results from these experiments were that virtually no trinitrobenzenesulfonate-reactive material appeared on the outer leaflet of the vesicle for 12 d. The $t_{\frac{1}{2}}$ for transbilayer movement of phosphatidylethanolamine in this system was estimated to be at least 80 d. This experiment is summarized in Figure 14.4.

This study also examined a crucial factor not addressed by the preceding investigations, namely, the role of chemical oxidation of phospholipids and its effects on the results. The conclusions were

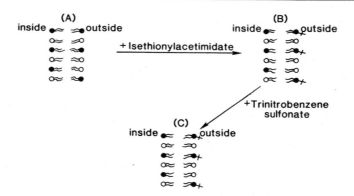

Figure 14.4. Measurement of the transbilayer movement of phosphatidylethanolamine. (A) Liposomes containing 10 mol % phosphatidylethanolamine (denoted in black) were reacted with isethionylacetimidate which (B) modifies the phosphatidylethanolamine in the outer leaflet. Subsequent treatment of the liposomes with trinitrobenzenesulfonate gives no reaction (C) and demonstrates that no additional phosphatidylethanolamine molecules appear at the outer leaflet. If transbilayer movement of lipid had occurred, then trinitrobenzenesulfonate-reactive material would have appeared at the outer leaflet. The $t_{\frac{1}{2}}$ for transbilayer movement was estimated to be 80 d (Roseman, Litman and Thompson 1975).

that nonspecific chemical oxidation can rapidly compromise the integrity of the vesicles and give faster rates of apparent transbilayer movement than occur otherwise. In retrospect it appears that the faster rates of translocation originally reported for spin-labeled vesicles may have been due to some lipid oxidation in these preparations.

The results from these experiments clearly demonstrate that transbilayer movement of phospholipids is a very slow, if not nonexistent process in model membranes. But what about biological membranes? As discussed earlier, experiments with procaryotes and eucaryotes have indicated that the enzymes for phospholipid synthesis are localized to the cytosolic face of the pertinent membranes, and it appears that transbilayer movement of phospholipids is required for new membrane synthesis. This requirement has been substantiated in experiments performed in bacteria, in membranes derived from endoplasmic reticulum, and in erythrocyte membranes.

Transbilayer Movement of Phospholipids: Observations of Biological Membranes

In experiments performed with *B. megaterium*, Rothman and Kennedy (1979) used trinitrobenzenesulfonate, under conditions where the probe did not enter the cell, to distinguish between phosphatidylethanolamine molecules located on the outer and inner sides of the cell membrane. This technique was coupled with pulse-chase experiments with [^{32}P]inorganic phosphate and [^{3}H]glycerol and demonstrated that newly synthesized phosphatidylethanolamine is found on the cytoplasmic surface of the cell membrane and is rapidly translocated to the outer leaflet of the membrane with a $t_{\frac{1}{2}}$ of 3 min at 37°C. Although the translocation is rapid, it does not occur coincident with synthesis, but rather, well after the molecule is synthesized. In addition, the translocation can occur in the absence of phosphatidylethanolamine synthesis. These findings indicate that lipid synthesis and translocation are two distinct events.

A second biological membrane that was examined for the transbilayer movement of phospholipids was the microsomal membrane (Zilversmit and Hughes 1977) (principally derived from endoplasmic reticulum) from rat liver. Like the cytoplasmic membrane of *B. megaterium*, this membrane system is actively involved in the synthesis of phospholipids. Animals were injected with either [^{14}C]choline or [^{32}P]inorganic phosphate to radiolabel hepatic phospholipids, and liver microsomes were prepared. Both short- and long-term labeling protocols were used in these studies, with identical results. Long-term labeling with [^{32}P]inorganic phosphate ensures uniform labeling of phospholipid pools on both sides of the membrane. The transbilayer movement of phospholipids was examined using phospholipid exchange proteins. The results from these experiments provided evidence that 85–95% of phosphatidylcholine, phosphatidylethanolamine, phosphatidylserine, and phosphatidylinositol were exchanged between labeled microsomes and excess acceptor membranes in 1–2 h. The labeled phospholipids exchanged with kinetics indicative of a single pool, and independent criteria demonstrated that membrane integrity was preserved throughout the experiment. These results implicate transbilayer movement of phospholipids and indicate that the rate of translocation is faster than can be detected with phospholipid exchange proteins. Estimates of the $t_{\frac{1}{2}}$ for translocation are 45 min. This value, however, must be an upper limit because the amount of exchange protein used could not exchange out the labeled phospholipid with a $t_{\frac{1}{2}}$ of less than 45 min.

A third system used to study transbilayer lipid movement was the mammalian erythrocyte. Although erythrocytes do not have the capacity to synthesize phospholipids, they can exchange phospholipids with serum lipoproteins. Renooij and co-workers (1976) therefore exchanged ^{32}P-labeled phospholipids from serum lipoproteins into the rat erythrocyte. They subsequently treated the red cells with phospholipase A_2, which selectively degrades the phospholipids in the outer leaflet of the bilayer. The labeled phospholipid pool available to the phospholipase A_2 was observed to decrease with time even though there was no change in the total amount of radiolabeled lipid in the cell. Apparently, the lipid had moved to an inaccessible domain of the membrane, presumably the inner leaflet of the bilayer. The $t_{\frac{1}{2}}$ of this process was determined to be 4.5 h.

Bloj and Zilversmit (1976) also have examined transmembrane movement in erythrocyte membranes. In their experiments rat red cells were radiolabeled with ^{32}P in vivo, isolated, lysed, and the membranes resealed to form ghosts. Radiolabeled phosphatidylcholine was exchanged out of the sealed ghosts using phosphatidylcholine exchange protein and excess acceptor membranes. The kinetics of labeled phosphatidylcholine exchange were biphasic, with 75% exchanging rapidly and 25% exchanging much more slowly. Making the assumptions that the rapidly exchanging pool of lipid represents the outer leaflet of the bilayer and the slowly exchanging pool was originally on the inner leaflet of the bilayer, the authors calculated the $t_{\frac{1}{2}}$ for transmembrane movement. They determined the $t_{\frac{1}{2}}$ for inside-to-outside transitions of phosphatidylcholine to be 2.3 h for the rat erythrocyte.

The results from the preceding experiments with biological membranes clearly provide strong evidence that the transbilayer movement of phospholipids occurs and that it can be very rapid. The experiments conducted with liposomes indicate that the opposite is true. What is the difference between model membranes and biological membranes? Quantitatively, the most significant difference would seem to be the presence of proteins. But is the presence of any membrane protein sufficient to effect transmembrane movement, or are there specific membrane proteins that manifest this function in biological membranes? At present, the answers to these questions are unknown, but two lines of experimentation with biological membranes suggest that transbilayer movement of phospholipid may be a specific process.

Enveloped viruses, such as influenza, bud from host-cell plasma membranes and carry a complement of cellular phospholipid in addition to the viral proteins. Rothman and co-workers (1976) prepared ^{32}P-labeled viruses and examined the transbilayer movement of phospholipids with phospholipid exchange proteins (and excess acceptor membranes) and phospholipase C. Approximately half of the phosphatidylcholine pool was available for exchange or degradation, depending on the probe. The balance of the phosphatidylcholine pool remained inaccessible for several days. Basing their calculations on the assumption that the inaccessible pool resides on the inner leaflet of the bilayer, the authors estimated the $t_{\frac{1}{2}}$ for transbilayer movement to be 10 d.

Sandra and Pagano (1978) prepared inverted (cytosolic side facing outward) plasma membrane visicles from mouse LM cells that had been labeled with [^{3}H]choline. Examination of the exchange of [^{3}H]phosphatidylcholine from the plasma membrane to excess acceptor vesicles demonstrated that there were two pools—one rapidly exchanging and one slowly exchanging. The rate-limiting step in the exchange of the latter was assumed to be the transbilayer movement of phosphatidylcholine. The $t_{\frac{1}{2}}$ for this process was estimated to be 88 h. Thus, these last two experiments provide clear evidence of biological membranes that do not exhibit rapid transbilayer movement of phospholipids.

The mechanism by which transbilayer movement of phospholipids occurs is unknown. The results from experiments conducted with model membranes and biological membranes indicate that it is not an unmediated spontaneous exchange of one phospholipid for another across the bilayer. From experiments with bacteria, red cells, and microsomal membranes it appears quite likely that membrane proteins facilitate this process. However, as with influenza virus and isolated plasma membranes, the presence of membrane proteins per se does not ensure the transbilayer movement of lipids. The results implicate specific proteins in this process.

The energetic requirements for transmembrane movement of phospholipids have been investigated by Langley and Kennedy (1979). Using *B. megaterium* and a trinitrobenzenesulfonate probe they demonstrated that the rapid transmembrane movement of newly synthesized phosphatidylethanolamine was unaffected by inhibitors of ATP synthesis and protein synthesis. Thus, the driving force for phospholipid translocation is independent of metabolic energy, lipid synthesis, and protein assembly into cell membranes.

As cited earlier in this chapter, *B. megaterium* exhibits an asymmetric distribution of phosphatidylethanolamine, with 30% on the external leaflet of the cell membrane and 70% on the internal leaflet. Since rapid transmembrane movement of phospholipids continues to occur under conditions of poisoned metabolism, Langley and Kennedy examined the effects on phosphatidylethanolamine asymmetry. Once again, the method they used was trinitrobenzenesulfonate modification of phosphatidylethanolamine. The results demonstrated that the asymmetric distribution of phosphatidylethanolamine was not only maintained but slightly enhanced in the presence of energy poisons. Stated in another way, the chemical gradient (asymmetry) of phosphatidylethanolamine did not collapse in the absence of metabolic energy. This finding suggests that the asymmetric distribution of phosphatidylethanolamine represents a stable, equilibrium state. The basis of the asymmetry then, may be the ionic interaction between phospholipids, proteins, and ions at the different leaflets of the bilayer. Whether the asymmetry of other biological membranes reflects an equilibrium state or a very slow rate of transmembrane movement after the asymmetry is established requires further experimentation to ascertain.

INTERMEMBRANE ASSEMBLY OF LIPIDS
Compositional Differences Among Membranes

Organisms with multiple membrane systems, such as Gram-negative bacteria, photosynthetic bacteria, and the eucaryotes, raise significant questions about the mechanisms of membrane biogenesis. In a "simple" organism such as *E. coli* there are two membrane systems: the inner or cytoplasmic membrane, and the outer membrane. The entire synthetic apparatus for phospholipid synthesis is located at the inner membrane. Consequently, there must exist a mechanism for exporting phospholipids from the inner membrane to the outer membrane. Comparison of the phospholipid composition of isolated inner and outer membranes reveals that there are significant quantitative differences in the amounts of phosphatidylethanolamine (enriched in outer membranes) and phosphatidylglycerol (enriched in inner membrane). Thus, any model for the assembly of the outer membrane should take this observation into account. In addition to an inner and outer membrane, Gram-negative photosynthetic bacteria, such as *R. sphaeroides*, elaborate intracytoplasmic membrane systems. The intracytoplasmic membranes are devoid of most phospholipid synthetic activities and hence require transport of phospholipid from the cytoplasmic membrane. Whether the mechanism of transport is similar to that required for phospholipid assembly into the outer membrane is presently unknown.

The basic questions of phospholipid transport from membranes that contain the lipid synthetic apparatus to membranes incapable of lipid synthesis is considerably amplified in eucaryotic cells. The principal membrane structures of eucaryotes are the endoplasmic reticulum, Golgi apparatus, lysosomes, mitochondria, and nuclear and plasma membrane. In animal cells the majority of the phospholipids and cholesterol are synthesized by the enzymes in the endoplasmic reticulum and exported to other target organelles. The net results are the phospholipid compositions shown in Table 14.1. From the data in the table it is apparent that the membranes of different organelles can have remarkably different lipid compositions. The most notable differences are the enrichment of cholesterol and sphingomyelin in the plasma membrane and their virtual absence from the mitochondria. Cardiolipin appears to be found exclusively in the mitochondria. The small amount of cardiolipin in the endoplasmic reticulum is attributable to mitochondrial contamination. Within the mitochondrion there are significant differences between the outer and inner membrane. In general, the outer membrane of the mitochondrion appears more similar to the endoplasmic reticulum than does the inner membrane. The biochemical mechanisms for lipid transport to the different organelles are unknown, as are the mechanisms by which differences in lipid composition are achieved and maintained. Any mechanism proposed for the interorganelle transport

Table 14.1. Lipid Compositions of Subcellular Organelles from Rat Liver

Phospholipid*	Endoplasmic reticulum		Mitochondrial membranes		Lysosomal membrane	Nuclear membrane	Golgi membrane	Plasma membrane
	Rough	Smooth	Inner	Outer				
Lysophosphatidylcholine	2.9	2.9	0.6	—	2.9	—	5.9	1.8
Sphingomyelin	2.4	6.3	2.0	2.2	16.0	6.3	12.3	23.1
Phosphatidylcholine	59.6	54.4	40.5	49.4	41.9	52.1	45.3	43.1
Phosphatidylinositol	10.1	8.0	1.7	9.2	5.9	4.1	8.7	6.5
Phosphatidylserine	3.5	3.9	1	1	—	5.6	4.2	3.7
Phosphatidylethanolamine	20.0	22.0	38.8	34.9	20.5	25.1	17.0	20.5
Cardiolipin	1.2	2.4	17.0	4.2	—	—	—	—
Phospholipid/protein (μmole P/mg)	0.33	0.47	0.34	0.46	0.21	—	—	0.37
Cholesterol/Phospholipid mol %	0.07	0.24	0.06	0.12	0.49	—	0.152	0.76

Sources: Data for all membranes except Golgi are from Colbeau, Machbaur, and Vignais (1971). Data for Golgi are from Keenan and Moore (1970).
* Values for individual lipids are percentage of phospholipid phosphorus.

of phospholipids should necessarily take differences in lipid composition into account.

Potential Mechanisms for Lipid Transport

At present there is no definitive biochemical evidence demonstrating the mechanisms by which phospholipids and cholesterol move from one membrane domain to another. It is useful to consider as many potential mechanisms for transport as possible in order to have a clear idea of whether experimental results support, refute, or provide no information about a particular mechanism. A schematic summary of these mechanisms is given in Figure 14.5.

Probably the simplest mechanism for lipid assembly is one of solubility and diffusion (mechanism A). Molecules such as free fatty acids, liponucleotides (such as CDP-diacylglycerol), and phosphatidic acid have appreciable solubility and could diffuse in a nonfacilitated manner among organelles. Although this mechanism alone probably would not be able to account for compositional variance, the

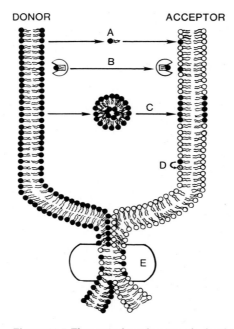

Figure 14.5. Theoretical mechanisms for lipid assembly into membranes. A, solubility and diffusion; B, soluble protein carrier; C, vesicle-mediated transport (Vesicle-target fusion may be spontaneous or protein mediated. The protein-mediated event could be catalytic or noncatalytic.); D, target organelle modification of transported lipid; E, donor-target organelle fusion (Donor-target fusion may be spontaneous or protein mediated. The protein-mediated event may be catalytic or noncatalytic.).

metabolism of these compounds in target organelles (mechanism D) could account for some compositional differences between donor and targe organelles.

The very limited solubility of the other phospholipids and cholesterol makes the aspect of a protein carrier an attractive mechanism for assembling new lipids into membranes. Such a mechanism (B) could be specific if there was a single carrier for each phospholipid. Some chemical interactions between the donor and acceptor membranes and a soluble carrier might ensure unidirectional movement of certain lipids and not others, thereby accounting for both net transport and specificity.

Alternate mechanisms might occur at a more macroscopic level and include physical contact and fusion between donor and acceptor organelles (mechanism E). The fusion events could be mediated by anionic phospholipids and divalent cations and be spontaneous, or they might be mediated by specific fusion proteins whose activity is tightly regulated by the cell. Proteins involved in fusion could function catalytically or noncatalytically. A plausible example of a catalytic fusion protein would be a membrane-bound phospholipase capable of generating fusogenic lipids (such as lysophosphatidylcholine). Noncatalytic fusion proteins might interact in a receptor-ligand fashion accompanied by a conformational change (possibly one requiring metabolic energy) that overcomes some of the energetic barriers of bilayer-bilayer fusion. If fusion between donor and acceptor organelles were to occur, one would anticipate (based upon the fluid mosaic model of membranes) that there would be a rapid intermixing of lipids and proteins between the organelles. Since target organelles are characterized by their unique protein and lipid compositions, it seems that a necessary corequisite of fusion phenomena would have to be the presence of a sorting mechanism (that is, a molecular filter), that allowed only certain molecules to move between donor and acceptor membranes.

An alternative to donor and target organelle fusion is the interposition of discontinuity between the two organelles via the formation of transport vesicles (mechanism C), which could provide an efficient means of ensuring the complete segregation of proteins and lipids between organelles. However, vesicular mechanisms for lipid transport still require the presence of molecular filters to ensure that only the correct molecules are exported. The potential mechanisms involved in vesicle–target organelle fusion could be either spontaneous or protein mediated (and either catalytic or noncatalytic) as already discussed.

One mechanism that should not be overlooked is the subsequent metabolism of lipids after their arrival at the target organelle (mechanism D). Phosphatidylglycerol and cardiolipin synthesis by the mitochondrion are good examples of this sort of posttransport modification. Whether selective degradation by phospholipases may also play a role in determining target membrane composition is not known.

PHENOMENA THAT MAY PLAY A ROLE IN LIPID ASSEMBLY
Morphology

Examination of cells by electron microscopy can often provide important clues to the nature of underlying biochemical phenomena. In the case of the Gram-negative bacteria whose inner membrane contains the lipid synthetic apparatus, morphological studies (Bayer 1968) have provided evidence of zones of adhesion between the inner membrane and the outer membrane. Although identification of such structures does not constitute a proof of the mechanism of lipid transport, it does indicate that notions of donor-target membrane fusion are feasible.

Research into interorganelle lipid transport and assembly in eucaryotic cells is usually focused on phospholipids synthesized in the endoplasmic reticulum (the location of the majority of the lipid synthetic apparatus) that are exported to target organelles. The relevant morphological questions are whether there are regions of continuity between the endoplasmic reticulum and other organelles, or whether the mechanisms of membrane transport are primarily vesicular. The majority of the morphological evidence implicates vesicular mechanisms for traffic between the endoplasmic reticulum and Golgi, and from the Golgi to the lysosomes and plasma membrane (Palade 1975). However, morphological arguments have been put forth for zones of continuity between the endoplasmic reticulum and the Golgi. The majority of the data argues for discontinuity between the mitochondrion and the endoplasmic reticulum, although there has been some scant evidence of continuity between the outer mitochondrial membrane and the endoplasmic reticulum. There has not been a clear demonstration of vesicle traffic to the mitochondrion. Thus, from the morphologist's viewpoint, vesicular traffic among the endoplasmic reticulum, Golgi, and plasma membrane is a potential mechanism for lipid transport. Whether lipid transport among these organelles is obligately connected to vesicle movement remains to be demonstrated. The other potential mechanisms for phospholipid transport, namely, solubility and diffusion, and via soluble protein carriers, are not easily investigated by the technique of electron microscopy; more information has been obtained by biochemical studies.

Lipid Binding Proteins

Investigations into the mechanism of lipid transport from the endoplasmic reticulum to the mitochondrion by Wirtz and Zilversmit

(1968) led to the identification of phospholipid exchange proteins. These proteins were originally shown to be capable of effecting phospholipid exchange between microsomes and mitochondria in vitro. The reaction proceeded in a manner consistent with exchange (that is, radioequilibration) of phospholipids and not net transfer of mass. These early studies culminated in the purification and characterization of a phosphatidylcholine exchange protein (Wirtz and Zilversmit 1970). Since the purification of this protein, several additional proteins have been purified and fully characterized. The well-characterized proteins thus far fall into three categories: (1) those specific for phosphatidylcholine; (2) those with a high activity with phosphatidylinositol and measurable activity with phosphatidylcholine and sphingomyelin; and (3) those that are nonspecific and capable of acting on sphingomyelin, cholesterol, and all phospholipids except diphosphatidylglycerol.

Phospholipid exchange protein activity has been detected in a wide variety of eucaryotic cells. At least one procaryote, *R. sphaeroides* has been shown to have lipid exchange activity (Cohen, Lueking, and Kaplan 1979). Because this organism elaborates intracytoplasmic membrane in an inducible fashion, it may prove useful in determining the coordinate regulation of membrane biogenesis and exchange protein activity.

Numerous assay systems have been developed to measure the activity of these proteins. One of the simplest and most convenient assays employs small unilamellar liposomes (donor membranes) containing the ^{3}H-labeled phospholipid of interest and trace amounts of [^{14}C]triolein or [^{14}C]cholesterol oleate, acceptor membranes such as mitochondria (usually from rat or beef liver), and a source of phospholipid exchange protein. Following a brief incubation, the mitochondria are readily separated from the liposomes by centrifugation, and the transfer of radiolabeled phospholipid is quantitated by liquid scintillation counting. The [^{14}C]triolein or [^{14}C]cholesterol oleate are routinely included in the assay systems because they do not function as ligands for phospholipid exchange proteins and hence can be used to assess any nonspecific binding between the liposomes and mitochondria.

When first isolated, phosphatidylcholine exchange proteins were demonstrated to contain 1 mol of lipid bound per mole of protein. This fact led to the notion that the proteins were always in the "loaded" state and hence could not effect net transfer of lipid mass. The question of whether the proteins can accomplish net transfer of mass is a separate issue from their ability to effect compositional modification of membranes. Clearly, judicious use of donor membranes of defined composition can be used to alter the lipid composition of acceptor membranes when the exchange protein has multiple specificities. A knowledge of the ability of phospholipid exchange proteins to carry phospholipids unidirectionally from one membrane domain to another is central to understanding their role in membrane biogenesis. Although many of the assay systems have focused upon either hetero- or homospecific exchange of phospholipids, recent experiments have demonstrated the net transfer of phospholipid mass.

Crain and Zilversmit (1980) have provided evidence for the net transfer of phospholipid mass from donor liposomes to apoproteins (from high-density lipoprotein) to high-density lipoprotein and to mitochondria stripped of their outer membrane. In each case the nonspecific protein was the only exchange protein capable of transferring lipid mass. A second demonstration of mass transfer was provided by Wirtz, Devaux, and Bienvenue (1980) using phosphatidylcholine exchange proteins and spin-labeled phosphatidylcholine. When the exchange protein was preloaded with spin-labeled phosphatidylcholine, a discrete spectrum characteristic of immobilized ligand was obtained. If the loaded exchange protein was incubated with micelles of lysophosphatidylcholine or phosphatidic acid liposomes, the ESR spectrum changed in a manner that indicated the probe entered the micelle or liposome. Since the exchange protein will not transfer either lysophosphatidylcholine or phosphatidic acid, the authors concluded that the protein had released the phosphatidylcholine analogue and remained in the empty state. This result indicated a net transfer of mass to the acceptor membrane.

Investigations of the activities of the exchange proteins have also focused upon the effects of membrane charge and composition upon lipid exchange. The introduction of negatively charged lipids such as phosphatidic acid or phosphatidylinositol into a membrane can markedly inhibit the ability of phosphatidylcholine exchange protein to transfer radiolabel from these membranes to acceptor membranes. The explanation given for this inhibition has been the high affinity of this protein for negatively charged membrane surfaces. A dramatic demonstration of the role of membrane composition in phospholipid exchange is seen when phosphatidylcholine exchange protein is incubated with donor membranes containing phosphatidylcholine, and acceptor membranes containing only phosphatidylethanolamine. Under these conditions no transfer of phosphatidylcholine is observed. These observations demonstrate that membrane lipid composition can greatly influence the activity of exchange proteins.

In addition to proteins which exchange phospholipids, there are two other classes of intracellular lipid-binding proteins. The newest identified class comprises those that transfer glycosphingolipids (Metz and Radin 1980). At present our knowledge about these is quite limited, although the activity appears distinct from the nonspecific phospholipid exchange proteins. The second group of proteins bind both fatty acids and sterols. These proteins have been investigated by a number of different research groups whose perspective on the protein has been from the point of view of the ligand. It has recently been realized (Ockner, Manning, and Kane 1982), however, that several of these ligand-binding activities—sterol carrier protein, Z-protein, and fatty acid binding protein—are attributable to the same protein. This family of sterol-fatty acid binding proteins is capable of augmenting the metabolism of both types of lipids in certain reactions, presumably by facilitating the interaction of enzyme and substrate. Quantitatively this class of proteins constitutes 4–5% of the soluble protein of liver. In metabolic situations where these proteins bind substrates for a given reaction (fatty acyl CoAs) but not the products of the reaction (phospholipid), their net effect is to transfer lipid mass into a growing membrane. Another potential function for these proteins may be to keep large amounts of free fatty acids and intermediates in sterol synthesis available for metabolism while at the same time preventing their partitioning into the phospholipid bilayer, where they may cause structural or permeability changes. Although these fatty acid binding proteins have not been found in procaryotes, the acyl carrier protein (Chapter 3) may well serve the same function.

A major research question concerning the phospholipid exchange proteins is, What is their role in membrane biogenesis? It is a curious fact that the discovery of exchange proteins grew out of inquiries into the mechanism of lipid transport between the endoplasmic reticulum and mitochondria. As a consequence of the original studies many exchange proteins have been purified, characterized, and sequenced. Now, however, the question has been turned around to ask, What is the function of these proteins in vivo? This question has been difficult to answer. Current approaches to address the question have (a) attempted to demonstrate enhanced exchange protein activity in rapidly growing cells and tissues, (b) analyzed lipid transport with analogues of phospholipids in intact cells and in in vitro exchange assays, and (c) attempted to inhibit lipid transport in cells concomitant with determining the effects of inhibitors upon in vitro phospholipid exchange.

The ability of these proteins to effect the net transfer of lipid mass only under somewhat specialized conditions still leaves open the question about their ability to do this in vivo. Although much emphasis has been placed upon the issue of the net transfer of mass by these proteins it may also be irrelevant to the in vivo function of these proteins. In the presence of other cellular mechanisms for transfer of lipid mass, the exchange proteins may play a more subtle role by affecting the compositional dissimilarity among organelles.

Target Organelle Modification

Another important consideration in membrane assembly is the role of enzymes within the target organelle that change the lipid composition by their action on imported lipid. The presence of phosphatidylglycerolphosphate synthase and phosphatase and cardiolipin synthase (Hostetler, van den Bosch, and van Deenen 1971) within the mitochondrion of the higher eucaryotes appears sufficient to account for the exclusive localization of cardiolipin within the mitochondrion. The presence of phosphatidylserine decarboxylase (Dennis and Kennedy 1972), also within the mitochondrion, seems sufficient to explain the paucity of phosphatidylserine within this organelle. Plasma membrane fractions prepared from rat liver or from cultured mouse 3T3 fibroblasts exhibit high specific enzyme activities for sphingomyelin synthase (Voelker and Kennedy 1982; Marggraf, Anderer, and Kanfer 1981) and may well account for the enrichment of this lipid in the plasma membrane fractions of various cells and tissues.

In summary, both morphological and in vitro biochemical studies have provided evidence for a variety of processes that may be involved in lipid assembly into cell membranes. The very nature of these studies has provided an isolated and static view of membrane assembly. A logical next step is to ask how these observations agree with experiments performed at the cellular level.

OBSERVATIONS OF MEMBRANE ASSEMBLY AT THE CELLULAR LEVEL
Procaryotes

The Gram-negative bacteria contain an inner cytoplasmic membrane and an outer membrane. As discussed earlier, substantial evidence demonstrates that the lipid synthetic apparatus is located at the inner membrane. The development of techniques for separating inner and outer membranes by Osborn and co-workers (1972) has provided the means to investigate the movement of phospholipids between these membranes.

Donohue-Rolfe and Schaechter (1980) have investigated the translocation of phospholipids between the inner and outer membranes of *E. coli*. Pulse-chase labeling of phosphatidylethanolamine revealed that the specific activity of this lipid was fivefold higher in the inner membrane than the outer membrane immediately following a 30-s pulse with [^{3}H]glycerol. During the chase period the specific activity of the outer membrane increased, while that of the inner membrane decreased. After several minutes the specific activities of both membranes asymptotically approached the same value, which indicated radioequilibration between the membranes. The $t_{\frac{1}{2}}$ for the translocation of phosphatidylethanolamine was determined to be 2.8 min. The translocation was independent of protein synthesis, lipid synthesis, and ATP synthesis. It appeared, however, to be dependent upon the cell's protonmotive force.

Thus, the driving force for lipid movement to the outer membrane does not appear to be the insertion of new lipid or protein into the membrane. Collapse of the cell's proton gradient might adversely affect a variety of process, but it seems rather unlikely that it would inhibit the action of a soluble (phospholipid exchange protein–like) carrier. The role in lipid transport of zones of adhesion between the inner and outer membranes (Bayer 1968) still remains obscure, but their putative role in other transport processes such as lipopolysaccharide assembly into the outer membrane make them appealing candidates for involvement in this process.

Another approach to the study of intermembrane lipid movement in Gram-negative bacteria has used liposome fusion to the outer membrane of *Salmonella typhimurium* coupled with measurement of phospholipid translocation to the inner membrane (Jones and Osborne 1977). Exogenously added phopholipids rapidly equilibrate between the inner and outer membranes. This is true not only for lipids that are normally found in the *Salmonella*, but also for such foreign lipids as phosphatidylcholine and cholesterol oleate. These results indicate that rapid nonspecific movement of phospholipids occurs between the two membranes. They also suggest that the compositional differences found between the lipids of the inner and outer membrane may reflect an equilibrium condition based upon protein-lipid interactions rather than a specific sorting mechanism.

Further evidence of the nonspecificity of this transport process comes from work with mutants of *E. coli* that are temperature sensitive for phosphatidylserine decarboxylase (Langley, Hawrot, and Kennedy 1982). These mutants accumulate large amounts of phosphatidylserine (normally present in trace amounts in procaryotes) at the expense of phosphatidylethanolamine. This phosphatidylserine synthesized at the inner membrane is rapidly and efficiently equilibrated with the outer membrane with a $t_{\frac{1}{2}}$ of 12 to 13 min. When these mutants are shifted to the permissive temperature, after they have accumulated phosphatidylserine, the active decarboxylase enzyme can metabolize the entire phosphatidylserine pool. This means that the phosphatidylserine that accumulated in the outer membrane was transported back to the inner membrane.

The apparent lack of specificity in the transport process is not true for all lipids. Mutants defective in the enzyme diacylglycerol kinase (Raetz and Newman 1979) accumulate substantial amounts of diacylglycerol in the inner membrane but do not export it to the outer membrane. At present it is not evident if the lipid accumulation at the inner membrane represents an inherent inability of the transport machinery to act upon diacylglycerol or an active sequestration of diacylglycerol molecules in domains segregated from the transport machinery.

Eucaryotes   An interesting combination of biochemical and cytological analyses of the intermembrane movement of lipids has been developed by Pagano, Longmuir, and Martin (1983). A fluorescent derivative of phosphatidic acid which contains (*N*-4-nitrobenzo-2-oxa-1,3-diazole)aminocaproic acid in the *sn*-2 position of the molecule is the tool that they used in these investigations.

1-Acyl-2-(*N*-4-nitrobenzo-2-oxa-1,3-diazole)aminocaproylphosphatidic acid

When liposomes containing the fluorescent lipid are presented to Chinese hamster lung fibroblasts at 2°C, they rapidly take up the fluorescent lipid and metabolize it to the diacylglycerol analogue. The distribution of this compound within the cell (still at 2°C) is remarkable. One observes areas of fine reticular and large punctate fluorescence. These fluorescent areas have been identified by independent criteria as endoplasmic reticulum and mitochondria, respectively. If the cells are warmed to 37°C, most of the cellular fluorescence is lost to the medium in the form of the fluorescent fatty acid. The residual fluorescent lipid within the cell is metabolized primarily to phosphatidylcholine and triacylglycerol. When metabolism of the analogue yields fluorescent triacylglycerols, these lipids are found primarily in intracellular droplets composed almost exclusively of the triacylglycerol analogue. The presence of the fluorescent moiety allows the visualization of this process in living cells at the level of light microscopy in addition to the chemical quantitation of metabolism by fluorescence spectrophotometry of the isolated lipids. The use of this and related fluorescent probes to study intermembrane transport of lipids is still being developed, but it seems to have great potential for application to fundamental questions about membrane biogenesis.

Experiments to compare directly the activities of phospholipid exchange proteins in vitro and in vivo were conducted by Yaffe and Kennedy (1983). These studies made use of the choline analogue *N*-propyl-*N*,*N*-dimethylethanolamine. This amino alcohol is rapidly incorporated into the cellular lipid of baby hamster kidney cells, yielding phosphatidylpropyldimethylethanolamine. The lipid is

$$CH_3CH_2CH_2 - \overset{\overset{\displaystyle CH_3}{|}}{\underset{\underset{\displaystyle CH_3}{|+}}{N}} - CH_2CH_2OH$$

N-propyl-N,N-dimethylethanolamine

apparently synthesized in the endoplasmic reticulum, and a fraction of the cellular pool is transported to the mitochondria. In this respect the lipid analogue is metabolized in a manner nearly identical with phosphatidylcholine. However, in vitro assays show that both crude and partially purified preparations of phosphatidylcholine exchange protein derived from baby hamster kidney cells are essentially incapable of acting upon the analogue. Thus, the in vivo results demonstrating interorganelle transfer are inconsistent with in vitro measurement of phospholipid exchange activity.

In separate experiments directed at the same question, Voelker (1983) has utilized the metabolism of serine to phosphatidylethanolamine in baby hamster kidney cells as an index of the intracellular movement of phospholipid. The rationale for this approach is that phosphatidylserine synthase is an enzyme of the endoplasmic reticulum, and phosphatidylserine decarboxylase is an enzyme of the mitochondria. The metabolism of serine to phosphatidylethanolamine thus requires synthesis of phosphatidylserine in the endoplasmic reticulum, transport to the mitochondria, and decarboxylation. Inhibitors of ATP production (NaN_3 and NaF) and protein synthesis inhibit the metabolism of serine to phosphatidylethanolamine without affecting the activity of phosphatidylserine synthase or phosphatidylserine decarboxylase. The results suggest that the transport of phosphatidylserine between the organelles is disrupted by the metabolic poisons. The activity of phosphatidylserine exchange protein is unaffected by these metabolic poisons. Thus, in this system, inhibition of lipid movement in vivo cannot be correlated with inhibition of lipid exchange in vitro.

The potential role of vesicular transport in membrane biogenesis has been examined directly by DeSilva and Siu (1981) in *Dictyostelium discoideum*. During the chemotactic migration stage of the life cycle of this organism, there is a burst of phospholipid synthesis and preferential insertion of newly synthesized phospholipids into the plasma membrane. The genesis of new plasma membrane is concomitant with an increase in the quantity of intracellular lipid-rich vesicles. These vesicles are greatly reduced in number in mutants that cannot aggregate and in wild-type cells that are in the preaggregation stage of their life cycle. Pulse-chase labeling with [³H]glycerol demonstrates that radiolabeled lipids can be chased from the vesicle fraction to the plasma membrane. In addition, the assembly of phospholipids into the plasma membrane can be blocked by the addition of colchicine. These results implicate vesicle trafficking in plasma membrane assembly and suggest that cytoskeletal elements may also be involved in lipid transport. Independent experiments studying phospholipid transfer in the sprouting neuron (Pfenninger and Johnson 1983) have suggested a similar mechanism of vesicle transport and involvement of cytoskeletal elements in expansion of cellular plasma membrane. The results obtained in these systems clearly indicate the importance of vesicle transport in the assembly of cell-surface membranes.

In addition to the phospholipids, another important membrane component is cholesterol. DeGrella and Simoni (1982) have investigated the transport of newly synthesized cholesterol from the endoplasmic reticulum to the plasma membrane of Chinese hamster ovary cells. The cells were pulse labeled with [³H]acetate and the plasma membrane fraction isolated and analyzed for the presence of labeled cholesterol. They determined that the transport process occurs with a $t_{\frac{1}{2}}$ of approximately 10 min, is temperature dependent, and proceeds with kinetics that are different from plasma membrane glycoprotein insertion. These investigators also found that the transport process could be partially inhibited by energy poisons.

In summary, the results from both noncellular and cellular investigation of lipid transport indicate that a multitude of factors may contribute to membrane assembly. Although there is not at present a definitive elucidation of the mechanism of lipid assembly for a single (target) organelle, the data strongly suggest that there may not be a unifying mechanism for all organelles. More specifically, there is circumstantial evidence for virtually all the proposed mechanisms in Figure 14.5. The idea of lipid solubility and diffusion from one membrane to another is suggested by the work with fluorescent phosphatidic acid. The redistribution of this lipid from the plasma membrane to the endoplasmic reticulum and mitochondria occurs at 2°C, a temperature at which phospholipid exchange proteins do not function (in vitro), vesicle trafficking is negligible, and most metabolic enzymes are barely active. These results are also consistent with regions of continuity among these organelles and two-dimensional diffusion within the bilayer. The data obtained with procaryotes favors continuity among donor and target membranes. The inability of diacylglycerol to be transported also suggests (operationally) the presence of molecular filters. The large body of work with phospholipid exchange proteins argues for some soluble carrier mechanisms.

Whether these proteins function in net transport or compositional alteration of membranes is still not clear. Experiments with phospholipid analogues and metabolic inhibitors implicate mechanisms other than exchange proteins but do not give an indication of what that mechanism might be. Specific experiments examining vesicular movement of lipids demonstrate that during rapid surface membrane expansion, lipid assembly via this route is substantial. It is not known, however, if all surface membrane lipids are obligately transported by this mechanism.

FUTURE DIRECTIONS
The future directions of the lipid aspect of membrane biogenesis need to become more highly focused. Several specific, basic questions in this area need to be answered. These include: (1) What is the relationship between protein synthesis and transport to a given organelle, and lipid transport? (2) What is the role of lipid synthesis in lipid transport? (3) Is metabolic energy required for lipid transport? (4) What are the roles of ionic and proton gradients in the transport process? (5) In cases where vesicle movement seems to be involved, are all the target organelle lipids transported via this mechanism? (6) What is the role of phospholipid exchange proteins in transport?

Some of these questions can be addressed in experimental systems that are widely employed, while others may require the use of less well known experimental systems. The *R. sphaeroides* system seems to offer some interesting advantages over other traditional procaryotes. Perhaps other bacteria with intracytoplasmic membranes (Hager, Goldfine, and Williams 1966) would also prove useful. In mammalian systems the alveolar type II cell (Mason, Dobbs, and Greenleaf 1977) may be a particularly useful tool. This cell stores large amounts of phospholipid in specialized organelles known as *lamellar bodies*. The biogenesis of the lamellar body may well reflect the amplification of a specific, constitutive lipid transport and membrane assembly process.

The most effective approach to the problem of lipid assembly into eucaryotic cell membranes is most likely to come from combined genetic and biochemical analyses. Significant achievements have been made in establishing mutants in lipid metabolism from mammalian cells (Raetz 1982). The most versatile eucaryote for studying the problem will probably prove to be *Saccharomyces cerevisiae*. The ease of manipulating the organism and the large base of genetic information make it especially suitable. A variety of phospholipid synthesizing mutants (Chapter 8) have been isolated, and this organism seems to have great potential for developing mutants in phospholipid transport.

PROBLEMS

1. The structure of cell membranes is a consequence of the hydrophobic effect and the amphipathic nature of phospholipids. What structure would you expect triacylglycerols to form? What structures would you predict for mixtures of triacylglycerols, phospholipids, and proteins in circulating lipoprotein particles?
2. Describe in detail how phospholipid exchange proteins can be used to examine the role of lipid diversity in plasma membrane function. What are some of the limitations of this technique?
3. Trinitrobenzenesulfonate and isethionylacetimidate are two useful reagents for studying asym-

metry and transmembrane movement of lipids. With which lipids do they react, and what are the products? With what other membrane components will these reagents react? Some amidinated phospholipids are only very poorly resolved from the parent phospholipid by thin-layer chromatography. If you had a mixture of amidinated and nonamidinated phospholipids (such as one might have after attempting to measure membrane asymmetry) how would you resolve them?
4. What are some of the important experimental considerations in determining the lipid asymmetry of an isolated organelle?

BIBLIOGRAPHY

Entries preceded by an asterisk are suggested for further reading.

Bayer, M. E. 1968. Areas of adhesion between wall and membrane of *Escherichia coli*. *J. Gen. Microbiol.* 53:395–404.

*Bell, R. M.; Ballas, L. M.; and Coleman, R. A. 1981. Lipid topogenesis. *J. Lipid Res.* 22:391–403.

Bell, R. M.; Mavis, R. D.; Osborn, M. J.; and Vagelos, P. R. 1971. Enzymes of phospholipid metabolism: Localization in the cytoplasmic and outer membrane of the cell envelope of *Escherichia coli* and *Salmonella typhimurium*. *Biochim. Biophys. Acta* 249:628–35.

Bloj, B., and Zilversmit, D. B. 1976. Asymmetry and transposition rates of phosphatidylcholine in rat erythrocyte ghosts. *Biochemistry* 15:1277–83.

*Bretscher, M. S., and Raff, M. C. 1975. Mammalian plasma membranes. *Nature* 258:43–49.

Cohen, L. K.; Lueking, D. R.; and Kaplan, S. 1979. Intermembrane phospholipid transfer mediated by cell free extracts of *Rhodopseudomas spheroides*. *J. Biol. Chem.* 254:721–28.

Colbeau, A.; Machbaur, J.; and Vignais, P. M. 1971. Enzymic characterization and lipid composition of rat liver subcellular membranes. *Biochim. Biophys. Acta* 249:462–92.

Crain, R. C., and Zilversmit, D. B. 1980. Net transfer of phospholipid by the nonspecific phospholipid transfer proteins from bovine liver. *Biochim. Biophys. Acta* 620:37–48.

DeGrella, R. F., and Simoni, R. D. 1982. Intracellular transport of cholesterol to the plasma membrane. *J. Biol. Chem.* 257:14256–62.

Dennis, E. A., and Kennedy, E. P. 1972. Intracellular sites of lipid synthesis and the biogenesis of mitochondria. *J. Lipid Res.* 13:263–67.

DeSilva, N. S., and Siu, C. H. 1981. Vesicle-mediated transfer of phospholipids to plasma membrane during cell aggregation in *Dictyostelium discoideum*. *J. Biol. Chem.* 256:5845–50.

Donohue-Rolfe, A. M., and Schaechter, M. 1980. Translocation of phospholipids from the inner to the outer membrane of *Escherichia coli*. *Proc. Natl. Acad. Sci. USA* 77:1867–71.

*Farquhar, M. G., and Palade, G. E. 1981. The Golgi apparatus (complex)—(1954–1981)—from artifact to center stage. *J. Cell Biol.* 91:775–1035.

Hager, P. O.; Goldfine, H.; and Williams, P. J. 1966. Phospholipids of bacteria with extensive intracytoplasmic membranes. *Science* 151:1543–44.

Hostetler, K. Y.; van den Bosch, H.; and van Deenen, L. L. M. 1971. Biosynthesis of cardiolipin in liver mitochondria. *Biochim. Biophys. Acta* 239:113–19.

Jones, N. C., and Osborne, M. J. 1977. Translocation of phospholipids between the outer and inner membranes of *Salmonella typhimurium*. *J. Biol. Chem.* 252:7405–12.

*Kader, J. C.; Douady, D.; and Mazliak, P. 1982. "Phospholipid transfer proteins." In *New com-*

prehensive biochemistry, ed. A. Neuberger and L. L. M. van Deenen. Vol. 4, *Phospholipids*, ed. J. N. Hawthorne and G. B. Ansell, 279–311. Amsterdam: Elsevier.

Keenan, T. W., and Morre, D. J. 1970. Phospholipid class and fatty acid composition of Golgi apparatus isolated from rat liver and comparison with other cell fractions. *Biochemistry* 9:19–24.

Kornberg, R. D., and McConnell, H. M. 1971. Inside-outside transitions of phospholipids in vesicle membranes. *Biochemistry* 10:1111–20.

Langley, K. E.; Hawrot, E.; and Kennedy, E. P. 1982. Membrane assembly: Movement of phosphatidylserine between the cytoplasmic and outer membrane of *Escherichia coli*. *J. Bact.* 152:1033–41.

Langley, K. E., and Kennedy, E. P. 1979. Energetics of rapid transmembrane movement and of compositional asymmetry of phosphatidylethanolamine in membranes of *Bacillus megaterium*. *Proc. Natl. Acad. Sci.* 76:6245–49.

Marggraf, W. D.; Anderer, F. A.; and Kanfer, J. N. 1981. The formation of sphingomyelin from phosphatidylcholine in plasma membrane preparations from mouse fibroblasts. *Biochim. Biophys. Acta* 664:61–73.

Mason, R. J.; Dobbs, L. G.; and Greenleaf, R. D. 1977. Alveolar type II cells. *Fed. Proc.* 36:2697–2702.

Metz, R. J., and Radin, N. S. 1980. Glucosylceramide uptake protein from spleen cytosol. *J. Biol. Chem.* 255:4463–67.

Ockner, R. K.; Manning, J. A.; and Kane, J. P. 1982. Fatty acid binding protein. *J. Biol. Chem.* 257:7872–78.

*Op den Kamp, J. A. F. 1979. Lipid asymmetry in membranes. *Ann. Rev. Biochem.* 48:47–71.

Osborn, M. J.; Gander, J. E.; Parisi, E.; and Carson, J. 1972. Mechanism of assembly of the outer membrane of *Salmonella typhimurium*. *J. Biol. Chem.* 247:3962–72.

Pagano, R. E.; Longmuir, K. J.; and Martin, O. C. 1983. Intracellular translocation and metabolism of a fluorescent phosphatic acid analogue in cultured fibroblasts. *J. Biol. Chem.* 258:2034–40.

Palade, G. 1975. Intracellular aspects of the process of protein secretion. *Science* 189:347–58.

Pfenninger, K. H., and Johnson, M. 1983. Membrane biogenesis in the sprouting neuron. I.

Selective transfer of newly synthesized phospholipid into the growing neurite. *J. Cell Biol.* 97:1038–42.

*Raetz, C. R. H. 1982. "Genetic control of phospholipid bilayer assembly." In *New comprehensive biochemistry*, ed. A. Neuberger and L. L. M. van Deenen. Vol. 4, *Phospholipids*, ed. J. N. Hawthorne and G. B. Ansell, 435–77. Amsterdam: Elsevier.

Raetz, C. R. H., and Newman, K. F. 1979. Diglyceride kinase mutants of *Escherichia coli*: Inner membrane association of 1,2-diglyceride and its relation to synthesis of membrane-derived oligosaccharides. *J. Bact.* 137:860–68.

Renooij, W.; van Golde, L. M. G.; Zwaal, R. F. A.; and van Deenen, L. L. M. 1976. Topological asymmetry of phospholipid metabolism. *Eur. J. Biochem.* 61:53–58.

Roseman, M.; Litman, B. J.; and Thompson, T. E. 1975. Transbilayer exchange of phosphatidylethanolamine for phosphatidylcholine and *N*-acetimidoyl phosphatidylethanolamine in single-walled bilayer vesicles. *Biochemistry* 14:4826–30.

Rothman, J. E., and Dawidowicz, E. A. 1975. Asymmetric exchange of vesicle phospholipids catalyzed by phosphatidylcholine exchange protein. *Biochemistry* 14:2809–16.

Rothman, J. E., and Kennedy, E. P. 1977. Asymmetrical distribution of phospholipids in the membrane of *Bacillus megaterium*. *J. Mol. Biol.* 110:603–18.

———. 1979. Rapid transmembrane movement of newly synthesized phospholipids during membrane assembly. *Proc. Natl. Acad. Sci. USA* 74:1821–25.

*Rothman, J. E., and Lenard, J. 1977. Membrane asymmetry. *Science* 195:743–53.

Rothman, J. E.; Tsai, D. K.; Dawidowicz, E. A.; and Lenard, J. 1976. Transbilayer phospholipid asymmetry and its maintenance in the membrane of influenza virus. *Biochemistry* 15:2361–70.

Sandra, A., and Pagano, R. E. 1978. Phospholipid asymmetry in LM cell plasma membrane derivatives: Polar head group and acyl chain distributions. *Biochemistry* 17:332–38.

*Tanford, C. 1973. *The hydrophobic effect. Formation of micelles and biological membranes*. New York: Wiley.

Verkleij, A. J.; Zwaal, R. F. A.; Roelofsen, B.; Comfurius, P.; Kastelijn, D.; and van Deenen, L. L. M. 1973. The asymmetric distribution of phospholipids in the human red cell membrane. *Biochim. Biophys. Acta* 323:178–93.

Voelker, D. R. 1983. Inhibition of phospholipid assembly into mammalian cell membranes. *Fed. Proc.* 42:1804.

Voelker, D. R., and Kennedy, E. P. 1982. Cellular and enzymic synthesis of sphingomyelin. *Biochemistry* 21:2753–59.

Wirtz, K. W. A.; Devaux, P. F.; and Bienvenue, A. 1980. Phosphatidylcholine exchange protein catalyzes the net transfer of phosphatidylcholine to model membranes. *Biochemistry* 19:3395–99.

Wirtz, K. W. A., and Zilversmit, D. B. 1968. Exchange of phospholipids between liver mitochondria and microsomes in vivo. *J. Biol. Chem.* 243:3596–3602.

———. 1970. Partial purification of phospholipid exchange protein from beef heart. *FEBS Lett.* 7:44–46.

Yaffe, M. P., and Kennedy, E. P. 1983. Intracellular phospholipid movement and the role of phospholipid transfer proteins in animal cells. *Biochemistry* 22:1497–1507.

Zilversmit, D. B., and Hughes, M. E. 1977. Extensive exchange of rat liver microsomal phospholipids. *Biochim. Biophys. Acta* 469:99–110.

CHAPTER 15

Assembly of Proteins into Membranes

Reinhart A. F. Reithmeier

ORGANIZATION OF MEMBRANE PROTEINS

Biological membranes are asymmetric; that is, one surface of the membrane is different from the opposite surface. For proteins, this asymmetry is absolute. All copies of the same intrinsic membrane proteins are oriented in the same way, while extrinsic membrane proteins are confined to one surface of the membrane or the other. This asymmetry suggests that biological membranes are not self-assembled structures. Indeed, reconstitution of membranes by mixing proteins and lipids usually results in a symmetrical membrane. How then does this asymmetry originate? The spontaneous transmembrane movement of membrane proteins occurs at a negligible rate. Once proteins are assembled into a membrane, their orientation with respect to the membrane remains fixed. Membrane protein asymmetry is therefore a result of the biosynthesis and assembly of membrane proteins.

Classification of Membrane Proteins

Membrane proteins may be classified into two groups based on the nature of their interaction with the membrane. *Extrinsic* or peripheral membrane proteins are not associated with the hydrophobic core of the lipid bilayer. These proteins are associated with the inner or outer surface of the membrane and are held on the membrane by their association with intrinsic membrane proteins or perhaps acidic phospholipids. Extrinsic membrane proteins can be readily removed from membranes by adjusting the ionic strength or pH of the medium. *Intrinsic* or integral membrane proteins are hydrophobically bonded to the lipid bilayer. These proteins can be solubilized from the membrane by detergents. Detergents are small amphiphilic molecules that disrupt lipid-lipid and lipid-protein interactions. Mild nonionic detergents such as Triton X-100 or $C_{12}E_8$ can substitute for endogenous lipids while maintaining subunit interactions and the native state of membrane proteins. Most, if not all, intrinsic membrane proteins span the lipid bilayer. This feature allows some intrinsic membrane proteins to act as transport proteins and to provide communication from one side of the membrane to the other.

Intrinsic membrane proteins can be further subdivided into two types. The first type, designated as *simple* membrane proteins, span the bilayer only once. Examples of simple intrinsic membrane proteins include glycophorin of the human red cell membrane (Figure 15.1), the membrane protein G of vesicular stomatitis virus (VSV), the μ_m chain of IgM, the major histocompatibility antigen (HLA), and the coat protein of M13 bacteriophage. These proteins

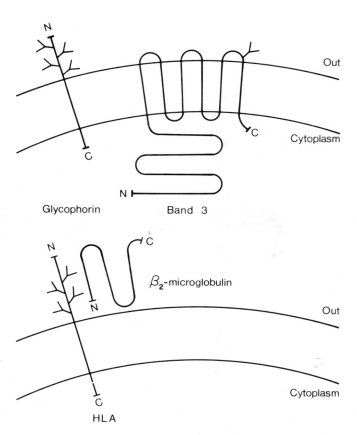

Figure 15.1. Models for the arrangement of various membrane proteins in their native membrane. N, amino terminus; C, carboxyl terminus; Y, carbohydrate.

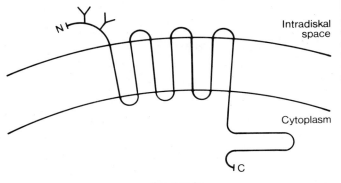

Rhodopsin

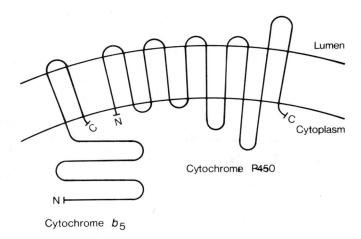

Cytochrome P450

Cytochrome b_5

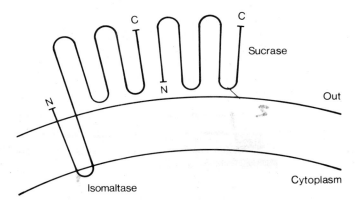

Sucrase

Isomaltase

Figure 15.1. (Continued)

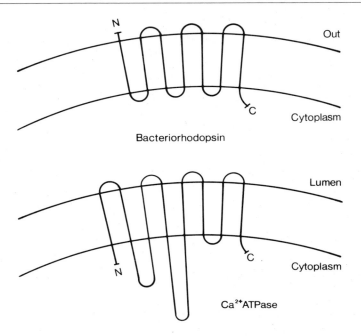

Figure 15.1. (Continued)

display cell-surface markers to the medium. The membrane-spanning segment is hydrophobic and is arranged as an α-helix. This portion of the protein serves to anchor the extracellular domain to the membrane. The portion of the protein just within the cytoplasm is highly charged, usually containing a number of closely spaced arginine or lysine residues. The portion of the protein exterior to the cell is also highly polar, usually containing carbohydrate residues. An α-helical arrangement of the polypeptide is favored in a hydrophobic environment such as the nonpolar region of a lipid bilayer, since this protein structure is maximally hydrogen bonded. Transfer of a hydrogen bond found in a protein backbone (C=O . . . H—N) into a nonpolar environment has a $\Delta G_{trans} = -1.4$ kcal/mol, while the non-hydrogen-bonded pair has an unfavorable $\Delta G_{trans} = 4.1$ kcal/mol. The acyl chains of phospholipid molecules are incapable of forming hydrogen bonds. Hence, the major forces holding intrinsic membrane proteins in the lipid are hydrophobic in nature. There are short-range van der Waals interactions as well.

The second type of intrinsic membrane protein is designated as *complex*, since it has a more complex folding pattern in the membrane. This folding ranges from two membrane-spanning segments arranged as a hairpin loop to an arrangement of seven or more membrane-spanning segments. Proteins with a single loop embedded in the bilayer include cytochrome b_5 and the isomaltase subunit of sucrase-isomaltase, a small-intestinal brush border membrane enzyme (Figure 15.1). The loop anchors the protein to the membrane. These loops are hydrophobic in nature and are probably arranged as two α-helices joined by a β-turn. It is unlikely that β-turns are within the lipid bilayer. A survey of the location of β-turns in *soluble* proteins has revealed that most reverse turns are found at the protein surface, not within the hydrophobic core of the folded protein. The carbonyl and amine groups in reverse turns are hydrogen bonded to water. Burying a reverse turn in a hydrophobic environment such as a lipid bilayer would not permit the formation of these hydrogen bonds.

Many membrane proteins, especially those that are involved in transport, contain a large number of membrane-spanning segments. Bacteriorhodopsin contains seven such sequences (Figure 15.1). These sequences are arranged as an α-helical bundle. Since there is ample opportunity in such a bundle of helices for the formation of ionic bonds such as salt bridges, these segments need not be as hydrophobic as simple intrinsic membrane proteins. The essential requirement is that those residues that are exposed to the lipid bilayer be hydrophobic. The interactions holding these helices together cannot be only hydrophobic, since the helices would be dissociated in the nonpolar bilayer. Ionic interactions which are stabilized in a hydrophobic environment may play an important role in stabilizing these assemblies. Similarly, the Na^+/K^+ ATPase, the Ca^{2+} ATPase, and Band 3, the anion transport protein of the red cell membrane, span the membrane a number of times (Figure 15.1). The orientation of the amino and carboxyl termini of membrane proteins with respect to the membrane varies. For example, the amino terminus and carboxyl terminus can be on opposite sides of the membrane, as in bacteriorhodopsin, or on the same side, as in isomaltase. The orientation of membrane proteins is determined by their assembly pathway.

The Energetics of Assembly of Membrane Proteins

Intrinsic membrane proteins are held in the membrane by virtue of the hydrophobic nature of the amino acid side chains that are inserted into the lipid bilayer. A quantitative measure of the hydrophobicity of a side chain can be made by considering the free energy of transfer of the amino acid side chain from water to a nonpolar solvent like dioxane or ethanol. The water-accessible surface area of the side chains is directly related to the ΔG of transfer (Figure 15.2). The free-energy difference for the transfer of an amino acid residue from a random-coil conformation in aqueous media to an α-helical conformation in a lipid bilayer can be calculated from the contributions of the accessible surface area value, the unsatisfied hydrogen bonds, and the energy required to neutralize charged groups (Table 15.1)

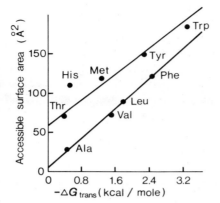

Figure 15.2. Water-accessible areas of amino acid side chains as a function of their free energy of transfer from water to ethanol or dioxane. Adapted from Schulz and Schirmer (1979).

Table 15.1. Estimated Free Energy Difference for the Transfer of an Amino Acid Residue from a Random-Coil Conformation in Aqueous Media to an α-Helical Conformation in a Hydrophobic Phase

Amino acid	ΔG_{trans} (kcal/mol)
Phe	−3.39
Met	−2.70
Ile	−2.51
Leu	−2.41
Val	−2.00
Trp	−2.00
Cys	−1.51
Ala	−1.00
Gly	0.0
Thr	0.91
Tyr	1.12
Ser	1.51
Glu	2.41
Asn	2.92
Pro	3.32
His	3.42
Lys	4.21
Glu	5.92
Asp	7.41
Arg	11.30

Source: The values are calculated from Von Heijne (1981).

(von Heijne and Blomberg 1979). The ΔG_{trans} values predict how readily an amino acid in a polypeptide chain will partition into a hydrophobic phase. The membrane-spanning portions of intrinsic membrane proteins must have a high negative ΔG_{trans} value for the sum of the amino acids in the hydrophobic lipid phase. Polar and charged side chains are usually found at the surface of soluble proteins. These groups can form hydrogen bonds with water. Burying isolated charged groups in a lipid bilayer is highly unfavorable (Table 15.1). Interestingly, the formation of a salt bridge between charged groups allows the bonded groups to dissolve in a hydrophobic solvent. The burying of a salt bridge in a hydrophobic environment has a ΔG_{trans} of ~ -15 kcal/mol. Polar residues unless neutralized will therefore not partition readily into the nonpolar portion of the lipid bilayer.

The problem of the assembly of proteins into membranes can be reduced to three fundamental questions:

1. How do hydrophilic amino acid residues pass across the lipid bilayer?
2. How does the complex folding pattern of membrane proteins originate?
3. What determines the subcellular location of the membrane protein?

Extrinsic membrane proteins that are associated with the cytoplasmic face of membranes do not have to cross the lipid bilayer during synthesis. These proteins are synthesized on membrane-free ribosomes and assemble onto the membrane after their synthesis is complete. For intrinsic membrane proteins, however, a considerable portion of their mass may occur on the extracellular or luminal side of membranes. This portion of the protein is exposed to water and contains a high proportion of polar residues as well as, perhaps, carbohydrate. The polar amino acid residues must be translocated across the hydrophobic lipid bilayer. This problem is one that is also faced by secreted proteins. As we shall see, simple intrinsic membrane proteins may be thought of as "partially secreted" proteins. This type of membrane protein follows a pathway similar to secreted proteins. Since the pathway for secretion has been well characterized, this is where we shall begin our discussion.

A number of similar models have been proposed to account for the folding pattern of intrinsic membrane proteins. The folding of membrane proteins, like that of soluble proteins, is dictated by their amino acid sequence. The origin of the folding pattern of well-characterized membrane proteins will be discussed. Eucaryotic cells contain a number of compartments specialized to carry out various functions. Many proteins of the plasma membrane, endoplasmic reticulum, sarcoplasmic reticulum, and lysosomes are synthesized on ribosomes bound to the endoplasmic reticulum and then routed to their final destination via coated vesicles and the Golgi apparatus. Most proteins found in mitochondria or chloroplasts are made on free polysomes and assembled posttranslationally. The various signals involved in the assembly, folding, and localization of membrane proteins will be discussed. Studies with bacteria such as *Escherichia coli*, which allow the ready use of mutants as an aid to dissecting the mechanisms of protein secretion and assembly, will be discussed.

SECRETION OF PROTEINS AND THE SIGNAL HYPOTHESIS

The biosynthesis and assembly of membrane proteins is a very active area of research. This topic may be considered to have started by consideration of protein secretion. The major developments in assembly are summarized in Table 15.2. The area has changed from morphological studies in the 1950s to molecular biological approaches that are used today.

The Palade Secretion Pathway

There are two classes of ribosomes in eucaryotic cells. *Free polysomes* are those which exist free of membrane in the cytoplasm, while *bound polysomes* are tightly associated with the cytoplasmic face of the endoplasmic reticulum. No structural differences in the ribosomal RNA or ribosomal proteins have been found between the two classes of ribosomes isolated by high-salt extraction. Thus, ribosomal subunits that were membrane bound can be subsequently found in the free population and vice versa.

Membrane-bound polysomes are associated with the endoplasmic reticulum membrane by two very different types of interactions. The first involves a binding, which is labile to high ionic strength, of the large ribosomal subunit to ribosome binding sites in the endoplasmic reticulum membrane. The second interaction involves an association of the nascent polypeptide with the endoplasmic reticulum membrane. Bound polysomes cannot therefore be removed from the endoplasmic reticulum membrane by high-salt treatment alone. Puromysin treatment, which terminates protein synthesis and releases the nascent polypeptide, in combination with high salt releases the polysome from the endoplasmic reticulum membrane.

Secreted proteins are made on membrane-bound polysomes. This was originally shown in a cell-free system for proteins such as albumin. This abundant plasma protein is made by liver and secreted into the bloodstream. The pathway for the secretion of proteins by the exocrine pancreatic cell was examined by Palade (1975) and

Table 15.2. A Brief History Lesson

1945	Endoplasmic reticulum discovered.
1955	Ribosomes discovered.
1960	Secreted proteins are synthesized on rough endoplasmic reticulum.
1964	Intracellular transport pathway for secreted proteins described.
1966	Secreted proteins are segregated into microsomal vesicles.
	Ribosomes are attached to the endoplasmic reticulum by their large subunits.
1967	Secreted proteins are transported through the Golgi apparatus.
1969	Cell-free translation shows secreted proteins are made on membrane-bound polysomes.
1971	Blobel and Sabatini propose that the region at the amino terminus of secreted proteins is recognized by a receptor in the endoplasmic reticulum.
1972	Singer and Nicholson outline fluid mosaic model.
	Milstein and co-workers show that signal sequence at the amino terminus of nascent polypeptide directs immunoglobulin to endoplasmic reticulum.
	Glucose-containing lipid-linked oligosaccharide is transferred to protein in cell-free preparation.
1973	Posttranslational assembly of membrane proteins proposed by Bretscher.
1974	Palade awarded Nobel Prize for physiology and medicine.
1975	Signal hypothesis outlined by Blobel and Dobberstein.
	Cell-free cotranslational assembly system described.
	Three-dimensional model of bacteriorhodopsin proposed on basis of electron diffraction.
1976	Mutant *E. coli* lipoprotein isolated.
	Gene-fusion techniques applied to secretion in *E. coli.*
	Rabbit reticulocyte cell-free translation described.
1977	Complete sequence of signal peptides of immunoglobulins reported.
	Inouye loop model proposed.
	Rothman and Lenard apply signal hypothesis to membrane proteins.
1978	Alteration in *E. coli* mutant prolipoprotein signal sequence characterized.
	Posttranslational uptake of chloroplast precursors demonstrated.
1979	Direct-transfer model outlined.
	Wickner proposes membrane trigger hypothesis.
1980	Signal recognition particle isolated.
1981	Helical hairpin hypothesis proposed.

his colleagues (1975). This work and other studies on the structural and functional organization of the cell led to the Nobel Prize for physiology or medicine in 1974. Using elegant electron microscopy techniques and cell subfractionation, Palade and co-workers traced

Table 15.3. Definitions

Signal sequence: The sequence of amino acids at the amino terminus of a
   nascent polypeptide that is required for signal recognition particle blockage
   of translation and that is cotranslationally cleaved by signal peptidase in the
   endoplasmic reticulum. A stable signal sequence is not cleaved. Internal
   signal sequences, which are not cleaved, are not located at the amino
   terminus. The properties of signal sequences are summarized in the text.
   The term *leader sequence* is avoided, since it has been used to identify a
   mRNA sequence preceding the initiator codon of the trp and other
   operons. This leader mRNA is translated to produce a small leader peptide.

Integration sequence: A hairpin loop of polypeptide that interacts directly with
   a lipid bilayer to allow integration of a polypeptide into a membrane. The
   term *insertion sequence* is avoided, since it has been used to describe small
   mobile genetic elements.

Stop-transfer sequence: A hydrophobic sequence followed by a highly polar,
   usually charged, region that prevents the further movement of the nascent
   polypeptide across the endoplasmic reticulum membrane. This sequence is
   equivalent to the transmembrane segment of simple intrinsic membrane
   proteins. Also called *halt sequence*.

---

the pathway of secreted proteins from the endoplasmic reticulum,
through the Golgi, and finally to the plasma membrane and the
extracellular medium via secretory vesicles.

Blobel Signal Blobel and Sabatini first proposed a hypothesis in 1971 (Table 15.2) to
Hypothesis account for the binding of polysomes which synthesize secreted
proteins to the endoplasmic reticulum. In this early model, they
suggested that proteins destined for secretion contain a "signal" at
the amino terminus of the nascent polypeptide (Table 15.3). They
proposed that this signal could be a unique sequence of amino acids
which is recognized by a binding factor on the endoplasmic reticulum
membrane that mediates attachment of the polysome.

The signal hypothesis was further elaborated by Blobel and Dobberstein (1975). A model to illustrate the signal hypothesis is shown
in Figure 15.3. The major features are the following:

1. Ribosomes exist in two forms, free and bound to the
   endoplasmic reticulum.
2. The translation of all proteins is initiated on free ribosomes.
   There is therefore no need for specialized ribosomes or pools
   of ribosomes.
3. The mRNA for secreted proteins contains a sequence of codons
   after the initiator codon (AUG) that code for a unique sequence

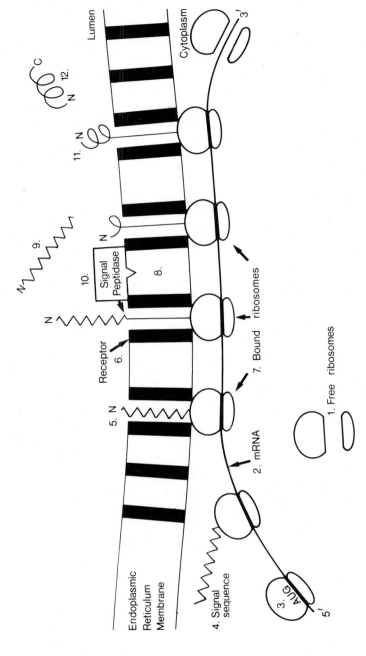

Figure 15.3. An early version of the signal hypothesis as proposed by Blobel and Dobberstein (1975). The hypothesis is described in detail in the text; the numbers in the figure correspond to the numbers in the section "Blobel Signal Hypothesis."

of amino acids at the amino terminus of the nascent polypeptide.

4. This sequence of amino acids is termed a *signal sequence* (Table 15.3). Since approximately 40 amino acids of the nascent polypeptide are buried within the ribosome, the signal sequence plus 40 amino acids must be synthesized before the signal is exposed.

5. The signal sequence directs the polysome to the endoplasmic reticulum membrane. The polysome is held in place on the membrane by the nascent polypeptide and by ribosome attachment to the endoplasmic reticulum.

6. The attachment of the polysome is mediated by a receptor present in the endoplasmic reticulum.

7. This receptor and/or other proteins form a multimeric hydrophilic channel that crosses the membrane. The large subunit of the ribosome is attached directly to the channel.

8. The nascent polypeptide is pushed through this channel as translation proceeds. The polypeptide is therefore sequestered into the lumen of the endoplasmic reticulum in a cotranslational fashion.

9. The signal sequence is removed before translation is complete. This is termed *cotranslational cleavage*.

10. The protease that removes the signal peptide is called *signal peptidase*. It is located on the luminal side of the endoplasmic reticulum membrane. For cleavage of the signal peptide to occur, the signal sequence plus enough amino acids to span the membrane (approximately 23) plus the approximately 40 residues buried in the ribosome must be synthesized.

11. Translation continues and the polypeptide is pushed into the lumen. Folding of the protein begins within the lumen.

12. Upon termination of translation, the ribosome dissociates and the subunits are recycled. The polypeptide is drawn through the channel as a result of the completion of the folding of the protein.

In Vitro Assembly System

The ability of a cell-free system to synthesize polypeptides has played a key role in elucidating the mechanism of assembly of membrane proteins. Two in vitro translation systems are widely used: the rabbit reticulocyte lysate and the wheat germ system. These systems contain ribosomes, translation factors, tRNA, amino acids, and a tRNA acylating system. To these systems one simply adds mRNA, some K^+ and Mg^{2+}, and amino acids—including a labeled amino acid, usually [^{35}S]methionine—and translation occurs (Figure 15.4).

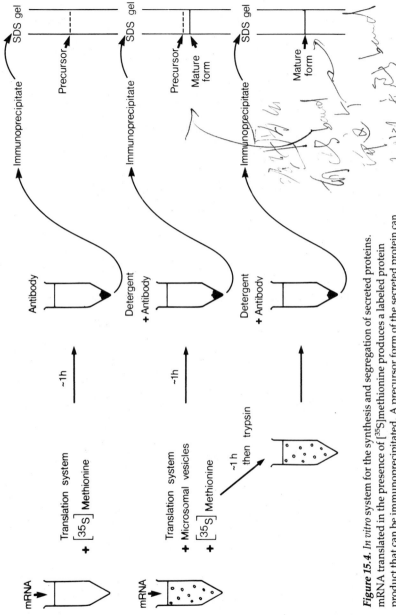

Figure 15.4. In vitro system for the synthesis and segregation of secreted proteins. mRNA translated in the presence of [³⁵S]methionine produces a labeled protein product that can be immunoprecipitated. A precursor form of the secreted protein can be identified by SDS gel electrophoresis. If translation takes place in the presence of microsomal vesicles, some of the precursor form will be processed to the mature form. Trypsin treatment of the vesicles after translation is complete will destroy any unassembled precursor. The mature form of the secreted protein will be resistant to degradation, since it is sequestered within the lumen of the microsomal vesicles.

reticulocyte
网织,红细胞)

The reticulocyte lysate contains endogenous mRNA (mainly for hemoglobin) that can be removed by nuclease treatment. The nuclease used is calcium dependent and can be subsequently inhibited by EGTA. Purified mRNA can then be added to the lysate and translated. The newly synthesized polypeptide contains radioactive methionine and can be resolved from other proteins by sodium dodecyl sulfate gel electrophoresis followed by autoradiography. A mixture of mRNAs can also be translated in the same system. The polypeptide of interest can be immunoprecipitated using specific antibody. The polypeptide can be identified by gel electrophoresis after the antibody-antigen complex is dissociated with sodium dodecyl sulfate.

For a study of the translation of secretory proteins, the basic translation system may be supplemented with endoplasmic reticulum membranes. These are closed vesicles with their cytoplasmic side facing out that are prepared from a microsome fraction of exocrine organs such as pancreas. These microsomes can bind nascent polypeptides of secretory and membrane proteins, cleave the signal sequence cotranslationally, and sequester the secretory polypeptide into the lumen of the endoplasmic reticulum or incorporate the polypeptide into the endoplasmic reticulum membrane, in the case of membrane proteins. If the completed polypeptide is completely sequestered, it is resistant to the action of proteases such as trypsin. This protease resistance is a convenient assay for the incorporation of the protein into the lumen of the endoplasmic reticulum. Microsomes may be added at different times after the initiation of translation to determine the timing of association of the nascent polypeptide with the endoplasmic reticulum membrane. Microsomes may also be added after translation is complete as an assay for posttranslational cleavage and assembly. The microsomes contain the signal peptidase and the enzymes necessary for the cotranslational addition of high-mannose oligosaccharide to form proteins that contain N-linked carbohydrate. Proteins synthesized in the absence of microsomal membranes will not be glycosylated and will have an intact signal sequence.

The Milstein Experiment: Secreted Proteins Are Made With an Amino-Terminal Signal Sequence

In 1972, Milstein and co-workers published an important paper (Table 15.2) dealing with the cell-free synthesis of immunoglobulin light chains. They found that the light-chain product synthesized in the cell-free reaction had a slightly higher molecular weight (an increase of 1500) than the mature light chain. The mRNA for the light chain was isolated from the microsomal fraction of a myeloma tumor cell line. Using microsomes they produced the mature light chain (Figure 15.5). However, if polysomes derived from the microsomes were added without microsomes, mainly the higher molecular weight form was produced. Since the light chain contains no carbohydrate, the difference could not be accounted for by addition of sugar residues. By peptide mapping, the investigators showed that the two forms differed at the amino terminus. The precursor form was the primary translation product, since it was labeled with [^{35}S]N-formyl-methionyl-tRNA$_F$. Posttranslational addition of microsomes to the cell-free system did not result in conversion of the precursor form to the mature form. The cleavage of the amino-terminal extension therefore occurs cotranslationally. Milstein and co-workers proposed that this amino-terminal region acts as a signal to bind the nascent polypeptide to the membrane, thereby indicating secretion.

Properties of signal sequences

1. Signal sequences are 15–25 amino acids in length.
2. Signal sequences are found at the amino terminus of nascent polypeptides.
3. The amino terminus of the signal contains the initiator methionine residue, which is followed by a short hydrophilic, usually charged region.
4. Signal sequences are very hydrophobic, the nonpolar amino acids being in the center of the sequence. If such a sequence were in a soluble protein, the hydrophobic portion would be buried within the core of the protein structure.
5. The cleavage mediated by the signal peptidase (signalase) occurs adjacent to small, uncharged amino acids such as glycine, serine, alanine, or cysteine.
6. Signal sequences are α-helical in conformation. An examination of the sequences of signal peptides has led to the prediction that they are in an α-helical conformation. This conformation would be favored if the peptide were in a hydrophobic environment. The regions of the signal sequence predicted to be in a β-sheet structure would also prefer to be α-helical in a hydrophobic environment. In an α-helical form, approximately 23 amino acids are required to span the membrane; the β-form requires only 11 amino acids. A synthetic signal sequence corresponding to preproparathyroid hormone has been examined by circular dichroism. This peptide had 27% α-helix, 43% β, and 30% random coil in water but 46% α-helix, 0% β, and 54% random coil in hexafluoro-2-propanol, a nonpolar solvent. Certain mutants of *E. coli* in the lam B and maltose binding protein have been localized to the signal sequences for these proteins. Mutant forms of these binding proteins in which proline or glycine,

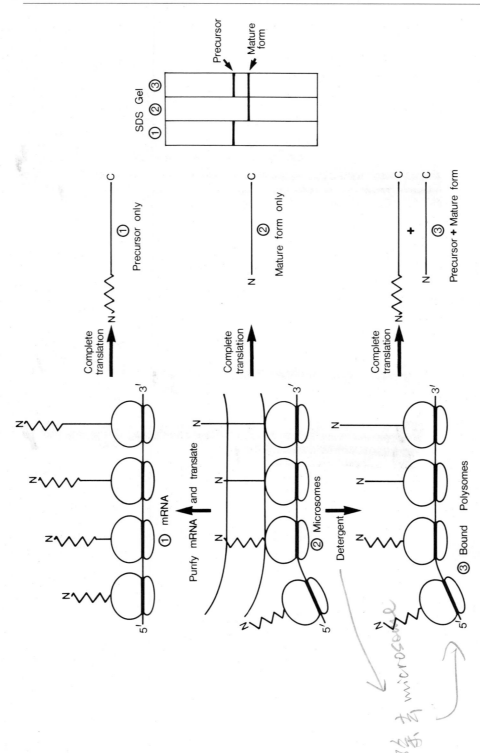

Figure 15.5. The Milstein experiment: (1) mRNA (purified from membrane-bound polysomes) when translated produces the precursor form of the immunoglobulin light chain. (2) Completion of translation using intact microsomes with bound polysomes produces the mature form. The microsomal membranes contain the signal peptidase and the translocation machinery for segregation of the secreted immunoglobulin. (3) Bound polysomes are produced by dissolving the endoplasmic reticulum membrane with detergent. Completion of translation produces both the precursor and mature forms. The precursor form contains an intact signal sequence, since the detergent removed the signal peptidase.

which interrupt α-helical structure, are substituted for the wild-type amino acids are not cleaved or exported.

7. There is a β-turn close to the cleavage site.
8. The signal peptide is removed intact by an endopeptidase reaction; however, its ultimate fate has not been determined.

The signal sequence of the secreted light chain of IG from myeloma M-321K has been sequenced and displays these features:

Met-Glu-Thr-Asp-Thr-Leu-Leu-Leu-Gly-Ala-Val-Leu-Leu-

Trp-Val-Pro-Gly-Ser-Thr-Gly-Asp-Ile-Leu-Thr. . .
$\overset{\uparrow}{20}$

The arrow indicates the site of cleavage by signal peptidase.

Signal Recognition Particle

The receptor for the signal sequence has been identified and purified (Walter and Blobel 1981). It is termed *signal recognition particle* (SRP) and has a molecular weight of 250,000 and a sedimentation coefficient of 11 S. It is composed of a single 7 S RNA molecule and 6 polypeptides with molecular weights of 72,000, 68,000, 54,000, 19,000, 14,000, and 9000. These proteins are immunologically unrelated, suggesting that they are not proteolytic degradation products of one another. The 7 S RNA found in HeLa cells has been sequenced. The sequence of 7 S RNA from rat Novikoff hepatoma is nearly identical. It is composed of 303 bases with a triphosphate at the 5′ terminus and contains a central block of 140 bases that are homologous to human middle-repeat Alu DNA. The Alu family of DNA sequences is an abundant (5% of the human genome) repetitive sequence, 300 base pairs in length, that is highly dispersed in the human genome.

The subcellular distribution of the 7 S RNA has been determined: 38% is membrane associated, 47% ribosome associated, and 15% free. More than 75% of the 7 S RNA is associated with the SRP. 7 S RNA has been found in liver, pituitary, placenta, and, interestingly, reticulocytes. The SRP can be removed from microsomes by high-salt extraction. SRP-depleted microsomes are active for synthesis of secretory proteins in reticulocyte lysates but not in the wheat germ system because the reticulocyte lysate contains significant amounts of SRP, whereas the wheat germ system does not. The purified SRP can be dissociated in the presence of EDTA; readdition of Mg^{2+} causes reassembly of the particle.

The SRP has some very interesting functions. First, SRP binds weakly to individual ribosomes ($K_D = 10^{-5}M^{-1}$). However, it binds very tightly to polysomes ($K_D = 10^{-8}M^{-1}$) containing nascent secretory proteins which contain signal sequences such as preprolactin.

SRP does not bind tightly to polysomes synthesizing hemoglobin. SRP treated with N-ethylmaleimide, a sulfhydryl reagent, does not bind to polysomes. This sulfhydryl reagent reacts with the 68,000- and 9000-dalton polypeptides and partially with the 54,000-dalton protein. Second, once SRP binds to polysomes, it inhibits the further synthesis of secretory proteins but not hemoglobin. This binding and inhibition require the synthesis of about 70 amino acids, resulting in exposure of the signal peptide. SRP therefore blocks further translation once the signal peptide is no longer contained within the ribosome. SRP may bind directly to the signal peptide, but this has yet to be demonstrated. Although the leucine analogue β-hydroxyleucine can be incorporated into a signal sequence, SRP does *not* bind to polysomes with nascent polypeptides containing β-hydroxyleucine. The role of the 7 S RNA in the function of SRP has yet to be elucidated; it may play an important role in assembly of SRP. Third, SRP mediates the binding of polysomes to the endoplasmic reticulum. This binding of the SRP-polysome complex to the endoplasmic reticulum is mediated by a docking protein present in the endoplasmic reticulum and results in a release of the translation block.

The synthesis of preprolactin has been examined in some detail. Complete synthesis of preprolactin in vitro takes about 10 min. The polysomes containing preprolactin mRNA and SRP associate with the endoplasmic reticulum within 1 min. Addition of salt-stripped (SRP-depleted) microsomal membranes to SRP-blocked polysomes relieves the inhibition of preprolactin synthesis any time after initiation of protein synthesis. SRP-blocked preprolactin polysomes contain an 8000-dalton form of preprolactin, which corresponds to about 70 amino acids. The signal sequence is 30 amino acids in length, and approximately 40 amino acids are buried in the ribosome. SRP therefore stops further protein synthesis when the entire signal sequence is exposed. Antibody against the 54,000-dalton polypeptide of SRP interferes with the release of the translation block, suggesting that the 54,000-dalton polypeptide may be involved in the attachment of SRP to the endoplasmic reticulum. The identification and characterization of the SRP has proven that protein translocation across the endoplasmic reticulum is a receptor-mediated process.

Docking Protein   The receptor in the endoplasmic reticulum for the SRP-polysome complex is a polypeptide with a molecular weight of 72,000. This protein is membrane bound and has been termed *docking protein*. Elastase treatment of SRP-depleted microsomes releases a water-soluble fragment of the docking protein (with a molecular weight of 60,000). Antibodies against this fragment have identified the parent 72,000-dalton molecule. The docking protein is not immunologically related to any of the SRP proteins. The purified docking protein fragment is capable of releasing the SRP-induced translation block. It is believed that the docking protein serves as a receptor for SRP-polysomes in the endoplasmic reticulum. The interaction of SRP-polysomes with the docking protein releases SRP from polysomes resulting in a continuation of translation.

Ribophorins Rough endoplasmic reticulum from rat liver contains two integral membrane proteins that are not found in smooth endoplasmic reticulum. These proteins have molecular weights of 65,000 and 63,000 and are designated as ribophorins I and II, respectively. There are one or two ribophorin molecules per ribosome. These proteins remain attached to ribosomes when the endoplasmic reticulum membrane is dissolved with detergent. The ribophorins can be cross-linked to ribosomes by chemical reagents. It has been suggested that ribophorins provide the attachment site for the large subunit of the ribosome to the endoplasmic reticulum.

Signal Peptidase The signal peptidase is an endopeptidase that is located on the luminal side of the endoplasmic reticulum membrane. It is an intrinsic membrane protein that can be solubilized with deoxycholate. The enzyme requires phospholipids for activity, and phosphatidylcholine is the most effective. In detergent, the peptidase is a large particle with a Stokes radius of 55 Å. Digestion of the intact endoplasmic reticulum with trypsin has no effect on the activity of the peptidase, since it resides on the luminal side of the endoplasmic reticulum membrane. Solubilized peptidase can cleave signal peptides posttranslationally. In vivo, however, the signal peptide is removed cotranslationally. The nascent polypeptide is inserted into the endoplasmic reticulum membrane, and the signal is removed once the sensitive bond is exposed to the luminal side of the membrane. A model showing the possible roles of signal recognition particle, docking protein, ribophorin, and signal peptidase is illustrated in Figure 15.6.

The Problem of Ovalbumin *Ovalbumin* is an egg white protein that is secreted by the chicken oviduct. It is not, however, made with a transient amino-terminal signal sequence. Only the initiator methionine residue is removed and replaced by an acetyl group. Polysomes synthesizing secreted proteins with amino-terminal signal sequences compete with nascent ovalbumin for binding to the endoplasmic reticulum membrane and

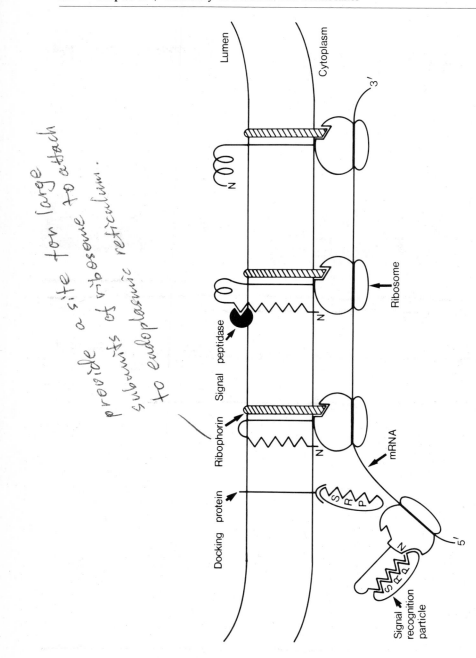

provide a site for a large
subunits of ribosome to attach
to endoplasmic reticulum.

Figure 15.6. Roles of signal recognition particle (SRP), docking protein, ribophorin, and the signal peptidase. Interaction of the SRP with the signal sequence (N~~) and the ribosome stops translation. The docking protein relieves this inhibition. The polysome is attached to the endoplasmic reticulum by interaction of the nascent polypeptide with the membrane and by binding of the large ribosomal subunit to ribophorin. The signal peptidase, located on the luminal side of the endoplasmic reticulum, cleaves the signal peptide.

sequester 俘陷

cotranslational sequestering within microsomes. These results suggest that ovalbumin contains the functional equivalent of a signal sequence. Ovalbumin therefore follows a pathway similar to that of other secreted proteins. SRP slows, but does not stop, the translation of ovalbumin.

The amino terminus of ovalbumin does not have the features of a signal sequence. It has been suggested that ovalbumin contains a stable, internal signal sequence. A peptide corresponding to residues 234–252 of mature ovalbumin has been seen to compete with preprolactin, a typical secretory protein, for binding to microsomal membrane. However, this competition was observed with high concentrations of this hydrophobic peptide. Synchronized translation experiments have shown that ovalbumin is glycosylated and sequestered within microsomes when the membranes are added before, but not after, 150 amino acids of ovalbumin have been synthesized. The nascent polypeptide chain of ovalbumin containing approximately 60 amino acids was sufficient to cause ovalbumin-synthesizing polysomes to become attached to microsomal membranes. These results suggest that the ovalbumin signal sequence is close to the amino terminus of the protein, yet it is retained in the mature protein.

Inspection of the amino acid sequence of ovalbumin reveals that there is a hydrophobic sequence (residues 27 to 45) that follows a charged sequence. This hydrophobic region may act as a signal sequence. It has been observed for the bacterial lipoprotein that cleavage of the signal sequence is not necessary for translocation and assembly into the membrane. The presence of a large hydrophilic sequence adjacent to the hydrophobic region may prevent cleavage. Alternatively, a mutation may have occurred that affects the cleavage site. Thus, we can conclude that proteolytic removal of a signal sequence is not required for the secretion of ovalbumin.

N-GLYCOSYLATION OF PROTEINS
Attachment of Carbohydrate

Many secreted and membrane proteins are glycosylated; an incredible diversity of sugar structures is possible. Carbohydrate plays an important role in recognition phenomena such as cell-cell interactions. The sugar residues can be attached in an *O*-linked or *N*-linked fashion. Little is known about the pathway of assembly of *O*-linked carbohydrate. The *N*-linked oligosaccharide is joined to asparagine residues (Hubbard and Ivatt 1981). The asparagine-linked carbohydrate contains the sequence

$$\text{-Asn-X-}\left(\begin{array}{c}\text{Ser}\\\hline\text{Thr}\end{array}\right)\text{-}$$

with a Carbohydrate group attached to Asn

Carbohydrate Dolichol

transport 不清楚

cytoplasmic Site

where X can be any amino acid. All *N*-glycosylated sites contain this sequence; however, not all Asn-X-Ser/Thr sequences are glycosylated. This sequence is predicted to be a β-turn located at the surface of the protein. The crystal structure of immunoglobulin has been determined; the carbohydrate is attached to a β-turn at the surface of the protein. The carbohydrate is transferred from a lipid-linked donor dolichol to the nascent polypeptide in the rough endoplasmic reticulum. *N*-Glycosylation is a cotranslational event. The donor of the carbohydrate moiety to the protein in mammalian cells has the structure

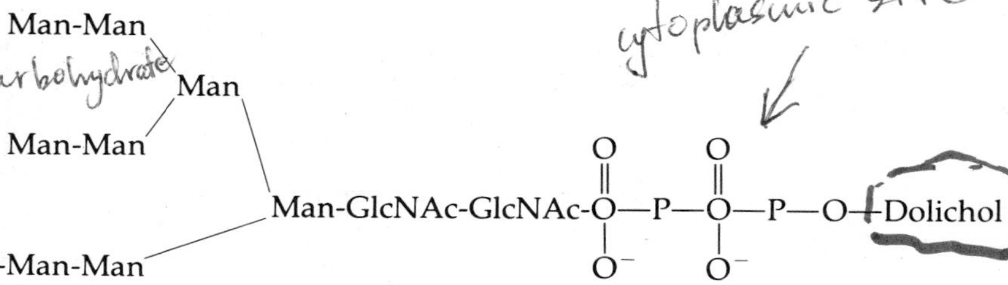

transferase

luminal

protein transferase

carbohydrate

cytoplasmic site

Man-Man
$\quad$ Man
Man-Man

Glc-Glc-Glc-Man-Man-Man

Man-GlcNAc-GlcNAc-O—P—O—P—O—Dolichol

$$\text{Man-GlcNAc-GlcNAc-O}-\overset{\overset{\displaystyle O}{\|}}{\underset{\underset{\displaystyle O^-}{|}}{P}}-O-\overset{\overset{\displaystyle O}{\|}}{\underset{\underset{\displaystyle O^-}{|}}{P}}-O-\text{Dolichol}$$

where Man is mannose, GlcNAc is *N*-acetylglucosamine, Glc is glucose. Dolichol is an isoprenoid compound with the following structure:

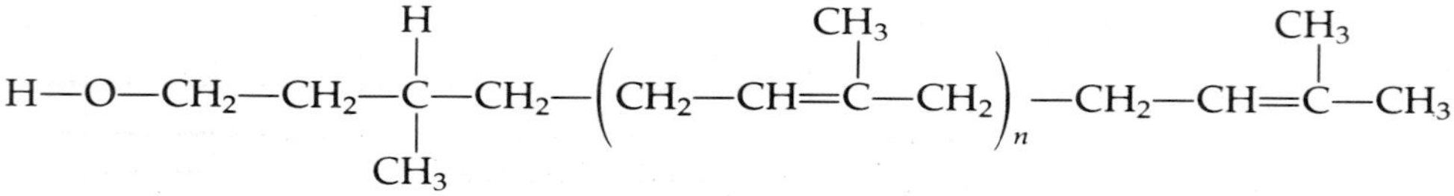

$$H-O-CH_2-CH_2-\overset{\overset{\displaystyle H}{|}}{\underset{\underset{\displaystyle CH_3}{|}}{C}}-CH_2-\left(CH_2-CH=\overset{\overset{\displaystyle CH_3}{|}}{C}-CH_2\right)_n-CH_2-CH=\overset{\overset{\displaystyle CH_3}{|}}{C}-CH_3$$

Dolichol

where *n* may be as high as 20. The donor is built up by addition of sugar residues from UDP-GlcNAc, GDP-Man (for the innermost five mannose residues), Man-P-dolichol (for the remaining mannose residues), and Glc-P-dolichol. The successive addition of these sugars to the dolichol occurs on the cytoplasmic face of the endoplasmic reticulum. The mechanism by which this massive carbohydrate structure crosses the endoplasmic reticulum membrane has not been determined.

不清楚 {

The enzyme that transfers the sugars from the donor to the nascent polypeptide is located on the luminal side of the endoplasmic reticulum. Secreted proteins such as ovalbumin, IGG heavy chain, and membrane protein G of VSV are glycosylated before termination of translation. Protein-linked carbohydrate is found on the luminal side of the endoplasmic reticulum. Glycosylation occurs only upon insertion of the nascent polypeptide into the endoplasmic reticulum. The acceptor sequence Asn-X-Ser/Thr must be exposed on the luminal side of the endoplasmic reticulum, since more than 45 amino acids must be added before glycosylation of the G-protein will occur. The transferase has recently been purified.

Processing of the Glycoprotein

Glycosylation may be considered as a two-step process. All *N*-linked glycoproteins initially contain the sugar structure $(Glc)_3(Man)_9(GlcNAc)_2$, whereas mature glycoproteins contain a large variety of other sugar structures. How then does the wide diversity of *N*-linked sugar structures originate? The initial sugar structure is first trimmed, then elongated. The pathway for processing the carbohydrate structure of the glycoprotein depends not on the initial sugar structure, since it is identical in all cases, but on the conformation of the folded polypeptide. The processing enzymes must therefore recognize not only the sugar residues but also the protein. The first stage of glycosylation is the cotranslational attachment of the sugar from the dolichol donor. The second stage is the posttranslational modification or processing of this structure.

In the processing pathway, the terminal glucose residues are removed quickly in the endoplasmic reticulum (Figure 15.7). No α-mannosidases have been localized in the endoplasmic reticulum. The $(Man)_9(GlcNAc)_2$ structure has been found in a mature protein, the unit-A glycopeptide of thyroglobin. Four mannose residues are removed in the Golgi (Figure 15.7). $(Man)_5(GlcNAc)_2$ sugar structures have been found in some mature glycoproteins. This sugar structure is termed *high mannose*. Endoglycosidase H cleaves the β-1,4 linkage between the two *N*-acetyl glucosamine residues of high-mannose structures:

$$
\begin{array}{c}
Man \\
\quad \searrow \\
\qquad Man \\
\qquad \nearrow \qquad\qquad \searrow \\
Man \qquad\qquad\qquad\qquad Man\text{-GlcNAc-GlcNAc-Asn} \\
\qquad \nearrow \\
\qquad Man
\end{array}
\qquad
\begin{array}{c}
endo\ H \\
\downarrow
\end{array}
$$

Sequential addition of further sugar residues to the high-mannose form creates complex glycoproteins (Figure 15.7). Complex oligosaccharides are not cleaved by endoglycosidase H. GlcNAc transferase I adds a single GlcNAc residue to $(Man)_5(GlcNAc)_2$ (Figure 15.7). Addition of this sugar allows another mannosidase, termed the *late mannosidase*, to act. Addition of another GlcNAc between the two

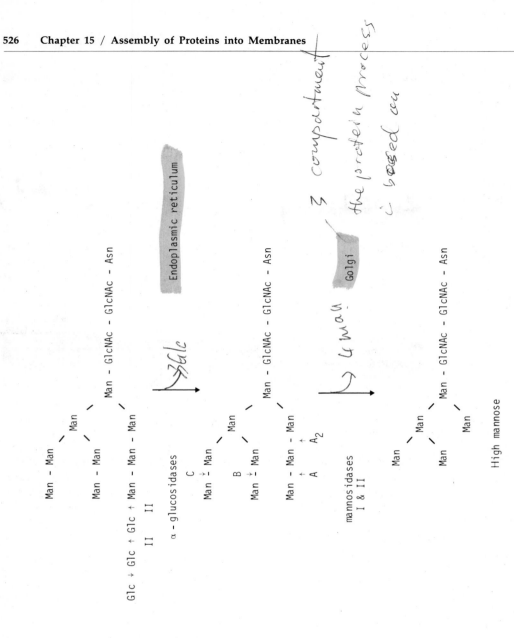

Endoplasmic reticulum

Golgi

Man – Man
         \
          Man
         /   \
Man – Man   Man – GlcNAc – GlcNAc – Asn
         /
Glc → Glc → Glc → Man – Man – Man
        II   II

α - glucosidases

                    C
                Man → Man
                      \
                       Man
                      /   \
                Man → Man   Man – GlcNAc – GlcNAc – Asn
                      B   /
                Man – Man – Man
                   ↑   ↑
                   A   A₂

mannosidases
I & II

Man
   \
    Man
   /   \
Man   Man – GlcNAc – GlcNAc – Asn
         /
       Man

High mannose

GlcNAc transferase I

handwritten notes: →3Glc   4 man   3 compartment   the protein process   i based on

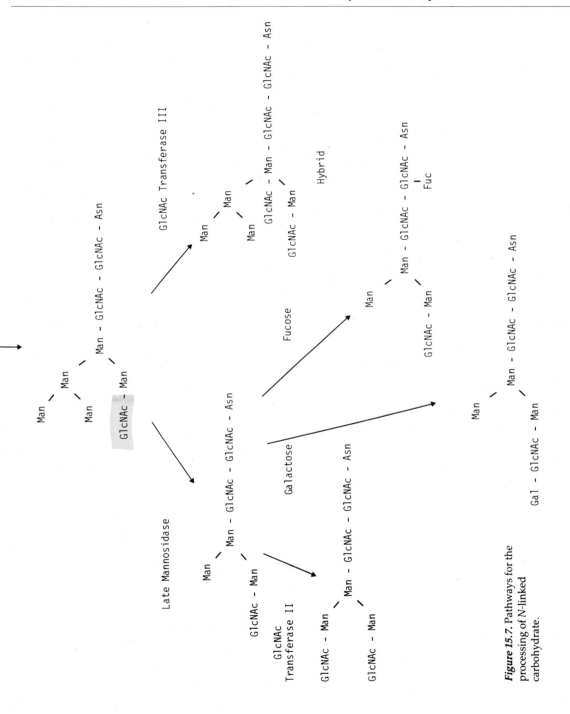

Figure 15.7. Pathways for the processing of *N*-linked carbohydrate.

arms (bisecting) by transferase III prevents removal of further mannose by the late mannosidase. This leads to a hybrid sugar structure with one arm being high mannose and the other complex. The $(GlcNAc)(Man)_3(GlcNAc)_2$ can be enlarged by addition of GlcNAc, galactose, or fucose (Figure 15.7), which in turn can be followed by addition of sialic acid residues.

BIOSYNTHESIS OF SIMPLE MEMBRANE PROTEINS IgM and the Relationship Between the Biosynthesis of Secreted Proteins and Simple Transmembrane Proteins

Immunoglobulin molecules are composed of two identical heavy chains linked by disulfide bonds to two identical light chains. The structure of the amino-terminal region has a variable structure and is responsible for antigen binding. The carboxyl terminus contains a constant region. Different classes of immunoglobins can be distinguished by the sequence of the constant region of their heavy chains. Immunoglobulins such as IgM can exist in two different forms—as a secreted antibody and as a cell-surface antigen. The secreted and membrane-bound forms of the heavy chains of the μ chain IgM are different only in their carboxyl-terminal sequences (Early et al. 1980), as shown here:

CHO
|
-Asp-Lys-Ser-Thr-Gly-Lys-Pro-Thr-Leu-Tyr-Asn-Val-Ser-Leu

Ile-Met-Ser-Asp-Thr-Gly-Gly-Thr-Cys-Tyr

Secreted form

-Asp-Lys-Ser-Thr-Glu-Gly-Glu-Val-Asn-Ala-Glu-Glu-Glu-Gly

Phe-Glu-Asn-Leu-Trp-Thr-Thr-Ala-Ser-Thr

Phe-Ile-Val-Leu-Phe-Leu-Leu-Ser-Leu-Phe

Tyr-Ser-Thr-Thr-Val-Thr-Leu-Phe-Lys-Val

Lys

Membrane-bound form

The membrane-bound form contains a sequence of hydrophobic acids (underlined) followed by a basic region composed of two lysine residues. This arrangement is characteristic of simple transmembrane proteins. The secreted form has a completely carboxyl-terminal region that includes an asparagine-linked oligosaccharide (CHO). What can account for this difference?

An examination of the gene structure provides the answer (Early et al. 1980). The gene for the IgM heavy chain is composed of eight exons (Figure 15.8). The first exon codes for the signal sequence. The next exon corresponds to the variable region. The type of IgM molecule that is produced is dependent on how the mRNA transcript is spliced together. A mRNA coding for the secreted protein results if the fourth exon remains intact. If, however, the region coding for the secreted C-terminus is removed and the exons coding for the membrane-associated form are attached to the third region of the mRNA, the membrane-bound form will be made.

The difference between the membrane-bound form and the secreted form of the same protein is in the manner in which the initial RNA transcript is spliced together. The same gene therefore codes for both the secreted and membrane forms of the protein. The difference in the structure of the two forms of the protein is that the membrane-bound form contains a stop-transfer sequence, which is described in the next paragraph (and Table 15.3). If this sequence is not present, the protein is secreted. Both forms of the molecule are synthesized with an amino-terminal signal sequence. Simple membrane proteins can be thought of as partially secreted proteins. Their synthesis follows a pathway similar to that of a secreted protein, both

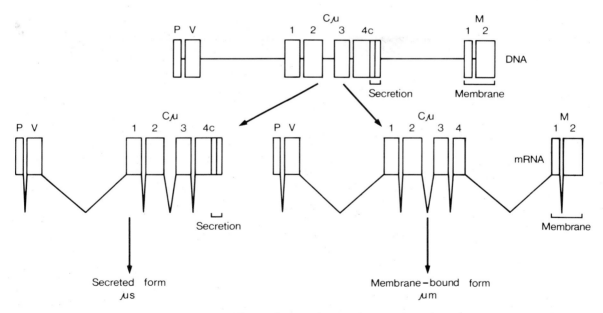

Figure 15.8. The production of secreted (μs) and membrane-bound (μm) forms of IgM is dictated by the splicing pattern of the mRNA transcript. The boxed regions indicate exons in the DNA. The mRNA is produced by removing the RNA between the boxed regions. The mRNA for the secreted form is derived from exons P (which codes for the signal sequence), V (variable region) Cμ1,2,3,4, and c. The mRNA for the membrane-bound form is derived from exons P, V, Cμ1,2,3,4, and M 1&2. Adapted from Early et al. (1980).

being made with amino-terminal signal sequences that are cotranslationally cleaved. Both secreted and simple transmembrane proteins are made on membrane-bound polysomes. Secreted proteins pass entirely through the endoplasmic reticulum membrane, while the membrane protein gets "stuck" in transit.

The transmembrane segments of simple membrane proteins act as stop-transfer sequences (Table 15.3). These sequences prevent the complete translocation of the protein across the membrane and lock the protein into the lipid bilayer. Stop-transfer sequences are long enough to span the bilayer as an α-helix. They are hydrophobic and are followed by a number of charged residues. These charged residues prevent the carboxyl end of the protein from entering the lipid bilayer. It has also been hypothesized that the hydrophobic stop-transfer sequence dissociates a multimeric protein tunnel in the endoplasmic reticulum membrane that allows the transmembrane movement of the hydrophilic portion of nascent polypeptide. Whether translocation of protein across the membrane is mediated by a protein channel remains to be shown.

VSV G Protein

The G-protein of the lipid enveloped virus, vesicular stomatitus virus (VSV), is an example of a simple membrane protein (Figure 15.9). This glycoprotein is made on membrane-bound ribosomes of the host cell and migrates to the plasma membrane, where it is incorporated into the plasma membrane of the host cell from which the virus buds (Katz et al. 1977). In the virus and the plasma membrane of the host cell, the protein is arranged with its amino terminus on the outside. There is a single transmembrane sequence and a 30–amino acid internal domain. The protein is cotranslationally inserted into the endoplasmic reticulum membrane with concomitant glycosylation.

Two types of experiments have been performed on the biosynthesis of the G-protein. First, pulse-chase experiments have involved Chinese hamster ovary cells infected with VSV. The cells were labeled with [^{35}S]methionine for 5 min and chased for various times. At 0 min of chase the G-protein was in form G_1 (with a molecular weight of 65,000). During the chase G_1 was converted to a *higher* molecular weight form G_2 (with a molecular weight of 67,000). This change was due to differences in the extent of glycosylation. Second, in cell-free translation experiments, viral mRNA in the absence of microsomes produced a more rapidly migrating form of the G-protein, G_0, that is unglycosylated. The G_0 form is a precursor of the mature G-protein with a 16–amino acid *N*-terminus signal sequence:

$$\overset{10}{\text{Met-Lys-Thr-Ile-Ile-Ala-Leu-Ser-Tyr-Ile-Phe-Cys-Leu-Val-Leu-Gly}}\overset{\downarrow}{\text{-Lys}}$$

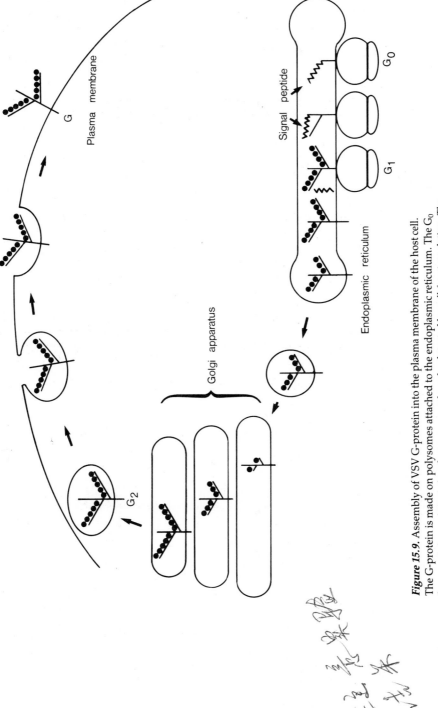

Figure 15.9. Assembly of VSV G-protein into the plasma membrane of the host cell. The G-protein is made on polysomes attached to the endoplasmic reticulum. The G_0 form contains an intact signal sequence and can be detected by cell-free translation. The signal peptide is removed before translation is complete. Carbohydrate is added cotranslationally to produce the G_1-form. This high-mannose form is trimmed and then elongated in the Golgi apparatus to produce the mature G_2-form. A fatty acid has also been added. The movement of the G-protein from endoplasmic reticulum to Golgi and Golgi to plasma membrane is mediated by coated vesicles.

[handwritten margin note: ≈1 in the presence → of microsome. cell free translation system can produce G₁]

G_0 is the primary translation product, since the methionine at the 1-position is donated by the initiator Met-tRNA$_{met}$. Cotranslational cleavage takes place between Gly-16 and Lys-17 (indicated by ↓). Lys-17 is the first residue of the mature G-protein. Deletion mutations in the signal peptide result in a mutant protein that is made on free polysomes.

If microsomal membranes are included in the cell-free translation system, the G-protein is in the form G_1. The microsomes must be present during protein synthesis for production of the G_1-form. G_1 is cotranslationally inserted into the endoplasmic reticulum membrane. Insertion takes place before approximately 80 residues have been synthesized. The amino-terminal domain is glycosylated and is sequestered within the microsome. The carboxyl-terminal 30–amino acid portion is exposed on the outside (cytoplasmic side) of the microsome. The signal sequence of G_0 is cotranslationally removed during assembly into the membrane, since G_1 has an amino-terminal sequence identical with that of the mature G-protein.

Inhibition of glycosylation of the G-protein with tunicamycin (which prevents the transfer of the lipid-linked carbohydrate $Glc_3Man_9GlcNAc_2$ to the acceptor protein) inhibits production of VSV from host cells. In these inhibited cells, the G_0 form is the main G-protein produced. Glycosylation may therefore be necessary for the proper assembly of this protein into the membrane. The initial glycosylation of G_0 to form G_1 occurs in the endoplasmic reticulum. The G_1-form of the protein is transferred via coated vesicles (vesicles coated with the protein clathrin that are involved in endocytosis and the movement of proteins from endoplasmic reticulum to Golgi and Golgi to plasma membrane) to the Golgi, where addition of further carbohydrate takes place to form G_2. G_2 then moves to the plasma membrane via a second set of coated vesicles. The G_1-form, which binds to Concanavalin A affinity columns, has the high-mannose (endoplasmic reticulum) form of carbohydrate that can be removed by endoglycosidase H. An additional posttranslational modification of G_1 takes place. The fatty acid palmitate is covalently attached to the protein shortly before processing of the oligosaccharide occurs. Little is known about the role of this covalently attached lipid, but it may play a role in attachment of the G-protein to membranes. The fatty acid is attached to a cysteine residue close to the carboxyl terminus. Site-specific mutagenesis at this site prevents addition of the fatty acid. Nevertheless, the assembly of G-protein into the plasma membrane and virus production are, however, normal. The transmembrane portion of the G-protein acts as a stop-transfer sequence. Deletion of this region of the protein results in a secreted mutant protein. The conclusion from these studies is that the G-protein of VSV follows a biosynthetic pathway similar to that of secreted proteins. This pathway may be summarized as follows:

$$G_0 \xrightarrow{\substack{\text{Endoplasmic} \\ \text{reticulum}}} G_1 \xrightarrow{\text{Golgi}} G_2 \xrightarrow{\substack{\text{Plasma} \\ \text{membrane}}} \text{Virus}$$

BIOSYNTHESIS OF COMPLEX MEMBRANE PROTEINS
Alternate Assembly Models

The original signal hypothesis can be used to describe the mechanism of protein secretion and the assembly of simple membrane proteins. Many membrane proteins, however, have a complex folding pattern in the membrane. A number of other models have been proposed to account for the assembly of simple and complex membrane proteins (Sabatini et al. 1982).

One such model is termed the *direct transfer model* (von Heijne and Blomberg 1979). The hydrophobic nature of signal sequences led to the suggestion that the signal sequence interacts directly with the lipid bilayer. Calculation of the hydrophobicity of signal sequences based on the transfer of an amino acid residue from a random-coil conformation in water to an α-helical conformation in a lipid bilayer indicated that signal sequences can insert into the membrane directly (Table 15.1). Therefore, no protein receptor is required for the binding of the nascent polypeptide chain to the endoplasmic reticulum. The tight binding of the ribosome to its receptor allows the nascent polypeptide to be driven through the membrane by the energy of translation. The protein is envisioned as being inserted as a loop (Figure 15.10). Once the initial segment is inserted in the membrane, an additional hydrophilic residue can be inserted into the bilayer from the cytoplasmic side as long as a hydrophilic residue is removed from the other side of the endoplasmic reticulum membrane. Translocation continues until a stop-transfer sequence appears and becomes associated with the membrane. This sequence is hydrophobic and interacts strongly with the membrane. The hydrophobic segment is followed by a basic region that does not enter the lipid phase because of its very positive ΔG_{trans} value.

A similar model was proposed earlier by Inouye, based on his laboratory's work on the assembly of the lipoprotein into the outer membrane of *E. coli*. In this *loop model*, the positively charged amino-terminus region of the signal peptide of the prolipoprotein associates with the negatively charged lipids of the inner surface of the cytoplasmic membrane (Figure 15.10). The hydrophobic portion of the signal sequence inserts into the bilayer, with the amino terminus always remaining on the cytoplasmic side of the membrane. This insertion takes place before synthesis of the prolipoprotein is completed. Cleavage of the signal peptide also takes place cotranslationally but after modification of a cysteine residue at the cleavage site

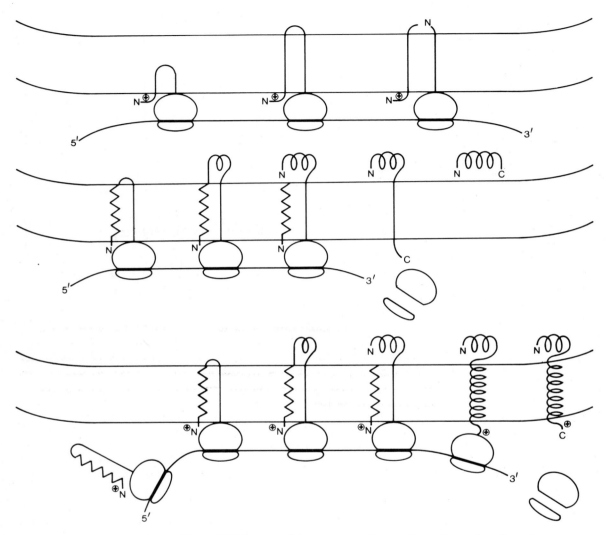

Figure 15.10. Loop models can account for secretion and assembly of membrane protein. Inouye loop model (top) accounts for assembly of *E. coli* lipoprotein (Di Rienzo, Nakamura, and Inouye 1978). The direct-transfer model and helical hairpin hypothesis can be applied to secretion (middle) and assembly of simple membrane proteins (bottom). In these models the protein is inserted as a hairpin loop. The signal peptide (N$\sim\!\!\sim$) is hydrophobic enough to make the ΔG_{trans} of the entire loop strongly negative. The portion of the loop within the bilayer is α-helical to maximize the number of hydrogen bonds. For secretion the polypeptide is pushed through the bilayer due to the strong binding of the ribosome to the endoplasmic reticulum and the energy of translation. The signal peptide is removed cotranslationally. Once translation is complete, the polypeptide is pulled across the membrane by the folding of the protein. The mechanism for assembly of a simple membrane protein is identical except for the presence of a stop-transfer sequence that prevents further translocation of the nascent polypeptide.

and exposure of the cleavage site to the signal peptidase present in the periplasmic space.

Blobel (1980) has expanded the original signal hypothesis to include the assembly of membrane proteins. Simple intrinsic membrane proteins contain, in addition to an amino-terminal signal sequence a hydrophobic stop-transfer or halt sequence. The complex folding pattern of many membrane proteins can be accounted for by integration or insertion sequences (Table 15.3). This sequence is a loop of polypeptide sufficient to span the lipid bilayer twice, with a high content of hydrophobic residues.

Engelman and Steitz proposed a similar model in 1981. They suggested that insertion of a polypeptide into a membrane is accomplished by a hairpin structure composed of two α-helices which partitions into membranes because of the hydrophobic nature of the sequence. In this model, called the *helical hairpin hypothesis*, the signal sequence is oriented with its amino terminus on the cytoplasmic side of the membrane. The function of the hydrophobic signal sequence is to allow more polar residues in the second half of the hairpin to be accommodated in the membrane. As long as the total ΔG_{trans} of the hairpin loop is strongly negative, it will insert into the lipophilic phase.

The identification of the SRP suggests that the signal sequence interacts initially with this receptor, the function of which is to stop translation and to direct the polysome to the endoplasmic reticulum membrane. The release of the translation block by the docking protein suggests that once the polysome binds to the endoplasmic reticulum the signal sequence is no longer bound to the signal recognition particle. The signal sequence may at this point insert into the lipid bilayer. The insertion of polypeptide into the membrane as a hairpin loop can account for the orientation and folding of many membrane proteins and will be used to illustrate the pathway for the assembly of simple and complex membrane proteins.

These hypotheses do not, however, provide a mechanism for the translocation of the amino terminus across the membrane without an amino-terminal signal sequence. It may be possible that association of the amino-terminal region with a subsequent region of the protein causes translocation across the membrane.

Figure 15.10 illustrates the mechanism of assembly of secreted simple and complex membrane proteins based on these loop models.

Posttranslational versus Cotranslational Assembly

The mechanism of integration of simple membrane proteins relies on the cotranslational insertion of protein into the endoplasmic reticulum membrane, removal of a signal peptide, and locking of the protein into place with a stop-transfer sequence. As we have seen with ovalbumin, secretion does not require removal of a signal sequence. Similarly, many membrane proteins are made without cleavable signal sequences. In addition, membrane proteins can associate with the endoplasmic reticulum membrane posttranslationally. Most proteins found within the mitochondria are taken up posttranslationally.

Biosynthesis of Cytochrome P-450 and Cytochrome b_5

The biosynthesis and assembly of cytochromes P-450 and b_5 into the endoplasmic reticulum membrane are of interest, since these proteins are found in the same membrane but follow very different biosynthetic routes. The arrangement of these proteins in the endoplasmic reticulum membrane is shown in Figure 15.1. Cytochrome P-450 has a molecular weight of approximately 50,000 and spans the membrane a number of times. The amino-terminal sequence of the mature protein, which begins with methionine, is extremely hydrophobic. In these respects it is similar to the signal sequences found in secreted and simple membrane proteins. The complete sequence of cytochrome P-450 has been obtained from cDNA obtained from isolated mRNA. The sequence has a number of hydrophobic stretches consistent with a complex folding pattern in the membrane.

Cytochrome P-450 is made on membrane-bound polysomes and is inserted directly into the endoplasmic reticulum membrane (Figure 15.11). The protein is then transferred in a step that is independent of protein synthesis to smooth endoplasmic reticulum. Since smooth and rough endoplasmic reticulum are contiguous, the translocation of cytochrome P-450 may be simply due to lateral movement of the molecules in the plane of the membrane. Cytochrome P-450 synthesized in a cell-free translation system has the same amino-terminal sequence as the mature protein. Cytochrome P-450 is inserted into the endoplasmic reticulum without cleavage of an amino-terminal signal sequence. Although the amino terminus of P-450 resembles a signal sequence, this portion of the molecule is retained in the mature molecule. This sequence may be termed a *stable signal sequence*. The amino terminus may be inserted as a hairpin loop and the term *integration sequence* (Table 15.3) may be applied.

Cytochrome b_5 is held in the membrane by a single loop of polypeptide located at the carboxyl terminus of the protein. This protein is synthesized exclusively on free polysomes and is therefore inserted into the membrane posttranslationally (Figure 15.11). This makes sense, since after the carboxyl-terminal residue is added to the protein, approximately 40 carboxyl-terminal amino acids are still buried in the ribosome. These terminal 40 amino acids include the portion of the protein that binds it to the membrane. This sequence is exposed only when the ribosomal subunit dissociates following

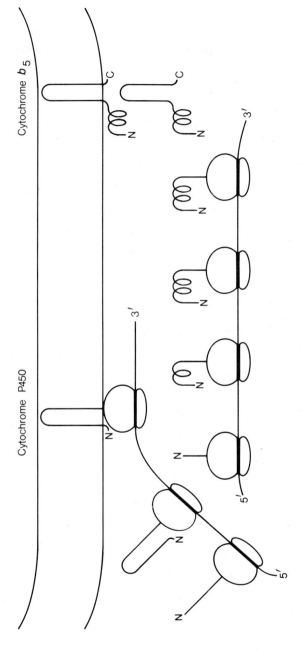

Figure 15.11. Models for the attachment of cytochromes P-450 and b_5 to the endoplasmic reticulum membranes. Cytochrome P-450 attaches to the endoplasmic reticulum as soon as the first transmembrane loop at the amino terminus is exposed from the ribosome (about 50 amino acids plus 40 within the ribosome). In contrast, cytochrome b_5 cannot attach until translation is complete, since the integration sequence which is located near the carboxyl terminus is buried within the ribosome.

termination of protein synthesis. The released polypeptide inserts spontaneously into the endoplasmic reticulum. The amino-terminal domain folds during protein synthesis into a native conformation. The purified protein has been shown to insert into lipid vesicles and to become functional. Cytochrome b_5 is found in most membranes of the cell; however, the endoplasmic reticulum is the major location of this protein.

Band 3, the Anion Transport Protein of the Human Erythrocyte Membrane

The Band 3 protein of the human red cell plasma membrane is a well-characterized example of an intrinsic membrane protein with a complex folding pattern in the membrane (Figure 15.1). Band 3 has a two-domain structure. The amino-terminal domain (with a molecular weight of 41,000) is located in the cytosol and is responsible for binding the cytoskeleton to the membrane. The carboxyl-terminal domain (with a molecular weight of 55,000) is embedded in the lipid bilayer and is responsible for anion transport. Little polypeptide is exposed on the exterior of the cell. Band 3 is glycosylated in the carboxyl-terminal domain, with the carbohydrate facing the cell exterior. Band 3 is made without an amino-terminal signal sequence, which is reasonable, since the amino terminus does not have to cross the membrane. Band 3 is made on membrane-bound polysomes and is cotranslationally inserted into the membrane. Approximately one-half of Band 3 must be synthesized before the nascent polypeptide chain is able to associate with the endoplasmic reticulum membrane. This association may be mediated by an insertion or integration sequence that corresponds to the transmembrane loop closest to the amino terminus (Figure 15.12). Additional loops of protein can insert into the membrane as they are made. Whether these loops insert directly into the lipid bilayer or interact initially with other components of the endoplasmic reticulum, such as SRP, has not been established.

Ca^{2+} ATPase and Calsequestrin

Like Band 3, the Ca^{2+} ATPase of the sarcoplasmic reticulum membrane of muscle cells has a complex folding pattern in the membrane (Figure 15.1). This protein is made on bound polysomes and is cotranslationally inserted into the endoplasmic reticulum. The protein is made without an amino-terminal signal sequence. The initiator methionine is acetylated and the first transmembrane loop is found very close to the amino terminus. This loop could act as an integration sequence to attach the nascent polypeptide to the endoplasmic reticulum. The Ca^{2+} ATPase, once inserted into the endoplasmic reticulum, could diffuse laterally to its final location in the sarcoplasmic reticulum, which is formed by extension of the endoplasmic reticulum.

Calsequestrin is found within the lumen of the sarcoplasmic

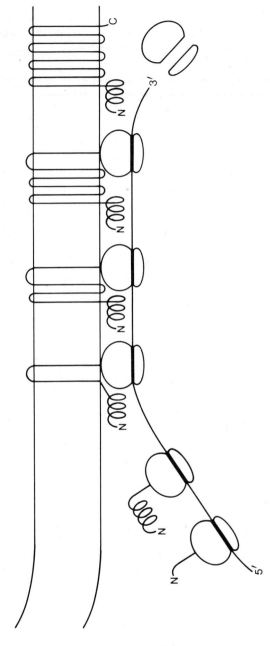

Figure 15.12. Model for the assembly of Band 3 into the endoplasmic reticulum membrane. The amino terminus of Band 3 (about 40,000 daltons) is located in the cytosol. The nascent polypeptide does not attach until the first transmembrane loop, located about halfway along the chain, is exposed from the ribosome. Further loops of polypeptide are inserted as they are exposed. Dissociation of the ribosome would leave the carboxyl terminus on the cytosolic side of the membrane.

reticulum. It is made on bound polysomes and is cotranslationally translocated into the endoplasmic reticulum. Calsequestrin is made with an amino-terminal signal sequence that is cotranslationally cleaved. This mechanism is identical with that for secreted proteins. Calsequestrin is glycosylated in the endoplasmic reticulum in a cotranslational fashion and may move to the Golgi, where the carbohydrate is trimmed. It is possible that calsequestrin never enters the Golgi but is processed in the endoplasmic reticulum.

Rhodopsin

Rhodopsin, the visual pigment of the eye, has been sequenced, and a folding pattern for the protein in the disc membrane has been proposed (Figure 15.1). Interestingly, the amino terminus of this protein is on the luminal side of the membrane, but the protein is made without an amino-terminal signal sequence. The amino-terminal methionine is donated by the initiator methionyl-tRNA and is blocked by a covalently linked fatty acid. The protein is made on bound polysomes and is glycosylated at residues 2 and 15. The covalently linked fatty acid and glycosylation may be involved in the translocation of the amino terminus across the membrane without the aid of a signal sequence.

Bacteriorhodopsin

The bacterial photoreceptor from *Halobacterium*, bacteriorhodopsin, is similar in some ways to the vertebrate protein rhodopsin. The protein crosses the membrane seven times with the amino terminus outside the cell and the carboxyl terminus facing the cytoplasm. Bacteriorhodopsin is made (Figure 15.1) with an amino-terminal extension:

Met-Leu-Glu-Leu-Leu-Pro-Thr-Ala-Val-Glu-Gly-Val-Ser-Gln-.

However, the amino-terminal extension does not resemble a signal sequence. How the amino terminus crosses the membrane is uncertain.

Assembly of Multisubunit Systems

Little attention has been focused on the biosynthesis and assembly of multisubunit membrane proteins. This area is very important, since most membrane proteins are made up of subunits, either identical or nonidentical. For example, the major histocompatibility antigen (HLA) protein is associated in the plasma membrane with β_2-microglobulin (Figure 15.1). HLA is made on membrane-bound polysomes with an amino-terminal signal sequence that is cotranslationally cleaved upon insertion of the protein into the endoplasmic reticulum. The β_2-microglobulin is synthesized with a 19-residue amino-terminal signal sequence. This peripheral membrane protein is cotranslationally processed and segregated across the microsomal membrane, where it associates with the HLA protein. It has been found that the β_2-microglobulin subunit does not associate with HLA until the HLA carbohydrate has been processed in the Golgi. The Golgi may be the site of assembly of multisubunit membrane proteins.

Sucrase-Isomaltase

The dimeric intestinal enzyme sucrase-isomaltase is synthesized as a single-chain precursor protein. The isomaltase subunit contains the membrane anchor at its amino terminus, which is located on the exterior of the microvillus membrane (Figure 15.1). The sucrase subunit, which is derived from the carboxyl terminus of the precursor, is held on the membrane by its association with the anchored isomaltase subunit. The one-to-one stoichiometry of this enzyme system is achieved by synthesizing a single precursor molecule that is subsequently cleaved to give equal numbers of the two subunits. It is envisioned that the sucrase-isomaltase is made with an amino-terminal signal sequence that inserts into the endoplasmic reticulum as a hairpin loop (Figure 15.13). This is followed by a second loop that has a hydrophobic strand followed by a hydrophilic strand. The rest of the polypeptide is extruded in a manner similar to that of secreted proteins. Cleavage of the signal sequence results in the location of the amino and carboxyl termini of the protein exterior to the cell. The protein is anchored by a single hairpin loop at the amino terminus.

Acetylcholine Receptor

A more extreme version of the problem of assembly of multisubunit membrane proteins systems is the assembly of the acetylcholine receptor. This integral membrane protein of known sequence is composed of four different polypeptides with the stoichiometry $\alpha_2\beta\gamma\delta$. Each of these subunits is translated from separate mRNA molecules on membrane-bound polysomes. All four polypeptides contain amino-terminal signal sequences that are co-translationally cleaved upon insertion into the membrane. The assembly of the acetylcholine receptor involves SRP. All subunits are cotranslationally glycosylated. The subunits may be inserted into the endoplasmic reticulum separately in a nonnative form, perhaps as homopolymers. In the membrane, there is a slow conformational change of the subunits to the native form. The subunits then come together to form an active complex which relies on specific protein-protein inter-actions involving the four differennt subunits. Exclusion of one subunit results in an inactive receptor. Unassembled subunits are rapidly degraded.

It has also been suggested that the subunits form homopolymers that exchange to form the active receptor. It is interesting to speculate that formation of the native complex requires modification of the

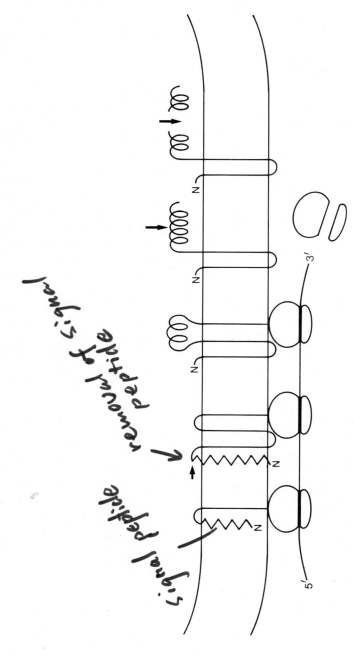

Figure 15.13. Model for the assembly of sucrase-isomaltase. The nascent polypeptide contains a signal sequence that allows the amino terminus to bind to the endoplasmic reticulum as a transmembrane loop. This is followed by a second loop (integration sequence) which is also inserted into the membrane. The signal peptide is removed cotranslationally. Protein synthesis continues pushing the polypeptide across the membrane. After synthesis is complete, the carboxyl terminus is pulled across the membrane by folding of the lumenal domain. Cleavage of the protein produces the sucrase and isomaltase subunits.

carbohydrate which occurs in the Golgi. An experiment involving inhibition of glycosylation with tunicamycin may provide an answer to this question.

IMPORT OF PROTEINS INTO ORGANELLES

Mitochondria are formed by the interplay of two genetics systems, the nuclear and mitochondrial. The vast majority of mitochondrial proteins are encoded on nuclear DNA. Their mRNA is translated on free polysomes, and the polypeptides are imported into the mitochondria posttranslationally. The pathway of assembly of mitochondrial proteins is clearly different from the translocation of proteins across the endoplasmic reticulum (Teintze and Neupert 1983).

The mitochondrion is bounded by two membranes which define a number of compartments: the outer membrane, the intermembrane space, the inner membrane, and the matrix. Proteins imported into mitochondria must find their way not only to the proper organelle but also to the proper site within the organelle. In addition, a number of the enzyme systems within the mitochondrion are multisubunit, with some subunits being encoded by the mitochondrial DNA and others by the nuclear DNA. The coordination of these two systems is a problem that the assembly system must face.

Experimental Systems

The biosynthesis and assembly of mitochondrial proteins have been studied in two ways. First, pulse-labeling studies have been used to detect precursors and follow their conversion in yeast cells and the fungus *Neurospora*. Second, the import of precursors into isolated mitochondria in a cell-free system has been demonstrated; this procedure has given us the most information. The binding of the protein precursor to mitochondria can be assayed directly; however, it is difficult to show that the precursor has been incorporated into the proper submitochondrial compartment and is functional.

The import of mitochondrial proteins can be broken down into a number of steps:

1. synthesis of the extramitochondrial precursor
2. binding to mitochondrial receptor
3. translocation
4. cleavage and processing
5. assembly

Synthesis of precursor. Mitochondrial proteins coded by nuclear DNA are synthesized in a precursor form on free polysomes. Many of these precursors are larger than the mature form, since they have amino-terminal extensions. These regions of the precursors are not at all like the signal sequences found on secretory and simple-membrane proteins. The amino-terminal extensions vary from a few amino acids up to about 100. The sequence of the precursor form of the *Neurospora* proteolipid subunit of the ATPase has been determined by gene sequencing. The precursor region is 66 amino acids in length and is hydrophilic, with a large number of basic residues. It is unlikely that the amino-terminal extensions of mitochondrial precursors play a direct role in binding to the mitochondrial receptor, since the exposure of the amino-terminal region of the precursor on the nascent polypeptide would lead to binding of the polysomes to mitochondria, resulting in cotranslational import; this is not observed. Moreover, a number of mitochondrial proteins, including the ADP–ATP translocator of the inner membrane, are made without an amino-terminal precursor region. The conformation of this protein before assembly into the mitochondrion is, however, different from the mature form, since the mature form, but not the precursor, will bind the inhibitor atractyloside. Cytochrome *c* is also made without an amino-terminal extension. The mature form, however, contains covalently attached heme. Possibly all mitochondrial proteins synthesized on free polysomes are made in a precursor form that differs in some significant way from the mature form.

Receptor binding. It is attractive to hypothesize that proteins destined for import into mitochondria are recognized by a specific receptor on the surface of this organelle. The detection and characterization of a receptor require the availability of significant amounts of the ligand. Large quantities of apo-cytochrome *c* can be chemically prepared from mature cytochrome *c*. Cytochrome *c* is localized on the outer surface of the inner membrane of the mitochondrion. The protein is synthesized on free polysomes without an amino-terminal extension. The completed polypeptide binds to a trypsin-sensitive site on the outer surface of the mitochondrion. The uptake of apo-cytochrome *c* can be inhibited by *deuterohemin*, an analogue of heme, which prevents the covalent attachment of heme to cytochrome *c*. The binding is saturable and can be competed with apo-cytochrome *c*. Mature cytochrome *c* does not displace the apo form.

The specificity of the receptor is further supported, since apo–cytochrome *c* does not inhibit the import of precursors to other mitochondrial proteins, including the ATPase proteolipid, holo-cytochrome *c* and the ADP–ATP translocator. It would not make sense to have a unique receptor for each mitochondrial protein; however, there may be a different receptor for proteins destined for each submitochondrial compartment.

The recognition of precursor forms of proteins destined for import into the mitochondrion suggests that the conformation and folding of the precursors are crucial in the uptake process. The presence of additional polypeptide at the amino terminus, the lack of a prosthetic group, or an aqueous environment rather than the native membrane environment would likely result in a precursor with a conformation different from the mature form. The function of a precursor form may be to produce a folding pattern of the protein which is different from that of the mature protein. The importance of folding has been emphasized by Wickner (1979) and is discussed in the section "Trigger Hypothesis".

Translocation. Once bound to a receptor, the mitochondrial protein precursors must be translocated to their proper mitochondrial compartment. The translocation pathways are at least as varied as the number of different locations within the mitochondria. The *porin protein*, which is located in the outer membrane and acts as a channel, may simply insert into this membrane. This protein is not made with an amino-terminal extension and does not require an electrochemical gradient across the inner membrane for assembly. No receptor for this protein has been detected.

Cytochrome *c* is found between the two mitochondrial membranes. It binds to a specific receptor and is posttranslationally translocated across the outer membrane. This protein is made without an amino-terminal extension and does not require a membrane potential for assembly (Figure 15.14).

Proteins that reside in the matrix must cross two membranes. There are sites of contact between the inner and outer mitochondrial membranes which may provide more direct access to the interior of the mitochondrion. Carbamoyl phosphate synthetase is a matrix enzyme that has been extensively studied. It is made on free polysomes with an amino-terminal extension of about 5,000 daltons. It is posttranslationally assembled into the mitochondrion, followed by cleavage of the amino-terminal extension.

Cleavage and processing. A matrix-associated protease has been identified that cleaves precursors of mitochondrial proteins to the mature form at neutral pH. The correct cleavage by this protease has been demonstrated for subunit V of cytochrome *c* oxidase. The protease requires Zn^{2+} and is inhibited by EDTA and sulfhydryl reagents. The protein acts on native precursors but not denatured ones. It is not inhibited by typical serine protease inhibitors. Partial purification of the protease has revealed that it probably exists as a dimer, with a subunit molecular weight of 59,000.

An example of the assembly of an inner membrane protein that involves cleavage by the protease is the F_1 portion of the ATPase which faces the matrix side of the inner mitochondrial membrane. It is composed of a number of subunits and is anchored to the inner

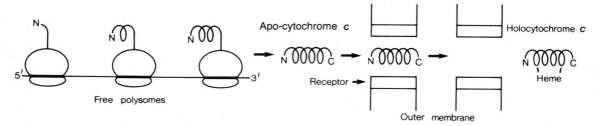

Figure 15.14. Assembly of mitochondrial cytochrome *c*. Apo-cytochrome *c* is synthesized on free polysomes and binds to a specific receptor located in the outer membrane of the mitochondrion. Translocation into the intermembrane space and attachment of heme completes the assembly of this protein.

membrane by a number of hydrophobic polypeptides. The α-, β-, and γ-subunits of the ATPase are synthesized on cytoplasmic polysomes as separate precursor subunits. This is most clearly shown by the fact that each precursor contains a methionine residue at its amino terminus that was donated by the initiator N-formylmethionyl-tRNA$_F$. These higher molecular weight precursors, when added to mitochondria under conditions that exclude protein synthesis, are taken up and processed to mature polypeptide.

Cytochromes c_1 and b_2 in yeast each undergoes two separate proteolytic cleavages before they are assembled into their first positions in the intramembranous space (Figure 15.15). Cytochrome b_2 is made as a 68,000-dalton precursor that enters the mitochondrion under the influence of the electrochemical potential. A proteolytic cleavage, carried out by the matrix protease, produces a 65,000-dalton intermediate that is associated with the outer surface of the inner membrane. The cytochrome b_2 is released into the intramembranous space by a second proteolytic cleavage. Flavin and heme are added before both cleavages have been completed (Figure 15.15). A similar pathway is followed by cytochrome c_1, although the second proteolytic cleavage leaves the mature cytochrome c_1 tightly associated with the outer surface of the inner mitochondrial membrane.

The incorporation of a number of mitochondrial proteins has been shown to require a membrane potential across the inner mitochondrial membrane. Proteins such as porin of the outer membrane and cytochrome c, which do not require an energized membrane for import, do not cross the inner membrane. Although cytochrome b_2 is located between the two mitochondrial membranes, it is processed by the neutral protease found within the mitochondrial matrix and requires a membrane potential for assembly. In many studies it has been shown that uncouplers of oxidative phosphorylation inhibit the

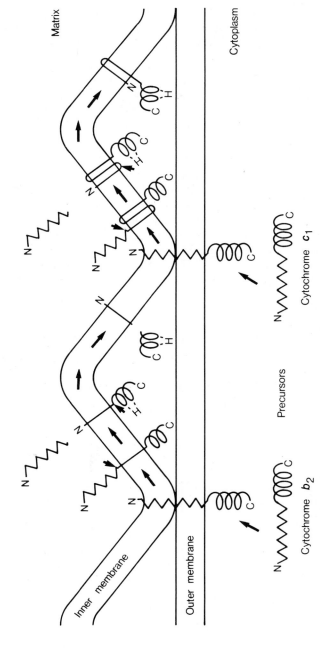

Figure 15.15. Assembly of mitochondrial cytochromes b_2 and c_1. Both proteins are made with amino-terminal extensions on free polysomes. Cytochrome b_2 binds to the mitochondrion, and the extension is removed by a protease located in the matrix. A second proteolytic cleavage releases the protein into the intermembrane space. The assembly of cytochrome c_1 is very similar. The second cleavage step, however, leaves this cytochrome attached to the outer surface of the inner mitochondrial membrane.

uptake and processing of mitochondrial proteins. Mitochondria containing high levels of ATP but a collapsed membrane potential fail to import these proteins.

Assembly. Very little is known about the assembly of proteins coded by mitochondrial DNA. Subunit II of cytochrome *c* oxidase is translated on mitochondrial ribosomes. These ribosomes are associated with the inner surface of the inner mitochondrial membrane. This protein is probably cotranslationally inserted into the inner mitochondrial membrane.

The assembly of the subunits of the chloroplast enzyme ribulose-1,5-biphosphate carboxylase has been studied in some detail. This enzyme is composed of eight copies each of large (55,000-dalton) subunits and an equal number of small (15,000-dalton) subunits. Interestingly, the large subunit is encoded by chloroplast DNA, while the small subunit is encoded by nuclear DNA. The stoichiometry between the two subunits is controlled by degradation of excess imported small subunit by a constitutive proteolytic degradation system. The lack of association of the small subunit with the large subunit makes it susceptible to degradation. Inhibition of chloroplast protein synthesis does not impair the synthesis and import of the small subunit, suggesting that syntheses of the two subunits are quite independent of one another. The small subunit is made on free polysomes as a precursor (with a molecular weight of 20,000) with an acidic amino-terminal extension. The uptake is dependent on light and is abolished by uncouplers of oxidative phosphorylation. Treatment of chloroplasts with proteases prevents uptake, suggesting that a receptor system may be involved.

Lysosomal Enzymes The biosynthesis of two lysosomal hydrolases, cathepsin D and β-glucuronidase, has been studied in a cell-free system. It was found that these proteins are made approximately 2000 daltons larger than the mature forms. The precursors are made on bound polysomes and are cotranslationally cleaved and sequestered by microsomal membranes. The segregation of cathepsin D across the endoplasmic reticulum requires a signal recognition particle. The biosynthesis of lysosomal enzymes involves a pathway similar to that of secreted proteins. Cultured liver cells that were treated with tunicamycin to prevent glycosylation produced the unglycosylated forms of cathepsin D and β-glucuronidase that were secreted. These studies suggest that the carbohydrate residues on lysosomal enzymes play an important role in directing the proteins to the proper subcellular compartment. Indeed, it was found that a significant amount of these hydrolytic enzymes was secreted under normal conditions. The secreted product differed in the carbohydrate structure and molecular weight from the lysosomal form. Further proteolytic processing takes place in the lysosome.

Considerable evidence has accumulated that mannose-6-phosphate residues play an essential role in directing hydrolases to lysosomes. Fibroblasts cultured from patients with I-cell disease secrete large amounts of lysosomal hydrolases rather than incorporating them into lysosomes. These secreted products lack the mannose-6-phosphate recognition marker found on lysosomal enzymes. Mannose-6-phosphate residues may serve as a sorting signal for lysosomal enzymes.

The peroxisomal enzymes catalase and uricase are not segregated by microsomal membranes and may be taken up by a posttranslational system similar to that of mitochondria.

BIOSYNTHESIS OF BACTERIAL CELL ENVELOPE PROTEINS
Assembly of Membrane Proteins in *E. coli*

The bacterium *E. coli* is surrounded by an envelope that is composed of a cytoplasmic (inner) membrane, a peptidoglycan layer, and an outer membrane. The periplasmic space is enclosed between the inner and outer membranes. *E. coli* ribosomes may be free in the cytoplasm or associated with the inner surface of the cytoplasmic membrane. The bound polysomes are attached to the inner membrane by the nascent polypeptide. Unlike eucaryotic bound polysomes, no salt-labile linkage is required for binding of bacterial ribosomes to the membrane. The bacterial cell faces a problem similar to that of the eucaryotic cell—proteins must be routed to the proper compartment. In the case of the periplasmic space, protein must be secreted across the cytoplasmic membrane. Outer-membrane proteins must cross the cytoplasmic membrane and then be assembled into the outer membrane. Receptors for outer-membrane proteins may be other proteins or lipopolysaccharide, a unique component of this membrane.

Ribosomes are attached to the cytoplasmic membrane of *E. coli* by the emerging nascent chains only. Since the ribosomes are not anchored to the membrane, and the energy of translation cannot push the nascent polypeptide through the membrane, translation would simply move the ribosomes away from the membrane. Removal of the periplasmic domain of nascent proteins does not release the ribosome. Hence, folding of the polypeptide in the periplasmic space is not required for the attachment of ribosomes. The uptake of secreted proteins into inside-out vesicles is not simply insertion into the lipid bilayer. Vesicles treated with protease lose their ability to take up proteins, suggesting that secretion requires the participation of translocation machinery. The nascent polypeptide is tightly attached to the translocation machinery which directs the nascent polypeptide across the inner membrane.

Table 15.4. Location of Bacterial Envelope Proteins

Site	Protein (gene)	
Inner membrane	M13 coat protein	(gene 8)
	Lac carrier	(Lac γ)
Periplasmic space	Alkaline phosphatase	(pho A)
	β-lactamase	(bla)
	Maltose-binding protein	(mal E)
Outer membrane	λ-receptor	(lam B)
	Porin	(omp A)
	Lipoprotein	(lpp)

The biosynthesis and assembly of a number of envelope proteins have been studied (Table 15.4). Proteins destined for secretion into the periplasmic space and assembly into the outer membrane are made with an amino-terminal signal sequence. Proteins of the inner membrane may be made as precursors, but no signal sequence has been found for an *E. coli* membrane protein to date. For example, the lactose carrier and the band-c subunits of the F_0 ATPase are made without a signal sequence, while penicillin-binding proteins are made as precursors.

Bacteriophage M13
Coat Protein

The biosynthesis and assembly of the coat protein of bacteriophage M13 into the cytoplasmic membrane of the host *E. coli* has been extensively studied (Wickner 1983). Bacteriophage M13 produces a nonlytic infection of *E. coli*. The virus contains no lipid, but each copy of DNA is enclosed in 2000 copies of the coat protein. The coat protein is inserted into the *E. coli* inner membrane with its amino-terminal 20 amino acids exposed to the periplasmic space, a 20-residue hydrophobic segment spanning the membrane, and a 10-residue carboxyl-terminal domain in the cytoplasm. The viral DNA becomes coated with this protein as it passes through the membrane. While in the cytoplasmic membrane, the coat protein resembles a simple intrinsic membrane protein.

The coat protein is made first in a procoat form on free polysomes with an amino-terminal extension of 23 amino acids:

$$\overset{10}{\text{Met-Lys-Lys-Ser-Leu-Val-Leu-Lys-Ala-Ser-Val-Ala-Val-Ala-Thr-}}$$

$$\underset{20}{\text{Leu-Val-Pro-Met-Leu-Ser-Phe-Ala}}\underset{\uparrow}{\text{-Ala-Glu-Gly-Asp-Asp-Pro-Ala-Lys-Ala-Ala-Phe-Asn-}}\ldots$$

This extension has all the characteristics of a signal sequence.

Pulse-chase experiments on the assembly of the coat protein have been difficult, since the conversion of procoat protein to coat protein is complete within a few seconds. Mutants defective in virus assembly have been employed. In these mutants, the majority of procoat protein is found in the cytoplasm. Pulse-chase experiments have shown a posttranslational conversion of cytoplasmic procoat protein to membrane-bound coat protein. Using these results, Wickner has proposed that coat protein is made on free polysomes, then is released as a water-soluble procoat protein which becomes membrane bound posttranslationally, and finally is converted to mature membrane coat protein (Figure 15.16). The conversion to coat protein and correct assembly require a membrane potential.

The gene for the signal peptidase of *E. coli* has been cloned and sequenced. The protein has been purified from an overproducing strain of *E. coli*. Interestingly, the purified peptidase cleaves not only procoat to coat, but also processes mammalian precursors for secreted proteins such as pre-β-lactamase and pre-IG kappa chain to the mature forms. Conversely, dog pancreas microsomal signal peptidase cleaves bacterial precursors such as M13 procoat protein. This phenomenon may be expected from the similarities of the signal sequences and suggests that the assembly mechanisms in bacteria and higher organisms are highly conserved. This feature has allowed the synthesis, processing, and secretion of human preproinsulin to be carried out by *E. coli* from a plasmid containing the human gene for preproinsulin, a breakthrough in biotechnology.

Wickner Trigger Hypothesis

Using primarily the results with coat protein, Wickner (1979) proposed a *membrane trigger hypothesis*. Wickner hypothesized that the function of the signal sequence is to change the folding pathway of the protein and perhaps to make a membrane protein precursor water soluble. In this hypothesis, the assembly of the precursor into the membrane is driven by the energy of protein folding. The precursor protein associates directly with the lipid bilayer and spontaneously inserts into the hydrophobic core of the membrane. There is no need for a protein pore in this hypothesis, as was suggested by Blobel. In addition, assembly of proteins into the membrane is also not necessarily cotranslational, since the completed precursor can associate with the membrane. Removal of the signal peptide renders the assembly process irreversible.

The Procoat Controversy

Two models have been proposed for the assembly of the M13 procoat protein into the inner membrane of *E. coli*. Wickner has suggested that the procoat is made on free polysomes as a water-soluble precursor and is posttranslationally assembled into the membrane. The assembly of the procoat protein into the membrane is thought to

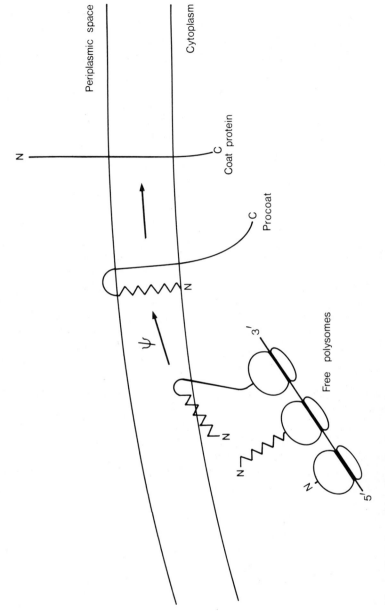

Figure 15.16. Assembly of bacteriophage M13 coat protein into the inner membrane of *E. coli*. The coat protein is made with an amino-terminal signal sequence. This procoat form associates with the inner membrane immediately after synthesis is complete. A membrane potential (ψ) drives the completed polypeptide across the membrane. The signal sequence of the procoat protein is removed by a signal peptidase located in the periplasmic space. The coat protein is held in place by a stop-transfer sequence.

occur spontaneously and without the need for translocation machinery. In contrast, Blobel has suggested that procoat protein is assembled in a manner similar to that of secreted protein. He has suggested that procoat protein is cotranslationally inserted into the inner membrane of *E. coli*, where cleavage of the signal sequence takes place. The experimental evidence that relates to this controversy is as follows:

1. Blobel and Wickner have demonstrated that M13 procoat can be assembled into membrane cotranslationally. Addition of membranes after procoat protein synthesis is complete does not permit processing or assembly of procoat. Procoat aggregates in solution and becomes incompetent for assembly, which may account for the lack of processing.

2. Wickner has used assembly mutants of *E. coli*. The assembly mechanism or receptor may be saturated by procoat protein, causing most of procoat protein to end up in the cytoplasm rather than in the membrane.

3. The procoat does not appear to be water soluble. Rather, it aggregates, sedimenting at 5 S, and in time will not assemble. It also seems unlikely that a hydrophobic signal peptide would render the procoat more water soluble.

4. Assembly is not driven by protein folding. Wickner has shown by the use of uncouplers of oxidative phosphorylation that a membrane potential is required for the proper assembly of procoat. If assembly of procoat is blocked by CCCP, an uncoupler of oxidative phosphorylation, the procoat is found exclusively in the membrane.

5. In a mutant where Glu-2 of the mature coat protein is replaced by Leu-2, the procoat is found in the membrane, not in the cytoplasm. However, the signal peptidase cleaves the mutant form as readily as the wild type. The reason for slow processing has not been established.

A model for assembly of procoat protein is illustrated in Figure 15.17. The procoat protein is 73 amino acids long. The signal peptidase is located on the periplasmic side of the cytoplasmic membrane. The signal sequence plus a membrane-spanning length (approximately 23 amino acids as an α-helix) must be exposed before the signal sequence can be removed. About 40 amino acids are buried within the ribosome of eucaryotes, and a similar number in procaryotes. Obviously, after the last amino acid has been added to procoat, only $73 - 40$, or 33 amino acids are exposed from the ribosome. This is too few to allow cleavage by periplasmic signal peptidase. Cleavage

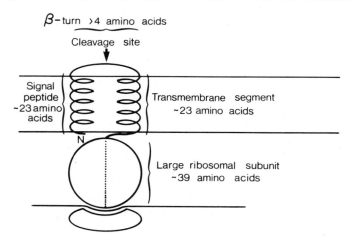

Figure 15.17. Possible arrangement of a nascent polypeptide in the plasma membrane of *E. coli* based on a helical hairpin model. Cleavage of the signal sequence would require the condensation of 23 + 4 + 23 + 40, or 90 amino acids (see text). The procoat protein cannot be cleaved until its synthesis is complete, since it is only 73 amino acids long.

of procoat protein therefore must be posttranslational. The ribosome must dissociate in order to allow the procoat protein to have access to the signal peptidase.

Periplasmic Proteins   Proteins destined for the periplasmic space must be translocated across the cytoplasmic membrane. Periplasmic proteins such as alkaline phosphatase, β-lactamase, and the maltose-binding protein are made with amino-terminal extensions. These proteins are made on membrane-bound polysomes that are attached to the cytoplasmic membrane by their nascent chain. However, the translocation of β-lactamase across the membrane does not occur immediately but requires that a large portion of the protein be synthesized before translocation occurs. It has been difficult to distinguish a completely cotranslational versus posttranslational mode of transfer.

Outer-Membrane Proteins   The outer membrane of *E. coli* contains several proteins present in large amounts. This membrane contains an unusual lipoprotein (Figure 15.18). The amino-terminal cysteine residue is blocked by a fatty acid, and the side-chain sulfhydryl is linked to a diacylglycerol. This protein is made with a 20-residue amino-terminal signal sequence:

$$\overset{\text{10}}{\text{Met-Lys-Ala-Thr-Lys-Leu-Val-Leu-Gly-Ala-Val-Ile-Leu-Gly-Ser-Thr-}}$$

$$\underset{\text{20}}{\text{Leu-Leu-Ala-Gly-}_{\uparrow}\text{Cys-Ser-Ser-Asn- } \ldots} \quad \text{(Asp)}$$

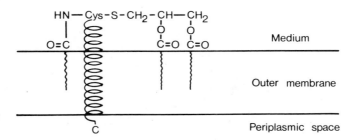

Figure 15.18. Model for the arrangement of the lipoprotein in the outer membrane of *E. coli*. O=C are fatty acyl groups.

The signal sequence is removed only after addition of diacylglycerol to the cysteine sidechain. Site-specific mutagenesis, in which the cysteine is replaced by a glycine residue, prevents addition of the diacylglycerol and also removal of the signal peptide. The signal peptidase characterized by Wickner apparently does not cleave the prolipoprotein; hence, another protease must exist. After removal of the signal sequence, a fatty acid is linked to the amino terminus. A mutation in the signal peptide which changes Gly-14 to Asp-14 prevents cleavage of the signal sequence. However, this mutant lipoprotein is assembled into the outer membrane. Proteolytic cleavage of the signal sequence is therefore not necessary for the correct assembly of this membrane protein.

The omp A protein of the outer membrane of *E. coli*, which functions as a pore, is made with an amino-terminal signal sequence. The intact protein contains 325 amino acids. The carboxyl terminus of this protein has been shortened by restriction-enzyme cutting of the gene for omp A. A shortened derivative containing 193 amino acids is still assembled into the outer membrane and is functional. The carboxyl-terminal region is obviously not required for proper assembly. However, an omp A protein with only 133 amino acids is not localized properly and is unstable and nonfunctional. Residues between numbers 133 and 193 from the amino terminus are required for the correct assembly of the omp A protein. A signal peptide alone is not sufficient to direct the polypeptide to its ultimate subcellular location.

Gene-fusion Studies

Elegant gene-fusion studies (Silhavy, Benson, and Emr 1983) have defined various domains of membrane proteins required for secretion and assembly in *E. coli*. The gene for the cytoplasmic protein β-galactosidase has been fused to DNA coding for various lengths of the lambda (λ) receptor of the outer membrane (Figure 15.19). The λ-receptor is made on bound polysomes with a 25–amino acid signal sequence. Fusion I contains only 3 amino acids of the λ-receptor. Not surprisingly, this protein is found in the cytoplasm.

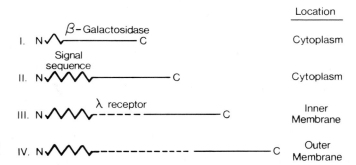

Figure 15.19. Protein products produced by fusion of the genes for the λ-receptor (see text for details).

Fusion II contains the entire signal sequence for the λ-receptor. This protein is also found in the cytoplasm. This result shows that the signal sequence alone is not sufficient for secretion and assembly of the hybrid protein. Fusion III contains the signal sequence plus 173 amino acids of mature λ-receptor. This protein is localized to the cytoplasmic membrane, jamming the export pathway for other secreted proteins, resulting in a lethal phenotype. A larger fusion containing 241 amino acids (fusion IV) is translocated to the outer membrane. These experiments show that the localization of proteins to their proper compartment requires other information in addition to the signal sequence.

Mutations in the signal sequence of the λ-receptor have shown that the hydrophobic region of the signal sequence is essential for translocation (Figure 15.20). A 12–amino acid deletion from the signal sequence prevents processing, as does a shorter 4–amino acid deletion (residues 10–13). Additionally, if residues 14, 15, 16, or 19 are changed to charged residues, processing is prevented (Figure 15.20). The hydrophobic core of the signal sequence probably exists as an α-helix. The 4–amino acid deletion in the signal sequence results in a nonfunctional signal. Secondary mutations in the 10–13 deleted sequence of Pro-9 to Leu-9 or Gly-17 to Cys-17 restore the function of the signal sequence. Proline and glycine are helix breakers. Therefore, replacement by leucine or cysteine increases the length of the hydrophobic core and promotes an α-helical conformation. These

Figure 15.20. Mutations in the signal sequence of the λ-receptor. The hydrophobic stretch is underlined. Altered amino acids are in brackets. The arrow indicates the cleavage site.

$$
\overset{}{\text{Met - Met - Ile - Thr - Leu - Arg - Lys - }}\underset{\text{(Ser)}}{\underline{\text{Leu - Pro - }\overset{10}{\text{Leu}}}}
$$

$$
\underline{\text{- Ala - Val - Ala - Val - Ala - Ala - Gly - Val - Met - }\overset{20}{\text{Ser}}}
$$
$$
'\text{(Asp)(Glu) (Glu) (Arg)}\qquad\text{(Arg)}
$$

$$
\underline{\text{- Ala - Gln - Ala - Met - Ala}}\overset{\downarrow}{\text{ - }}\text{Val - Asp - Phe - . . .}
$$

studies have shown that a signal sequence of defined length and conformation is required for export.

The ultimate location of the protein is not defined by the signal sequence alone. The signal sequence for β-lactamase, a periplasmic enzyme, was fused to the amino terminus of phoE, an outer-membrane protein. The polypeptide was assembled into the outer membrane in spite of being made with a signal sequence for a periplasmic protein.

FUTURE DIRECTIONS The application of molecular biological approaches such as gene splicing and deletion will lead to a better understanding of the roles played by various protein domains in their assembly into membranes (Yost, Hedgpeth, and Lingappa 1983). Reconstitution studies with the purified components of the assembly process will lead to a better understanding of the function of each component in the pathway. Structural studies on purified components, such as the signal recognition particle, are essential to our understanding of the assembly process. Further studies on the role and fate of signal sequences are required. The mechanisms by which proteins are routed to their correct compartments have not been well established. The role of the sugar residues of glycoproteins is particularly relevant to this problem. Mutants of bacteria and yeast will be of great value in characterizing the assembly pathways in these organisms. Identification and characterization of the translocation machinery in bacteria and subcellular organelles and their comparison with the endoplasmic reticulum mechanism may reveal common themes or mechanisms for assembly.

PROBLEMS

1. Using the ΔG_{trans} values in Table 15.1, calculate the energy holding the most hydrophobic 23–amino acid transmembrane segment of μM of IgM in the membrane.
2. From your knowledge of the structure and organization of Band 3 and glycophorin (Figure 15.1) in the human red cell membrane, outline a model for the biosynthesis and assembly of these two types of intrinsic membrane proteins.
3. Compare Blobel's signal hypothesis and Wickner's trigger hypothesis. Use the example of the M13 coat protein to distinguish these two mechanisms.
4. Support the view that simple intrinsic membrane proteins are "partially secreted" proteins. Use the biosynthesis of the G-protein of VSV to illustrate your argument.

BIBLIOGRAPHY

Entries preceded by an asterisk are suggested for further reading.

*Blobel, G. 1980. Intracellular protein topogenesis. *Proc. Natl. Acad. Sci. USA* 77:1496–1500.

Blobel, G., and Dobberstein, B. 1975. Transfer of proteins across membranes. *J. Cell Biol.* 67:835–51.

*Clément, J. M. 1983. On the insertion of proteins into membranes. *Biochimie* 65:325–38.

558 Chapter 15 / Assembly of Proteins into Membranes

Di Rienzo, J. M.; Nakamura, K.; and Inouye, M. 1978. The outer membrane proteins of gram-negative bacteria: Biosynthesis, assembly, and functions. *Ann. Rev. Biochem.* 47:481–532.

Early, P.; Rogers, J.; Davis, M.; Calame, K.; Bond, M.; Wall, R.; and Hood, L. 1980. Two mRNAs can be produced from a single immunoglobulin μ gene by alternative RNA processing pathways. *Cell* 20:313–19.

*Engelman, C. M., and Steitz, T. A. 1981. The spontaneous insertion of proteins into and across membranes: The helical hairpin hypothesis. *Cell* 23:411–22.

Hubbard, S. C., and Ivatt, R. S. 1981. Synthesis and processing of asparagine-linked oligosaccharides. *Ann. Rev. Biochem.* 50:555–83.

Katz. F. N.; Rothman, J. E.; Knipe, D. M.; and Lodish, H. F. 1977. Membrane assembly: Synthesis and intracellular processing of the vesicular stomatitis viral glycoprotein. *J. Supramolec. Struct.* 7:353–70.

Milstein, C.; Brownlee, G. G.; Harrison, T. M.; and Mathews, M. B. 1972. A possible precursor of immunoglobulin light chains. *Nature New Biol.* 239:117–20.

Palade, G. 1975. Intracellular aspects of the process of protein synthesis. *Science* 189:347–58.

*Rothman, J. E., and Lenard, J. 1977. Membrane asymmetry. *Science* 195:743–53.

Sabatini, D. D.; Kreibich, G.; Morimoto, T.; and Adesnik, M. 1982. Mechanisms for the incorporation of proteins in membranes and organelles. *J. Cell Biol.* 92:1–22.

Silhavy, T. J.; Benson, S. A.; and Emr, S. D. 1983. Mechanisms of protein localization. *Microbiol. Rev.* 47:313–44.

Schulz, G. E., and Shirmer, R. H. 1979. *Principles of protein structure.* New York: Springer-Verlag.

Teintze, M., and Neupert, W. 1983. "Biosynthesis and assembly of mitochondrial proteins." In *Cell membranes: Methods and reviews,* vol. 1, ed. E. Elson, W. Frazier, and L. Glaser, 89–115. New York: Plenum.

von Heijne, G. 1981. On the hydrophobic nature of signal sequences. *Eur. J. Biochem.* 116:419–22.

*von Heijne, G., and Blomberg, C. 1979. Transmembrane translocation of proteins: The direct transfer model. *Eur. J. Biochem.* 97:175–81.

Walter, P., and Blobel, G. 1981. Translocation of proteins across the endoplasmic reticulum. *J. Cell Biol.* 91:557–61.

*Wickner, W. 1979. The assembly of proteins into biological membranes: The membrane trigger hypothesis. *Ann. Rev. Biochem.* 48:23–45.

Wickner, W. 1983. M13 coat protein as a model of membrane assembly. *Trends in Biochem. Sci.* 8:90–94.

Yost, C. S.; Hedgpeth, J.; and Lingappa, V. R. 1983. A stop transfer sequence confers predictable transmembrane orientation in a previously secreted protein in cell-free systems. *Cell* 34:759–66.

Answers to Problems

CHAPTER 2

1. Outer-monolayer lipids would be expected to be more impermeable due to their relatively saturated composition. However, the negative surface potential of the inner monolayer would be expected to increase Na^+ permeability and reduce Cl^- permeability. These permeabilities may be measured by trapping radioactive varieties of Na^+ and Cl^- in LUV systems and monitoring efflux.

2. Integral proteins would not be able to move past the intrabilayer inverted lipid cylinder barrier. Lipids in the outer monolayer of apical or basolateral domains would be retained in these regions. Lipids able to move across the bilayer would be able to move to the inner monolayer, diffuse past the barrier, and subsequently flip back out to the outer monolayer of the other domain.

3. More-unsaturated acyl chains impart a greater cross-sectional area in the acyl region (increased cone shape). This could be compensated for by increasing the DGluDG species, which have a larger head group.

4. You would expect a rapid initial K^+ efflux facilitated by valinomycin, which would give rise to a membrane potential (inside negative) where $\Delta \psi = -59 \log[K_i^+]/[K_0^+]$, balancing the interior K^+ chemical potential. Subsequent efflux would be much slower, as K^+ efflux would then be coupled to Na^+ influx.

5. Factors which induce H_{II} organization will result in fusion. For unsaturated phosphatidylethanol-amine-phosphatidylserine LUVs, these factors are the addition of Ca^{2+}, a lowering of pH, an increase in the cholesterol content, or an increase in the temperature.

CHAPTER 3

1. The biosynthesis of fatty acids and phospholipids will immediately cease. Lipid A acylation will also stop, since the 3-hydroxymyristic acid in this lipid is derived from fatty acid biosynthesis. Lipoprotein production will be normal since the acyl groups attached to the amino terminus are derived from phospholipids, which will still be in abundant supply after cerulenin treatment.

2. The addition of glucose will result in MDO biosynthesis, which would be absent without the addition of glucose. MDO production would cause turnover of the sn-glycerol-1-phosphate moiety of phosphatidylglycerol. The dgk mutation will result in the accumulation of diacylglycerols in the medium only in the presence of glucose and hence MDO biosynthesis. MDO biosynthesis is ionic-strength controlled; therefore, at low ionic strength, MDO production is high, as is phosphatidylglycerol turnover and diacylglycerol accumulation. The converse is true at high ionic strengths.

3. The effect of depriving a *gpsA* mutant of the glycerol phosphate supplement is the same as removing the glycerol supplement from a *plsB* strain: Phospholipid biosynthesis is arrested, long-chain acyl-ACPs accumulate, and the membrane density increases due to the continued synthesis of membrane protein in the absence of phospholipid production.

4. The most likely enzyme to be affected is 3-ketoacyl-ACP synthase II, which is responsible for the production of *cis*-vaccenate and temperature control. The identity of the new mutant as a *fabF* allele would be established if the new mutation maps at the *fabF* locus and is complemented by the wild-type allele in merodiploid strains.

5. Acyl-ACP should be almost completely denatured at pH 9.5; therefore, the thioesterase would appear to be specific for the denatured conformation of ACP. This point could be established by the following.
   a. Using acyl-deuterio-ACP as a substrate. This ACP derivative is less stable to conformational change, and the pH optimum of the thioesterase using this substrate should be lower.
   b. Adding agents that stabilize acyl-ACP to pH-induced hydrodynamic expansion, which should inhibit the thioesterase activity. Spermidine and Mg^{2+} are two agents discussed in the text.
   c. Examining the pH optimum of the enzyme using a substrate analogue that does not have a complex tertiary structure. Two good candidates would be an acyl-phosphopantetheine peptide or acyl-CoA.

6. The inhibition of 3-hydroxydecanoyl-ACP dehydrase (*fabA*) will result in the inhibition of unsaturated fatty acid biosynthesis, the accumulation of disaturated phospholipid molecular species, and the eventual cessation of cell growth. The addition of oleic acid to the medium would abolish these effects by providing an exogenous unsaturated fatty acid for phospholipid biosynthesis. The *fabAUp* mutant would be resistant to 3-decynoyl-*N*-acetylcysteamine due to the overproduction of 3-hydroxydecanoyl-ACP dehydrase. In fact, resistance to 3-decynoyl-*N*-acetylcysteamine was how the *fabAUp* strain was isolated.

CHAPTER 4

1. 2-Bromopalmitoyl-CoA is an inhibitor of carnitine palmitoyl-transferase I but not of carnitine palmitoyltransferase II.

2. Malonyl-CoA inhibits carnitine palmitoyltransferase I and consequently the oxidation of palmitic acid. It has no effect on the oxidation of octanoate because short-chain and medium-chain fatty acids can directly enter the mitochondrial matrix, where they are activated by short-chain or medium-chain acyl-CoA synthetase.

3. In the reaction catalyzed by acyl-CoA oxidase, electrons are transferred directly to oxygen and therefore do not yield ATP. Hence, the oxidation of palmitic acid by peroxisomes and mitochondria will yield $129 - 5 = 124$ ATP in contrast with the oxidation by mitochondria alone, which yields 129 ATP. Efficiency: mitochondria alone, 100%; peroxisome plus mitochondria, 96%.

4. The *E. coli* mutant that grows only on decanoate is a *fadL* mutant defective in the uptake of long-chain fatty acids. Medium-chain fatty acids like decanoate, in contrast with long-chain fatty acids, can enter the bacterium in the absence of the *fadL* gene product and will be oxidized because the enzymes of fatty acid oxidation are constitutively expressed in both mutants.

5. In uncoupled mitochondria, NADPH is not formed because the transhydrogenase is inactive. The absence of NADPH causes an inhibition of the oxidation of docosahexaenoylcarnitine only, because NADPH is required for the removal of the double bonds extending from even-numbered carbon atoms. Dehydrogenation of glutamate by glutamate dehydrogenase yields NADPH, which stimulates the oxidation of docosahexaenoylcarnitine.

CHAPTER 5

1. (a) The concentrations of its substrates and products are far from thermodynamic equilibrium. (b) Maximum velocity of the enzyme, as measured in cell extracts under optimal conditions, is usually the slowest of all enzymes in the pathway. (c) The concentration of the product of

the enzyme, malonyl-CoA, increases when flux through the pathway increases. If the catalytic efficiency of the next enzyme in the pathway, fatty acid synthase, were regulated, malonyl-CoA concentration would decrease when flux is increased. (d) Acetyl-CoA carboxylase catalyzes the first committed step in the pathway for fatty acid synthesis. In most pathways, the first committed step is regulated. For an exception, see phosphatidylcholine biosynthesis in Chapter 8. (e) There is no known physiologically relevant regulation of the catalytic efficiency of ATP-citrate lyase or fatty acid synthase, the other two enzymes of the cytosolic part of the pathway.

2. Acetyl-CoA carboxylase requires a tricarboxylate anion such as isocitrate for catalytic activity. Long-chain fatty acyl-CoAs such as stearoyl-CoA bind to the enzyme and inhibit the activation caused by tricarboxylate anions.

3. Enzymes which catalyze reactions where the concentrations of the substrates and products are near thermodynamic equilibrium cannot be regulatory. If the activity of such an enzyme is stimulated, flux in both the forward and reverse directions will be stimulated to the same extent; therefore, unidirectional flux will not be influenced significantly.

4. The β-ketoacyl synthase reaction requires the participation of acyl groups donated by the acyl carrier peptide and the active catalytic center for β-ketoacyl synthase. According to the model proposed by Wakil and co-workers, these two sites are on different subunits of fatty acid synthase. Since all intermediates are enzyme-bound, the overall synthesis of fatty acids cannot proceed without an active β-ketoacyl synthase reaction.

5. Glucagon acts in both hepatocytes and adipocytes to increase the intracellular concentration of cAMP. Increased levels of cAMP in hepatocytes will lead to (a) inhibition of the synthesis of acetyl-CoA carboxylase, leading to a decrease in its concentration and activity; (b) phosphorylation and inhibition of the activity of acetyl-CoA carboxylase; (c) inhibition of the activities of phosphofructokinase and pyruvate kinase by a combination of allosteric and covalent modification mechanisms. This, in turn, will lead to inhibition of the flow of pyruvate to citrate via

the pyruvate dehydrogenase. As a consequence, less citrate will be produced in the cytosol. Less citrate means less activation of acetyl-CoA carboxylase. Any citrate present will be even less effective with the phosphorylated enzyme. (d) Increased cAMP in adipose tissue will lead to the activation of hormone-sensitive lipase, increased hydrolysis of triacylglycerol to fatty acids and glycerol, and an increased plasma concentration of unesterified fatty acids. The increase in free fatty acids in plasma will elevate intracellular unesterified fatty acids and fatty acyl-CoA in the liver. An increased concentration of long-chain fatty acyl-CoA will inhibit acetyl-CoA carboxylase, the inhibitor being more potent when the enzyme is phosphorylated.

6. Insulin and epinephrine cause increased phosphorylation of different tryptic peptides in the acetyl-CoA carboxylase molecule. Phosphorylation of one site is postulated to increase activity, while phosphorylation of the other site inhibits activity.

CHAPTER 6

1. For saturated fatty acids, bond angles are arranged to form a linear acyl chain. A cis double bond alters acyl chain linearity by about 30°, and the spatial width is nearly tripled. In contrast, a trans bond alters the angle of the chain only slightly and increases the spatial width by less than 25% compared with the saturated analogue. The intermediate nature of a trans acid in decreasing London–van der Waals forces between adjacent acyl chains is reflected by its relative melting point and solubility. Thus, interactions of trans acids in biological systems might be expected on a physical basis to be between those of saturated and cis acids. However, this is not always the case, and bond position along the acyl chain can be at least as important an influence as bond configuration.

2. Fatty acyl-CoA derivatives form micelles at concentrations above $30\mu M$. Thus, in the absence of albumin, micelles are formed which either are inactive as substrate or inhibit condensation. Albumin appears to prevent micelle formation at $50-100\mu M$ acyl-CoA while permitting monomer

interaction and higher rates of condensation. Albumin may also restrict hydrolysis of the 16:0 thioester in the reaction medium. While condensation is generally rate limiting to overall elongation, it is actually lowered in the absence of albumin. This suggests that condensation may be underestimated, as conversion of β-keto acyl-CoA back to the primer is not prevented by albumin inhibition of β-ketothiolase (Sprecher 1981).

3. The position of the double bond in $16:1(n - 10)$ is Δ^6, in contrast with the Δ^9 position in the $16:1(n - 7)$ isomer predominating in most tissues. Quantitative analysis of these two isomers in isolated lipid extracts could involve gas chromatography (before or after cleavage of the double bond). Positional isomers of the fatty acyl methyl esters could be separated using a high-efficiency capillary column, although care must be used in distinguishing bond positions differing by only one or two carbon positions. Following oxidation or reductive ozonolysis of the double bond in the total 16-carbon monoene fraction, another alternative would be chromatographic analysis of the chain length of fragments (usually the portion from the double bond to the carboxyl end is most reliably quantitated, particularly for monoene chains). If formation of the $(n - 10)$ isomer in brain were by the Δ^6 desaturase involved in polyunsaturated fatty acid synthesis, *in vitro* assays with $[1-^{14}C]16:O$-CoA (or fatty acid and cofactors) would be inhibited by competition from exogenous unlabeled $18:2(n - 6)$ or $18:3(n - 3)$; an inhibitor of Δ^9 desaturation, such as the cyclopropene acid, sterculic acid, would not reduce formation of $16:1(n - 10)$. Such differential inhibition of synthesis of the $(n - 10)$ and $(n - 7)$ isomers of brain 16:1 has been demonstrated (Cook 1979).

4. Conversion of $18:3(n - 3)$, $18:2(n - 6)$, and $18:1(n - 9)$ to more unsaturated acids involves an initial Δ^6 desaturation. With a normal balance of these acids, competition for Δ^6 desaturase (18:3 > 18:2 > 18:1) allows for adequate formation of $18:3(n - 6)$, and ultimately $20:4(n - 6)$, with little or no detectable desaturation of $18:1(n - 9)$ and subsequent $20:3(n - 9)$ formation. However, restricted intake of the essential fatty acids results in formation and accumulation of $20:3(n - 9)$ from $18:1(n - 9)$ and a marked

decrease in $20:4(n - 6)$. In these circumstances the triene:tetraene ratio (or more accurately, the ratio of $20:3(n - 9)$ to $20:4(n - 6)$) is a reasonable index of the deficiency. However, in cases where the activity of Δ^6 desaturase is restricted, for example, by low enzyme level, inhibition, or cofactor restrictions, neither trienes or tetraenes would be formed. A resultant functional deficiency would not be reflected in the ratio and, furthermore, would not be corrected by administration of $18:2(n - 6)$ alone.

5. Under normal circumstances, dietary $18:2(n - 6)$ is converted to $20:4(n - 6)$ in three steps involving Δ^6 desaturation to $18:3(n - 6)$, two-carbon chain elongation to $20:3(n - 6)$, and Δ^5 desaturation to $20:4(n - 6)$. Generally, the intermediates $18:3(n - 6)$ and $20:3(n - 6)$ do not accumulate appreciably. Competition for Δ^6 desaturase, with $18:2(n - 6)$ being more effective than $18:1(n - 9)$, usually prevents formation of $18:2(n - 9)$ and further elongation and desaturation to $20:3(n - 9)$. However, in the absence of $18:2(n - 6)$ competition, $18:2(n - 9)$ and $20:3(n - 9)$ would be formed from $18:1(n - 9)$. Thus, in the patient, a severely restricted intake of essential fatty acids results in decreased $18:2(n - 6)$, $20:3(n - 6)$, and $20:4(n - 6)$ and increased $18:2(n - 9)$ and $20:3(n - 9)$. This imbalance could be corrected by supplementing the intravenous feeding with some form of lipid emulsion (for example, Intralipid) containing essential fatty acyl chains. By serial sampling (with great caution for an infant) and gas chromatographic analyses of serum lipid acyl chain composition, a return to normal could be assessed. The $20:3(n - 9)$ to $20:4(n - 6)$ ratio (triene:tetraene) should decrease from greater than 1 to a normal level of below 0.2—if $20:3(n - 9)$ were detectable at all under normal conditions.

CHAPTER 7

These are brief outlines that try to summarize the more important asepcts.

1. Phosphatidate can be synthesized by the sequential esterification of the two hydroxyls of glycerolphosphate by glycerolphosphate acyl-

transferase and lysophosphatidate acyltransferase (Figure 7.2). This reaction can take place on the cytoplasmic side of the endoplasmic reticulum or on the inner surface of the outer mitochondrial membrane. The glycerol phosphate acyltransferases in these two sites are different enzymes and have different properties (Table 7.2). Alternatively, phosphatidate can be produced by the acylation of dihydroxyacetone phosphate by a specific acyltransferase in the peroxisomes or by the glycerol phosphate acyltransferase of the endoplasmic reticulum (Table 7.2). Acyldihydroxyacetone phosphate is then converted to lysophosphatidate by a reductase that uses NADPH, and lysophosphatidate is then acylated to phosphatidate (Figure 7.2). The main route for the synthesis of phosphatidylethanolamine, phosphatidylcholine, and triacylglycerol is normally thought to involve the phosphatidate that is formed in the endoplasmic reticulum by the esterification of glycerol phosphate. The formation of phosphatidate in mitochondria can compete with β-oxidation for acyl-CoA esters and help to regulate fatty acid metabolism. Some of the phosphatidate formed in mitochondria can be used for the synthesis of acidic phospholipids, for example, diphosphatidylglycerol. Acyldihydroxyacetone phosphate can be used for the synthesis of the alkyl- and alkenyl-lipids. The importance of the peroxisomes in glycerolipid synthesis is not understood. The physiological significance of glycerol phosphate relative to dihydroxyacetone phosphate as a direct precursor for glycerolipid synthesis is also not known.

2. Short- and medium-chain fatty acids are more readily hydrolyzed from triacylglycerols than are long-chain fatty acids. The short- and medium-chain acids are often located at the 3-position of natural acylglycerols. They can be hydrolyzed by gastric lipase (Figure 7.4) and absorbed directly from the stomach. The gastric lipase is particularly important in young animals. Most of the remaining digestion of the short- and medium-chain acids, and the bulk of the digestion of the long-chain acids, takes place in the lumen of the small intestine. Pancreatic lipase hydrolyzes fatty acids at the 1- and 3-positions, and some limited hydrolysis (about 15%) of 2-monoacylglycerols

is catalyzed by a nonspecific hydrolase (Figure 7.4). The products of lipolysis dissociate from the oil droplet (Figure 7.5). The long-chain fatty acids and monoacylglycerols accumulate in the micellar phase, whereas the short-chain acids partition into the aqueous phase. The uptake of the short-chain acids can take place against a concentration gradient, whereas the uptake for the long-chain fatty acids is energy-independent. The latter relies on re-esterification to maintain an inward diffusion gradient. Long-chain fatty acids are rapidly activated and are preferentially esterified, mainly by the monoacylglycerol pathway (Figure 7.6). The remaining long-chain acids are incorporated into triacylglycerols by the glycerol phosphate, or dihydroxyacetone phosphate, pathways, or they can be incorporated into phospholipids and cholesterol esters. These complex lipids are assembled into chylomicrons, which are transported by the lymphatic route (Figure 7.7). The short-chain acids are discriminated against by the enzymes of esterification, and monoacylglycerols containing them can be readily degraded in the enterocytes. Consequently, short- and medium-chain acids are transported in their unesterified forms bound to albumin in portal blood. Decanoic and dodecanoic acids are found in both portal blood and chylomicrons (Table 7.3).

3. In general, the enzymes involved in triacylglycerol synthesis in adipose tissue are controlled reciprocally with those of lipolysis (Figures 7.11 and 7.12). Their activities are decreased by the catecholamines, and this effect is antagonized by insulin. In the liver, the control is more complex. When insulin concentrations are high, hepatic triacylglycerol synthesis increases to esterify the fatty acids derived from de novo synthesis. β-Oxidation decreases (Figure 7.9), and the enzymes of esterification compete more effectively for acyl-CoA. (See also the answer to Problem 4.) However, when lipolysis in adipose tissue increases because of low availability of insulin and an increased action of stress hormones, the liver receives an increased fatty acid supply (Figure 7.13). The potential of the liver to synthesize triacylglycerols is increased mainly through changes in phosphatidate phosphohydrolase activity (Table 7.4). This enzyme becomes functionally active if the supply of fatty acids

is more than is required to support the increased rate of β-oxidation. (See also the answer to Problem 5.)

4. The high concentrations of absorbed glucose increase the concentrations of circulating insulin relative to glucagon and the catecholamines. Concentrations of cAMP in the liver fall, and fatty acid synthesis is stimulated. The high malonyl-CoA concentrations inhibit β-oxidation (Figure 7.9), and the acyl-CoA that is formed by de novo synthesis is efficiently incorporated into triacylglycerol which is secreted in VLDL. Although some of this can be hydrolyzed by muscle tissue, the majority of the fatty acids derived from the triacylglycerols of VLDL are obtained by adipose tissue, since its lipoprotein lipase activity is increased by insulin. Glucose uptake by adipose tissue is also stimulated by insulin and the glucose can be used to provide the glycerol backbone of triacylglycerols or for fatty acid synthesis (Figure 7.11). The low concentration of cAMP in the adipocytes stimulates the activity of the enzymes responsible for triacylglycerol synthesis and causes the inactivation of hormone-sensitive lipase (Figure 7.12). Consequently, there is a net storage of triacylglycerol.

5. The low concentrations of insulin relative to glucagon, catecholamine, and corticotropin increase cAMP concentrations in adipocytes, leading to the phosphorylation of the hormone-sensitive lipase and to its activation (Figure 7.12). Stored triacylglycerol is broken down, and fatty acids are released into the blood. Some of these are taken up by muscle tissue and used to produce energy by β-oxidation (Figure 7.13). Most of the remaining fatty acids go into the liver, in which the high cAMP and low malonyl-CoA concentrations promote β-oxidation and ketogenesis. Ketones are secreted by the liver and are used by brain and muscle as a form of energy (Table 7.6; Figure 7.13). The remaining fatty acids are esterified mainly to triacylglycerols and phosphatidylcholine. This process is facilitated by the high activity of phosphatidate phosphohydrolase (Table 7.4) and the activation and translocation of the phosphohydrolase and CTP:phosphocholine cytidylyltransferase to the membrane-associated compartment, which

occurs when fatty acid availability is high (Figure 7.10; Chapter 8). The effects of cAMP in displacing the phosphohydrolase and cytidylyltransferase from the membranes is overcome by the high fatty acid concentrations (Table 7.5; Chapter 8). The increased rate of triacylglycerol synthesis can produce a fatty liver. Triacylglycerol and phosphatidylcholine are also secreted by the liver as VLDL. The hydrolysis of the triacylglycerol from the VLDL occurs mainly in heart and skeletal muscle, where the activity of lipoprotein lipase is increased by glucocorticoids. Uptake of fatty acids from VLDL by adipose tissue is prevented by the lack of insulin, which decreases its lipoprotein lipase activity. This decreased uptake by adipose tissue coupled with an increased secretion of VLDL by the liver explains the hypertriglyceridemia that is seen in ketotic diabetes.

CHAPTER 8

1. No! By plotting the radioactivity in only N-methyl-phosphatidylethanolamine, you have ignored all the N-methyl-phosphatidylethanolamine that was formed during the incubation and subsequently converted to dimethyl-phosphatidylethanolamine and phosphatidylcholine. A considerably more accurate estimation would be obtained if the radioactivity in N-methyl-phosphatidylethanolamine, N,N-dimethyl-phosphatidylethanolamine, and phosphatidylcholine was determined and the data substituted in the following equation:

$$\text{nmol } N\text{-Methyl-PE} = \frac{\text{dpm in } N\text{-methyl-PE} + \frac{1}{2} \text{ dpm in dimethyl-PE} + \frac{1}{3} \text{ dpm in PC}}{\text{Specific radioactivity of AdoMet}}$$

Plotting 1/(nmol N-methyl-phosphatidylethanolamine) against 1/(concentration of S-adenosylmethionine) will give an apparent K_m value for S-adenosylmethionine in the conversion of phosphatidylethanolamine to N-methyl-phosphatidylethanolamine (Audubert and Vance 1983).

2. First, incubate the hepatocytes with [CH$_3$-^{3}H] choline for a short time period (30 min). This will label the major precursor of phosphatidylcholine—phosphocholine—to the same extent in all cells. Next, change the medium to one containing unlabeled choline with or without forskolin. Harvest the cells immediately and at various subsequent times (30, 60, 90 min). Determine the label appearing in phosphocholine and phosphatidylcholine after thin-layer chromatography. The observation of significantly fewer counts in phosphatidylcholine in the forskolin-treated cells would suggest an inhibition of phosphatidylcholine biosynthesis. Furthermore, if most of the precursor counts were in phosphocholine, and after 90 min more radioactivity was present in the phosphocholine of forskolin-treated cells than in the untreated cells, the results would suggest an inhibition of the cytidylyltransferase reaction by forskolin.

3. The culture medium contains no ethanolamine. Hence, the levels of the aqueous precursors of phosphatidylethanolamine (ethanolamine, phosphoethanolamine, and CDP-ethanolamine) are very low in these cells. Consequently, the supply of ethanolamine is rate limiting for phosphatidylethanolamine biosynthesis via the CDP-ethanolamine pathway. The small amount of [^{3}H]ethanolamine added is rapidly incorporated into the cells. Since the concentration of the aqueous precursors is very low, the [^{3}H]ethanolamine is not significantly diluted. This results in a rapid passage of radioactivity through the intermediates into phosphatidylethanolamine.

4. In hepatocytes, the [^{3}H]serine is incorporated into phosphatidylserine via base exchange. The phosphatidylserine is decarboxylated to phosphatidylethanolamine in the mitochondria. The phosphatidylethanolamine can be converted into phosphatidylcholine by the methylation pathway on the microsomes of the hepatocytes. HeLa cells also make phosphatidylserine, which can be decarboxylated to phosphatidylethanolamine. However, the methylation pathway is virtually inactive in these cells; thus, very little radioactivity from [^{3}H]serine is incorporated into phosphatidylcholine.

5. *Tetrahymena* has large amounts (25%) of the phosphatidylethanolamine analogue that contains 2-aminoethylphosphonic acid, which arises from phosphoenolypyruvate and would not be labeled with ethanolamine. Second, the phosphono bond would not be cleaved by phospholipase D. Hence, the labeled phosphatidylethanolamine would be cleaved by this phospholipase and would release all the radioactivity, but the phosphono analogue would remain unaltered and be recoverd after thin-layer chromatography as if it were undegraded phosphatidylethanolamine.

CHAPTER 9

1. The contaminated PAF sample should be treated with phospholipase A$_2$ (venom) to obtain 1-alkyl-2-lyso-*sn*-glycero-3-phosphocholine, which should then be separated from the unreacted 3-alkyl-2-acetyl-D-glycero-1-phosphocholine isomer by thin-layer chromatography and reacetylated using acetic anhydride to produce the natural PAF.

2. An ethanolamine and/or a choline plasmalogen phospholipid (1-alk-1-enyl-2-acyl-*sn*-glycero-3-phosphoethanolamine or 1-alk-1-enyl-2-acyl-*sn*-glycero-3-phosphocholine). A serine plasmalogen would also qualify as an answer.

$$H_2COCH=CHR$$
$$\underset{\substack{\| \\ O}}{R'COCH}$$
$$\underset{\substack{| \\ OH}}{H_2COPOCH_2CH_2NH_2}$$

Ethanolamine plasmalogen

The (+) or (−) reactions indicate the following: (a) the presence of alk-1-enyl moiety, (b) the presence of acyl moieties, (c) a phospholipid with an *sn*-2 acyl moiety, (d) the absence of an acyl group at the *sn*-1 position of a glycerolipid,

(e) the presence of alkyl, alk-1-enyl, or acyl (reduced to alcohol) moieties, (f) the absence of a ketone and/or aldehyde functional group, (g) the presence of an acyl and/or alk-1-enyl moiety, (h) a phospholipid containing a choline, ethanolamine, or serine moiety, and (i) a phospholipid containing a phosphobase. The formation of an isopropylidine derivative after $LiAlH_4$ reduction indicates that two adjacent hydroxyl groups are present.

3. a. 0; both tritiums on C-1 are lost when the fatty alcohol is oxidized to the fatty acid.
   b. 1; the entire fatty alcohol chain including the oxygen is incorporated into the alkyl moiety by alkyldihydroxyacetone-P synthase.
   c. 0.5; when the alk-1-enyl moiety is synthesized by desaturation of the alkyl moiety of an intact phospholipid (1-alkyl-2-acyl-*sn*-glycero-3-phosphocholine), one of the hydrogens is abstracted by the alkyl desaturase from C-1 and the other from C-2 of the alkyl chain; the two tritiums lost in the desaturation reaction are the pro-*S* atoms. Thus, since both tritiums are present on C-1 and none on any of the other carbons of the alcohol, only one half of the ^3H and none of the ^{14}C would be lost from the doubly labeled hexadecanol.

4. Tritium at C-1 of hexadecanol. Both tritiums would be lost upon oxidation of the alcohol to the acid.

CHAPTER 10

Questions of this nature obviously can have a multitude of answers that are feasible, although some are more probable or significant than others. Those listed here are viewed by the author as the most obvious.

1. a. The use of phospholipids labeled at specific sites provides the most definitive method of establishing the positional specificity of a phospholipase. An example follows.

   Substrate: 1-[^3H]acyl-2-[^{14}C]acyl-glycerophosphorylbase
   Products of
   phospholipase A_1: [^3H] fatty acid and [^{14}C] acyl-glycerophosphorylbase

   phospholipase A_2: [^{14}C] fatty acid and [^3H] acyl-glycerophosphorylbase

   phospholipase B: [^{14}C] fatty acid and [^3H] fatty acid

   phospholipase C: [^3H] and [^{14}C] diacylglycerol
   phospholipase D: [^3H] and [^{14}C] phosphatidate

   These products are separated by thin-layer chromatography.
   Try some other combinations of radiolabeling to establish positional specificity.

   b. Since the activity of the enzyme is sufficiently high, an assay such as titration or a colorimetric reaction is desirable. These techniques are much more rapid and less expensive than the radioisotope procedure.

2. In order to establish substrate specificity, one must use phospholipids with different polar head groups under a variety of conditions. These conditions include:

   1. different polymorphic states of the lipid (bilayer, micelle, hexagonal array)
   2. changing surface charge of the lipid aggregate
   3. changing solute composition (such as metal-ion content or pH)
   4. changing fluidity and packing of phospholipid

   In general, the acyl composition does not appear to influence specificity appreciably, other than by influencing the physical state of the lipid. The effect of acyl chains on enzymatic activity can be determined if described in monolayer or mixed-micelle systems.

3. The two methods of choice for kinetic analysis of phospholipases are the monolayer and mixed-micelle (detergent-phospholipid) systems. The substrate concentration in monolayers is regulated by the surface pressure; therefore, the effect of pressure must be considered. When mixed micelles are used, both the ratio of phospholipid to detergent and the total amount of both detergent and phospholipid in a constant ratio must be determined. The binding of the enzyme to the micelle, the maximum velocity, and the affinity constant for the enzyme binding

to the substrate molecule (Michaelis complex), can then be determined. (See Verheij, Slotboom, and deHaas (1981) and Dennis (1983) for more complete descriptions of kinetic analysis.)

4. A number of factors could account for the inability of a phospholipase to attack a membrane. Some factors that could be examined are:

    1. molecular packing and fluidity of phospholipids in the membrane
    2. phospholipid composition of the membrane
    3. presence of an inhibitor(s) in membrane
    4. inaccessibility of membrane lipid (glycoprotein or glycolipid coating)

CHAPTER 11

1. PGI_2 is produced primarily by nucleated cells such as endothelial cells and smooth muscle cells. These cells synthesize new PGH synthase protein to replace enzyme which has been irreversibly inactivated by aspirin. In contrast, the major source of TxA_2 production is platelet cells, which lack protein biosynthetic activity. In order for further TxA_2 synthesis to occur, new platelets have to be formed. The lifetime of a platelet is roughly 4–5 d. Thus, it takes 4–5 d for TxA_2 synthesis to return to normal.

2. LTA_4 contains an epoxide linkage which is extremely unstable in the presence of acid, but which is relatively stable in base. Conversion of LTA_4 to its methyl ester permits one to extract this compound into organic solvents without acidifying the aqueous phase.

3. A cis methylene-interrupted double-bond system with double bonds at $n - 6$ and $n - 9$ of the fatty acid is the major structural requirement for the activity of soybean lipoxygenase. Most fatty acids having this structural group are substrates for soybean lipoxygenase. Soybean lipoxygenase will not function on fatty acid substrates which lack double bonds at these positions or have trans double bonds at these positions.

4. One explanation is that there are two sets of prostaglandin receptors in collecting tubule cells—one type which is coupled to adenylate cyclase and which can interact only with PGE_2 and a second type which is involved in heterolo-gous desensitization of adenylate cyclase and which can interact with both PGE_2 and $PGF_{2\alpha}$.

CHAPTER 12

1. Sphinganine synthetase involves the condensation of palmitoyl-CoA and the pyridoxal phosphate–bound intermediate from serine to yield 3-oxosphinganine. Pyridoxal phosphate reacts with serine to give a Schiff base intermediate that loses a proton to form another intermediate, believed to be the reactive form involved in the condensation reaction.

2. Golgi membrane receptors are postulated to be necessary for the active transport of sugar nucleotides from the cytosol to the lumen of the Golgi apparatus. Cotransport of nucleoside monophosphates from the lumen suggests a typical antiport mechanism.

3. Most plant glycosphingolipids contain inositol phosphate, which has never been found in a glycosphingolipid of higher animals.

4. Treatment of the brain ganglioside called GD1a with bacterial neuraminidase yields GM1 ganglioside. The sialic acid attached to the internal galactose residue is not hydrolyzed from the glycolipid by this enzyme, presumably because of steric hindrance.

5. a. Using two purified galactosidases with proven specificities, one for α-Gal linkages and one for β-Gal linkages, you could incubate the glycolipid with each enzyme overnight (at the appropriate pH optimum and in a suitable detergent-buffer mixture) and spot an aliquot of the two mixtures on a TLC plate. One mixture will contain a less-polar glycolipid (greater mobility on TLC) and will appear at a higher R_F value when the plate is sprayed with a color-producing reagent. The difference in TLC mobility between the parent glycolipid and the product will indicate how many sugar residues were removed by comparison with reference standards such as $GbOse_3Cer$, LacCer, and GlcCer. If all three galactose residues were removed, the product will have the same R_F as GlcCer, and the anomeric configuration of all three galactose residues will be identified (α or β depending

upon which enzyme was active). If one or two galactose residues were removed, treatment of the product (recovered from TLC or by column chromatography) with the other galactosidase should further hydrolyze the intermediate, and so forth.

If at least 100 μg of pure glycolipid have been isolated, ^{1}H-NMR can be used to establish the anomeric configuration of each sugar.

b.

Gal III $\longrightarrow$

$$
\begin{array}{l}
\text{CH}_2\text{OAc} \\
\text{HC—OMe} \quad 117 \\
\text{AcO—CH} \\
\text{MeO—CH} \quad 233 \\
\quad 161 \\
\text{HC—OAc} \\
\text{CH}_2\text{OMe} \quad 45
\end{array}
$$

Gal II $\longrightarrow$

$$
\begin{array}{l}
\text{CH}_2\text{OAc} \\
\text{HC—OAc} \\
189 \quad \text{MeO—CH} \\
\quad \text{MeO—CH} \quad 205 \quad 233 \\
161 \\
\text{HC—OAc} \\
\text{CH}_2\text{OMe} \quad 45
\end{array}
$$

c. No, a fucose residue is a necessary part of the blood group B–active antigenic determinant.

6. The sulfotransferase involved in sulfatide biosynthesis is located on the luminal aspect of Golgi membranes. Transfer to the GalCer in an exogenously added micelle is impossible, therefore, since the micelle cannot penetrate the Golgi membrane. Addition of a detergent makes the membrane "leaky," and the enzyme and substrate can then interact.

7. Lactosylceramide is converted to GbOse$_3$Cer in the presence of UDP-Gal and a specific galactosyltransferase; the product is an intermediate in the biosynthesis of Gb5b. Glycosphingolipids of the lacto series are synthesized from LacCer via an N-acetylglucosaminyl transferase.

8. GM1 and, perhaps, GD1b (depending upon whether a neuraminidase cleaves the NeuAc

(α2-8)NeuAc glycosidic linkage of GD1b). Further catabolism is not possible because of the deficiency of β-galactosidase activity.

9. Current theories about the action of oncogenes, and descriptive studies of glycosphingolipid composition of uninfected and virus-transformed cells, suggest that the transcription of glycosyltransferase genes might be altered, leading to products that are not present in the uninfected cells. It is equally possible that translation of mRNA might be different, and in vitro translation might be compared. This is an ambitious project, however, since very few of the glycosyltransferases have been purified, and monoclonal antibodies needed for such experiments are not available. Recent work on the effect of certain oncogenes on protein kinases has suggested that increased covalent modification of certain enzymes may be a concomitant of viral transformation. Studies of the specific activities of glycosyltransferases might be revealing, but these experiments might also require monoclonal antibodies.

CHAPTER 13

1. The epidemiological finding is consistent with the concept that the development of atherosclerosis is due to the balance of the amounts of both the cholesterol-loading lipoproteins (there is an increased content of VLDL and LDL in the plasma of those at increased risk for atherosclerosis) and cholesterol-depleting lipoproteins (HDL). Remember, though, that only a fraction of HDL is apparently of major significance for removal of cholesterol from cells, that LCAT plays a key role in maintaining effective efflux of cholesterol, and that cholesterol efflux *in vitro* becomes saturated at very low plasma concentration of apo A-1-HDL. That is, a further increase in apo A-1-HDL will not alter the rate of cholesterol efflux.

2. The major evidence for a direct role of the apo B,E receptor in the suppression of cholesterogenesis is that cholesterol accumulates in tissues when this receptor is genetically active. However, since normally the receptor does not

seem to be involved in suppression of cholesterogenesis, it is unlikely that the apo B,E receptor plays a simple role, either in normal metabolism or in atherogenesis in the vascular bed. It is more likely that the abnormal composition of LDL in this disease, now reported by many laboratories, plays a more direct role.

3. Pure cholesterol does not affect cholesterogenesis; only oxidized cholesterol (in pure or lipoprotein form) or oxysterols acting via a specific intracellular oxysterol carrier protein appear to be able to mediate cholesterol synthesis. The effects of LDL on cellular cholesterol metabolism have all been reproduced with lipoprotein-free oxysterols.

4. First, it suggests that both the apo B,E receptor and the distinct apo E receptor proposed for the hepatocyte membrane must react with the same region of the apo E sequence. Second, this supports the argument made in this chapter that perhaps the only significant role of the apo B,E receptor in normal physiology is to clear lipoproteins containing apo E.

5. The existence of a saturable process is insufficient evidence for a receptor pathway—even cholesterol transfer between synthetic lipid vesicles has these properties (McLean and Phillips 1981). Likewise, cholesterol transfer probably depends mainly on the cholesterol content (that is, the chemical potential of cholesterol) in the acceptor and need not imply ligand-receptor interaction. Loss of function following action of cell-surface proteases could indicate a receptor function if the protease action did not change the thickness of the diffusion barrier at the cell surface. The use of chemical modification of the ligand in the case of cholesterol transfer may be ambiguous if aggregation of the ligand occurred, since for thermodynamic reasons, smaller particles will be better acceptors of cholesterol than larger particles. The problem is a very difficult one technically. For recent experimental data, see Oram, Brinton and Bierman (1983) and McLean and Phillips (1981).

6. The mutant, receptor-deficient cells might lose cholesterol less efficiently to their medium than normal cells; under these conditions, the defect might lie in the absence of normal up-regulation, rather than in a primary deficiency of the receptor gene. The point has apparently not been investigated experimentally.

7. (a) A reversible enzyme would require that factors changing the level of enzyme protein would probably have little effect on the balance of substrate and product. (b) While ACAT has not been isolated, and cannot be quantitated at present in cells, inhibitors of its activity (for example, progesterone) significantly lower cholesteryl ester levels of cholesterol-loaded cells. Since the level of cholesteryl esters in most cells is normally very small, it is most likely that the level of ACAT, not of a distinct cholesteryl ester hydrolase, is rate limiting when cells become loaded with cholesteryl esters, at least experimentally (Figure 13.7).

8. It is possible that 26-hydroxylation might precede 7-hydroxylation; since 12-α-hydroxylase acts only with substrate containing an intact cholestane side chain, formation of 26-hydroxycholesterol would commit to chenodeoxycholic acid formation. However, the 26-hydroxylase activity in humans is almost entirely mitochondrial, so this pathway probably has little importance. The most likely alternative is that cholic and chenodeoxycholic acids are derived from different cellular pools of cholesterol; the point is discussed further by Bjorkhem et al. (1983).

9. (a) The most likely reason is that this lipoprotein is normally the most important carrier of free cholesterol from cell membranes to plasma LCAT. The concentration of this HDL is increased secondarily to the LCAT deficiency, perhaps because of the increase in free-cholesterol content of plasma lipoprotein in this situation. (b) Since net transport of free cholesterol from normal cell membranes to plasma requires LCAT, LCAT-deficient plasma would not be expected to catalyze any net flux of cholesterol; this is what has been observed experimentally.

10. Since plasma mevalonate levels appear to reflect rates of cholesterogenesis, change to a vegetarian diet (that is, a low cholesterol diet in this context) should lead to an increase in total-body cholesterogenesis and hence an increase in plasma mevalonate levels. The experiment would be an interesting one.

CHAPTER 14

1. Triacylglycerols are not amphipathic, so they would be expected to form structures that absolutely minimize the interaction with water as a consequence of the hydrophobic effect. In most cases this will be a completely separate oil phase. A good example of this in biological systems is seen in the structure of the adipocyte, where the triacylglycerols exist as a large fat droplet. In lipoprotein molecules the predicted structure would be a fat droplet surrounded by a protein and phospholipid "skin" whose hydrophilic moieties interact with the aqueous environment (Chapter 13).

2. The properties of the nonspecific exchange protein allow one to use mass action to substitute one phospholipid for another or for cholesterol. The requirements for performing such experiments are purified nonspecific exchange protein, purified membranes whose phospholipid composition is known, a rapid reliable assay for a specific membrane function (for example, a membrane enzyme, or an amino acid transport system), a simple means for separating the exchange protein from liposomes, and liposomes of defined composition. For example, to increase the amount of phosphatidylinositol in a membrane, the reaction would be conducted using liposomes with 50–60 mol % phosphatidylinositol. A 10-fold excess of total liposome lipid over acceptor membrane lipid would ensure efficient transfer. Radiolabeled phosphatidylinositol in the donor membranes could be used to quantitate the amount transferred. The inclusion of a nonexchangeable lipid such as [^{14}C]triolein can be used to assess nonspecific binding. After incubation with the exchange protein, the liposomes and target membranes can be separated and the lipid transfer quantitated. Finally, the functional assay can be performed and the effects of altering membrane phospholipid composition assessed. Some of the drawbacks of this technique are (1) The time required to accomplish exchange makes measurement of labile membrane functions difficult. (2) If the membranes are closed vesicles and do not exhibit transbilayer movement of phospholipids, then only one side of the bilayer can be altered. This physical property of the membrane needs to be determined *beforehand*. (3) There can be some ambiguity in assigning changes in function to the increase of a specific phospholipid as opposed to the removal of a specific phospholipid from the membrane.

3. These reagents modify the phospholipids containing primary amines, phosphatidylserine, and phosphatidylethanolamine. The products are N-trinitrophenylphosphatidylethanolamine, N-trinitrophenylphosphatidylserine, N-acetimidoylphosphatidylethanolamine and N-acetimidoylphosphatidylserine. The student should draw these structures. These reagents will react with any primary amine, so the other reactive species in cell membranes are the amino group of lysyl residues and the amino group of amino termini in polypeptides. These reagents do not react with arginyl residues and do not form stable covalent adducts of sulfhydryls. One method for resolving amidinated from nonamidinated phosphatidylethanolamine is to treat the mixture with trinitrobenzenesulfonate. Only the nonamidinated molecules are reactive. The resultant N-trinitrophenylphosphatidylethanolamine is easily separated from N-acetimidoylphosphatidylethanolamine by thin-layer chromatography.

4. The important criteria to consider are the following:
   a. Purity of the membrane preparation. Does the membrane preparation represent only one organelle, or is it contaminated with a variety of other cell membranes?
   b. Closure. Are the membranes sealed units with a discrete outside and inside? Leaky membrane preparations can allow access of probes to both sides of the membrane, compromising the results.
   c. Homogeneous orientation. Are all the membranes right side out or inside out? This question requires independent documentation of vesicle sidedness.
   d. Impermeant reagents. Can the probes cross the membrane, or will they act on only one side of the bilayer?
   e. Complementary analysis. Do inside-out and right-side-out vesicles give the same result with the same probe? Do two different probes give the same result?

f. Cryptic domains. Is there evidence that non-reactive pools are on the other side of the bilayer and not sequestered in nonreactive pools that are independent of bilayer distribution?

g. Uniform labeling. If in vivo radiolabeling preceded the analysis then the membrane lipids should be radiolabeled in proportion to their mass distribution.

CHAPTER 15

1. The 23–amino acid sequence ending with the Phe-4 residue from the carboxyl terminus has a $\Delta G_{trans} = -22.01$ kcal/mol.

2. Band 3 is made on bound polysomes without an amino-terminal signal sequence. The nascent polypeptide will insert into the endoplasmic reticulum membrane once the first transmembrane loop is exposed from the ribosome. Further loops will integrate into the membrane as they are synthesized. The carbohydrate will be attached cotranslationally. Glycophorin is also made on bound polysomes but with an amino-terminal signal sequence that is cotranslationally removed. The glycophorin is cotranslationally glycosylated, and translocation across the endoplasmic reticulum membrane is prevented by a stop-transfer sequence. This pathway would involve SRP and resembles the biosynthesis of secreted proteins.

3. According to the signal hypothesis, the coat protein is made in a precursor form on bound polysomes. The signal sequence binds to a receptor in the inner membrane of *E. coli* and is cotranslationally removed. The translocation of the coat protein is prevented by a stop-transfer sequence. According to the trigger hypothesis, the coat protein is made on free polysomes as a procoat protein. The function of the signal sequence is to alter the folding pathway of the coat protein. The procoat binds to the inner membrane and spontaneously inserts into the membrane. Cleavage of the procoat by the signal peptidase renders the process irreversible.

4. The G-protein is made with a signal sequence that is similar to the signal sequences found on secreted proteins. Like secreted proteins, the G-protein is made on bound polysomes and is cotranslationally inserted into the endoplasmic reticulum membrane. The signal peptide is cleaved cotranslationally, and carbohydrate is added before translation is complete. The complete translocation of the G-protein is prevented by a stop-transfer sequence. Deletion of this sequence would result in a secreted product. The G-protein moves from the endoplasmic reticulum through the Golgi to the plasma membrane via coated vesicles. This pathway is identical with that followed by secreted proteins.

Index

Alkylglycerol phosphotransferase.
See ATP:alkylglycerol
phosphotransferase
Alkylglycols, alkyl cleavage
enzyme, 286
Alkyldihydroxyacetone phosphate,
219, 278–83, 285, 294
Alkyldihydroxyacetone phosphate
oxidoreductase, 278–79, 285
Alkyldihydroxyacetone phosphate
synthase, 278–79, 281–283, 294
Alkyl ether lipids,
biosynthesis, 216, 219
Alkyl lipids
analysis, 275–78
biological function, 295–96
biosynthesis, 282–84
catabolism, 285–89
platelet activating factor, 289–93
regulation of metabolism, 293–95
structures, 271–75
1-Alkyl-2-lysoglycerol-3-phosphate,
284, 291
1-Alkyl-2-lysoglycerol-3-phosphate
acetyltransferase, 291, 294
1-Alkyl-2-lysoglycero-3-
phosphocholine, 277–79,
292–93
1-Alkyl-2-lysoglycero-3-
phosphocholine
acyltransferase, 292–93
1-Alkyl-2-lysoglycero-3-
phosphoethanolamine,
277–79
1-Alkyl-2-lysophospholipids, alkyl
cleavage enzyme, 286
Alkyl monooxygenase. *See* Alkyl
cleavage enzyme
1-Alkyl-2-propionylglycerol, 291
Aluminum oxide, 34
Ambiquitous enzymes, 232
Amino acids, free energy, protein
conformation, 508–10
Aminoacyl-tRNA and phospholipid
synthesis, 100
γ-Aminobutyric acid, effect of sterols
on, 14
Aminoethylphosphonate
transaminase, 268
Aminoethylphosphonic acid,
266–68
Amphiphilic amines, phosphatidic
acid phosphohydrolase, 221
Analbuminemia, 438
Anaphylaxis, slow-reacting
substance, 349, 356, 358–59
Angiotensin
arachidonate release, 331
on phosphatidate
phosphohydrolase, 231

Anion transport protein. *See* Band 3
protein
Antheridiol, 3, 14
Antibodies
to docking protein, 520
to lipoprotein lipase, 442
to lipoproteins, 439
Antidiuretic hormone, 345, 347–48
Antigens I and i,
glycosphingolipids,
395–96
Apical domains, 67
Apo B,E receptor
and atherosclerosis, 454–55,
468–70
catabolism of LDL, 454–58
catabolism of VLDL, 454
cholesterol balance, 417–18,
449–50
in dysbetalipoproteinemia, 470
genetic deficiency, 454–55, 469
hypothesis, 419–22
membrane fluidity, 8
suppression by oxysterols,
421–22
turnover of, 455
Apo E receptor, 450
chylomicron remnants, 452–53
in dysbetalipoproteinemia, 470
VLDL remnants, 454
Apolipoprotein. *See* Apoprotein
Apoprotein A-1
cofactor for LCAT, 437–39
in HDL, 427–28, 447–48
in intestine, liver, 431–32
Apoprotein A-2, 427–28, 447–48
Apoprotein A-4, 431
Apoprotein B, 427–28
in β-VLDL, 461
deficiency, 457
in LDL, 454–58
secretion by intestine, 430–31, 433
secretion by liver, 429–30, 433
in VLDL, 446, 453–54
Apoprotein C-2, 427–28
chylomicron degradation, 452
deficiency of, 427
and lipoprotein lipase, 226–27,
443–44, 309
secretion by liver, 430–31
VLDL degradation, 454
Apoprotein D, 427–28, 437
Apoprotein E, 427–28
apo B,E receptor, 454–58
in β-VLDL, 461
deficiency, 457
secretion by liver, 430–31
secretion by macrophages, 431
Apoproteins
of plasma lipoproteins, 426–28

secretion in chylomicrons,
430–31
secretion in VLDL, 429–30
Arachidic acid, 190
Arachidonate cascade
definition, 325–26, 330
relation to phospholipases, 301,
316–17, 320–21
Arachidonic acid
in cats, 198
hydroosmotic effect, 347–48
precursor of eicosanoids, 325–53,
356–58
release by phospholipases, 301,
317, 320–21
release by plasmalogenase, 288
release in PAF-cycle, 292–93,
295–96
structure, 190
zinc deficiency, 208
Archaebacteria, 19
glycerolipids, 84
phytanoyl ether lipids, 84, 113
Argininosuccinate lyase, regulation
240
Argininosuccinate synthetase,
regulation, 240
Arylsulfatase, 395
Asialoganglioside GM2, 398
Asparagine, glycoprotein assembly,
523–24
Aspirin
effect on cyclooxygenase, 336–37
effect on prostaglandin synthesis,
326
Asthma leukotrienes, 359
Asymmetry
membrane lipids, 26, 32
membrane proteins, 40
Atherogenesis. *See* Atherosclerosis
Atherosclerosis
endothelial damage, 467–68
lack of prostaglandin I₂, 342
receptor theory, 468–70
role of ACAT in, 419
role of cholesterol in, 406, 449,
455–56
triacylglycerols, 214
unsaturated fatty acids, 209–10
ATP–ADP exchange, effect of
cholesterol, 17
ATP–ADP translocase, inhibition by
atractyloside, 378
ATP:alkylglycerol
phosphotransferase, 284
ATP:citrate lyase
cDNA, 174
in fatty acid synthesis, 148, 150,
153
in 3T3-L1 cells, 175–77